Monitoring Coastal Marine Ecosystems Under Climate Change

K. K. Basheer Ahammed ·
Arvind Chandra Pandey · Bikash Ranjan Parida ·
I Wayan Gede Astawa Karang
Editors

Monitoring Coastal Marine Ecosystems Under Climate Change

Machine Learning and Artificial Intelligence Techniques

Springer

Editors
K. K. Basheer Ahammed
Department of Marine Science
Faculty of Marine Science and Fisheries
Udayana University
Denpasar, Indonesia

Arvind Chandra Pandey
Department of Geoinformatics
School of Natural Resource Management
Central University of Jharkhand
Ranchi, Jharkhand, India

Bikash Ranjan Parida
Department of Geoinformatics
School of Natural Resource Management
Central University of Jharkhand
Ranchi, Jharkhand, India

I Wayan Gede Astawa Karang
Department of Marine Science
Faculty of Marine Science and Fisheries
Udayana University
Denpasar, Indonesia

ISBN 978-3-032-09136-9 ISBN 978-3-032-09137-6 (eBook)
https://doi.org/10.1007/978-3-032-09137-6

This Springer imprint is published by the registered company Springer Nature Switzerland AG
The registered company address is: Gewerbestrasse 11, 6330 Cham, Switzerland

Contents

Remote Sensing and Geographic Information Science for Monitoring and Managing Coastal and Marine Ecosystems

Long-Term Shoreline Change Detection Along the Point Calimere Coastal Stretch of Tamil Nadu Using Landsat Data and GIS-Based DSAS Technique

S. Kaliraj, Reji Srinivas, S. Om Prakashraju, and P. S. Fathimathu Sahala

Abstract Assessing shoreline change is essential for identifying coastal accretion and erosion and understanding coastal morphology and dynamics. Coastal erosion represents a significant hazard to coastal communities, particularly in areas increasingly prone to coastal disasters. This study investigates shoreline changes along the Point Calimere coast in Nagapattinam district, Tamil Nadu, focusing on decadal changes from 2000 to 2024. The Digital Shoreline Analysis System (DSAS) was utilized to assess these changes using satellite imagery from Landsat 7—Enhanced Thematic Mapper Plus (ETM+) and Landsat 9—Operational Land Imager (OLI), both with a spatial resolution of 30 m, acquired during the pre-monsoon periods of 2000 and 2024. DSAS effectively analyzes shoreline changes through key metrics, including Net Shoreline Movement (NSM), End Point Rate (EPR), and Shoreline Change Envelope (SCE). The long-term shoreline dynamics (2000–2024) at 100 m transects along the coastal stretches showed the remarkable changes due to erosion and accretion processes, and the EPR is measured at the average erosion rate of −0.20 m/yr (found in 48.22% of transects), and an average accretion rate of 0.25 m/yr (31.18% of transects), respectively, with the uncertainty value of +/− 1m. The maximum erosion rate is −18.80 m/yr at the transect ID 943, in segment 3 (Point Calimere – Kodiyakarai), meanwhile the maximum accretion rate is at 17.29 m/yr at the transect ID 1771, in segment 5 (Manamelkudi). Notably, the 20.60% of transects in moderate change (seasonal erosion/accretion) with an average rate of 0.12 m/yr, and indicates a predominant trend of coastal erosion along the Point Calimere stretch. These metrics provide crucial insights into coastal erosion and accretion dynamics, which are essential for effective coastal zone management and planning. The findings

S. Kaliraj (✉) · R. Srinivas · P. S. Fathimathu Sahala
National Centre for Earth Science Studies (NCESS), Ministry of Earth Sciences (MoES), Government of India, Thiruvananthapuram, Kerala, India
e-mail: s.kaliraj@ncess.gov.in

S. Om Prakashraju
Department of Geology, Anna University, Chennai, Tamil Nadu, India

P. S. Fathimathu Sahala
Ministry of Earth Science (MoES) Research Fellow (MRFP), IITM, Pune, India

NCESS Research Scholar, Faculty of Marine Sciences, CUSAT, Kochi, Kerala, India

K. K. Basheer Ahammed et al. (eds.), *Monitoring Coastal Marine Ecosystems Under Climate Change*, https://doi.org/10.1007/978-3-032-09137-6_1

from DSAS offer valuable data for the development of strategies aimed at mitigating coastal erosion and promoting sustainable coastal development.

Keywords Shoreline change · Coastal erosion · GIS · DSAS · EPR · NSM · SCE · Point Calimere · Tamil Nadu

1 Introduction

The coastal shoreline is a dynamic interface between the land and sea, supporting diverse ecosystems and human settlements (Kaliraj et al. 2013a; Basheer Ahammed and Pandey 2019; Rani et al. 2020; Ciecholewski 2024; Dike et al. 2024). Approximately 40% of the global population lives within the proximity of coastal areas, highlighting the immense economic, environmental, and social significance of these regions (Zambrano-Medina et al. 2023). Coastal areas are governed by intense human activities (tourist activities, settlements, infrastructure for shoreline protection, construction of ports and dams, coastal mining) and natural processes such as coastal erosion, accretion, and inundation (Kaliraj et al. 2013b; Awad and El-Sayed 2021). Both anthropogenic and natural activities result in shoreline changes, coastal landscape development, urbanization, and resource use disputes, further leading to environmental degradation and increased vulnerability (Sekar et al., 2024). Shorelines are dynamic, undergoing spatial and temporal evolution in response to tidal flows, sea-level variations, storm surges, and land tectonics (Kundu and Mandal 2024; Sekar et al. 2024). The states of shoreline seen around the world include erosion (land being eroded—sediment reduction), accretion (land being built up—sediment accumulation), and equilibrium, i.e., steady shoreline (Kaliraj et al. 2016, Zambrano-Medina et al. 2023). Shoreline constantly changes, mandating seasonal, short-term, and long-term monitoring and management (Kaliraj et al. 2017; Mujabar and Chandrasekar 2013; Basheer Ahammed and Pandey 2021).

Coastal erosion is a pressing global concern, especially for countries like India, where it presents a significant challenge. With a coastline spanning 7517 km and approximately 35 percent of the population residing within 100 km of the coast (Ajai 2012), India faces considerable vulnerability to coastal hazards. The country's coastline includes nine coastal states, two union territories, and 66 coastal districts. Tamil Nadu, a crucial coastal state, has a coastline of 1076 km along the Bay of Bengal and the Indian Ocean, accounting for about 15% of India's total coastal length. The coastal region of Tamil Nadu is particularly susceptible to natural hazards, with coastal erosion being one of the most severe issues, impacting numerous coastal villages (Sekar et al. 2024). High population density and coastal erosion, coupled with sea-level rise, profoundly impact coastal areas, mainly because much of the population relies heavily on coastal resources. Rising sea levels increase the susceptibility of coastlines to various hazards, making low-lying coastal areas particularly vulnerable. This vulnerability manifests through saltwater intrusion into estuaries and coastal aquifers, which can compromise freshwater resources and degrade critical habitats such as mangroves and tidal flats (Sekar et al. 2024; Basheer Ahammed and Pandey 2021; Zambrano-Medina et al. 2023).

Remote sensing techniques provide reliable data for monitoring shoreline changes, effectively capturing the dynamic interface between land and water. When integrated with Geographic Information Systems (GIS), these techniques offer more accurate and consistent information derived from satellite data (Kaliraj et al. 2016; Kılar 2023). To support detailed shoreline analysis, the US Geological Survey (USGS) developed the Digital Shoreline Analysis System (DSAS), a freely accessible software tool within the ArcGIS platform. DSAS enables the delineation of shorelines and the assessment of their fluctuations through statistical measures (Baig et al. 2020; Basheer Ahammed and Pandey, 2022). This tool is invaluable for conducting comprehensive assessments of shoreline dynamics using satellite imagery, offering critical insights into long-term trends in shoreline movement (Dey et al. 2021; Magesh et al. 2014; Seenipandi, 2020; Sundar and Sannasiraj, 2009). Key statistical methods, including End Point Rate (EPR), Net Shoreline Movement (NSM), and Shoreline Change Envelope (SCE), can be effectively calculated using DSAS in combination with remote sensing imagery and GIS. In this study, satellite imagery and DSAS's statistical tools were employed to assess shoreline changes along the Point Calimere coastal stretch in Tamil Nadu.

2 Study Area

The study area is located on the Coromandel Coast of the Bay of Bengal in the state of Tamil Nadu, India. The geographical extent of the study area covers a latitude of 9° 33′ 52.41″ N–10° 47′14.76″ N and a longitude of 78° 53′45.59″ E—79°50′42.28" E with the coastal length of 250 km. Figure 1 shows the geographical location of the study area. The study area covers Point Calimere (Kodiyakarai coast), well-known for its wetlands (Ramsar site) and unique coastal configuration structure. Administratively, the coastal stretch extends into Karaikkal, Nagapattinam, Thiruvarur, Thanjavur, and Pudukkottai districts. The coastal stretch is characterized by a plain topography with a gentle slope descending toward the coast. The major river flowing through this region is the Coleroon, which features recent alluvial formations. The aquifer system in Point Calimere is characterized by a reasonably thick, discontinuous, confined layer of fresh groundwater, which is overlaid by saline water toward the coast. The water table in this area ranges from 1 to 8 m in depth. Dunes predominantly cover the landscape between Nagapattinam and Vedaranyam, while the cultivated soils in this region are generally sandy in texture. The district receives rainfall under the influence of both the southwest and northeast monsoons.

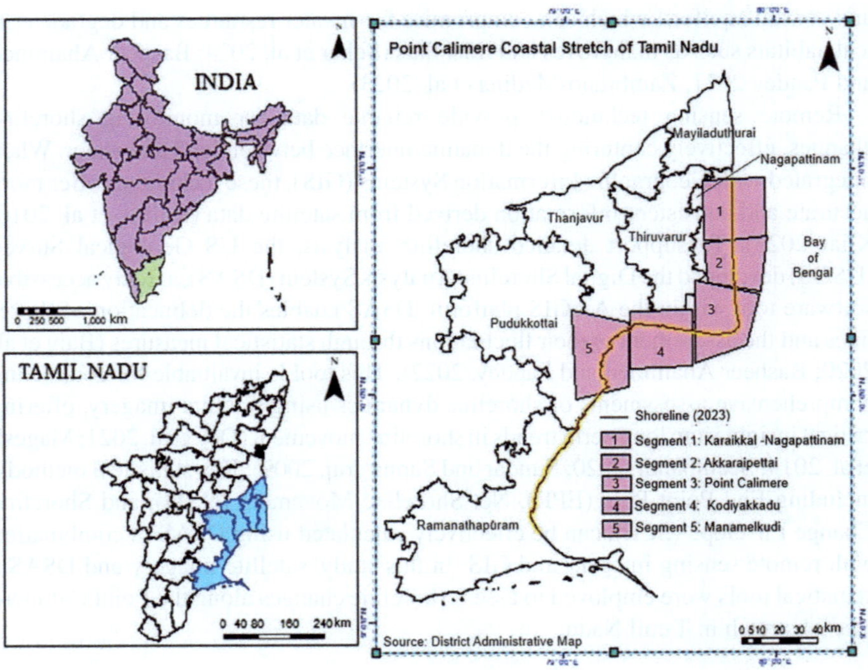

Fig. 1 Location map of the study area

3 Materials and Methods

The shoreline changes along the Point Calimere coastal stretch were determined using satellite imagery from Landsat 7 ETM + dated September 10, 2000, and Landsat 9-OLI dated June 14, 2024. Landsat images were collected from the USGS-Earth Explorer portal using minimal cloud cover (< 10%) to ensure precise shoreline extraction. The specific details of the selected images are presented in Table 1.

The methodology employed in this study is summarized in Fig. 2. Initially, the Landsat images underwent radiometric and atmospheric corrections to enhance data quality and accuracy. The images were then systematically pre-processed to improve their spectral and spatial characteristics, enabling a clearer distinction between land and water features. Shoreline extraction was performed using the Normalized Difference Vegetation Index (NDVI) and the Normalized Difference Water Index (NDWI), which facilitated the identification of the shoreline—the distinct boundary between

Table 1 Characteristics of satellite images used in the study

Sl. No.	Satellite/Sensor	Resolution	Date of acquisition	Spectral bands	Path/Row
1	Landsat 7 ETM +	30 m	September 2000	8	142/53
2	Landsat 9-OLI	30 m	June 2024	8	142/53

land and water. This process also encompassed the delineation of open water bodies, wetlands, and river mouths (estuaries) within the study area. In the NDVI analysis, index values range from − 1 to 1, indicating the vegetation and non-vegetation features; similarly, the NDWI index values indicate the difference between land and water features, wherein higher values indicate the presence of water and lower values indicate non-water features, such as vegetation or bare soil (Singh et al. 2023). The ratio of VNIR bands of the Landsat 7 and 9 images is used to compute the NDVI and NDWI, enhancing the identification of water bodies. The NDVI and NDWI equations are described as follows (McFeeters 1996).

$$\mathrm{NDVI} = (\mathrm{NIR} - \mathrm{R}) / (\mathrm{NIR} + \mathrm{R}) \tag{1}$$

In this analysis, the NDVI equation is executed on the Landsat 7—ETM + image as NDVI = (Band 4—Band 3)/(Band 4 + Band 3), and on the Landsat 8–9, NDVI

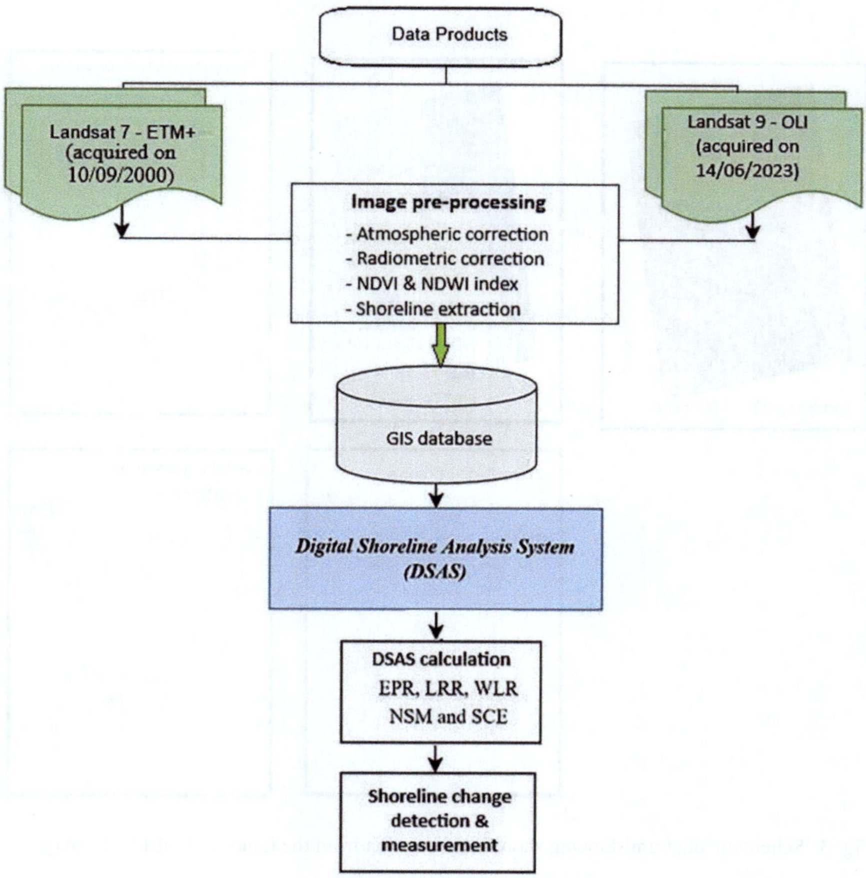

Fig. 2 Methodology flow chart showing the shoreline change detection using DASAS

= (Band 5—Band 4)/(Band 5 + Band 4). NDVI provides a single band product with an index value between − 1 and + 1, whereas < 0 indicates the occurrence of water bodies (coastal water, wetlands, and backwater).

Similarly, the NDWI equation is expressed as,

$$NDWI = (Green - NIR) / (Green + NIR) \tag{2}$$

In this analysis, the equation represented for Landsat 7 data is NDWI = (Band 2—Band 4)/(Band 2 + Band 4), and for Landsat 8 data is NDWI = (Band 3—Band 5)/(Band 3 + Band 5). NDWI values greater than 0.5 indicate the presence of water bodies, while values less than 0.2 suggest vegetation and built-up areas. The NDWI results effectively distinguish land from water, enabling the extraction of shorelines for the years 2000 and 2024 along the Point Calimere coastal stretch. Figure 3 illustrates the schematic steps involved in shoreline extraction from the Landsat 9-OLI False Colour Composite (FCC) image.

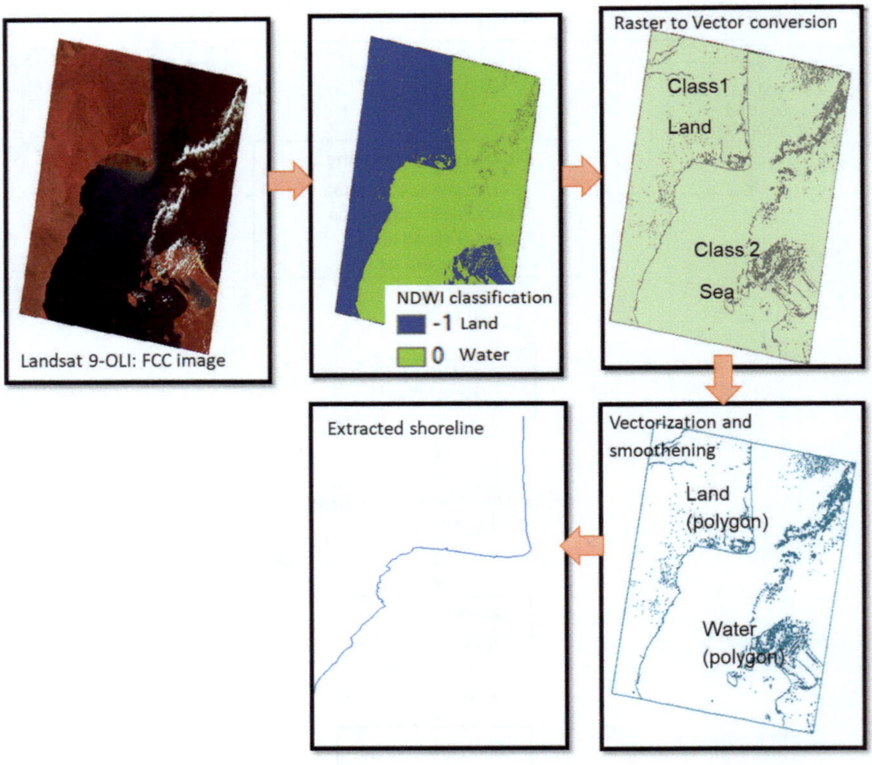

Fig. 3 Schematic diagram showing shoreline extraction from the Landsat 9-OLI FCC image

DSAS-Based Shoreline Change Analysis

The Digital Shoreline Analysis System (DSAS) was employed to assess shoreline changes using extracted shoreline data from the years 2000 and 2024. DSAS computes rate-of-change statistics for time-series shoreline datasets (Kılar 2023; Singh et al. 2023). It utilizes a transect-based method, generating transects perpendicular to a user-defined baseline and intersecting multiple shoreline positions across different periods (Oyedotun 2014; Quang et al. 2021). These intersection points between transects and shorelines serve as the basis for calculating change rates (Quang et al. 2021). In this study, three key parameters were used: End Point Rate (EPR), Net Shoreline Movement (NSM), and Shoreline Change Envelope (SCE). EPR provides the rate of shoreline change in meters per year (m/yr), while NSM and SCE represent the total distance of shoreline movement in meters (Ankrah 2024).

EPR is calculated by dividing the Net Shoreline Movement (NSM) by the time interval between the earliest and latest shoreline positions. This method estimates the average annual rate of shoreline change by quantifying the total distance of shoreline movement over the elapsed time.

$$\text{EPR} = \text{Distance of shoreline change(NSM)} \big/ \text{number of years of shoreline change} \tag{3}$$

EPR estimates the general rate of shoreline change between two shoreline positions, while the Linear Regression Rate (LRR) is applied when more than two shoreline positions are available, providing a more accurate and statistically robust estimate. Shoreline change characteristics are further described using NSM and SCE. NSM measures the absolute distance of shoreline shift between the oldest and most recent shoreline positions along each transect without considering the rate of change. In contrast, SCE represents the maximum extent of shoreline movement, calculated as the distance between the shoreline's furthest and nearest positions from the baseline across the entire period. Using DSAS, shoreline changes from 2000 to 2024 were analyzed to estimate both the distance and rate of shoreline movement. These outputs were then spatially visualized using GIS techniques to evaluate shoreline dynamics along the Point Calimere coastal stretch.

3.1 Uncertainty Analysis

The results derived from the DSAS include both positive and negative values, which indicate shoreline accretion and erosion, respectively. However, DSAS outputs are subject to certain uncertainties that may affect the accuracy of the results. These uncertainties can arise from several sources, including the spatial resolution of satellite images, image registration errors, digitization inaccuracies, variations in tidal levels at the time of image acquisition, and elevation-related discrepancies (Gopinath et al. 2023).

4 Results and Discussion

The DSAS measures the shoreline change using two estimations, namely distance and rate, using the statistical approaches of SCE, NSM, and EPR. The SCE and NSM measure the distance of the shoreline shift at a transect within the time interval. However, the EPR measures the rate of shoreline change. The analysis of shoreline changes in Nagapattinam District, Tamil Nadu, India, was conducted using the Digital Shoreline Analysis System (DSAS) version 6.0.168, covering a period from 2000, to 2024. This comprehensive analysis employed rigorous methodological standards, including a 90% confidence interval and a 2 m shoreline threshold. Five distinct rate types were evaluated: Shoreline Change Envelope (SCE), Net Shoreline Movement (NSM), End Point Rate (EPR), Linear Regression Rate (LRR), and Weighted Linear Regression (WLR). A total of 2,676 transects were analyzed, providing a robust dataset for assessing shoreline changes. The long-term shoreline dynamics (2000–2024) at 100 m transects along the coastal stretches showed the remarkable changes due to erosion and accretion processes, and the EPR is measured at the average erosion rate of −0.20 m/yr (found in 48.22% of transects), and an average accretion rate of 0.25 m/yr (31.18% of transects), respectively, with the uncertainty value of +/− 1m. The maximum erosion rate is -18.80 m/yr at the transect ID 943, in segment 3 (Point Calimere – Kodiyakarai), meanwhile the maximum accretion rate is at 17.29 m/yr at the transect ID 1771, in segment 5 (Manamelkudi). Notably, the 20.60% of transects in moderate change (seasonal erosion/accretion) with an average rate of 0.12 m/yr. The cumulative measurement of EPR and NSM is given in Table 2.

The Shoreline Change Envelope (SCE) quantifies the maximum extent of shoreline movement over time, helping to identify zones of erosion, accretion, and relative stability. It is calculated by measuring the distance between the most landward and most seaward shoreline positions observed during the study period, encompassing all intermediate shoreline locations. Figure 4 illustrates the SCE variability across the shoreline and highlights areas vulnerable to significant change. As a non-directional metric, SCE provides an overall measure of shoreline fluctuation. The average SCE

Table 1 The cumulative measurement of EPR and NSM for 2000–2024

Shoreline Dynamics (2000–2024)	EPR (m/yr)			NSM (m)		
	Erosion	Accretion	Moderate change	Erosion	Accretion	Moderate change
Average rate (for all 5 segments)	−0.20 m/yr	0.25 m/yr	0.12 m/yr	−14.62 m	+23.68 m	4.15 m
% of Area	48.22%	31.18%	20.60%	52.64%	36.31%	11.05%
Maximum rate	−18.80 m/yr	17.29 m/yr	12.04m/yr	−47.93 m	+33.56 m	28.62 m
Transect ID	ID 943	ID 1771	ID 724	ID 943	ID 1771	ID 724–726

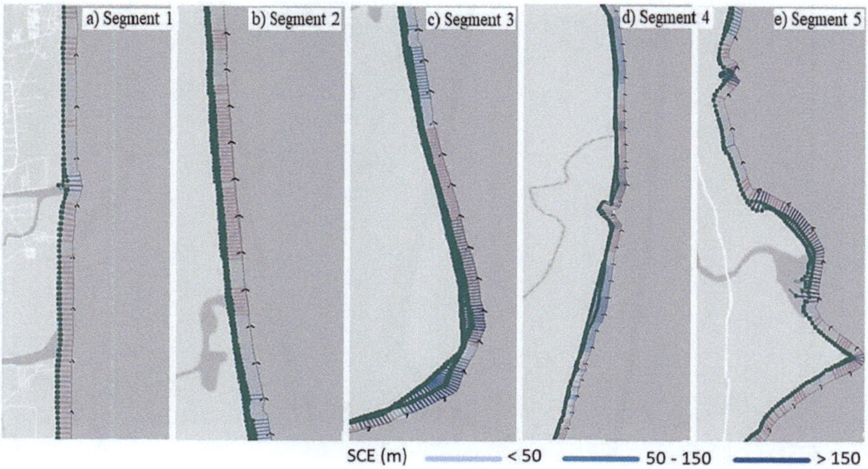

Fig. 4 Shoreline change envelope (SCE) rate: **a** Karaikkal–Nagapattinam, **b** Akkaraipettai–Nagapattinam, **c** Point Calimere, **d** Kodiyakkadu, and **e** Manamelkudi

across all transects is 40.24 m. The maximum SCE was recorded at Transect ID 943 with 427.93 m, while the minimum value of 0.01 m was observed at Transect ID 415, indicating considerable spatial variability in shoreline dynamics within the study area.

Net Shoreline Movement (NSM) represents the total linear distance between the oldest and most recent shoreline positions along a transect. Unlike the End Point Rate (EPR), which expresses the rate of shoreline change per year, NSM quantifies the absolute shoreline displacement over the analyzed period. Figure 5 indicates the spatial variability of NSM at the transect level, which calculates the net change in shoreline position from 2000 to 2024, showing an average distance of − 4.62 m, indicating a general trend of erosion. A significant portion of the transects, about 60.69%, exhibited negative distances corresponding to erosion, with the most severe erosion being − 427.93 m at transect ID 943. Conversely, 39.31% of the transects showed positive distances, indicating accretion, with the highest accretion measured at 393.56 m at transect ID 1771.

The End Point Rate (EPR), which calculates the rate of shoreline change (m/year) using only the earliest and latest shoreline positions in the time series, averaged − 0.20 m per year across all transects. Figure 6 shows the EPR rate that indicates the site of erosion and accretion in all segments. This negative average rate highlights an overall trend toward erosion over the analyzed period. Despite this, none of the transects exhibited statistically significant erosion or accretion rates. The highest recorded erosion rate was − 18.80 m per year at Transect ID 943, while the maximum accretion rate reached 17.29 m per year at Transect ID 1771.

The Linear Regression Rate (LRR) estimates the rate of shoreline change (m/year) at each transect by fitting a least-squares regression line through all available shoreline positions. The slope of this line indicates the average rate of change over

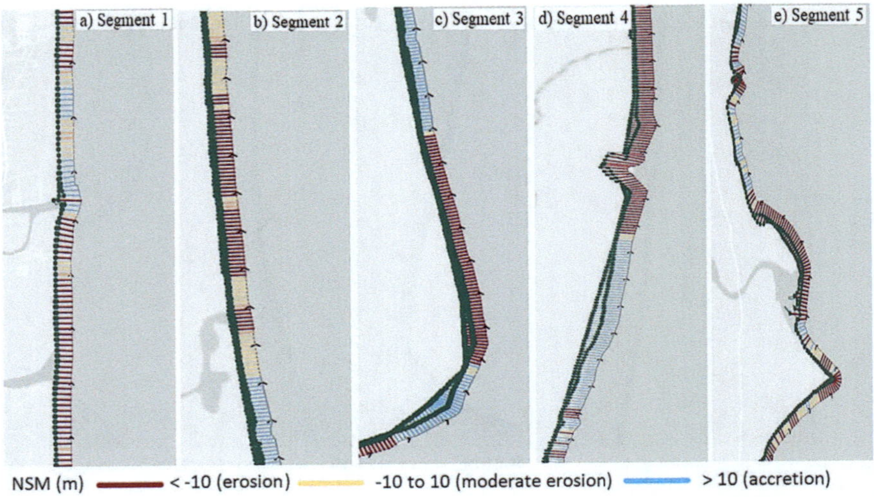

Fig. 5 Net shoreline movement (NSM) rate: **a** Karaikkal–Nagapattinam, **b** Akkaraipettai–Nagapattinam, **c** Point Calimere, **d** Kodiyakkadu, and **e** Manamelkudi

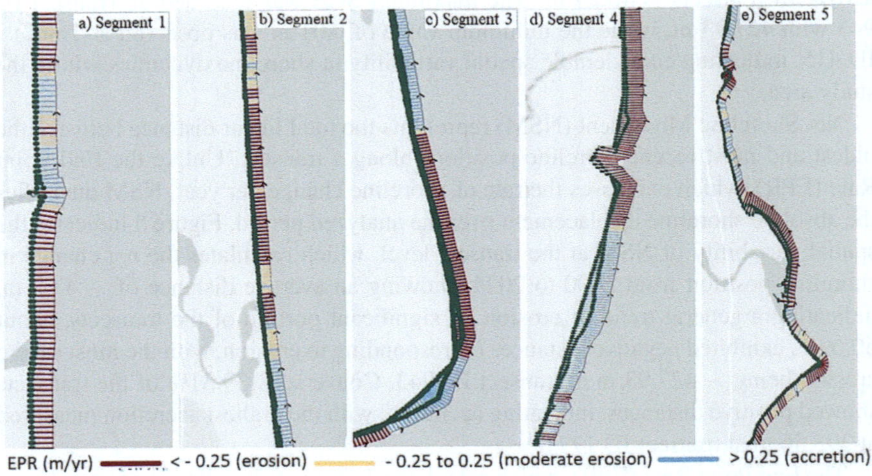

Fig. 6 End Point Rate (EPR) rate: **a** Karaikkal–Nagapattinam, **b** Akkaraipettai–Nagapattinam, **c** Point Calimere, **d** Kodiyakkadu, and **e** Manamelkudi

the study period. LRR accounts for trends over multiple years, providing a more comprehensive analysis of long-term shoreline behavior. LRR is a robust method for assessing long-term shoreline trends, offering a more stable and statistically sound rate of change than methods based on fewer data points. The LRR is measured at the average rate of 1.89 m/yr., with a Confidence Interval (CI) value of 9.14 m/yr. (average of across transects), Moreover, the reduced 'n' (number of independent transects) is 31.65. Significantly, the uncertainty of the shoreline is considered as ± 1.63 m/yr., and the average rate of uncertainty is ± 1.89 m/yr. Moreover, the weighted linear regression (WLR) is estimated at the average rate of 1.97 m/year, which shows the Confidence Interval (CI) value at 9.04 m/year (average across transects) and the uncertainty value at ± 1.61 m/year, and the average rate with uncertainty is at 1.97 m/year.

The analysis revealed that the maximum accretion rate was 48.35 m/year at Transect ID 951, with an average accretion rate of 7.76 m/year. In contrast, the maximum erosion rate reached − 55.42 m/year at Transect ID 2783, and the average erosion rate was estimated at − 4.20 m/year. The study, covering 23 years (2000–2024), offers critical insights into shoreline dynamics. Results indicate a predominant trend of erosion, affecting 60.69% of the transects, with an average End Point Rate (EPR) of − 0.20 m/year. Meanwhile, 39.31% of the transects exhibited accretion. This comprehensive assessment underscores the highly dynamic nature of the shoreline and emphasizes the urgent need for sustained monitoring and effective coastal zone management to mitigate the impacts of both erosion and accretion.

5 Conclusion

The coastal region of Nagapattinam District in Tamil Nadu, India, undergoes significant changes due to both natural processes and human activities. The shoreline in Nagapattinam is highly variable; some areas face severe erosion while others experience accretion. These changes are influenced by monsoon wave action, sea-level rise, and sediment supply. The 2004 Indian Ocean tsunami had a profound impact on the Nagapattinam shoreline, causing extensive erosion and altering the coastal morphology. The recovery from this event has been slow, with ongoing erosion in some areas due to modified sediment dynamics and disrupted coastal processes. Coastal infrastructure, including sea walls, breakwaters, and land reclamation projects, has significantly affected shoreline dynamics. While these structures are designed to protect against erosion, they often disrupt natural sediment transport, increasing erosion in adjacent areas.

Effective management of the Nagapattinam shoreline requires an integrated approach that takes into account both natural processes and human impacts. Strategies for Integrated Coastal Zone Management (ICZM) should include ongoing monitoring, community involvement, and adaptive management to address the dynamic nature of the coastline. Engaging local communities in coastal management is essential for sustainable shoreline conservation. Community-based initiatives can raise awareness, promote sustainable practices, and protect coastal ecosystems and livelihoods. In conclusion, addressing the challenges of shoreline erosion and accretion in Nagapattinam requires a comprehensive and adaptive approach. Integrating scientific research with sustainable management practices and community participation makes creating a balanced and resilient coastal environment in Nagapattinam possible.

Acknowledgements The corresponding author is thankful to the Director, National Centre for Earth Science Studies (NCESS), Ministry of Earth Sciences (MoES), Government of India, Thiruvananthapuram, for providing continuous support and encouragement to complete this successful research work. The authors thankful to the Editors and anonymous reviewers for their insightful comments and suggestions. The authors acknowledge the IMD, GSI, SOI, and USGS-Earth Explorer for providing the necessary data sources.

References

Ajai NS (2012) Coastal zones of india, n.d. space application centre (ISRO), Ahmedabad 1, pp 1–609

Ankrah J (2024) Shoreline change and coastal erosion: an analysis of long-term change and adaptation strategies in coastal Ghana. Geo-Mar Lett 44(3):12. https://doi.org/10.1007/s00367-024-00774-5

Awad M, El-Sayed HM (2021) The analysis of shoreline change dynamics and future predictions using automated spatial techniques: case of El-Omayed on the Mediterranean coast of Egypt. Ocean Coast Manag 205:105568

Baig MRI, Ahmad IA, Shahfahad N, Tayyab M, Rahman A (2020) Analysis of shoreline changes in Vishakhapatnam coastal tract of Andhra Pradesh, India: an application of digital shoreline analysis system (DSAS). Ann GIS 26(4):361–376. https://doi.org/10.1080/19475683.2020.1815839

Basheer Ahammed KK, Pandey AC (2019) Shoreline morphology changes along the eastern coast of India, Andhra Pradesh by using geospatial technology. J Coast Conserv 23:331–353

Basheer Ahammed KK, Pandey AC (2021) Geospatial application for coastal morphology changes along the sand mining coast: a case study on Alappad, Kerala. In: Recent technologies for disaster management and risk reduction: sustainable community resilience & responses, pp 465–479

Basheer Ahammed KK, Pandey AC (2022) Assessment and prediction of shoreline change using multi-temporal satellite data and geostatistics: a case study on the eastern coast of India. J Water Clim Change 13(3):1477–1493

Ciecholewski M (2024) Review of segmentation methods for coastline detection in SAR images. Arch Comput Methods Eng 31(2):839–869. https://doi.org/10.1007/s11831-023-10000-7

Dey M, Sakthivel SP, Jena BK (2021) A shoreline change detection (2012–2021) and forecasting using digital shoreline analysis system (DSAS) tool: a case study of dahej coast, gulf of khambhat, Gujarat, India. Indonesian J Geogr 53(2). https://doi.org/10.22146/ijg.56297

Dike EC, Ameme BG, Efeovbokhan O (2024) Shoreline position trends in the Niger delta: analyzing spatial and temporal changes through Sentinel-1 SAR imagery. Geomatics Nat Hazards Risk 15(1). https://doi.org/10.1080/19475705.2024.2346150

Gopinath G, Thodi MFC, Surendran UP, Prem P, Parambil JN, Alataway A, Al-Othman AA, Dewidar AZ, Mattar MA (2023) Long-term shoreline and islands change detection with digital shoreline analysis using RS data and GIS. Water 15(2):244. https://doi.org/10.3390/w15020244, https://doi.org/10.1016/j.ocecoaman.2021.105568

Kaliraj S, Chandrasekar N, Magesh NS (2013a) Evaluation of coastal erosion and accretion process along the Southwest coast of Kanyakumari, Tamil Nadu using geospatial techniques. Arab J Geosci. ISSN: 1866-7538, https://doi.org/10.1007/s12517-013-1216-7

Kaliraj S, Chandrasekar N, Magesh NS (2013b) Impacts of wave energy and littoral currents on shoreline erosion/accretion along the southwest coast of Kanyakumari, Tamil Nadu using DSAS and geospatial technology. Environ Earth Sci. https://doi.org/10.1007/s12665-013-2845-6

Kaliraj S, Chandrasekar N, Ramachandran KK (2016) Mapping of coastal landforms and volumetric change analysis in the southwest coast of Kanyakumari south India using remote sensing and GIS techniques. Egypt J Remote Sens Space Sci. https://doi.org/10.1016/j.ejrs.2016.12.006

Kaliraj S, Chandrasekar N, Ramachandran KK, Srinivas Y, Saravanan S (2017) Coastal landuse and land cover change and transformations of Kanyakumari coast India using remote sensing and GIS. Egypt J Remote Sens Space Sci. https://doi.org/10.1016/j.ejrs.2017.04.003

Kılar H (2023) Shoreline change assessment using DSAS technique: a case study on the coast of Meriç Delta (NW Türkiye). Reg Stud Mar Sci 57:102737. https://doi.org/10.1016/j.rsma.2022.102737

Kundu K, Mandal JK (2024) Shoreline change detection and future prediction of Sundarban delta using remote sensing data and digital shoreline analysis system. J Indian Soc Remote Senss 52(3):485–503. https://doi.org/10.1007/s12524-024-01833-1

Magesh NS, Chandrasekar N, Krishnakumar S (2014) Evaluation of coastal environmental changes along the Nagapattinam coast, Tamil Nadu, India. Environ Monit Assess 186(8):5209–5220

McFeeters SK (1996) NDWI by McFeeters. Remote Sens Environ 25:687–711

Rani M, Kaliraj S, Sufia R, Kumar P, Haroon S (2020) Remote sensing of ocean and coastal environments, 1st edn. In: Earth Observation Series (Elsevier Publications USA). ISBN: 978-0-12-819604-5, https://doi.org/10.1016/C2019-0-00225-7

Mujabar PS, Chandrasekar N (2013) Shoreline change analysis along the coast between Kanyakumari and Tuticorin of India using remote sensing and GIS. Arab J Geosci 6(3):647–664. https://doi.org/10.1007/s12517-011-0394-4

Oyedotun TD (2014) Shoreline geometry: DSAS as a tool for historical trend analysis. Geomorphol Tech 3(2.2):1–12

Quang DN, Ngan VH, Tam HS, Viet NT, Tinh NX, Tanaka H (2021) Long-term shoreline evolution using DSAS technique: a case study of Quang Nam province, Vietnam. J Mar Sci Eng 9(10):1124. https://doi.org/10.3390/jmse9101124

Seenipandi K (2020) Coastal erosion and mitigation strategies in Tamil Nadu. J Coastal Res 36(2):213–224

Sekar LG, Androws X, Annaidasan K, Kumar A, Kannan R, Muthusankar G, Balasubramani K (2024) Assessment of shoreline changes and associated erosion and accretion pattern in coastal watersheds of Tamil Nadu, India. Nat Hazards Res 4(2):231–238. https://doi.org/10.1016/j.nhres.2023.09.008

Singh S, Meraj G, Kumar P, Singh SK, Kanga S, Johnson BA, Prajapat DK, Debnath J, Sahariah D (2023) Decoding chambal river shoreline transformations: a comprehensive analysis using remote sensing, GIS, and DSAS. Water 15(9):1793. https://doi.org/10.3390/w15091793

Sundar V, Sannasiraj SA (2009) Post-tsunami coastal erosion and recovery in Nagapattinam District, Tamil Nadu, India. J Coast Res 25(1):11–20

Zambrano-Medina YG, Plata-Rocha W, Monjardin-Armenta SA, Franco-Ochoa C (2023) Assessment and forecast of shoreline change using geospatial techniques in the Gulf of California. Land 12(4):782. https://doi.org/10.3390/land12040782

Estimation of Ashtamudi Lake Water Salinity and Its Spatiotemporal Variability Using Time-Series Sentinel-2A MSI Data and Remote Sensing Indices

S. Kaliraj and P. S. Fathimathu Sahala

Abstract Ashtamudi Lake, a tropical backwater located on the southwest coast of India, is connected to the Arabian Sea and experiences significant spatiotemporal variations in Lake Water Salinity (LWS). This study leverages Sentinel-2A MSI imagery to examine LWS variability in Ashtamudi Lake over eight months, covering both monsoon and non-monsoon seasons from 2018 to 2019. An empirical algorithm was developed to estimate LWS (PSU) from Sentinel-2A MSI bands. The results show that LWS ranges from 6.68 to 25.09 PSU during the monsoon, rising to 7.75–29.98 PSU in the non-monsoon season. LWS peaks at 29.98 PSU in February at the lake's mouth, driven by saline seawater inflow and tidal mixing. Monsoon freshwater inputs from river discharge and precipitation notably lower Lake Water Salinity (LWS) in nearshore and inland areas. Cross-validation with in situ data reveals a strong correlation, with a linear regression (R^2) value of 0.994 and a root mean square error (RMSE) of 0.448, demonstrating the high accuracy of LWS estimates derived from Sentinel-2A MSI images.

Keywords Lake water salinity · Sentinel-2A · Ashtamudi lake · Lake environment · Remote sensing

S. Kaliraj (✉) · P. S. Fathimathu Sahala
National Centre for Earth Science Studies (NCESS), Ministry of Earth Sciences (MoES), Government of India, Thiruvananthapuram, Kerala, India
e-mail: s.kaliraj@ncess.gov.in

P. S. Fathimathu Sahala
Ministry of Earth Science (MoES) Research Fellow (MRFP), DESK, IITM, Pune, India

NCESS Research Scholar, Faculty of Marine Sciences, CUSAT, Kochi, Kerala, India

K. K. Basheer Ahammed et al. (eds.), *Monitoring Coastal Marine Ecosystems Under Climate Change*, https://doi.org/10.1007/978-3-032-09137-6_2

1 Introduction

Lake Water Salinity (LWS) is a dynamically changing physicochemical property of lake water that varies across both spatial and temporal scales. It plays a critical role in influencing other water quality parameters such as Lake Water Temperature (LWT), chlorophyll-a (Chl-a), and nutrient levels, especially in lakes located near coastal regions (Barre et al. 2008; Font et al. 2010; Dierssen and Kaylan 2012; Akhil et al. 2016; Kaliraj et al. 2019a, b). Generally, the salinity of lake water is defined as the concentration of salts, such as sodium and chloride, within the lake. It is measured in Practical Salinity Units (PSU), where 1 PSU is equivalent to 1 g of salt per kg of water, based on the electrical conductivity properties of the lake water (Dehadri, 1970; McLusky, 1989; Prabhu et al. 2000; Soundarapandian et al. 2009; Smitha et al. 2014; Momin et al. 2016). For context, the average salinity of the global ocean is approximately 35.5 PSU, but this can range from less than 15 PSU in estuaries, where freshwater mixes with seawater, to over 40 PSU in hypersaline environments like the Dead Sea (Lewis 1980; UNESCO 1981; Mantyla 1987).

Monitoring and assessing lake water properties, including physical, chemical, and biological constituents, through in situ sampling methods provides highly accurate results. However, these methods are limited in their ability to capture the spatiotemporal variability of lake water properties efficiently (Gabarro et al. 2004; Boutin and Martin 2006; Brown et al. 2008; Geiger et al. 2013; Abe and Ebuchi 2014; Bhaskar and Jayaram 2015; Dinnat et al. 2019). Remote sensing technology, on the other hand, provides a powerful platform for the efficient assessment of water properties across diverse water bodies, including rivers, lakes, reservoirs, coastal regions, and oceans (Ellison et al. 1998; Sulochana and Muniyandi 2005; Ahn et al. 2008; Fournier et al. 2015; Klemas 2011; Lagerloef 2014; Durack 2015; Eladawy et al. 2017).

Multi-temporal satellite imagery is a valuable tool for monitoring and assessing Lake Water Salinity (LWS), Lake Water Temperature (LWT), and chlorophyll-a (Chl-a), which are crucial for understanding oceanic and climate dynamics (Burrage and Wesson 2008; Sharma et al. 2010a, b; Kaliraj et al. 2014a, b; Durack 2015; Rajabi et al. 2017). These images provide synoptic coverage, enabling water quality evaluation across spatial and temporal scales at both regional and global levels (Boutin and Martin 2006; Dierssen and Kaylan 2012). Satellites such as LANDSAT, Sentinel, MODIS, and Aquarius are capable of measuring LWS with an accuracy of 0.2–0.4 PSU, which aligns closely with in situ measurements from ARGO floats, RAMA, and OMNI buoys (Delcroix and McPhaden 2002; Brown et al. 2008; Font et al. 2010; Marghany et al. 2010; Abe and Ebuchi 2014; Bhaskar and Jayaram 2015; Dinnat et al. 2019). Satellite-derived data for LWT and LWS effectively capture surface conditions (depth = 0) but have limitations in deeper water measurements, such as those obtained by ARGO instruments, which operate at depths greater than 5 m. The NASA Aquarius mission (June 2011–August 2015), designed to measure sea surface salinity (SSS) globally, also contributed to LWS estimation in other water bodies. While the Aquarius data provides valuable insights, its coarse spatial resolution of 150 km and temporal resolution of 7 days limit its effectiveness for detailed studies of lakes, coastal, and estuarine environments. (Abe and Ebuchi 2014; Seelanki et al. 2018). Most satellite missions targeting salinity have focused on oceanic applications, with their derived products being too coarse for inland and coastal water studies (Marshall and Elliot, 1998; Kerry and Black, 2001; Damotharan et al. 2010; Ratheesh et al. 2013; Lin et al. 2019).

Despite some limitations, multispectral images from satellites like Landsat and MODIS are extensively used to estimate LWS, LWT, and chlorophyll-a (Chl-a) using linear regression algorithms. These satellite-derived estimates have shown strong correlations ($R^2 > 0.81$) with in situ sampling data, confirming their validity and usefulness in water quality assessments (Marghany et al. 2010; Salleh et al. 2013; Xiang et al. 2017; Kaliraj et al. 2019a, b). This satellite and ground-based data integration enhances our capacity to monitor and understand the dynamic changes in aquatic environments.

LWS products, derived from satellite images, facilitate routine monitoring of Lake Water Salinity. They are used for scientific applications such as tracking tropical salinity variability, seawater plumes, and climate change impacts. Researchers often use MODIS images to accurately retrieve sea surface temperature (SST) and sea surface salinity (SSS) from seawater columns at high resolution through statistical models and artificial neural network (ANN) algorithms. These methods frequently encounter data and methodological limitations, especially in coastal and lake environments where water properties can vary greatly (Ratheesh et al. 2013; Marghany et al. 2010; Salleh et al. 2013; Mishra et al. 2015; Xiang et al. 2017).

The use of multispectral images from the Landsat satellite series, including TM, ETM+, and OLI, has proven effective for measuring SSS and SST in coastal and ocean waters, achieving acceptable accuracy levels through various mathematical algorithms. These data are essential for studies on coastal and ocean dynamics due to their high spatial and temporal resolution, as well as consistent data quality. Similar approaches have been successfully applied to Sentinel data for studying lake environments (Kaliraj et al. 2019a, b). Building on these methods, the present study uses multi-temporal Sentinel-2A MSI images to estimate Lake Water Salinity and its monthly variability in Ashtamudi Lake.

2 Study Area

The coastal waters of Southern India exhibit significant variations in seawater salinity, which greatly influence coastal hydrodynamics and air-sea interaction processes. These salinity variations are crucial for understanding the climatic and monsoon patterns in the region. Ashtamudi Lake, an estuary connected to the Arabian Sea, is directly affected by these physicochemical and climatic factors that influence sea salinity (Kaliraj 2016). The salinity of Ashtamudi Lake is influenced by the freshwater input from the Kallada River, which flows into the lake. The Kallada River originates near Ponmudi in the Kulathupuzha hills of the Western Ghats and is formed by the merging of three tributaries: Kulathupuzha, Chenthurnipuzha, and Kalthuruthipuzha. This river travels approximately 121 km through pristine forests before reaching the Ashtamudi wetland at Peringalam, near Kollam. As Kerala's deepest estuary, Ashtamudi Lake reaches a maximum depth of 21 feet at the confluence (Fig. 1).

Geographically, Ashtamudi Lake is located between 8° 53′ 19.083″ N–9° 1′ 11.5″ N latitude and 76° 32′ 6.157″ E–76° 40′ 15.006″ E longitude. The lake extends about 61.42 km^2, with an average annual rainfall of 2400 mm, contributing to a yearly discharge of 3.375 km^3 of water. The region experiences temperatures

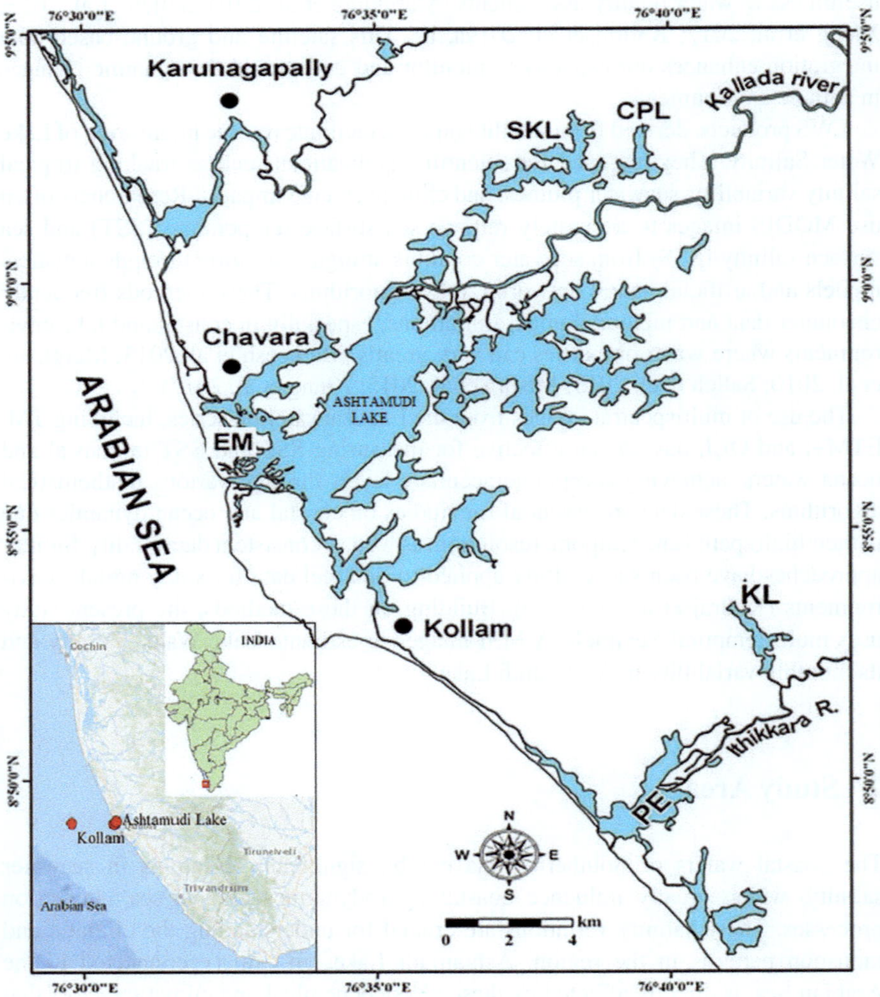

Fig. 1 The location map of the Study area. AE—Ashtamudi Lake; SKL—Sasthamkotta lake; CPL—Chelupola lake; EM—Estuarine mouth; PE—Paravur estuary; KL—Kotta Lake; SKSB (in inset map)—South Kerala Sedimentary Basin

ranging from a minimum of 25.5 °C to a maximum of 27.5 °C. The climate is usually hot and humid in April and May, while it is cooler in December and January. The water column of the lake is characterized by freshwater from the Kallada River and seawater from the Arabian Sea. The surrounding coastal landforms include laterite uplands, estuaries, dunes, and Quaternary and Tertiary sediments. Quaternary sediments in this area consist of recent marine and fluvial alluvium, while Tertiary sediments include laterite, sandstones, and clays from the Varkala formation. The coastal water system also receives discharges and runoff from the Kallada River's drainage networks (Kaliraj and Chandrasekar 2012; Kaliraj et al. 2015).

The depth of Ashtamudi Lake ranges from 1 to 10 m in the shallow areas near the shore and surf zones, increasing to depths of 20–30 m at a distance of 5–6 km offshore (Kaliraj et al. 2013; 2014a, b, 2017). These depths and water properties are subject to seasonal variations in hydrodynamics, such as changes in wind and currents. Monitoring and assessing Lake Water Salinity (LWS) and its monthly variability using multi-temporal Sentinel-2A MSI images is vital for understanding the hydrodynamic characteristics of the ocean and coastal environments in the study area.

3 Materials and Methods

Multi-temporal Sentinel-2A MSI images, sourced from the Copernicus Open Access Hub, were utilized to estimate Lake Water Salinity (LWS) and assess its monthly variability in the Ashtamudi Lake water column over eight months. The study covered the pre-monsoon (April–May 2018), monsoon (October–November 2018), and post-monsoon (December 2018–March 2019) seasons. LWS was mapped using a least squares algorithm, enabling detailed analysis of the spatial and seasonal variability patterns within the lake environment. The specifications of the Sentinel-2A MSI images, including the bands used for the study, are detailed in Table 1.

Figure 2 illustrates the flowchart outlining the systematic methodology for estimating Lake Water Salinity (LWS) and its monthly variability using Sentinel-2A MSI imagery. The analysis involved several key steps: initially, the Sentinel-2A multispectral bands were stacked, and land areas were masked using ENVI software to isolate the relevant water bodies. Subsequently, the ISODATA unsupervised classification technique was applied to distinguish land from the water column. Additional masking procedures were implemented to remove land, clouds, and intertidal zones, ensuring accurate boundary delineation between water and non-water areas. The least squares algorithm was then employed using the Raster Calculator module in ArcGIS to estimate LWS accurately. Given that the Sentinel-2A multispectral bands have varying spatial resolutions (10, 20, and 60 m), a resampling technique was used to achieve a uniform 10 m × 10 m pixel scale for LWS estimation. The LWS was calculated in Practical Salinity Units (PSU) using Eq. 1 through the Map Algebra tool in ArcGIS 10.6 software.

Table 1 Specifications of the Sentinel-2A MSI (L2A product)

Band	Wavelength (nm)	Spatial resolution (m)	Preference for the study
Band 1—coastal aerosol	443	60	Excluded
Band 2—blue	490	10	Used in the study
Band 3—green	560	10	
Band 4—red	665	10	
Band 5—vegetation red edge	705	20	Excluded
Band 6—vegetation red edge	740	20	
Band 7—near infrared (NIR)	783	20	
Band 8—Near infrared (NIR)	842	10	
Band 8A—near infrared (NIR)	865	20	Used in the study
Band 9—water vapor	945	60	Excluded
Band 10—cirrus	1375	60	Used in the study
Band 11—Shortwave infrared	1610	20	
Band 12—shortwave infrared	2190	20	

$$\begin{aligned}\text{LWS(PSU)} = [&14.256 - 240.163 * \text{Band 4} - 72.533 * \text{Band 8}A \\ &+ 124.700 * \text{Band 2} + 191.266 * \text{Band 3} + 36.044 * \text{Band 10} \\ &-11.117 * \text{Band 11} - 39.789 * \text{Band12}]\end{aligned} \tag{1}$$

where LWS denotes the salinity in Practical Salinity Units (PSU), and the *Bands* represent the specific Sentinel-2A MSI multispectral bands utilized in the analysis. The numerical coefficients in the regression equation were empirically derived from in situ salinity measurements. Figure 1 illustrates the spatial distribution of the 40 sampling points used to obtain in situ salinity data for Ashtamudi Lake. This systematic methodology facilitates a comprehensive LWS assessment, capturing spatial and seasonal variability within the study area. To validate the accuracy of the salinity estimates derived from Sentinel-2A MSI imagery, the results were cross-verified with in situ observations at the 40 sampling locations. Two statistical approaches were employed for validation: (i) Linear regression analysis (LRA), which quantifies the strength of the correlation, and (ii) root mean square error (RMSE) analysis, which evaluates the predictive accuracy of the model.

The RMSE, a widely accepted statistical measure, is calculated using the following equation, Eq. 2.

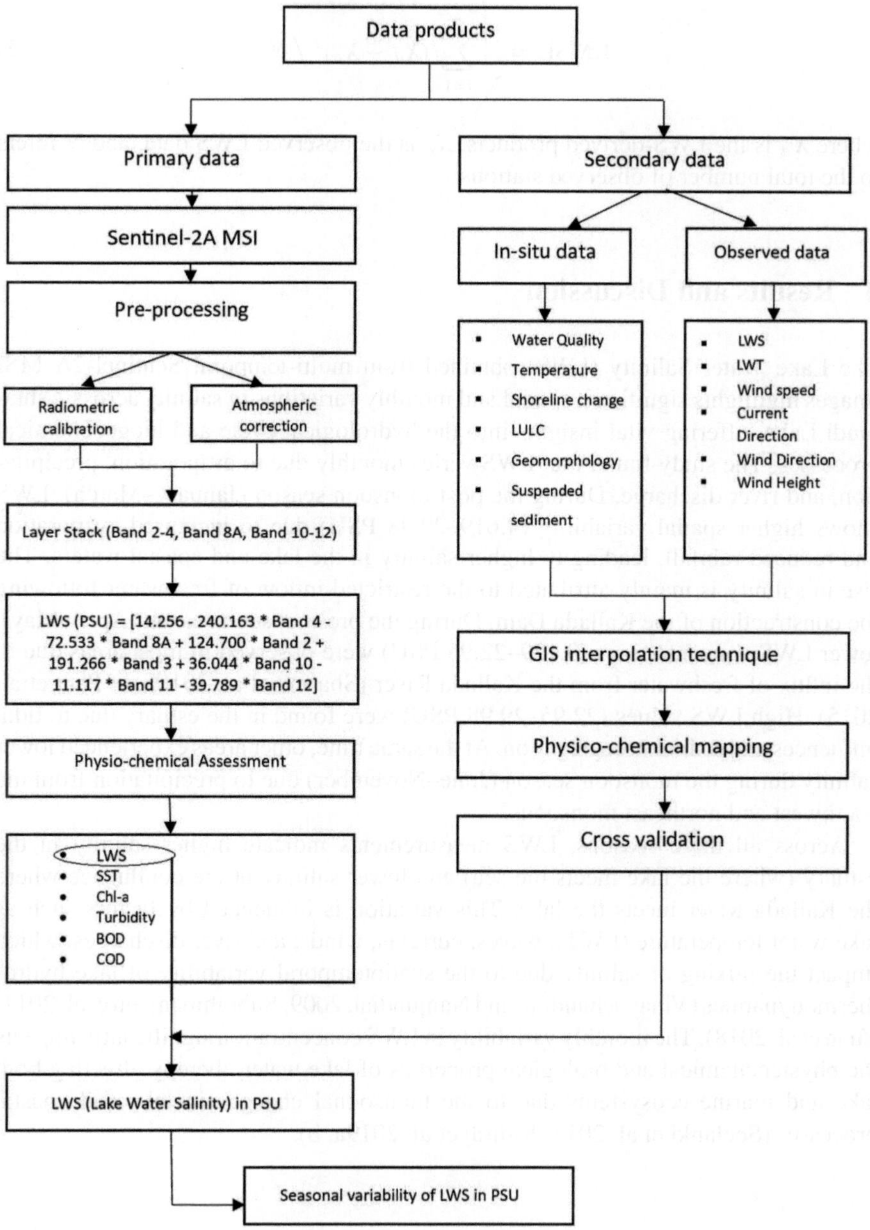

Fig. 2 Flowchart showing the method of LWS estimation using Sentinel-2A MSI data

$$\mathrm{RMSE} = \sqrt{\sum_{i=1}^{n} (X_d - X_o)^2 / N} \tag{2}$$

where X_d is the LWS-derived products, X_o is the observed LWS data, and N refers to the total number of observed stations.

4 Results and Discussion

The Lake Water Salinity (LWS) obtained from multi-temporal Sentinel-2A MSI images highlights significant spatial and monthly variations in salinity across Ashtamudi Lake, offering vital insights into the hydrological cycle and biogeochemical processes. The study found that LWS varies monthly due to evaporation, precipitation, and river discharge. During the post-monsoon season (January–March), LWS shows higher spatial variability (4.619–29.98 PSU) due to increased evaporation and reduced rainfall, leading to higher salinity in the lake and coastal waters. The rise in salinity is mainly attributed to the restricted inflow of freshwater following the construction of the Kallada Dam. During the pre-monsoon season (April–May), lower LWS concentrations (3.129–22.95 PSU) were observed in most areas due to the influx of freshwater from the Kallada River (Sharma et al. 2010a, b; Pant et al. 2015). High LWS values (22.95–29.98 PSU) were found in the estuary due to tidal influences and sediment segregation. At the same time, other areas experienced lower salinity during the monsoon season (June–November) due to precipitation from the southwest and northeast monsoons.

Across all three seasons, LWS measurements indicate higher salinity in the estuary (where the lake meets the sea) and lower salinity at the confluence where the Kallada River meets the lake. This variation is influenced by factors such as lake water temperature (LWT), waves, currents, winds, and river discharges, which impact the mixing of salinity due to the spatiotemporal variability of lake hydrothermodynamics (Vinayachandran and Nanjundiah 2009; Subrahmanyam et al. 2011; Yuan et al. 2018). The monthly variability in LWS concentration significantly impacts the physicochemical and biological properties of lake water, thereby affecting both lake and marine ecosystems due to the monsoonal changes in lake and coastal processes (Seelanki et al. 2018; Kaliraj et al. 2019a, b).

4.1 LWS Spatial Variability During Northeast Monsoon (October–November 2018)

Unlike most other Indian states, Kerala experiences two distinct monsoon periods: the Southwest monsoon, the primary monsoon from June to September, and the Northwest monsoon, a secondary contributor to Kerala's annual precipitation. During the Northwest monsoon, the monthly variation of salinity (LWS) in Ashtamudi Lake is estimated to range from 6.68 to 25.09 PSU. Notably, the lake water along the streams of the Kallada River exhibits higher LWS concentrations, reaching up to 24.53–25.09 PSU, compared to other regions. This elevated salinity is attributed to the leaching of fertilizers from nearby paddy fields and plantations, which, due to excessive runoff, introduce chemicals into the river water (Ratheesh et al. 2013; Mishra et al. 2015; Sengupta et al. 2016). From Fig. 3, it is evident that LWS spatial variability in lake water during monsoon (October–November) is found higher in the stream of Kallada River than in the middle of the lake due to the influence of northeast monsoon on lake hydrodynamic processes. The lake water, especially along the Kallada River stream, shows a higher average LWS concentration of 24.81 PSU. In contrast, the estuary water has lower salinity due to the discharge of freshwater from several river outlets.

4.2 LWS Spatial Variability in Pre-monsoon (April–May)

During the pre-monsoon period (April–May), the LWS in Ashtamudi Lake is generally lower, ranging from 3.129 to 7.55 PSU, significantly less than in other seasons, where the average LWS is higher at 25.465 PSU. Figure 3 illustrates the spatial variability of LWS during the pre-monsoon. The offshore water column, located away from the mouth of the lake, shows lower salinity levels (3.129–7.55 PSU) due to reduced tidal influence in these areas. Conversely, certain pockets within the lake and along the Kallada River stream exhibit higher salinity levels, ranging from 22.95 to 27.98 PSU. Elevated salinity is also observed in dried-up streams and tributaries due to increased evaporation rates (> 27.98 PSU). It is often noted that water from Ashtamudi Lake contributes to the salinity of the nearby Arabian Sea (Ratheesh et al. 2013; Kaliraj et al. 2014a, b; Lin et al. 2019). Some areas within the lake exhibit lower salinity levels (< 22 PSU), likely due to variations in monsoonal processes.

4.3 LWS Spatial Variability in Post-monsoon (December–March)

Figure 3 illustrates the spatial variability of Lake Water Salinity (LWS) during the post-monsoon period (December–March). During this period, the average LWS in the broader expanse of Ashtamudi Lake is around 25.2475 PSU. Most offshore waters, including those from the Kallada River stream and the farther portions of the lake, exhibit lower salinity levels (4.619–10.49 PSU). This reduced salinity is attributed to the annual precipitation collected in the Kallada River tributaries, which flows into Ashtamudi Lake throughout the post-monsoon period. However, higher salinity levels are observed in the nearshore regions, particularly in the estuary. This

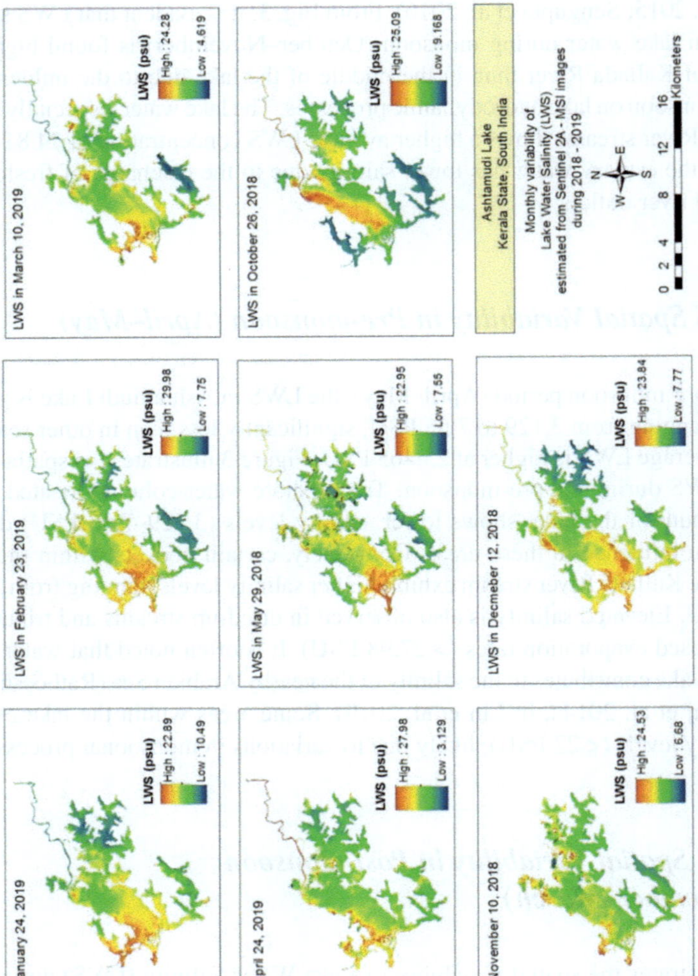

Fig. 3 Monthly variability of lake water salinity in the Ashtamudi Lake during the periods of 2018–2019

increase in salinity is due to the interaction between lake water and the Lake Water Temperature (LWT), which causes significant evaporation and accumulates salts with each successive post-monsoon period.

Additionally, the interaction between lake water and Arabian Sea waves contributes to the elevated salinity near the lake's mouth (Pant et al. 2015; Yuan et al. 2018). A few nearshore pockets exhibit even higher LWS concentrations, reaching up to 29.98 PSU. Such elevated salinity levels, comparable to the Sea Surface Salinity (SSS) of the Arabian Sea, are atypical for a lake. These high levels are likely caused by significant evaporation driven by elevated Sea Surface Temperature (SST) and atmospheric temperatures, in addition to the mixing of seawater from the Arabian Sea into the estuary.

4.4 LWS Monthly Variability in 2018–2019

The monthly variability of LWS in Ashtamudi Lake is captured using Sentinel-2A MSI images over eight months during 2018–2019. Figure 3 illustrates the fluctuations in LWS concentrations across the lake. Due to excessive cloud cover during the southwest monsoons, satellite images for June to September could not be processed. LWS measurements collected from 40 stations reveal the spatial and temporal variability of salinity across the monsoon (October–November 2018), pre-monsoon (April–May 2018), and post-monsoon (December 2018–March 2019) periods.

Table 2 presents the monthly variability of Lake Water Salinity (LWS) for 40 stations along Ashtamudi Lake. The mean LWS is 16.76 PSU, with a maximum of 21.24 PSU at station C1 and a minimum of 1.63 PSU at station KE4. In the offshore regions (braided branches of the lake), the LWS varies between 7.24 and 14.38 PSU, showing minimal spatial and temporal variation during pre- and post-monsoon periods. The LWS variability displays a standard deviation (SD) of 0.05–1.94 for most stations, indicating stable salinity levels in the nearshore (Estuary) regions throughout the seasons. In contrast, offshore regions (further from the middle of the lake and Kallada River streams) experience more variability across different seasons. Higher variability in local water salinity (with standard deviations ranging from 2.16 to 3.07) is observed at specific stations, including C4–C8, KE7–KE9, M8, and Ks7. These fluctuations in salinity are influenced by factors such as monsoon jets, surface salinification, and hydrodynamic processes, including upwelling, deep mixing, and river runoff discharge.

In offshore waters, lower LWS values are due to higher Lake Water Temperature (LWT) and reduced wave and coastal water impact in these expansive lake areas. Central lake areas show lower LWS because of precipitation mixing into the Kallada River during monsoons. Higher LWS levels can be attributed to domestic discharges, industrial wastes, slaughterhouse wastes, etc. LWS is generally lower on the western side of the nearshore than on the eastern side.

Between 2018 and 2019, LWS in shallow lake waters increased from monsoon to post-monsoon (summer), reaching values between 17.23 and 27.04 PSU and peaking

Table 2 Lake water salinity distribution in the Ashtamudi Lake

ID	Location		Salinity in Monsoon (PSU)		Salinity in Non Monsoon (PSU)	
	Longitude	Latitude	Observed (In Situ)	Estimated (Sentinel 2A)	Observed (In Situ)	Estimated (Sentinel 2A)
M1	76.552417	8.931333	20.36	19.48	22.22	21.87
M2	76.546389	8.935722	20.75	20.83	22.22	22.32
M3	76.551028	8.938944	20.67	21.32	22.56	22.31
M4	76.561556	8.931583	20.77	20.21	21.79	20.87
M5	76.570167	8.940056	19.98	19.07	21.47	21.47
M6	76.560861	8.948389	19.89	18.63	22.16	22.16
M7	76.571222	8.958028	20.35	19.83	21.99	21.99
M8	76.583417	8.955111	20.75	20.28	22.13	22.54
M9	76.583417	8.990139	20.12	20.12	21.32	21.45
M10	76.592944	8.97775	20.96	20.96	22.08	22
M11	76.596306	8.980417	18.68	18.68	21.3	21.57
M12	76.60425	8.971111	21.24	21.24	21.45	22
M13	76.606389	8.976778	20.78	20.78	21.81	21.5
C1	76.537556	8.963	21.24	21.87	22.11	21.69
C2	76.547667	8.954389	20.37	20.37	21.79	21.79
C3	76.546278	8.943722	20.7	19.44	21.33	21.65
C4	76.565	8.916667	19.37	19.37	21.12	21.32
C5	76.566639	8.918667	19.69	19.69	20.95	21.05
C6	76.57525	8.927083	19.66	19.66	22.38	23.48
C7	76.583722	8.930639	20.15	20.54	22.38	22.48
C8	76.590111	8.936028	20.68	20.12	22.38	22.45
C9	76.599667	8.936611	20.47	20.78	22.38	22.38
KE 1	76.572889	8.905361	7.26	7.56	23.45	23.45
KE2	76.582417	8.913444	7.22	7.22	24.17	24.17
KE 3	76.587111	8.918278	2.89	2.89	23.38	23.38
KE 4	76.582889	8.8985	1.63	1.63	22.43	22.43
KE 5	76.582528	8.904528	3.63	3.63	23.05	23.18
KE 6	76.586917	8.9035	4.2	5.6	25.09	25.42
KE 7	76.602056	8.910917	4.6	4.64	24.36	24.36
KE 8	76.608556	8.922917	5.18	4.53	24.64	24.74
KE 9	76.615833	8.929083	5.34	5.34	25.24	25.52
Ks1	76.604583	8.959833	19.36	19.36	21.07	21.23
Ks2	76.611833	8.963222	20.14	19.58	22.54	22.65
Ks3	76.619389	8.974806	20.03	20.03	22.71	22.71
Ks4	76.628889	8.983944	20.24	20.24	22.18	22.18
Ks5	76.63475	8.989556	20.42	20.42	22.21	22.21
Ks6	76.639583	8.965972	20.76	20.76	22.13	22.13
Ks7	76.637444	8.968222	19.65	20.65	21.5	21.5
Ks8	76.651861	8.965944	20.08	20.08	21.28	21.28
Ks9	76.655583	8.969361	20.17	20.17	22.14	22.14

at < 29.98 PSU during the northeast monsoon (October–November). Conversely, offshore LWS remained below 14 PSU and relatively homogeneous across seasons. Seasonal hydrodynamic changes affect LWS in lake and coastal waters, particularly in the estuary, where site-specific factors like precipitation and freshwater influxes also influence LWS. In Ashtamudi Lake, the salinity of the lake water fluctuates due to seasonal factors such as river discharge, freshwater influx, and precipitation, particularly affecting the lake's water column. In contrast, the salinity of offshore waters is more influenced by evaporation and precipitation levels.

5 LWS Validation Using Linear Regression Analysis and RMSE Estimation

The Lake Water Salinity (LWS) of Ashtamudi Lake is estimated using Sentinel-2A MSI images at a resolution of 10 × 10 m per pixel. This LWS-derived data, which spans three seasons over eight months, is validated against in situ LWS observations using two statistical methods: (i) Linear regression analysis (LRA) and (ii) root mean square error (RMSE).

Figure 4 illustrates the cross-validation of LWS using linear regression and RMSE between Sentinel-2A MSI-derived LWS and observed LWS data for 2018–2019. The linear regression analysis reveals a strong positive correlation with an R^2 value of 0.994, indicating a high level of agreement with observed LWS data across all seasons. Table 3 presents the statistical metrics for linear regression and RMSE, showcasing the accuracy of LWS-derived products against in situ measurements.

The RMSE value of ± 0.138 PSU indicates a strong correlation between Sentinel-2A MSI image-derived LWS products and observed LWS data. This low RMSE indicates the high accuracy of the LWS estimates from Sentinel-2A MSI images. Furthermore, the least squares algorithm outperforms the second-order polynomial algorithm, delivering more precise LWS estimates with minimal residual error.

The relationship between products derived from LWS data and the observed data shows a strong positive correlation, indicated by an R^2 value of 0.994 and a root mean square error (RMSE) of ± 0.296 PSU. This model exhibits high accuracy and can effectively retrieve LWS information using multispectral bands from Sentinel, Landsat, MODIS, and ASTER satellite data.

6 Conclusions

This study highlights the significant spatial and temporal variability in Lake Water Salinity (LWS) across Ashtamudi Lake, revealing critical insights into the lake's hydrological dynamics and biogeochemical processes. Utilizing multi-temporal Sentinel-2A MSI images, the study highlights several factors influencing Lake Water Salinity (LWS), including evaporation, precipitation, and river discharge. The highest salinity levels were observed during the post-monsoon season, primarily due to evaporation and reduced freshwater inflow following the construction of the Kallada Dam. Conversely, lower LWS values during the pre-monsoon period are attributed to increased freshwater input from the Kallada River. The study found that salinity is highest in the estuary, where the lake meets the sea, and lower at the confluence of the Kallada River, highlighting the role of tidal influences, winds, and hydro-thermodynamic variability. These findings emphasize the significant impact of LWS variability on the physicochemical and biological properties of both lake and coastal ecosystems, particularly in response to monsoonal changes. Using satellite remote sensing, particularly Sentinel-2A MSI imagery, proved effective for monitoring LWS

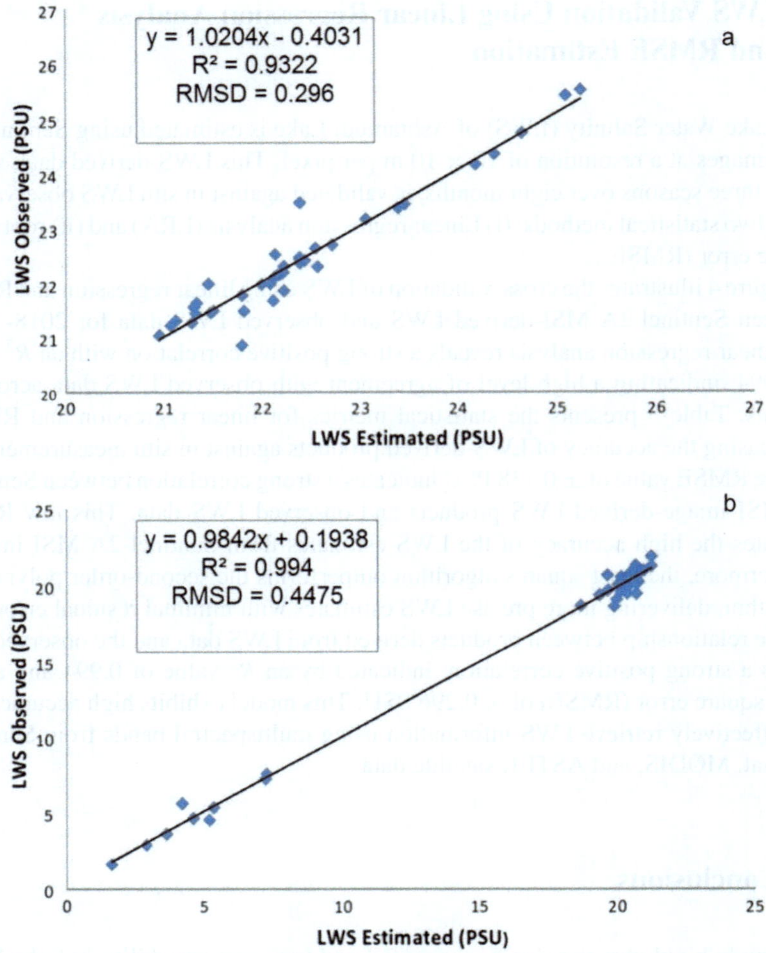

Fig. 4 Linear regression and RMSE estimation using LWS observed and LWS derived (Sentinel-2A MSI) for the periods of 2018–2019; monsoon and non-monsoon

Table 3 LWS correlation indices for monsoon and non-monsoon

	R^2	Root Mean Square Difference (RMSD)
Monsoon	0.994	0.4475
Non Monsoon	0.9322	0.296

variability, offering a reliable alternative to traditional measurement methods. This approach not only enhances understanding of the lake's salinity dynamics but also supports the broader goal of assessing the health and productivity of estuarine environments, which are highly sensitive to changes in salinity. The study underscores the importance of monitoring salinity as a critical environmental factor influencing estuarine and coastal ecosystems. The findings advance oceanographic and climate research while informing sustainable management strategies for conserving vital water resources and protecting ecosystem health.

Acknowledgements The corresponding author thanks the Director, National Centre for Earth Science Studies (NCESS), Ministry of Earth Sciences (MoES), Government of India, Thiruvananthapuram, for continuous support and encouragement to complete this successful research work. The corresponding author thanks Dr. K. Maya, Scientist—G, Group Head, BGG, NCESS, for continuous support. The authors thank the anonymous reviewers for their insightful comments and suggestions. The authors acknowledge the IMD, GSI, SOI, and EU-Copernicus Data Space Ecosystem for providing the necessary data sources.

References

Abe H, Ebuchi N (2014) Evaluation of sea-surface salinity observed by Aquarius. J Geophys Res Oceans 119(11):8109–8121. https://doi.org/10.1002/2014JC010094

Abowei JFN (2010) Salinity, dissolved oxygen, pH and surface water temperature conditions in Nkoro River, Niger Delta, Nigeria. Adv J Food Sci Technol 2(1):36–40

Ahn YH, Shanmugam P, Moon JE, Ryu JH (2008) Satellite remote sensing of a low salinity water plume in the East China Sea. Ann Geophys 26:2019–2035

Akhil VP, Lengaigne M, Vialard J, Durand F, Keerthi MG, Chaitanyi AVS, Papa F, Gopalakrishna VV, Montegut C (2016) A modeling study of processes controlling the Bay of Bengal sea surface salinity interannual variability. J Geophys Res Oceans 121(12):8471–8495. https://doi.org/10.1002/2016JC011662

Barre HMJ, Duesmann B, Kerr YH (2008) SMOS: the mission and the system. IEEE Trans Geosci Remote Sens 46:587–593

Bhaskar TVSU, Jayaram C (2015) Evaluation of Aquarius sea surface salinity with Argo sea surface salinity in the tropical Indian ocean. IEEE Geosci Remote Sens 12(6):1292–1296. https://doi.org/10.1109/LGRS.2015.2393894

Boutin J, Martin N (2006) ARGO upper salinity measurements: perspectives for L-band radiometer calibration and retrieved sea surface salinity validation. IEEE Geosci Remote Sens Lett 3:202–206

Brown MA, Torres F, Corbella I, Colliander A (2008) SMOS calibration. IEEE Trans Geosci Remote Sens 46:646–658

Burrage D, Wesson J (2008) Deriving sea surface salinity and density variations from satellite and aircraft microwave radiometer measurements: application to coastal plumes using STARRS. IEEE Trans Geosci Remote Sens 46(3):765–785

Damotharan P, Perumal NV, Perumal P (2010) Seasonal variations of physico-chemical characteristics of point Calimere coastal waters (south east coast of India). Middle East J Sci Res 6(4):333–339

Dehadri PV (1970) Changes in the environmental features of the Zuari and Mandovi estuaries in relation to tides. Proc Indian Acad Sci B72:68–80

Delcroix T, McPhaden M (2002) Interannual sea surface salinity and temperature changes in the western Pacific warm pool during 1992–2000. J Geophys Res 107:SRF 3-1–SRF 3-17. https://doi.org/10.1029/2001JC000862

Dierssen H, Kaylan R (2012) Remote sensing of ocean color. In: Earth System Monitoring. Springer Science, New York, NY, USA, pp 8952–8975

Dinnat EP, Le Vine DM, Boutin J, Meissner T, Lagerloef (2019) Remote sensing of sea surface salinity: comparison of satellite and in situ observations and impact of retrieval parameters. Remote Sens 11:750. https://doi.org/10.3390/rs11070750

Durack PJ. Ocean salinity and the global water cycle. Oceanography 28:20–31. https://doi.org/10.5670/oceanog.2015.03

Eladawy A, Nadaoka K, Negm A, Abdel-Fattah S, Hanafy M, Shaltout M (2017) Characterization of the northern Red Sea's oceanic features with remote sensing data and outputs from a global circulation model. Oceanologia 59:213–237

Ellison W, Balana A, Delbos G, Lamkaouchi K, Eymard L, Guillou C, Prigent C (1998) New permittivity measurements of seawater. Radio Sci 33(3):639–648

Font J, Camps A, Borges A, Martin NM, Boutin J, Reul N, Kerr YH, Hahne A, Mecklenburg S (2010) SMOS: the challenging sea surface salinity measurement from space. Proc IEEE 98:649–665

Fournier S, Chapron B, Salisbury J, Vandemark D, Reul N (2015) Comparison of space-borne measurements of sea surface salinity and colored detrital matter in the Amazon plume. J Geophys Res Oceans 120:3177–3192. https://doi.org/10.1002/2014JC010109

Gabarro C, Vall-closer M, Font J, Camps A (2004) Determination of sea surface salinity and wind speed by L-band microwave radiometry from a fixed platform. Int J Remote Sens 25(1):111–128

Geiger F, Grossi D, Trembanis C, Kohut T, Oliver J (2013) Satellite-derived coastal ocean and estuarine salinity in the Mid-Atlantic. Cont Shelf Res 63:S235–S242

Kaliraj S, Chandrasekar N (2012) Geo-processing model on coastal vulnerability index to explore risk zone along the south west coast of Tamilnadu, India. Int J Earth Sci Eng 5(5):1138–1147

Kaliraj S, Chandrasekar N, Magesh NS (2014b) Impacts of Wave energy and Littoral currents on Shoreline erosion/accretion along the southwest coast of Kanyakumari, Tamil Nadu, using DSAS and geospatial technology. Environ Earth Sci. https://doi.org/10.1007/s12665-013-2845-6

Kaliraj S, Chandrasekar N, Simon Peter T, Selvakumar S, Magesh NS (2015) Mapping of coastal aquifer vulnerable zone in the southwest of Kanyakumari, South India, using GIS-based DRASTIC model. Environ Monit Assess. https://doi.org/10.1007/s10661-014-4073-2

Kaliraj S, Chandrasekar N, Ramachandran KK (2016) Mapping of coastal landforms and volumetric change analysis in the southwest coast of Kanyakumari, South India, using remote sensing and GIS techniques. Egypt J Remote Sens Space Science. https://doi.org/10.1016/j.ejrs.2016.12.006

Kaliraj S, Chandrasekar N, Ramachandran KK, Srinivas Y, Saravanan S (2017) Coastal landuse and land cover change and transformations of Kanyakumari coast, India, using remote sensing and GIS. Egypt J Remote Sens Space Sci. https://doi.org/10.1016/j.ejrs.2017.04.003

Kaliraj S, Chandrasekar N, Ramachandran KK, Selvakumar S (2019b) Seawater intrusion vulnerability in the coastal aquifers of southern India—an appraisal of the GALDIT model, parameters' sensitivity, and hydrochemical indicators. Environ SciPollut Res 26:9755–9784. https://doi.org/10.1007/s11356-019-04401-0

Kaliraj S, Chandrasekar N, Magesh NS (2013) Geospatial techniques were used to evaluate coastal erosion and accretion processes along the southwest coast of Kanyakumari, Tamil Nadu. Arab J Geosci. https://doi.org/10.1007/s12517-013-1216-7

Kaliraj S, Chandrasekar N, Magesh NS (2014a) Multispectral image analysis of suspended sediment concentration along the Southern coast of Kanyakumari, Tamil Nadu, India. J Coast Sci 1(1):63–71. ISSN: 2348–6740

Kaliraj S, Chandrasekar N, Ramachandran KK (2019a) Phytoplankton (chl-a) Biomass seasonal variability in the Gulf of Mannar, South India: a remote sensing perspective. In: Kumar P et al. (eds) Applications and challenges of geospatial technology. https://doi.org/10.1007/978-3-319-99882-4_5

Kaliraj S (2016) Geospatial analysis of coastal geomorphological vulnerability on the west coast of Kanyakumari using Remote sensing Technology. Unpublished Ph.D. Thesis, Manonmaniam Sundaranar University, Tirunelveli, pp 83–110

Klemas V (2011) Remote sensing of sea surface salinity: an overview with case studies. J Coast Res 27(5):830–838. https://doi.org/10.2112/JCOASTRES-D-11-00060.1

Madhusoodana Kurup B, Black K, Thomas KV (2001) Developing a district coastal management plan and training for environmental improvement. Ashtamudi Estuary, Indian Fisheries concept

Lagerloef G (2014) Sea surface salinity. In: Njoku E (ed) Encyclopedia of remote sensing, vol XXV. Springer, 939p

Lewis, E.L. (1980) The Practical Salinity Scale 1978 and its antecedents. IEEE J. Ocean. Eng., OE-5(1): 3–8.

Lin X, Qiu Y, Cha J, Guo X (2019) Assessment of Aquarius sea surface salinity with Argo in the Bay of Bengal. Int J Remote Sens. https://doi.org/10.1080/01431161.2019.1615654

Mantyla AW (1987) Standard seawater comparisons updated. J Phys Ocean 17:543–548

Marghany M, Hashim M, Cracknell AP (2010) Modelling sea surface salinity from MODIS satellite data. Springer, Berlin, Heidelberg. ICCSA 2010, Part I, LNCS 6016, pp 545–556

Marshall S, Elliot M (1998) Environmental influences on the fish assemblage of the Humber estuary, UK. Estuar Coast Shelf Sci 46(2):175–184

McLusky DS (1989) The estuarine ecosystem 2nd edn. Chapman and Hall, New York, p 214

Mishra AK, Dwivedi S, Shrivastava A (2015) High resolution simulation of the salinity variability in the Bay of Bengal and Arabian Sea during the years 1998–2014 using an ocean circulation model. Discovery 39(180):173–179. ISSN: 2278--5469

Momin IM, Mitra AK, Mahapatra DK, Ragagopal EN (2016) Review of recent evaluation of sea surface salinity with satellite estimates in the tropical Indian Ocean. J Appl Remote Sens 10. https://doi.org/10.1117/1.jrs.10.046008

Nasnolkar CM, Shirodhkar PV, Singbal SVS. Mar Sci 25:120–124

Palanichamy S, Rajendran A (2000) Heavy metal concentration in seawater and sediments of Gulf of Mannar and Palk Bay, South east coast of India. Indian J Mar Sci 29:116–119

Pant V, Girishkumar MS, Bhaskar TVSU, Ravichandran M, Papa F, Thangaprakash VP (2015) Observed interannual variability of near-surface salinity in the Bay of Bengal. J Geophys Res Oceans 120(5):3315–3329. https://doi.org/10.1002/2014JC010340

Prabhu VA, Rajkumar M, Perumal P (2000) Seasonal variations in physico-chemical characteristics of Pichavaram mangroves, southeast coast of India. J Environ Biol 29(6):945–960

Rajabi S, Hasanlou M, Safari AR (2017) Spatial and temporal analysis of sea surface salinity using satellite imageryin Gulf of Mexico. Int Arch Photogrammetry, Remote Sens Spat Inf Sci XLII-4/W4

Ratheesh S, Sharma R, Sikhakolli R, Kumar R, Basu S (2013) Assessing sea surface salinity derived by Aquarius in the Indian Ocean. IEEE Geosci Remote Sens 11(4):719–722. https://doi.org/10.1109/LGRS.2013.2277391

Salleh D, Shattri M, Biswajeet P, Lawal B, Ahmad M (2013) Potential fish habitat mapping using MODIS-derived sea surface salinity, temperature and chlorophyll-a data: south China sea coastal areas, Malaysia. Geocarto Int 28:546–560

Seelanki V, Sreenivas P, Prasa KVSR (2018) Impact of Aquarius sea-surface salinity assimilation in improving the ocean analysis over Indian Ocean. Mar Geodesy. https://doi.org/10.1080/01490419.2017.1422817

Sengupta D, Raj GNB, Ravichandran M, Lekha JS, Papa F (2016) Near-surface salinity and stratification in the north Bay Of Bengal from moored observations. Geophys Res Lett 43(9):4448–4456. https://doi.org/10.1002/2016GL068339

Sharma R, Agarwal N, Momin IM, Basu S, Agarwal VK (2010b) Simulated sea surface salinity variability in the tropical Indian Ocean. J Climate 23(24):6542–6554. https://doi.org/10.1175/2010JCLI3721.1

Sharma R, Agarwal N, Momin IM, Basu S, Agarwal VK (2010) Simulated sea surface salinity variability in the tropical Indian Ocean. J Climate 23(24):6542–6554

Smitha R, Bhasha M, Sujit B, Raj K, Rashmi S (2014) Assessment of satellite-derived sea surface salinity in the Indian Ocean. IEEE Geosci Remote Sens Lett. https://doi.org/10.1109/LGRS.2012.2207943

Soundarapandian P, Premkumar T, Divakaran GK (2009) Studies on the physico-chemical characteristics and nutrients in the Uppanar estuary of Cuddalore, south east coast of India. Curr Res J Biol Sci 1(3):102–105

Subrahmanyam B, Murty VSN, Heffner DM (2011) Sea surface salinity variability in the tropical Indian Ocean. Remote Sens Environ 115(3):944–956. https://doi.org/10.1016/j.rse.2010.12.004

Sulochana B, Muniyandi K (2005) Hydrographic parameters of Gulf of Mannar and Palk Bay during a year of abnormal rainfall. J Mar Biol Assoc India 47(2):198–200

UNESCO (1981) The practical salinity scale 1978 and the international equation of state of seawater 1980. Techn Pap Mar Sci 36:25pp

Vinayachandran PN, Nanjundiah RS (2009) Indian Ocean sea surface salinity variations in a coupled model. Clim Dyn 33:245–263. https://doi.org/10.1007/s00382-008-0511-6

Xiang Y, Bei X, Xiangyang Y, Yebao L, Buli W, Cui B, Xin L (2017) Retrieval of remotely sensed sea surface salinity using MODIS data in the Chinese Bohai Sea. Int J Remote Sens 38:7357–7373

Yuan X, Salama M, Su Z (2018) An observational perspective of sea surface salinity in the southwestern Indian Ocean and its role in the south Asia summer monsoon. Remote Sens 10(12):1930. https://doi.org/10.3390/rs10121930

Assessment of Soil Physicochemical Characteristics and Heavy Metals Distribution in Coastal Urban Proximal Sites of Ernakulam District, Kerala, Using Multivariate Statistics and GIS Techniques

S. Kaliraj, C. N. Shaginimol, M. S. Parvathi, and P. S. Fathimathu Sahala

Abstract The study assessed the physicochemical characteristics, including pH, organic carbon, nitrogen, phosphorus, and the concentrations of heavy metals (Zn, Cr, Pb, Mn, Cu, Ni, and Cd) in the coastal lowland areas of Ernakulam district, Kerala. The findings indicate that the concentrations of heavy metals in the soil samples ranged as follows: chromium (Cr) from 0.073 to 0.311 mg/g, manganese (Mn) from 0.113 to 0.364 mg/g, nickel (Ni) from 0.017 to 0.065 mg/g, zinc (Zn) from 0.037 to 1.136 mg/g, cadmium (Cd) from 0.012 to 0.073 mg/g, copper (Cu) from 0.021 to 0.036 mg/g, and lead (Pb) from 0.039 to 0.053 mg/g. The concentration of heavy metals in soil samples was observed in the following descending order: Zn > Mn > Cr > Pb > Ni > Cd = Cu. Soil contamination was evaluated using the geo-accumulation index (I_{geo}), contamination factor (C_f), and Pollution Load Index (PLI). The average I_{geo} values were Cr-21.51, Mn-18.29, Ni-8.33, Zn-9.32, Cd-58.38, Cu-9.60, and Pb-6.87. Elevated I_{geo} values for Cd and Cr were noted at certain sampling stations. The PLI values for the sample stations (S1-S8) were 4.31, 3.60, 3.29, 4.21, 2.81, 2.32, 3.25, and 1.57, respectively. High contamination levels were observed for Cd and Cr at specific stations based on the C_f values. The spatial distribution of heavy metals was mapped using interpolation techniques. The results

S. Kaliraj (✉) · P. S. F. Sahala
National Centre for Earth Science Studies (NCESS), Ministry of Earth Sciences (MoES), Government of India, Thiruvananthapuram, Kerala, India
e-mail: s.kaliraj@ncess.gov.in

C. N. Shaginimol
School of Fishery Environment, Kerala University for Fisheries and Ocean Studies, Kochi, Kerala, India

M. S. Parvathi
Bishop Kurialacherry College for Women, Amalagiri, Kerala, India

P. S. F. Sahala
Ministry of Earth Science (MoES) Research Fellow (MRFP), DESK, IITM, Pune, India

NCESS Research Scholar, Faculty of Marine Sciences, CUSAT, Kochi, Kerala, India

K. K. Basheer Ahammed et al. (eds.), *Monitoring Coastal Marine Ecosystems Under Climate Change*, https://doi.org/10.1007/978-3-032-09137-6_3

revealed that zones with high concentrations of Cd were located near industrial dumping areas. Conversely, the highest Pb, Mn, Cr, Ni, Cu, and Zn concentrations were found in the southeastern region adjacent to the effluent dumping zones. The ecological risk assessment indicated significant risks associated with these metals in sediment samples.

Keywords Soil properties · Heavy metals contamination · Physicochemical properties · Pollution load index · Urban coast · Ernakulam

1 Introduction

Soil is vital in terrestrial ecosystems, primarily by supporting plant growth. It serves as a reservoir for essential nutrients and moisture required to produce forage and plant species. Soil layers, especially in the coastal lands, provide vital nutrients to the biota that comprise a complex mixture of organic matter, minerals, gases, and various organisms (Ali et al. 2015; Shaginimol et al. 2022). The specific composition of these constituents directly influences the soil's physical, chemical, and biological properties. Key physicochemical properties of soil include pH, salinity, texture, moisture, temperature, electrical conductivity, nitrogen, phosphorus, potassium, and organic matter (Tale and Ingole 2015; Tripathy et al. 2018; Salaj et al. 2022; Kaliraj et al. 2023). These properties are paramount in agriculture, directly affecting soil fertility and plant growth.

pH is a crucial indicator of nutrient availability in the soil, including essential elements such as Iron (Fe), Manganese (Mn), Zinc (Zn), and Copper (Cu). It balances the nutrients available to plants and those retained in the soil matrix (Tale and Ingole 2015). Moisture is another crucial factor, as the absorption of nutrients by plants is mainly dependent on its presence. Temperature also significantly influences soil processes and is subject to seasonal variations. An increase in temperature can accelerate chemical reactions, decrease the solubility of gases, and lower the soil pH (Shirbhate and Malode 2012; Lalitha et al. 2021). Nitrogen, phosphorus, and potassium are essential nutrients for plant growth. However, the continuous application of fertilizers to supply these nutrients can introduce trace elements into the soil, potentially leading to a significant enhancement in the soil mineral content. Human activities, such as agricultural practices, can redistribute heavy metals in a specific area, contributing to soil contamination and altering its composition (El Behairy et al. 2022). These parameters must be frequently monitored to understand significant disturbances and human interventions that may impact soil health and fertility.

Over time, soil continuously undergoes modification due to natural processes such as weathering, climate change, and human activities. These processes and various biotic activities contribute to the role of soil as a sink for the chemicals generated by both natural and anthropogenic actions (Jonathan et al. 2004; Kaliraj et al. 2017). Agricultural activity using fertilizers and pesticides, mining, vehicular emissions, and industrial activities caused soil contamination issues there. Soil contamination by the intervention of anthropogenic activities is an established phenomenon worldwide. Heavy metals in trace amounts are natural constituents of Earth's crust, and exceeding the limit results in environmental pollution. Elevated concentrations of heavy metals in soil are a major concern due to their high toxicity and long-term environmental persistence (Ali et al. 2015). Numerous studies have reported high concentrations of heavy metals in surface sediments (Fatoki and Mathabatha 2004; Jayaraju et al. 2009; Pande and Nayak 2013). Urban soils, in particular, interact with various pollutants and are more susceptible to anthropogenic activities (Bux et al. 2022). Multivariate statistical analysis, especially principal component analysis (PCA), plays a key role in analyzing pollution sources and understanding the distribution of heavy metals in the environment (Kaliraj et al. 2019; Anaman et al. 2022). PCA is recognized as an effective tool for identifying potential sources of heavy metals, while Geographic Information System (GIS) mapping provides valuable spatial insights for pinpointing these sources. The integration of PCA with GIS mapping significantly enhances the ability to accurately identify, allocate, and visualize potential sources of heavy metal contamination across the study area (Kaliraj et al. 2016).

The main objectives of this study are to identify the physicochemical properties of soil and evaluate the spatial variability of these elements across the study area. Additionally, the study aims to measure the concentrations of heavy metals (Zn, Cu, Ni, Mn, and Cd) and assess their spatial distribution. The study also seeks to define contamination levels and identify pollution sources using PCA, providing a comprehensive understanding of heavy metal distribution and its environmental implications.

2 Study Area

Ernakulam district, located in central Kerala, India, is known for its diverse natural and geographical features. It is bordered by the Trichur district to the north, Idukki to the east, Kottayam and Alappuzha districts to the south, and the Lakshadweep Sea to the west. The district spans coordinates between 9° 47′ 13″ N–10° 10′ 44″ N latitude and 76° 10′ 05″ E–77°05′24″ E longitude. Kochi, the district headquarters, is a prominent port city on India's west coast, often called the "Queen of the Arabian Sea." The district is primarily drained by the Periyar River, its tributaries in the north, and the Muvattupuzha River in the south. The coastal areas are strongly influenced by their proximity to the Arabian Sea, with diverse geomorphic features such as backwaters, coastal plains, estuaries, and lagoons. The geology of Ernakulam's coastal regions includes alluvial plains, coastal dunes, and estuarine deposits, reflecting its varied geological processes. The district experiences a wet monsoon-type climate, with three distinct seasons—monsoon (June–September), post-monsoon (October–January), and pre-monsoon (February–May)—and receives an average annual rainfall of 3359.2 mm, with 71% contributed by the monsoon (Jayaprakash, 2002). The temperature in Ernakulam district ranges from a mean monthly maximum of 28.1–31.4 °C and a minimum of 23.2–26 °C. The hottest months are March and

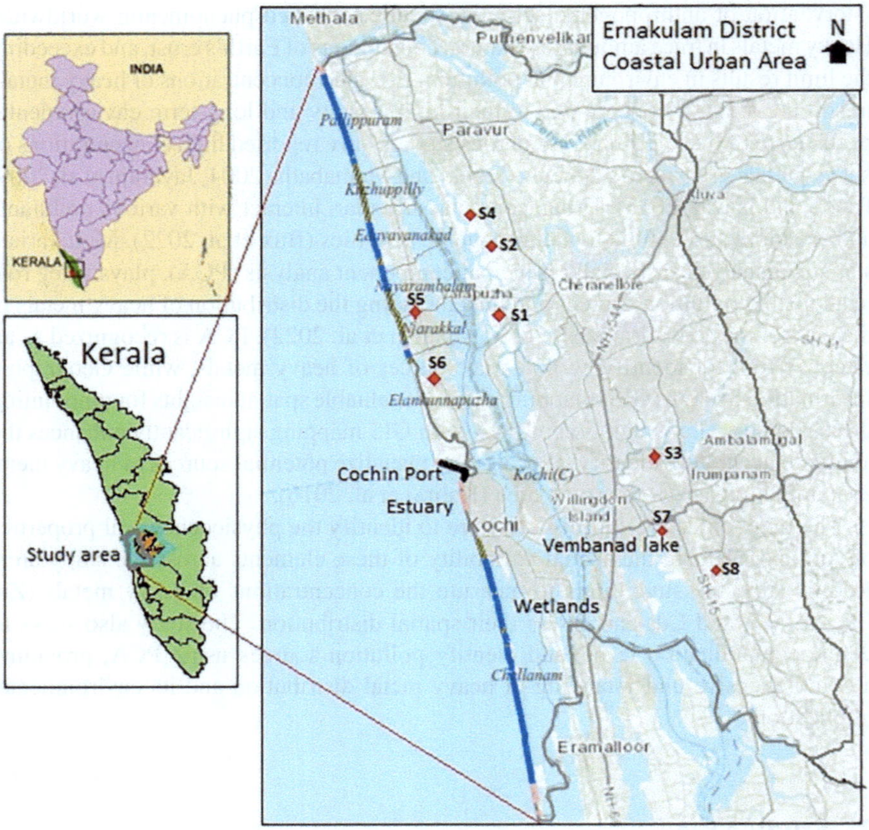

Fig. 1 Location of the study area showing soil sampling sites

April, while the coolest are December and January. The district's diverse topography supports various soil types, including laterite, alluvial, red loam, coastal sandy soil, and clay soil, with laterite being the most prevalent. Land use in the district consists of settlements (75%), cultivable land (12%), water bodies (13%), and marsh areas (12%). The ecological stability and aquatic productivity of the coastal estuarine ecosystem are closely linked to the environmental quality, significantly impacting the local population's economic viability (Fig. 1).

3 Laboratory Analyses

In this study, eight soil and sediment samples were collected from the coastal region of Ernakulam district in January 2023 for geochemical analysis. Samples were obtained using a stainless steel spatula, stored in zip-lock bags, and georeferenced using GPS and GIS tools. The samples were placed in watch glasses and oven-dried in the laboratory at 80 °C for 48 h. Following drying, the samples were homogenized using the coning and quartering technique, and extraneous materials such as stones and roots were manually removed. The samples were then finely ground using an agate mortar and pestle, sieved through a 230-micron mesh, and the resulting powders were stored in labeled sample bottles. These prepared samples were analyzed for physicochemical properties, including pH, salinity, organic carbon, phosphorus, nitrogen, heavy metal concentrations, and contamination indices.

The concentration of heavy metals and their oxides was determined using a Bruker S4 Pioneer sequential wavelength-dispersive X-ray fluorescence (WD-XRF) spectrometer. This instrument has a precision goniometer and seven analyzing crystals (OVO-55, PET, LiF 200, LiF 220, Ge, ADP, and InSb) to detect various elements accurately. It features a 4 kW Rhodium (Rh) X-ray tube, dual collimators (0.23° and 0.46°), and a 60-position automatic sample changer for high-throughput analysis. An argon/methane flow proportional counter and a scintillation counter facilitate elemental detection. The system utilizes SPECTRA plus software for both qualitative and quantitative elemental analysis. Major elements were analyzed using a single X-ray tube setup. In contrast, trace elements required additional excitation conditions involving multiple tubes and specific crystal combinations to achieve optimal sensitivity and resolution.

Approximately 5 g of homogenized wet sediment was placed in a pre-weighed, cleaned crucible for trace metal analysis. The samples were dried at 110 °C for 1 h to remove moisture. Subsequently, the dried samples were heated in a muffle furnace at 900 °C for 2 h to eliminate organic matter, calcium carbonate, interstitial water, and other volatile components. The crucible and sample were weighed before and after each heating step. After furnace treatment, the samples were cooled to room temperature in a desiccator to prevent moisture absorption before final weighing. The loss on ignition (LOI) was calculated based on the weight difference, providing an estimate of the particulate organic matter content in the samples.

$$\text{LOI}\% = \frac{w_2 - w_3}{w_2 - w_1} \times 100 \tag{1}$$

where w_1 = weight of crucible, w_2 = weight of crucible plus oven-dry sample, and w_3 = weight of crucible plus oven-dry sample after ignition.

4 Electrochemical Analysis

To measure pH, approximately 20 g of sediment was mixed with 100 mL of distilled water to achieve a uniform suspension, and the pH was then measured using a pH meter. The pH scale ranges from 1 to 14, with values 1 to 6 indicating acidity, 7 being neutral, and 8 to 14 indicating alkalinity. Salinity and conductivity were measured within one hour using 20 g of sediment and 100 mL of distilled water, with a conductivity cell immersed in the suspension. Soil salinity, the concentration of salts in the soil, is critical to soil fertility and plant growth. High electrical conductivity suggests the presence of significant ion concentrations (cations and anions), which can limit agricultural productivity.

5 Contamination Geochemistry

The level of heavy metal contamination in the Ernakulam district, derived from natural and anthropogenic sources, was assessed through an extensive analysis of sediment samples collected from the study area. The relative pollution levels at the sampling sites were determined using four key measurements: Contamination Factor (C_f), Degree of Contamination (C_{deg}), Geoaccumulation Index (I_{geo}), and Pollution Load Index (PLI). These metrics provided a detailed evaluation of the extent of heavy metal pollution in the region.

5.1 Geoaccumulation Index

The Geoaccumulation Index (I_{geo}) is a quantitative measure developed in the late 1970s to assess the degree of pollution in aquatic sediments. European studies have widely utilized it to evaluate trace metal contamination by comparing current metal concentrations with pre-industrial soil levels. The I_{geo} is calculated using Muller's (1969) formula:

$$I_{\text{geo}} = \log_2\left[C_n / (1.5 \times B_n)\right] \tag{2}$$

where C_n represents the concentration of a given element in the tested soil, and B_n is the average shale value as described by Turekian and Wedepohl (1961). The constant factor of 1.5 accounts for possible variations in baseline data due to lithogenic effects (Perumal et al. 2021; VineethKumar et al. 2020). According to Salomons and Forstner (1984), the I_{geo} classification includes the following: $I_{\text{geo}} > 5$ for extreme contamination, 4–5 for robust to extreme contamination, 3–4 for solid contamination, 2–3 for moderate to substantial contamination, 1–2 for moderate contamination, 0–1 for uncontaminated to moderate contamination, and < 0 for uncontaminated conditions.

5.2 Contamination Factor (C_f) and Degree of Contamination (C_{deg})

The soil contamination was assessed using the contamination factor and degree of contamination. The contamination factor represents the level of metal contamination in the soil and is calculated using the formula:

$$C_f = \left(C_{\text{heavy metals}} / C_{\text{background}}\right) \tag{3}$$

where $C_{\text{heavy metals}}$ are the metal concentration in the polluted soil, and $C_{\text{background}}$ is the natural background value of that metal. According to Hakanson (1980), four categories of contamination factors are defined: $C_f < 1$ indicates low contamination, $1 \leq C_f \leq 3$ indicates moderate contamination, $3 \leq C_f \leq 6$ indicates considerable contamination, and $C_f > 6$ indicates very high contamination.

The degree of contamination (C_{deg}) is the sum of the contamination factors for all elements examined, providing a comprehensive measure of environmental contamination. Four classes of contamination degrees are recognized: $C_{\text{deg}} < 6$ indicates low contamination, $6 < C_{\text{deg}} < 12$ indicates moderate contamination, $12 < C_{\text{deg}} < 24$ indicates considerably high contamination, and $C_{\text{deg}} > 24$ indicates high contamination. (Perumal et al. 2021).

5.3 *Pollution Load Index (PLI)*

The Pollution Load Index (PLI) assesses an area's pollution level and sediment toxicity. (Vineethkumar et al. 2020). It provides a comprehensive measure of contamination by considering the combined impact of multiple metals. The PLI is calculated using the formula:

$$\text{PLI} = \left[C_{f1} \times C_{f2} \times C_{f3} \times \cdots \times C_{fn}\right]^{1/n} \tag{4}$$

where C_f indicates the contamination factor, n for several metals, C_{f1} represents the contamination factor of the first location, C_{f2} represents the contamination factor of the second location, etc. PLI is classified as follows: a PLI value of ≤ 1 indicates no metal pollution, while a PLI value of > 1 indicates the presence of pollution (Tomlinson et al. 1980).

6 Total Organic Carbon (TOC) Analysis

TOC was determined using a catalytic oxidation combustion method at a high temperature of 720 °C, during which organic carbon is oxidized and converted into carbon dioxide (CO_2). Prior to analysis, finely powdered samples were treated with 2N hydrochloric acid (HCl) to eliminate Total Inorganic Carbon (TIC), followed by drying. A small aliquot (5–10 mg) of the acid-treated and dried sample was placed in a tin capsule and analyzed using a non-dispersive infrared (NDIR) detector, which quantifies the CO_2 released during combustion.

7 GIS Analysis

The GIS-based IDW technique, which utilizes in situ sampling data, is widely used to map the spatial distribution of various soil properties, offering a more detailed and quantitative understanding of soil characteristics. Accurately assessing the spatial variability of soil properties is essential for sustainable agricultural ecosystem management and reliable environmental modeling. This study measures soil properties and spatial variability, explicitly focusing on the spatial distribution of pH, salinity, phosphorus, organic carbon, nitrogen, and heavy metals.

8 Statistical Analysis

Multivariate analysis examines relationships among multiple variables and identifies associations among observed variables that may suggest underlying controlling factors and sources (Zhou et al. 2024). Principal component analysis (PCA), a method within multivariate statistical analysis, models new unobserved, latent variables (principal components) using the correlation structure among observed variables. In this study, PCA was employed to explore correlations between different elements and identify the sources of contamination.

9 Results and Discussion

This study investigates the physicochemical properties and heavy metal contamination of soils in the Ernakulam district. Soil samples were collected and analyzed to assess these characteristics. The results are presented in two sections: the first section addresses the physicochemical properties of the soil, while the second focuses on the spatial distribution of heavy metals—Zn, Cr, Pb, Mn, Cu, Ni, and Cd—using the GIS-based Inverse Distance Weighted (IDW) interpolation method.

9.1 *Spatial Distribution of Physicochemical Soil Characteristics*

Table 1 summarizes the results of the physicochemical parameters analyzed for the study area, offering a detailed overview of the soil characteristics. Soil pH and salinity, which remain relatively stable across seasons, directly and indirectly, influence the behavior of chemical elements in the soil. In the coastal area of Ernakulam district, soil pH ranged from 5.9 to 6.5, indicating mildly acidic conditions, likely resulting from a lack of acid-neutralizing minerals. Soil pH and salinity levels were higher at station S6 (Vypin South), near the coastal plain, possibly due to saltwater intrusion, where seawater infiltrates the soil and groundwater, increasing salinity and pH due to sodium chloride (Fig. 2).

The study also found that organic carbon content was higher in stations S1 (Edapally) and S4 (Paravur). High organic matter in soil is beneficial for agricultural practices as it reduces erosion; conversely, poor organic matter content can increase erosion (Kekane et al. 2015). Higher organic matter may also lower pH levels (Kekane et al. 2015). Nitrogen and phosphorus, essential nutrients for plant growth, showed varied distribution among the samples. Nitrogen, vital for amino acids, proteins, chlorophyll, and DNA, was higher in sample S3, while phosphorus, critical for cellular functions and better available in soils with high organic matter and minimal leaching (Tale and Ingole 2015), was higher in samples S1 and S3. The spatial distribution of soil properties was analyzed using the GIS-based Inverse Distance Weighting (IDW) technique in ArcGIS 10.8, with a 10 × 10 m per pixel resolution. Figures 3 and 4 illustrate the spatiotemporal characteristics of the physicochemical soil properties.

Table 1 Soil properties in coastal urban areas of Ernakulam district, Kerala, India

Sample ID	Location	pH	Salinity (PPT)	Organic Carbon (%)	Nitrogen (%)	Organic phosphorus (%)
S1	Edapally	5.9	1.0	0.74	0.02	46.8
S2	Alangad	6.1	1.0	0.06	0.04	36.2
S3	Thripunithura	5.9	1.0	0.43	0.08	48.3
S4	Paravur	6.2	1.0	0.91	0.03	15.7
S5	Vypin north	6.4	1.0	0.21	0.04	24.1
S6	Vypin south	6.8	1.0	0.24	0.02	24.4
S7	Palluruthy	6.3	2.0	0.04	0.06	32.6
S8	Mulamthuruthy	6.4	1.0	0.57	0.06	52.2

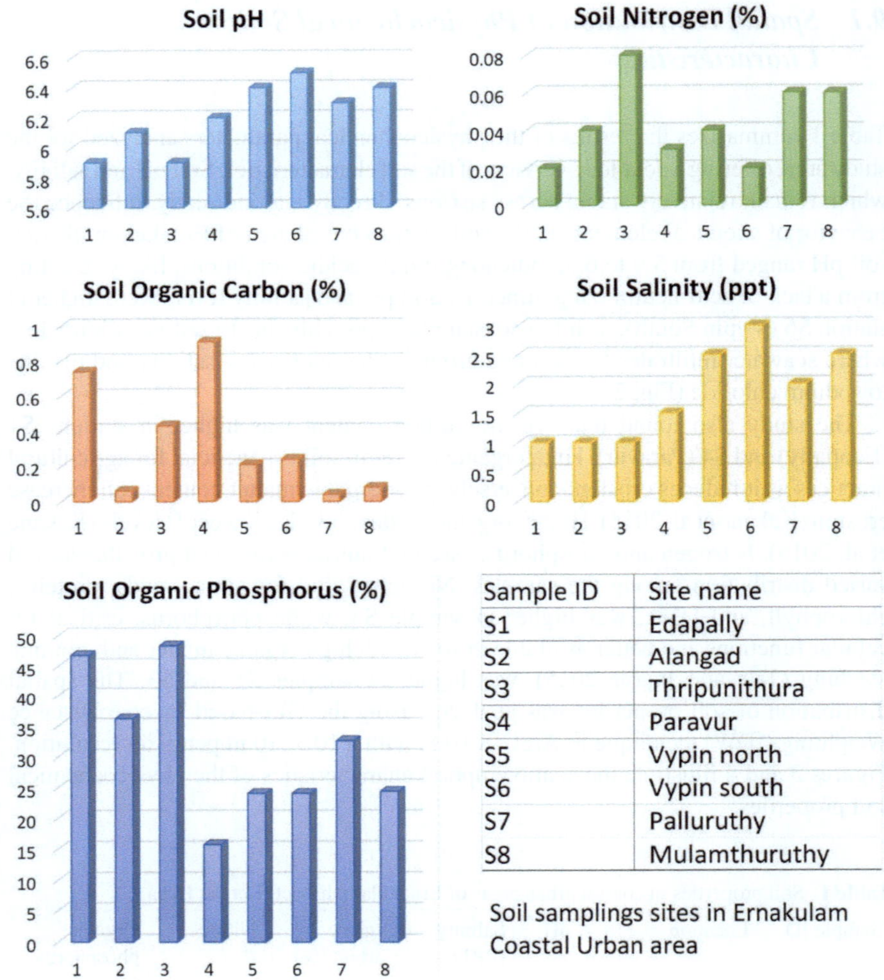

Sample ID	Site name
S1	Edapally
S2	Alangad
S3	Thripunithura
S4	Paravur
S5	Vypin north
S6	Vypin south
S7	Palluruthy
S8	Mulamthuruthy

Soil samplings sites in Ernakulam Coastal Urban area

Fig. 2 The graph indicates the difference in the values of soil physicochemical characteristics

9.2 Spatial Distribution of Heavy Metal Concentrations

The heavy metal concentration of the soil and sediments in mg/g from the study area is shown in Table 2. The spatial distribution of heavy metal concentration, namely Zn, Cr, Pb, Mn, Cu, Ni, and Cd, using the GIS-based IDW technique, is depicted in Figs. 5, 6, 7, and 8, which varies from place to place due to the impacts of various geoenvironmental factors.

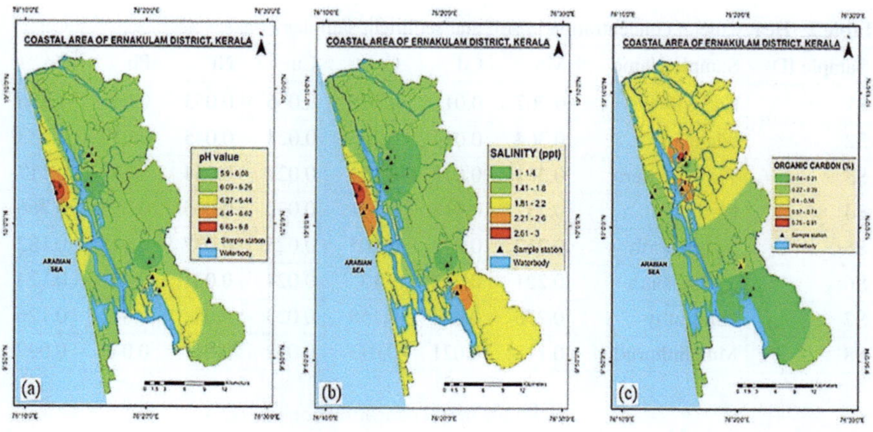

Fig. 3 Spatial distribution of **a** pH content, **b** salinity (PPT), and **c** organic carbon (%)

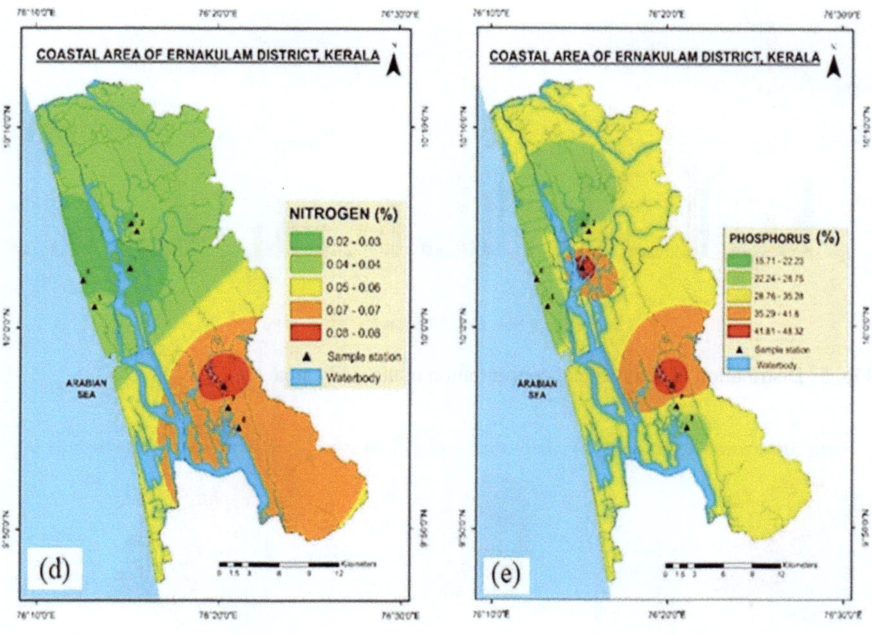

Fig. 4 Spatial distribution of **d** Nitrogen (%) and **e** phosphorous (kg/L)

Multivariate statistical analysis was conducted to identify the sources of various elements in the soil. The results of the principal component analysis (PCA) for total element concentrations are presented in Table 3. Following rotation, three principal

Table 2 Heavy metal concentration in soil and sediment samples

Sample ID	Sample Name	Mn	Cd	Cr	Cu	Ni	Pb	Zn
S1	Edapally	0.267	0.017	0.275	0.036	0.053	0.046	1.136
S2	Alangad	0.364	0.014	0.265	0.024	0.045	0.039	0.623
S3	Thripunithura	0.301	0.073	0.151	0.021	0.044	0.053	0.117
S4	Parur	0.23	0.018	0.311	0.032	0.065	0.052	0.766
S5	Vypin north	0.212	0.019	0.205	0.025	0.037	0.045	0.182
S6	Vypin south	0.221	0.012	0.13	0.024	0.031	0.051	0.124
S7	Palluruthy	0.278	0.064	0.158	0.023	0.05	0.042	0.126
S8	Mulamthuruthy	0.113	0.021	0.073	0.022	0.017	0.044	0.037

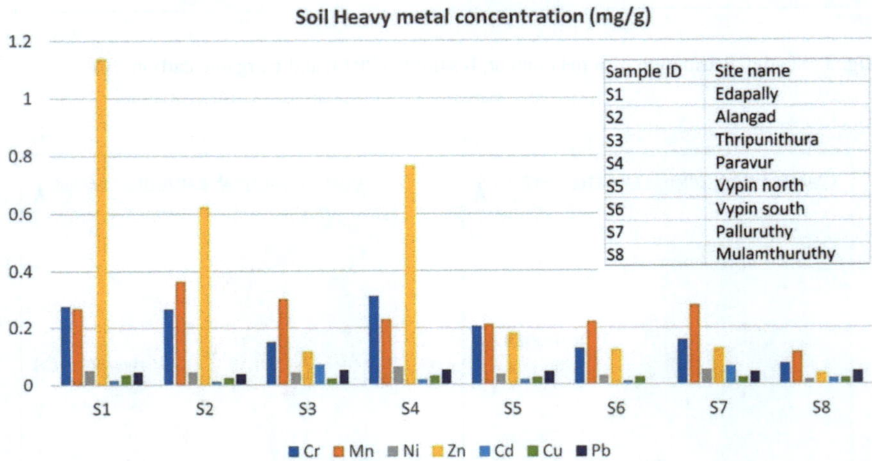

Fig. 5 Distribution of heavy metal concentration in the study area

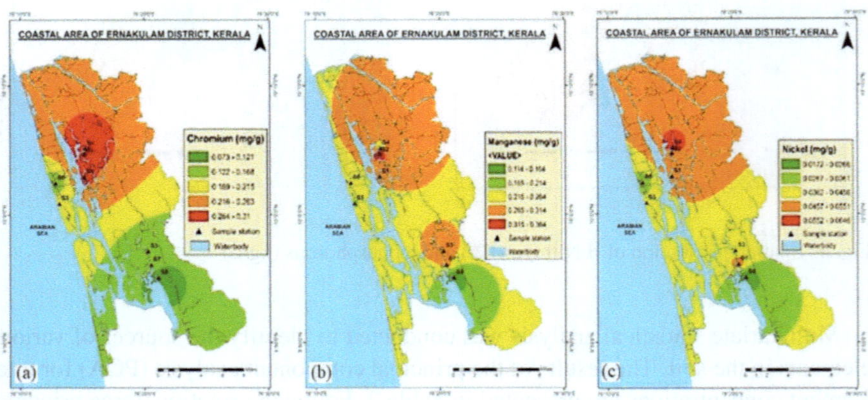

Fig. 6 Spatial distribution of heavy metal concentration in the study area **a** chromium (mg/g), **b** manganese (mg/g), and **c** nickel (mg/g)

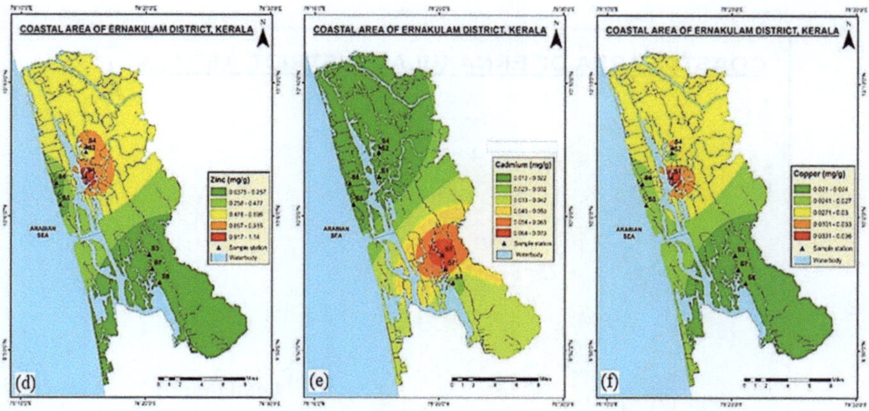

Fig. 7 Spatial distribution of heavy metal concentration in the study area **d** zinc (mg/g), **e** cadmium (mg/g), and **f** copper (mg/g)

components (PCs) with eigenvalues greater than one were extracted. PCA effectively reduced the dimensionality of the original dataset to three components, collectively explaining 91.36% of the total variance. The first principal component (PC1) accounts for 51.91% of the variance and exhibits strong loadings with Mn (1.373), Cd (0.90326), Cr (2.74601), Cu (2.47502), and Zn (2.69812). The second component (PC2), mainly associated with Cd, Mn, Ni, and Cu, explains 22.59% of the variance. The third component (PC3) strongly correlates with Pb and Mn, with a high loading of 0.80446, accounting for 16.86% of the variance. The observed correlation between elements such as Mn and Cu suggests a combination of geogenic and anthropogenic sources, as distinct component separation was not evident. Overall, the analysis indicates that Cd and Zn are predominantly derived from anthropogenic activities, particularly agricultural practices involving agrochemicals, which contribute to elevated concentrations of these elements in the soil (Fig. 9).

Cadmium (Cd) is a non-essential trace element widely distributed in the environment, with its concentrations in soils and groundwater influenced by both natural processes and anthropogenic activities. Elevated Cd levels are of particular concern due to their carcinogenic potential in humans, posing significant risks to food safety and drinking water quality. Major anthropogenic sources of Cd include mining activities, atmospheric deposition from combustion emissions, and Cd-containing fertilizers. Naturally, Cd occurs in minerals such as sulfides, carbonates, and phosphorites, further contributing to its environmental presence. In the study area, the primary sources of zinc (Zn) are ammonium phosphate fertilizers combined with zinc sulfate ($ZnSO_4$).

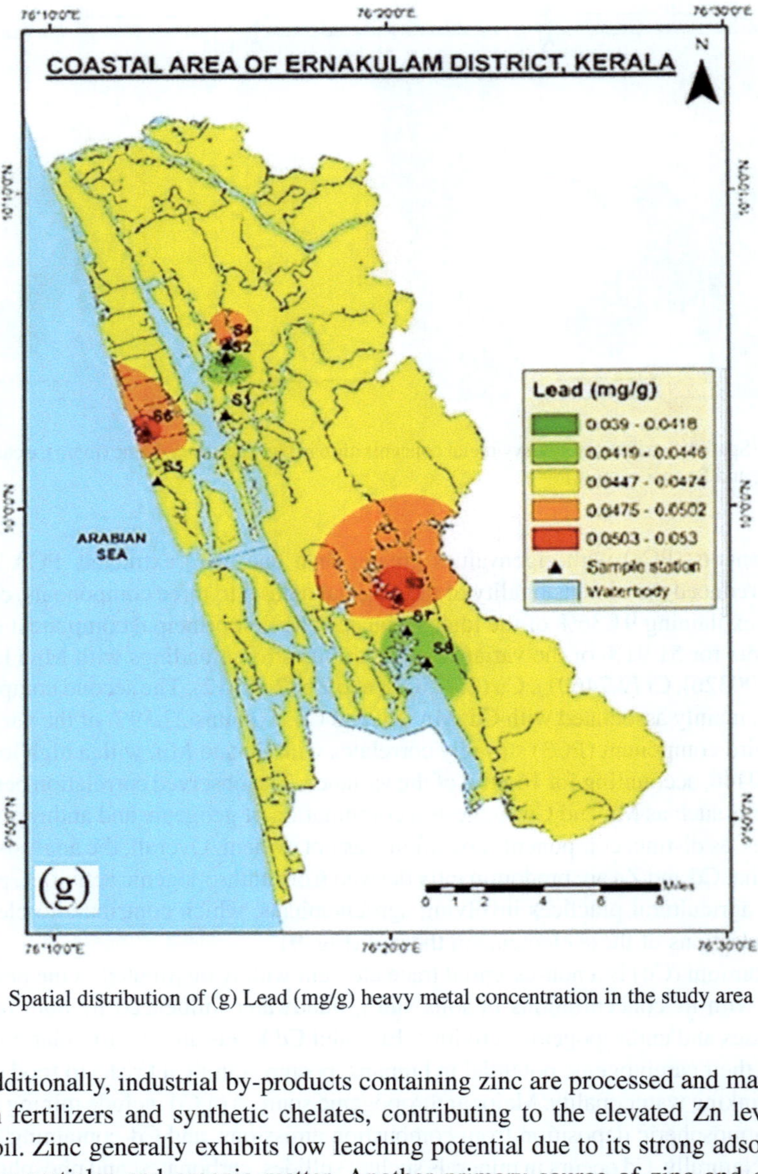

Fig. 8 Spatial distribution of (g) Lead (mg/g) heavy metal concentration in the study area

Additionally, industrial by-products containing zinc are processed and marketed as Zn fertilizers and synthetic chelates, contributing to the elevated Zn levels in the soil. Zinc generally exhibits low leaching potential due to its strong adsorption to clay particles and organic matter. Another primary source of zinc and copper (Cu) in the study area is the application of biosolids and manures, widely used to boost agricultural productivity. Animal wastes, particularly poultry and cattle manure, are commonly applied as slurries to highland crops. Although these manures are effective fertilizers, they can also introduce Zn and Cu into the soil, as these metals are frequently added to animal feed as growth promoters. The repeated application of such manures over time may result in a significant accumulation of these metals in the soil.

Table 3 Total variance and matrix of principal components analysis (Value in bold text exhibits strong +/− correlation of variables within PC1, PC2 and PC3)

Parameter	PC1	PC2	PC3
Mn	**1.373**	**1.71633**	**− 0.80446**
Cd	**− 0.90326**	**2.06307**	0.45312
Cr	**2.74601**	0.10653	− 0.1486
Cu	**2.47502**	**− 0.8565**	0.45196
Ni	**2.40723**	**1.02976**	0.3949
Pb	0.08193	0.18709	**2.09202**
Zn	**2.69812**	− 0.42989	− 0.11814
Percentage of total variance	51.91	22.59	16.86
Cumulative percentage	51.91	74.50	91.36

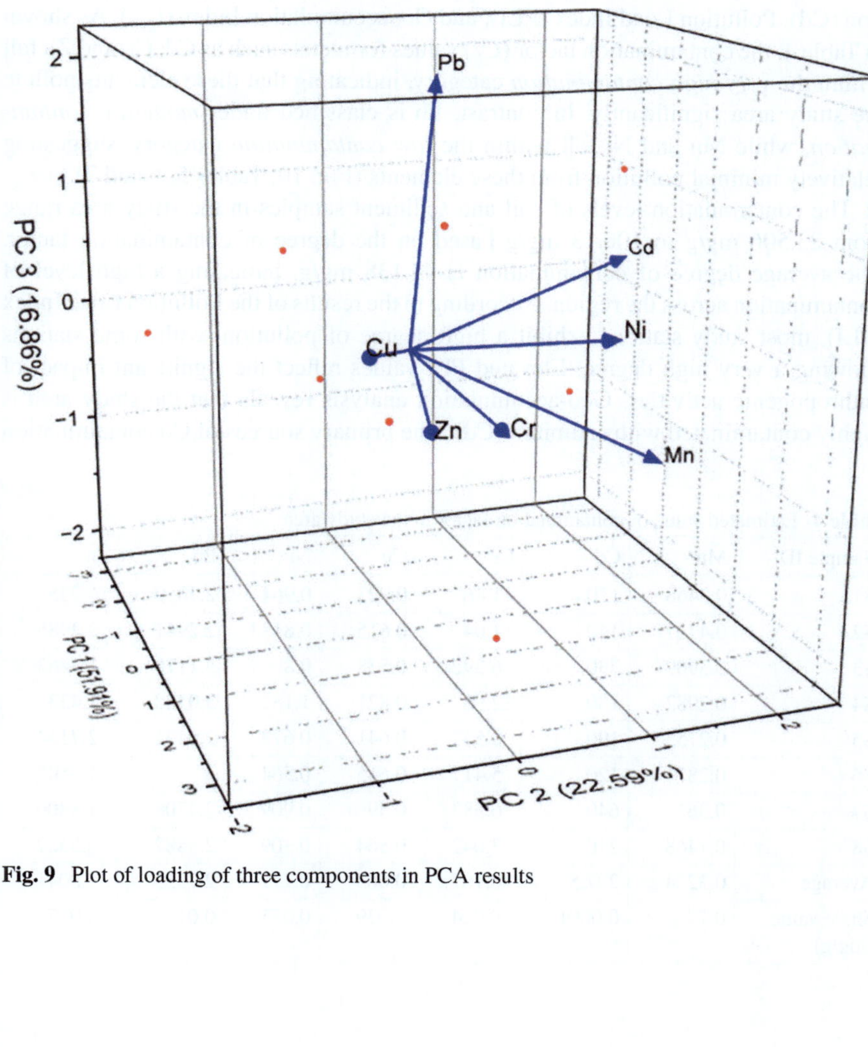

Fig. 9 Plot of loading of three components in PCA results

Source identification of metal ions in surface soils during both monsoon and non-monsoon seasons indicates that geogenic sources are predominant, as reflected by the three principal components derived for each season. However, evidence of source mixing is also apparent, likely influenced by rainwater-induced leaching in the study area. This process can substantially alter the distribution of metals in both surface and subsurface soils, with previous studies suggesting that such leaching may have long-term effects on soil metal concentrations.

9.3 Soil Contamination Indices

This section presents the results of the pollution assessment indices applied to soil and sediment samples, including the Contamination Factor (C_f), Degree of Contamination (Cd), Pollution Load Index (PLI), and Geoaccumulation Index (I_{geo}). As shown in Table 4, the contamination factor (C_f) values for metals such as Cd, Cr, and Zn fall within the *very high contamination* category, indicating that these elements pollute the study area significantly. In contrast, Pb is classified under *moderate contamination,* while Mn and Ni fall within the *low contamination* category, suggesting relatively minimal pollution from these elements (Fig. 10; Tables 5, 6 and 7).

The contamination levels of soil and sediment samples in the study area range from 23.506 mg/g to 106.13 mg/g based on the degree of contamination factor. The average degree of contamination is 45.138 mg/g, indicating a high level of contamination across the region. According to the results of the Pollution Load Index (PLI), most study stations exhibit a high degree of pollution, with some stations showing a very high degree. Elevated PLI values reflect the significant impact of anthropogenic activities. Geo-accumulation analysis reveals that the study area is highly contaminated with cadmium (Cd). The primary sources of Cd contamination

Table 4 Estimated value of contamination factor in the study area

Sample ID	Mn	Cd	Cr	Cu	Ni	Pb	Zn
S1	0.3468	170	11.46	0.923	0.964	2.7059	16.955
S2	0.4727	140	11.04	0.615	0.818	2.2941	9.2985
S3	0.3909	730	6.292	0.538	0.8	3.1176	1.7463
S4	0.2987	180	12.96	0.821	1.182	3.0588	11.433
S5	0.2753	190	8.542	0.641	0.673	2.6471	2.7164
S6	0.287	120	5.417	0.615	0.564	3	1.8507
S7	0.361	640	6.583	0.59	0.909	2.4706	1.8806
S8	0.1468	210	3.042	0.564	0.309	2.5882	0.5522
Average	0.3224	297.5	8.167	0.663	0.777	2.7353	5.8041
Shale value (mg/g)	0.77	0.0001	0.024	0.039	0.055	0.017	0.067

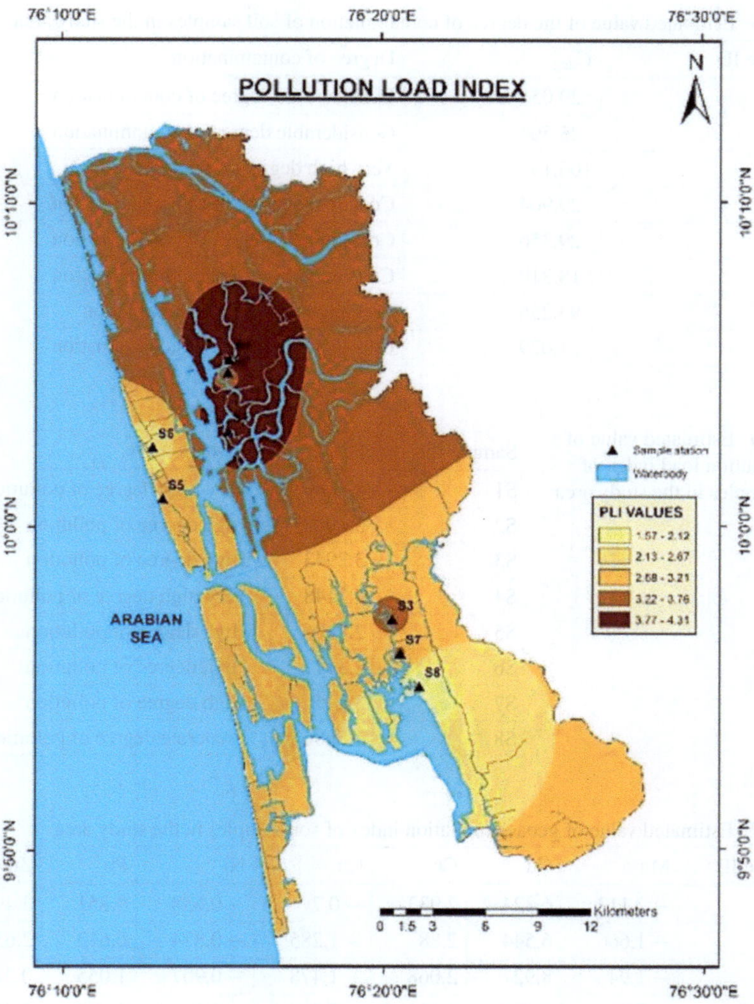

Fig. 10 Spatial distribution of pollution load index (PLI)

include waste discharge from industries such as the Cochin Shipyard and fishing harbors, mixing of agrochemicals and manure containing Cd and Zn from agricultural fields, rusting and paint removal from fishing boats, activities within fishing zones, and dense waterways associated with populated areas. The mobilization potential of heavy metals in sediments follows the order of Cd > Cr > Zn > Pb > Mn > Cu > Ni.

Table 5 Estimated value of the degree of contamination of soil samples in the study area

Sample ID	C_{deg}	Degree of contamination
S1	29.05	Considerable degree of contamination
S2	23.506	Considerable degree of contamination
S3	106.13	Very high degree of contamination
S4	29.964	Considerable degree of contamination
S5	29.356	Considerable degree of contamination
S6	18.819	Considerable degree of contamination
S7	93.256	Very high degree of contamination
S8	31.029	Considerable degree of contamination

Table 6 Estimated value of the pollution load index of soil samples in the study area

Sample ID	PLI	Remark
S1	4.3086	Very high degree of pollution
S2	3.6008	High degree of pollution
S3	3.2943	High degree of pollution
S4	4.2148	Very high degree of pollution
S5	2.8106	High degree of pollution
S6	2.3177	High degree of pollution
S7	3.245	High degree of pollution
S8	1.5686	Moderate degree of pollution

Table 7 Estimated value of geoaccumulation index of soil samples in the study area

Sample ID	Mn	Cd	Cr	Cu	Ni	Pb	Zn
S1	− 2.113	6.824	2.933	− 0.7	− 0.638	0.851	3.499
S2	− 1.666	6.544	2.88	− 1.285	− 0.874	0.613	2.632
S3	− 1.94	8.927	2.068	− 1.478	− 0.907	1.055	0.219
S4	− 2.328	6.907	3.111	− 0.87	− 0.344	1.028	2.93
S5	− 2.446	6.985	2.51	− 1.227	− 1.157	0.819	0.857
S6	− 1.666	6.544	2.88	− 1.285	− 0.874	0.613	2.632
S7	− 2.055	8.737	2.134	− 1.347	− 0.722	0.72	0.326
S8	− 3.353	7.129	1.02	− 1.411	− 2.279	0.787	− 1.44

10 Discussion

Using GIS techniques, assessing soil physicochemical properties and heavy metal distribution in the coastal urban proximal areas of Ernakulam District, Kerala, has offered valuable insights into the region's environmental quality and potential risks. Eight soil samples were collected from the study area to analyze various parameters.

The geochemistry of the area was studied using X-ray fluorescence (XRF), and contamination geochemistry was calculated from the XRF data. The application of GIS technology has been instrumental in spatially visualizing the distribution patterns of heavy metals and their correlation with soil characteristics. This spatial approach enables a better understanding of the underlying environmental processes and supports informed decision-making regarding land use planning, environmental management, and remediation strategies. By integrating geochemical data with GIS tools, it is possible to identify potential contamination hotspots and prioritize areas for intervention. The results show that the soil salinity and pH levels at sampling station S6 (Vypin South) are higher than in other locations. The soil properties at S6 have been significantly influenced, possibly due to its proximity to the coast and the impact of saltwater intrusion. A comprehensive analysis of soil parameters and heavy metal concentrations reveals that anthropogenic activities, urbanization, and coastal proximity have contributed to notable variations in soil properties and metal distributions across the study area.

The Pollution Load Index (PLI) indicates high pollution levels at sampling stations S1 and S4. The Geoaccumulation Index (Igeo) further shows very high cadmium (Cd) contamination levels in the study area. Principal component analysis (PCA) determined that cadmium (Cd) and zinc (Zn) primarily originated from anthropogenic sources. Various environmental factors influence the movement and distribution of these metals. As agrochemicals represent a significant source of Cd and Zn in the study area, ongoing agricultural practices may gradually increase the concentrations of these elements in the soil over time.

11 Conclusion

The study reveals the site-specific characteristics of soil physicochemical properties and heavy metals along the coastal urban proximity in the Ernakulam district of Kerala, India. The findings emphasize the urgent need for sustainable urban development practices and effective pollution control measures in coastal metropolitan areas. Addressing the elevated levels of heavy metals in specific locations is essential to safeguard human health and ecosystem integrity. Implementing measures such as proper waste management, promoting green spaces, and adopting eco-friendly construction practices can help minimize further soil degradation and prevent metal accumulation. However, it is crucial to recognize that this assessment represents a specific point in time. Ongoing monitoring is necessary to track changes in soil quality and metal concentrations over time. Future research is encouraged to investigate heavy metal contamination sources in greater detail, assess potential impacts on local flora and fauna, and examine long-term trends in soil health. These findings can help stakeholders develop sustainable strategies to ensure the health and resilience of these regions' environments and communities.

Acknowledgements The corresponding author expresses sincere gratitude to the Director, National Centre for Earth Science Studies (NCESS), Ministry of Earth Sciences (MoES), Government of India, Thiruvananthapuram, for the continuous support and encouragement provided throughout the successful completion of this research work. Special thanks are also extended to Dr. K. Maya, Scientist-G and Group Head, BGG, NCESS, for her support. The authors are grateful to the anonymous reviewers for their valuable comments and suggestions, which significantly improved the quality of the manuscript. The authors also acknowledge the IMD, GSI, SOI, USGS-Earth Explorer, and the EU-Copernicus Data Space Ecosystem for providing essential data sources.

References

Ali Z, Kazi AG, Malik RN, Naz M, Khan T, Hayat A, Kazi AM (2015) Heavy metal built-up in agricultural soils of Pakistan: sources, ecological consequences, and possible remediation measures, pp 23–42. https://doi.org/10.1007/978-3-319-14526-6_2

Anaman R, Peng C, Jiang Z, Liu X, Zhou Z, Guo Z, Xiao X (2022) Identifying sources and transport routes of heavy metals in soil with different land uses around a smelting site by GIS-based PCA and PMF. Sci Total Environ 823:153759. https://doi.org/10.1016/j.scitotenv.2022.153759

Bux RK, Haider SI, Batool M, Solangi AR, Shah Z, Karimi-Maleh H, Sen F (2022) Assessment of heavy metal contamination and its sources in urban soils of district Hyderabad, Pakistan, using GIS and multivariate analysis. Int J Environ Sci Technol 19(8):7901–7913. https://doi.org/10.1007/s13762-021-03691-7

El Behairy RA, El Baroudy AA, Ibrahim MM, Mohamed ES, Rebouh NY, Shokr MS (2022) Combination of GIS and multivariate analysis to assess the soil heavy metal contamination in some arid zones. Agronomy 12(11):2871. https://doi.org/10.3390/agronomy12112871

Fatoki O, Mathabatha S (2004) An assessment of heavy metal pollution in the East London and Port Elizabeth harbours. Water SA 27(2). https://doi.org/10.4314/wsa.v27i2.4997

Hakanson L (1980) An ecological risk index for aquatic pollution control.a sedimentological approach. Water Res 14(8):975–1001. https://doi.org/10.1016/0043-1354(80)90143-8

Jayaraju N, Sundara Raja Reddy BC, Reddy KR (2009) Metal pollution in coarse sediments of Tuticorin coast, Southeast coast of India. Environ Geol 56(6):1205–1209. https://doi.org/10.1007/s00254-008-1220-5

Jonathan M, Ram-Mohan V, Srinivasalu S (2004) Geochemical variations of major and trace elements in recent sediments, off the Gulf of Mannar, the southeast coast of India. Environ Geol 45(4):466–480

Kaliraj S, Chandrasekar N, Ramachandran KK (2016) Mapping of coastal landforms and volumetric change analysis in the southwest coast of Kanyakumari, South India Using Remote Sensing and GIS Techniques. Egypt J Remote Sens Space Sci. https://doi.org/10.1016/j.ejrs.2016.12.006

Kaliraj S, Chandrasekar N, Ramachandran KK, Srinivas Y, Saravanan S (2017) Coastal landuse and land cover change and transformations of Kanyakumari coast, India Using Remote Sensing and GIS. Egypt J Remote Sens Space Sci. https://doi.org/10.1016/j.ejrs.2017.04.003

Kaliraj S, Chandrasekar N, Ramachandran KK, Selvakumar S (2019) Seawater intrusion vulnerability in the coastal aquifers of southern India—an appraisal of the GALDIT model, parameters' sensitivity, and hydrochemical indicators. Environ Sci Pollut Res. https://doi.org/10.1007/s11356-019-04401-0

Kaliraj S et al (2023) Remote sensing indices-based soil properties measurement—a case study of the Thamirabarani River basin, south India. In: Dharumarajan et al. (eds) Remote sensing of soils: mapping, monitoring, and measurement. Elsevier, USA. ISBN: 978-0-44-318773-5, https://doi.org/10.1016/B978-0-443-18773-5.00030-2

Kekane SS, Chavan RP, Shinde DN, Patil CL, Sagar SS (2015) A review on physico-chemical properties of soil. Int J Chem Stud 3(4):29–32

Lalitha M, Kumar KSA, Nair KM, Dharumarajan S, Koyal A, Shivanand K, Kaliraj S, Hegde R (2021) Evaluating pedogenesis and soil Atterberg limits for inducing landslides in the Western Ghats, Idukki District of Kerala, South India. Nat Hazards 106:487–507. https://doi.org/10.1007/s11069-020-04472-0

Muller G (1969) Index of geoaccumulation in sediments of the Rhine river. GeoJournal 2:108–118

Pande A, Nayak GN (2013) Understanding distribution and abundance of metals with space and time in estuarine mudflat sedimentary environment. Environ Earth Sci 70(6):2561–2575. https://doi.org/10.1007/s12665-013-2298-y

Perumal K, Antony J, Muthuramalingam S (2021) Heavy metal pollutants and their spatial distribution in surface sediments from Thondi coast, Palk Bay, South India. Environ Sci Eur 33(1):63. https://doi.org/10.1186/s12302-021-00501-2

Salaj SS, Ramesh D, Suresh Babu DS, Kaliraj S, Chandrasekar N (2022) Appraisal of urban growth impacts on seawater intrusion vulnerability using GIS-based modified GALDIT-U model: a case study of Kozhikode coastal stretch, Kerala, South India. J Appl Remote Sens 16(1):012014. https://doi.org/10.1117/1.JRS.16.012014

Salomons W, Forstner U (1984) Metals in the hydrocycle. Springer, Berlin, Heidelberg, New York, Tokyo. https://doi.org/10.1007/978-3-642-69325-0

Shaginimol CN, Manojkumar B, Kaliraj S (2022) Spatial distribution of heavy metals in surface water from coastal areas of Ernakulam district, Kerala, Southwest coast of India. Ecol Environ Conserv 28:S151–S159. https://doi.org/10.53550/EEC.2022.v28i07s.026.

Shirbhate N, Malode SN (2012) Municipal solid waste management: a survey and physicochemical analysis of contaminated soil from sukali compost and landfill depot, batkuli road, Amravati. GJBB 1(2):215–219

Tale KS, Ingole S (2015) A review on the role of physico-chemical properties in soil quality. Chem Sci Rev Lett 4(13):57–66

Tomlinson DL, Wilson JG, Harris CR et al (1980) Problems in the assessment of heavy-metal levels in estuaries and the formation of a pollution index. Helgolander Meeresunters 33:566–575. https://doi.org/10.1007/BF02414780

Tripathy BR, Kaliraj S, Sajjad H, Joshi PK, Chaudhary BS, Kumar P (2018) Monitoring of seasonal variability and movement of suspended sediment concentrations along the Thiruvananthapuram coast, southern India, using the Landsat OLI sensor. Ocean Sci 14:1085–1092. https://doi.org/10.5194/os-14-1085-2018

Turekian KK, Wedepohl KH (1961) Distribution of the elements in some major units of the earth's crust. Geol Soc Am Bull 72:175–192. https://doi.org/10.1130/0016-7606(1961)72[175:DOTEIS]2.0.CO;2

Vineethkumar V, Sayooj VV, Shimod KP, Prakash V (2020) Estimation of pollution indices and hazard evaluation from trace elements concentration in coastal sediments of Kerala, southwest coast of India. Bull National Res Centre 44(1):198. https://doi.org/10.1186/s42269-020-00455-0

Zhou Z, Li D, Huh D, Xie M, Mun Y (2024) A simulation study of the performance of statistical models for count outcomes with excessive zeros. Stat Med 43(24):4752–4767. https://doi.org/10.1002/sim.10198

Evolution of Coastal Erosion in the Gulf of California During the Anthropocene Epoch and Climate Change

Yedid Guadalupe Zambrano-Medina, Marco A. Mora-Ramírez, Evangelina Avila-Aceves, Wenseslao Plata-Rocha, Cuauhtémoc Franco-Ochoa, Lidia Yadira Perez-Aguilar, and Gibrán Humberto Manjarrez-Pérez

Abstract Coastal zones are sensitive and vulnerable areas because of the combined interaction of physical, biological, and human activities. In this sense, climate change and human factors increase the negative effects of coastal erosion, which represent a significant threat to the integrity of these ecosystems by changing their ecological services and functions. The latter has changed the way weather patterns behave along the coast, particularly in the Gulf of California, localized in northwest Mexican, where rising sea levels, stronger waves, storm surges, and expansion of human activities have led to increased coastal erosion. Furthermore, coastal urban and tourism development, exploitation of natural resources, and unsustainable fishing practices in the Anthropocene era have contributed to the recent degradation of local marine

Y. Guadalupe Zambrano-Medina · E. Avila-Aceves · W. Plata-Rocha
Facultad de Ciencias de La Tierra y El Espacio, Universidad Autónoma de Sinaloa, Culiacán 80013, Sinaloa, México

Y. Guadalupe Zambrano-Medina · M. A. Mora-Ramírez
Facultad de Ciencias Químicas, Benemérita Universidad Autónoma de Puebla, Puebla de Zaragoza 72570, Puebla, México

C. Franco-Ochoa
Facultad de Ingeniería Civil, Universidad Autónoma de Sinaloa, Culiacán 80013, Sinaloa, México

G. Humberto Manjarrez-Pérez
Facultad de Ingeniería y Tecnología, Universidad Autónoma de Sinaloa, Mazatlán 82017, Sinaloa, México

Y. Guadalupe Zambrano-Medina (✉)
Posgrado en Tecnologías Agrícolas Limpias, Instituto de Ciencias, Benemérita Universidad Autónoma de Puebla, Puebla de Zaragoza 72570, Puebla, México
e-mail: yedidzambrano@uas.edu.mx; yedidzambranoposdoc@gmail.com

L. Yadira Perez-Aguilar
Facultad de Informática Culiacán, Universidad Autónoma de Sinaloa, Culiacán 80013, Sinaloa, México

E. Avila-Aceves · W. Plata-Rocha · L. Yadira Perez-Aguilar
Laboratorio Nacional CONAHCYT de Tecnologías de la Información Geoespacial Para los Sistemas Socioecológicos Resilientes (LaNCTIGeSSR), Toluca de Lerdo 50110, México

K. K. Basheer Ahammed et al. (eds.), *Monitoring Coastal Marine Ecosystems Under Climate Change*, https://doi.org/10.1007/978-3-032-09137-6_4

and coastal ecosystems. This chapter analyzes the negative impact on coastal ecosystems and the pattern of degradation caused by erosion in the Gulf of California. We performed a comprehensive review of the current state of the Gulf and performed a detailed analysis of the rates of coastal change using the Digital Shoreline Analysis System (DSAS) method, employing the weighted linear regression (WLR) statistic. The results show that the overall rate of change along the Gulf coast is 1.7 m/yr. Seventy percent of the coast has an erosion rate of 2.9 m/yr. The researchers identified the most prominent erosion hotspots on the northern coast of Baja California, the northern coast of Sonora, and along the coast of Sinaloa.

Keywords Coastal erosion · Shoreline · Anthropocene · Climate change · Gulf of California

1 Introduction

Coastal zone monitoring is an important task in national and international development because the coast is one of the most dynamic processes on earth and is the main socioeconomic environment worldwide (Dwivedi et al. 2023). The coastal or littoral zone makes up an area with a critical dynamic equilibrium that is sensitive to any alteration (Bardají et al. 2009). It is a maritime-terrestrial spatial strip of complex variable width and constitution. It harbors a diversity of environments and resources that interact with each other, creating ecosystems with unique environmental, climatic, geomorphological, and hydrological characteristics on the planet (Laignel et al. 2023; Plomaritis et al. 2018).

Knowing their behavior is fundamental for protecting the environment, as well as to avoid loss of life and property, to have a greater security in ports and to avoid the decrease of coastal land resources that arises mainly from erosion. In coastal areas, significant problems occur because of the modification of the natural sedimentary system causing a variation in the coastline (Merlotto et al. 2008), with erosive and/or cumulative tendency because of geomorphological modifications of the landscape, caused by physical processes such as tides, waves, cyclones, coastal sediment transport, and barriers, whether natural or anthropogenic, which intervene in different time domains as they can occur in a few hours or thousands of years.

Shoreline studies, which define the physical interface of land and water, serve as a basic tool for integrated coastal management, and many scientific, coastal management, and engineering studies need to analyze them (Boak and Turner 2005; Merlotto et al. 2013). These scientific studies of coastal erosion management are based on analyzing long-term or short-term shoreline changes (Bera and Maiti 2019) depending on the aim of the research. Using traditional land survey techniques can make it difficult and sometimes impossible to delineate the shoreline. Advances in remote sensing and GIS overcome these difficulties in shoreline position extraction by proposing automated methods on shoreline, such as image processing algorithms like edge detection and image segmentation techniques (Aedla et al. 2015).

Currently, this type of shoreline analysis is performed using Geographic Information Systems (GIS) based on mapping and quantitative analysis of erosion and accretion (Ahmad and Lakhan 2012; Barbaro et al. 2019; Jonah et al. 2016; Kallepalli et al. 2017; Lee et al. 2019; Zhang et al. 2019). The Digital Shoreline Analysis System (DSAS) (Thieler et al. 2009) is one of the most widely used GIS tools for analyzing shoreline changes in order to understand the rates of change of erosion and/or accretion.

Hydrodynamic forces such as sea level, wave conditions, storm surges increasingly influenced by climate change, as well as river flow are the main drivers of changes in coastal areas (Ranasinghe 2016), and the coast of the Gulf of California in Mexico is no exception to this trend. The Gulf of California is a paradigmatic example of how human activities and climate change are changing coastal processes in the Anthropocene era. In this context, a combination of anthropogenic and climatic factors that together are altering coastal stability and resilience in the region has intensified coastal erosion. This chapter discusses how the Anthropocene, marked by unprecedented human influence on natural systems and global climate change have exacerbated coastal erosion processes in the Gulf of California. Urbanization, coastal development, and infrastructure construction have changed natural sedimentation and erosion processes, while climate change has introduced new challenges, such as sea level rise and more extreme weather events. These combined factors are accelerating the degradation of Gulf coasts, underscoring the urgency of implementing coastal management strategies that consider both human influences and climate change to preserve these crucial ecosystems.

2 Gulf of California

The geography of the Gulf of California defines it as a deep, narrow, elongated body of water in a NW–SE orientation, with the arid Baja California peninsula to the west and the deserts of Sonora and Sinaloa to the east (Fig. 1). This results in a more continental climate compared to the nearby California Current (Bernal et al. 2001).

The Gulf of California, also known as the Sea of Cortes, is a region of unique scientific interest because of its geological complexity and climatic diversity. Between the Baja California peninsula and the states of Sinaloa and Sonora in northwestern Mexico, active tectonic processes, a rich sedimentary history, and climatic patterns that vary from tropical to subtropical shape this dynamic coastal system. The formation of the Gulf of California is associated with a complex tectonic system, which includes the separation of the Baja California peninsula from the North American continent. Geological faults and basin subsidence have created a coastal configuration characterized by cliffs, beaches, deltas and bays that are highly susceptible to erosion (Dorsey et al. 1995; Ledesma-Vázquez et al. 2009). The region is part of the San Andreas fault system, which implies significant seismic activity and a constantly changing coastal landscape, especially in the Upper Gulf. Areas near the San Andreas fault and other minor faults have high degrees of geologic instability,

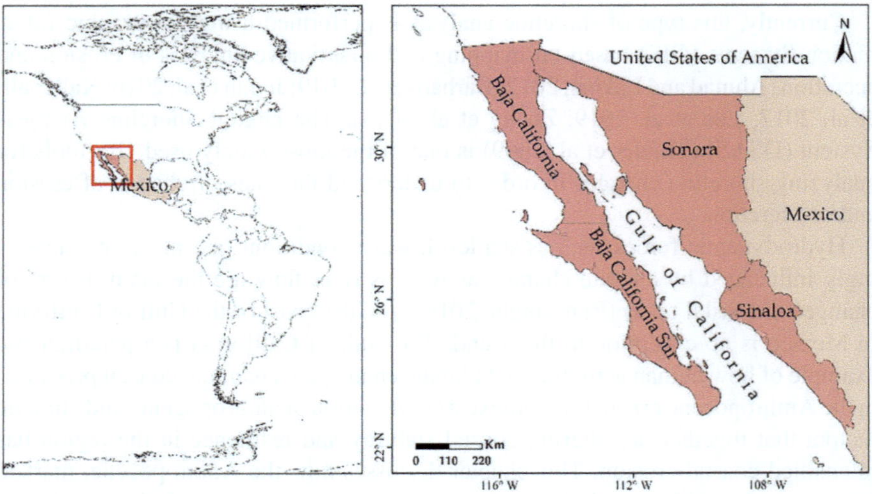

Fig. 1 The study area is in the Gulf of California

which may contribute to accelerate erosion in certain coastal segments (Pérez-Cruz 2013). This dynamic tectonic environment, together with the climatic variability and anthropogenic pressure present in the Anthropocene era, defines a highly vulnerable coastal system.

2.1 *Climate and Oceanographic Dynamics*

Its location influences the Gulf of California's climate in the transition between tropical and subtropical zones, resulting in wide climatic variability. It is highly susceptible to hurricanes and tropical storms, which contribute significantly to coastal erosion through storm surges and extreme waves (Ezcurra et al. 2019). Two major systems: the North American monsoon, which brings seasonal rains during the summer, and dominates weather patterns in the region, the Santa Ana winds, which occur mainly in winter (Anderson et al. 2000a, b; Zamora et al. 2014). These climatic factors have a direct impact on coastal erosion processes by altering wave patterns, precipitation, and sediment transport. During the summer, the monsoon influence causes torrential rains that increase surface runoff and fluvial erosion in the watersheds that drain into the Gulf of California. In contrast, Santa Ana winds generate dry conditions and strong winds that intensify wind erosion in coastal areas.

The Gulf of California has an arid climate, with 50–200 mm of annual rainfall and summer temperatures reaching 40 °C. In the upper Gulf, temperatures drop considerably in winter, with frost and even snowfall, while the lower Gulf is prone to tropical storms during the summer. Rainfall is scarce in the upper Gulf, concentrated mainly between October and May, while weak winds from the southeast dominate atmospheric wind forcing during the summer and strong winds from the northwest in winter (Sans 2018). This strongly influenced the ocean circulation forcing, along with tidal cycles and air-sea heat exchange, which defines the circulation pattern in the Gulf of California (Lluch-Cota et al. 2010).

Tidal currents in the Gulf of California are particularly strong due to the geomorphological configuration of the narrow gulf, resulting in a significant exchange of water with the Pacific Ocean. The astronomical tide is predominantly semi-diurnal, with variations in the tidal range: micro-tidal (< 2 m) in the lower Gulf, meso-tidal (2–4 m) in the middle Gulf, and macrotidal (> 4 m) in the upper Gulf (Servicio Mareográfico Nacional-UNAM 2024). Wave heights are around 0.30 m in the upper Gulf, 0.60–0.90 m in the middle Gulf, and 1.50–1.80 m in the lower Gulf (Kasper-Zubillaga et al. 2007).

2.2 *Ecological, Conservation and Population Importance*

The Gulf of California is home to one of the most diverse marine regions on the planet, with 244 islands, islets and coastal areas that are of great ecological importance. Various international designations, such as their recognition as a UNESCO World Heritage Site and including several sites on the list of wetlands of international importance under the Ramsar Convention, protect many of these areas. These designations highlight their global relevance and emphasize the need to protect them from harmful human activities. Protected areas in the Gulf of California not only preserve critical habitats for marine species but also contribute to the stability of local fisheries, which depend on healthy marine ecosystems. National and international conservation programs work toward mitigating threats to these areas, such as overfishing, pollution, unregulated coastal development, and climate change (Wasim et al. 2025). The creation of marine reserves and natural protected areas has been a key strategy to maintain the biodiversity and ecosystem services provided by the gulf.

The Gulf of California is not only ecologically valuable, but also of great importance to the human communities that inhabit its shores. Local populations depend heavily on the Gulf's natural resources for their livelihoods, especially through fishing, tourism and agriculture. Fishing is one of the main economic activities in the region, providing income and food for thousands of families. However, the intensive use of these resources has also led to significant challenges, such as overexploitation of species and degradation of critical habitats.

In addition, coastal development and urbanization have increased the vulnerability of coastal structures and communities to extreme weather events, such as hurricanes and storm surges. Coastal erosion, exacerbated by climate change, also poses a growing threat to infrastructure, housing, and tourist areas in the region. For example, storm surges and sea level rise are increasingly present risks that affect both human populations and ecological areas in the gulf. Adaptation and mitigation strategies are crucial to protect both human communities and the gulf's ecosystems.

Strategies include using barriers to reduce erosion, restoring coastal habitats, and planning sustainable urban development.

3 Contribution of the Anthropocene and Climate Change to Coastal Erosion

Today's coasts face a double threat: on the one hand, the growing trend of human activities in the Anthropocene era and the impacts of climate change, which is reflected in the intensification of coastal erosion, and the Gulf of California is no exception. Urbanization, coastal development and infrastructure construction have altered natural sediment and erosion processes. In addition, climate change has introduced new challenges, such as sea level rise and extreme weather events. Coastal degradation underscores the urgent need to implement coastal management strategies that consider both human influences and climate change to preserve these ecosystems.

Discerning between changes because of natural processes, human activities, and climate change on the coast at the regional or global level is very complicated because these factors (Fig. 2) interact with each other in time and space (Gómez-Pazo et al. 2019). However, natural processes and human activities in coastal areas are the major factors linked to the modern dynamics of the coastal zone (Bera and Maiti 2019). We need to understand clearly that multiple factors intervene in the coastline, and attributing a change in any fragment of the coastline to only a few of them is difficult. However, knowing the variability of the shoreline and its critical points of erosion or accretion, it is easier to identify which natural or anthropogenic actions causing a change in the shoreline.

The interaction between anthropogenic and climatic factors has created a vicious cycle that exacerbates coastal erosion in the Gulf of California. For example, the degradation of coastal ecosystems because of urbanization and infrastructure construction has reduced the natural buffering capacity of the coasts against sea level rise and storms in the Gulf of California. This climate change intensifies the

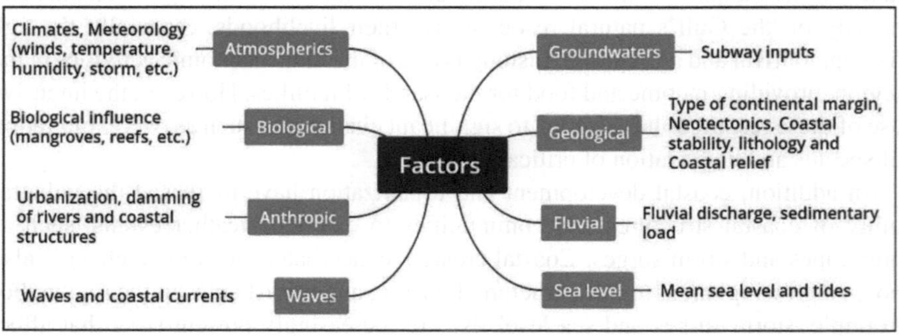

Fig. 2 Factors controlling the characteristics of coastal environments

Fig. 3 Erosion on Altata beaches in Sinaloa coasts

pressure on its already vulnerable coasts, accelerating the erosion process as shown in Fig. 3, where erosion is already so severe that it affects residential buildings on the Altata beach on the coast of Sinaloa.

The combination of unplanned coastal development, together with the impacts of climate change, highlights the need for integrated coastal zone management that considers both natural dynamics and human interventions. Long-term planning and adaptation measures are crucial to mitigate the combined effects of these forces on the coasts of the Gulf of California.

3.1 Impact of Urbanization and Coastal Development/ Coastal Infrastructure

Urbanization and coastal development in the Gulf of California have a significant impact on coastal erosion. As human populations have grown in the region, especially in urban areas such as La Paz, Cabo San Lucas, Guaymas, and Mazatlán, the construction of infrastructure, hotels, and tourism developments have increased considerably (Fig. 4). These anthropogenic activities have changed coastal landscapes, altering the natural processes of sedimentation and erosion. Disorderly expansion in coastal areas is possibly the fundamental way in which humans affect the environment. This process begins with the destruction or alteration of land, followed by the construction of settlements and their subsequent development. This geomorphological modification acts negatively and sometimes accentuates the increase of erosion on the coastline (Rangel-Buitrago 2009).

The removal of coastal vegetation and dunes to make way for urban infrastructure has reduced the natural ability of shorelines to resist erosion. Vegetation, such

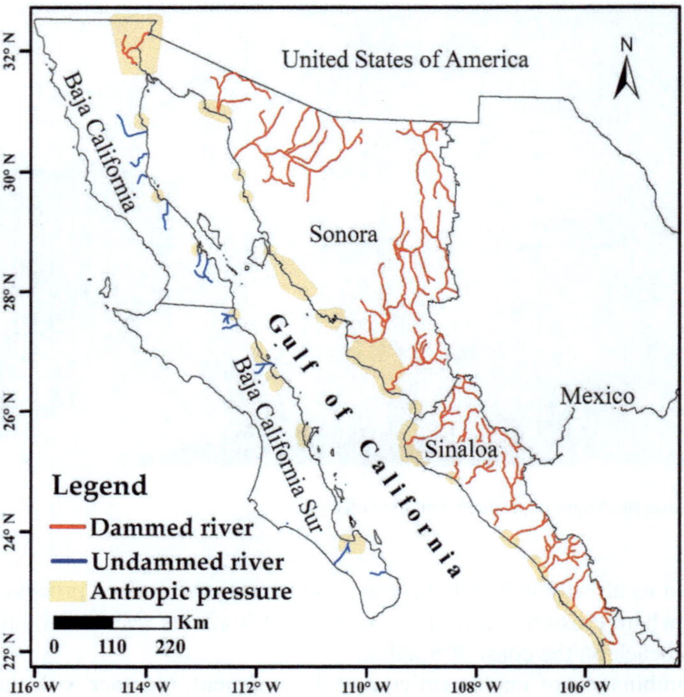

Fig. 4 Distribution of anthropic pressures and location of rivers with dams in the coastal region

as mangroves and seagrasses, plays a significant role in stabilizing sediment and protecting shorelines from erosion. However, urban expansion and the construction of roads, ports, and dams have degraded these ecosystems, exacerbating the vulnerability of coastal to erosion. There are many human actions that erode processes, and when coastal are affected, natural factors increase their erosive action to different degrees depending on the magnitude of the intervention, hindering the natural reconstruction of beaches and eroding (Merlotto et al. 2008). Sometimes recovery is irreversible or drastic changes in socioeconomic behavior are required to preserve the natural environment or mitigate the consequences of changes already made.

Therefore, anthropogenic influence on changes in the coastal plays an important role in its configuration, causing it to be more vulnerable to imminent climate change. In the Gulf of California, the construction of dams, the high population density near or on the beaches, as well as the main economic activities, such as tourism, agriculture and aquaculture on the coast, are important modifiers of sediment dynamics and ultimately affect the variability of the coastline eroding. Similarly, the construction of protective structures such as walls and breakwaters in order to balance coastal processes, when not well designed, become another trigger for morphological changes and erosion caused by wave energy, which is increasing at rates of 0.4% per year globally because of the increase in ocean temperature and these change sediment transport processes (Reguero et al. 2019).

3.2 *Climate Change: Sea Level Rise, Sea Surface Temperature, and Extreme Weather Events*

Climate change in the Anthropocene era arises from changes that are directly or indirectly caused by human activity. Carbon dioxide emissions are the key driver of global climate change (Fig. 5). Before the industrial revolution, global emissions were low. In 1950, the world emitted 6 billion tons of CO_2. By 1990, it had almost quadrupled, reaching over 20 billion tons. Emissions have continued to explode. We now emit approximately over 35 billion tons each year. The increase in greenhouse gas emissions contributes to temperature variation and the deterioration of the ozone layer. This has led to the intensification of this atmospheric phenomenon, which has recently increased because of various human activities, such as the use of electrical energy, fossil fuel industries, oil and gas, pollution from transportation, and the misuse of waste, among others (Dawahidi et al. 2019).

The Intergovernmental Panel on Climate Change concludes that the warming of the climate system is indisputable, and that humans have clearly influenced this phenomenon, affecting physical and biological systems and the (IPCC 2022). Proof of this is the increase in air and ocean temperature, melting of the poles, droughts, extreme weather events, erosion, floods, and the rise in global sea level. The latter being the most important climate indicator, which is largely driven by global temperature changes. When the average temperature rises, the global mean sea level rises (Nicholls and Cazenave 2010). Scientists have detected these effects throughout the climate system, with biophysical changes and natural processes being present in

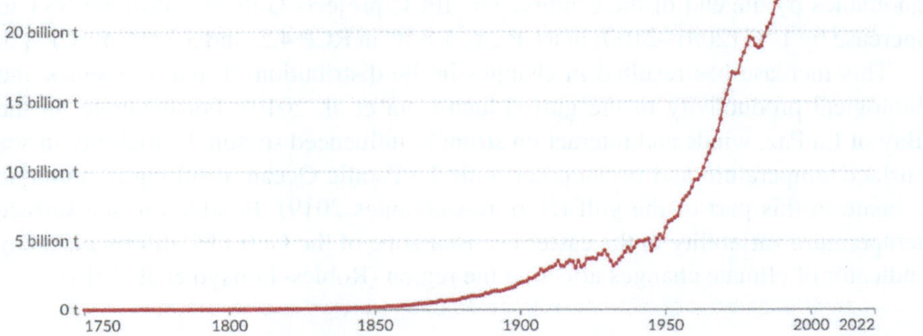

Fig. 5 Global CO_2 emissions from fossil fuels. *Source* (Ritchie 2022)

the planet's evolution. However, today, they mainly attribute these effects to anthropogenic climate change. They use this term because evidence shows that human activities in the Anthropocene era have recently caused an increase in the Earth's atmospheric temperature. At present, the distribution of the intensity of climate change affects is affecting all countries, but unevenly.

3.2.1 Sea Surface Temperature

Global warming associated with anthropogenic climate change impacts both the mean SST and the thermal and atmospheric processes that affect ocean circulations (Robles-Tamayo et al. 2018). One of the most important variables to understand how climate change in the Anthropocene era has affected the Gulf of California is sea surface temperature (SST), which is very important in coastal dynamics studies. The Gulf has experienced a significant increase in water temperature during the last decade, with researchers linking it to both global warming and the Niño-Southern Oscillation (ENSO) phenomena (Bernal et al. 2001; López Martínez et al. 2023), and analyzes of SST in the Gulf of California, such as the one developed by Soto Mardones et al. (1999). In his research on the SST of the Gulf of California using satellite images for the period 1983–1996, found that the SST decreases towards the northern part of the Gulf of California, finding that the variability increases year after year, and also noted the appearance and development of warm zones during the winter, which he related to anticyclonic gyres, as well as cold zones during the summer, which he attributed to cyclonic gyres in this geographic area. In the central and southern regions, there are greater amplitudes during winter on the continental side.

According to Lluch-Cota et al. (2010) and Soto Mardones (1999), the surface circulation and temperature in the Gulf of California are predominantly seasonal, showing a cyclonic surface circulation in summer and anticyclonic in winter. Ojeda-Ruiz et al. (2020) analyzed SST using remote sensing data (1985–2018) and detected higher anomalies in 1997 with a value of 2.9 °C, above average and increasing positive anomalies by the end of the century. The IPCC projects Gulf of California SST to increase by 1 °C (2081–2100) in RCP 2.6, 1.8 °C in RCP 4.5, and 3.2 °C in RCP 8.5.

This increase has resulted in changes in the distribution of marine species and biological productivity in the gulf (Lluch-Cota et al. 2010). For example, in the Bay of La Paz, winds and interaction strongly influenced seasonal variability in sea surface temperature and ocean color with the Pacific Ocean, resulting in a unique climate in this part of the gulf (Herrera-Cervantes 2019). In addition, sea surface temperature variability in the eastern coastal zone of the Gulf of California is a key indicator of climate changes affecting the region (Robles-Tamayo et al. 2018).

3.2.2 Sea Level Rise

The recent rise in sea level is one of the most important manifestations of climate change. Several factors caused it, mainly by the melting of the poles and the thermal expansion of seawater because of ocean warming. Local and regional factors, including Earth's crust movement, coastal subsidence, groundwater extraction, ocean circulation changes, atmospheric pressure (storms), and water temperature (Páez-Osuna et al. 2016) influence it.

The coast is vulnerable to sea level rise, is not static and is subject to erosion and accumulation processes. The number of sediments that reach the coast influences these processes and by the actions of climatic agents. Therefore, the coast is a dynamic element and is one system most affected by climate change, especially by sea level rise. This reaction of the coast depends on phenomena that characterize a particular zone, its typology and the characteristics of its backdrop (Kallepalli et al. 2017; Olivas and González 2018). This sea level rise will cause much more rapid coastal retreat with severe effects in low-lying areas. However, the precise regional rates depend on factors, such as local wind energy, seawater temperature, and tectonic uplift/sink, among others (Nicholls and Cazenave 2010; Reguero et al. 2019). There are simple response models that approximate how the coast will respond to sea level rise. The most commonly used method is the simple two-dimensional mass conservation principle known as the (Bruun 1962).

Systematic observation of sea level began after the launch of TOPEX / Poseidon in late 1992. Since then, the recording of sea level measured by satellite altimetry has led to improvements in the understanding of sea level change on regional and global scales. Currently, the global mean sea level is rising at rates of 3.6 mm according to AVISO data from 1993 to 2024 with an acceleration of 0.11 ± 0.05 mm/yr (Fig. 6), while wave energy is increasing 0.4% per year (Reguero et al. 2019). However, sea level rise (SLR) is not homogeneous in all oceans, for example, in the Pacific Ocean, it is 2.88 mm/yr, in the Gulf of Guinea on the Atlantic coast of Cameroon it is about 3.4 mm/yr (Fotsi et al. 2019), 2.2 mm/yr in the South Atlantic Bight in the United States (Passeri et al. 2014), and only 0.8 mm/yr on the Keelung coast in Taiwan (Tseng et al. 2010). The IPCC projections in the fifth report estimate that sea level rise will be between 0.25 m and 1.0 m in the year 2100.

In the Gulf of California, according to Páez-Osuna et al. (2016), the regional mean sea level trend estimated with satellite altimetry data and, with tide gauges within the Gulf, is 0.8 ± 0.8 mm and 2.5 ± 1.1 mm per year, respectively. Their study showed that sea level rise (1993–2015) has a geographical gradient, i.e., the upper Gulf has a higher mean trend (2.0 ± 0.4 mm/year), while in the middle Gulf it is lower (0.1 ± 0.3 mm/yr) and in the lower Gulf, the mean trend shows a transition from the Gulf entrance of 0.5 ± 0.4 mm/yr.

Fig. 6 Mean sea level from altimetry. *Source* (AVISO + Satellite Altimetry Data 2024)

3.2.3 Extreme Weather Events

The Gulf of California is in a transition zone between tropical and subtropical climate regimes, at the convergence of the North Equatorial Current and the North Pacific Gyre. Therefore, its coasts face exposure to multiannual phenomena such as the El Niño-Southern (ENSO) and North Pacific Decadal Oscillation Index (NPDO) (Lluch-Cota et al. 2013), as well as other phenomena such as the North American Monsoon and Cold Fronts. The effects of tropical cyclones (hurricanes and depressions) and strong northerly winds, known as Northers, also play an important role in the morphological changes of the Gulf of California coast as they impact it directly.

El Niño-Southern Oscillation (ENSO) is the most important signal of interannual variability in the Gulf of California. Its magnitude and regional manifestation depend mainly on the strength of the events (Lluch-Cota et al. 2010). Researchers observed the first evidence of an ENSO event in the Gulf of California in 1957, although the two largest ENSO events in recent times were recorded in 1982–1983 and 1997–1998 (Páez-Osuna et al. 2016). The North Pacific Decadal Oscillation Index (NPDO) is another important phenomenon to consider. PDO and ENSO control the SST variability patterns of the Lower Gulf of California, reinforcing their effects during the ENSO-positive and NPDO-negative periods, while mitigating them after the last NPDO-positive transition. An increase in wind intensity, especially in winter since the 1950s, although this effect is mitigated by the long-term warming of the tropical Pacific (Bernal et al. 2001). Therefore, the Pacific Ocean, through the Gulf of California, influences changes in water temperature and salinity, which relate to annual variations in sea level. In addition, precipitation is mainly influenced by summer rains associated with the ENSO phenomenon.

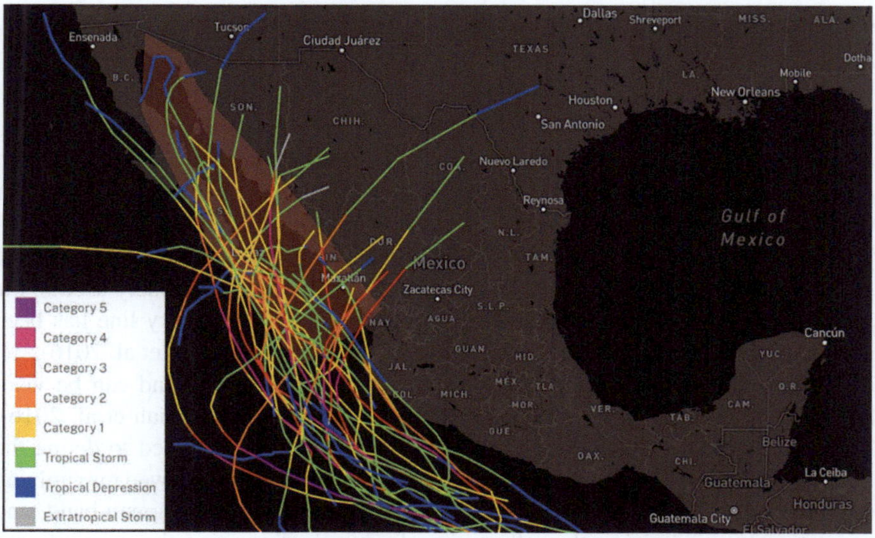

Fig. 7 Tracks of tropical storms that passed along the GC region from 1980 to 2024. *Source* (National Oceanic and Atmospheric Administration 2024)

It is important to consider cold fronts, which occur from September to March, with northwesterly winds blowing over the Gulf of California. (Mart 2018; Ponce Bastidas 2015). The North American Monsoon covers the period from July to September and is related to the onset of rains, tropical cyclones, the formation of a low-pressure center, and the shift from winter northwesterly winds to southerly or even southeasterly winds (Anderson et al. 2000a, b). Therefore, the Gulf faces extreme meteorological phenomena, such as tropical cyclones that depend on the phenomena. Figure 7 shows the trajectories of tropical cyclones, hurricanes, and depressions that have impacted the Gulf from 1980 to 2024.

Sinaloa and Baja California Sur are the states with the highest number of tropical cyclones that make landfall and have a direct impact on the coast (Farfán et al. 2014; Franco-Ochoa et al. 2020). For example, Category 1 Hurricane Manuel severely affected Sinaloa in September 2013, causing extensive damage to infrastructure and agriculture, as well as human losses. Category 3 Hurricane Willa made landfall in Sinaloa on October 23, 2018, causing torrential rains, flooding, and significant damage to infrastructure and housing, especially in coastal areas. Nora, which impacted Sinaloa in August 2021 as a tropical storm, caused heavy rains and flooding in several regions of the state, causing significant damage. Pamela, a Category 1 hurricane, affected the coast of Sinaloa on October 13, 2021, leaving behind heavy rains, flooding, and property damage in several communities. In addition, Tropical Storm Ileana, in 2024, generated high waves and coastal flooding in the southern Gulf of California. These events underscore the vulnerability of the Gulf of California coasts to tropical cyclones and hurricanes, especially the coasts of Sinaloa and southern Baja California Sur, and the importance of being prepared to mitigate their effects.

4 Erosion on the Coast of the Gulf of California

By applying geospatial and statistical techniques, realized analysis of coastal erosion over the last four decades in the Gulf of California, providing a deeper understanding of erosion in this region. For the study, we used Landsat satellite images to extract the coastline between 1980 and 2020. We downloaded these images for October to reduce the effects of seasonal variations in mean sea level and wave energy. The wet/dry line served as an approximation of the high-water line (HWL) in the study. The wet/dry line closely approximates the HWL, which is widely used in the investigation of shoreline changes (Oyedotun 2016), as the wet/dry line has been shown to approximate HWL closely (Barbaro et al. 2019; Kermani et al. 2016).

HWL is defined as the maximum of the last maximum tide and can be visually discernible in the satellite image (Boak and Turner 2005; Jonah et al. 2016). Therefore, the normalized difference water index (NDWI) was used to define the boundary between land and water (McFeeters 1996). This index allowed maximizing the reflectance of both water by working with wavelengths in the green band, and vegetation by minimizing the reflectance of water bodies using the near infrared (NIR) band (Ashtekar et al. 2019; McFeeters 1996; Mondejar and Tongco 2019).

The United States Geological Survey (USGS) developed Digital Shoreline Analysis System (DSAS) version 5.0, which we used to analyze shoreline changes and calculate rates of shoreline change from multiple historical shoreline positions. To apply DSAS, we created a geo-database and incorporated the obtained shorelines, as well as generated the baseline and transects every 100 m. The statistical model of weighted linear regression (WLR) was got, getting a margin of error for the units of rate of change of 0.7 m/yr considering that the Landsat images have a resolution of 30 m per pixel in contrast to the period analyzed. Considering these uncertainties incorporated into the model, we obtained the results of the rate of shoreline change from WLR with a confidence level of 95% (Henderson et al. 2018). WLR will consider all shorelines in the analyzed period and their associated uncertainties assigning a higher weight to transect intersection points with lower uncertainty value to get the best fit line (Baig et al. 2020) by using the following equation:

$$\omega = \frac{1}{e^2}$$

where ω is defined as a function of the variance in the measurement uncertainty (e), and the weighting value ω is linked to the shore data before examining its rate of change.

The results of the shoreline analysis for the entire Gulf of California region with WLR reveal an overall rate of change of − 1.7 m/yr. However, this figure includes both erosion and accretion processes. This chapter focuses on presenting the results in terms of erosion rates, which are the major topic of discussion. Seventy percent of the Gulf of California coastline shows an erosion rate of 2.9 m/yr, while 30% showed accretion with a rate of 1.8 m/yr (Fig. 8). The northern coasts of BC and Sonora are the most notable erosion points with erosion rates of 2.7 and 5.2 m/yr (Table 1). The construction of over 20 dams on the Colorado River since 1930, which reduced water and sediment input to almost zero, is the main attribution for the erosion (Carriquiry and Sánchez 1999; Carriquiry et al. 2001). The Gulf of California has both semi-diurnal and diurnal tides, with significant fluctuations in sea level, ranging from 6.95 m in San Felipe to 10 m in the Colorado River delta. Therefore, in this area, oceanic forcing has become the main regulator of sediment dispersion, transport, and deposition, replacing the fluvial processes that previously predominated (Azcúnaga 2005; Ledesma-Vázquez et al. 2009).

Along the coast of Sinaloa, the damming of its rivers largely contributed to erosion occurring at a rate of – 2.4 m/yr (Table 1) similar to that obtained by (Zambrano-Medina et al. 2023). And to the fact that its coasts are subject to extreme meteorological phenomena and that it also has various anthropogenic activities such as

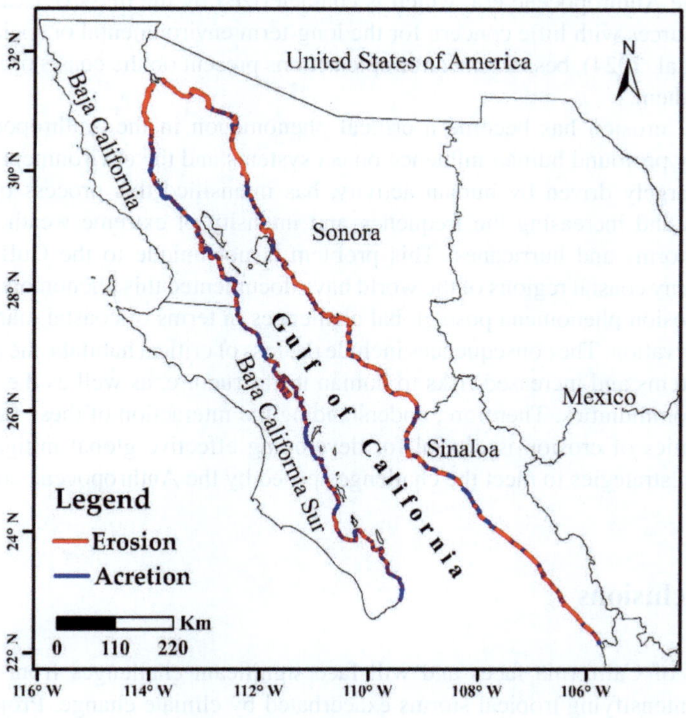

Fig. 8 Evolution of coastal erosion in the Gulf of California

Table 1 There is statistical data available for WLR

Statistics (m/yr)	Sinaloa	Sonora	BC	BCS	GC
Average rate	– 1.4	– 3.6	– 1.7	– 0.4	– 1.7
Average of all erosional rates	– 2.4	– 5.2	– 2.7	– 1.0	– 2.9
Average of all accretion rates	3.7	2.0	0.8	0.8	1.8

breakwaters, jetties, and anthropogenic pressure arising from coastal cities such as the city of Mazatlán which has experienced a high erosion of 2.4 m/yr, similar to that reported by Oyedotun et al. (2018), which reported a rate of $- 1.9 \pm 0.9$ m/yr. Playa Las Glorias in Sinaloa is one of the most eroded beaches presenting erosion rates of − 5 m/yr coinciding with Jiménez-Illescas et al. (2019). Finally, the coast of Baja California Sur maintains stability with an erosion rate of – 1 m/yr (Table 1), and we can attribute this result to wave energy and frequencies of tropical cyclones. However, anthropogenic pressure is also relevant in some areas of the coast.

The erosion in the Gulf Coast of California primarily affects areas near the mouths of dammed rivers, densely populated areas, and land used for agriculture and aquaculture. However, understanding it as an isolated phenomenon or attributing it to a single cause is not possible. It results from a complex network of interactions between geological processes, ocean dynamics, climate and, to a large extent, extractives in the Anthropocene era, which is characterized by the massive extraction of these resources with little concern for the long-term environmental or social impact (Olvera et al. 2024), besides the anthropic actions present on the coasts or indirectly affecting them.

Coastal erosion has become a critical phenomenon in the Anthropocene era, marked by profound human influence on ecosystems and the environment. Climate change, largely driven by human activity, has intensified this process by raising sea levels and increasing the frequency and intensity of extreme weather events, such as storms and hurricanes. This problem is not unique to the Gulf of California. Many coastal regions of the world have documented this phenomenon, where similar erosion phenomena pose global challenges in terms of coastal management and conservation. The consequences include the loss of critical habitats, the alteration of ecosystems and increased risks to human infrastructure, as well as the displacement of communities. Therefore, understanding the interaction of these factors and the dynamics of erosion is crucial for developing effective global mitigation and adaptation strategies to meet the challenges posed by the Anthropocene era.

5 Conclusions

The Gulf of California faces and will face significant challenges from sea level rise and intensifying tropical storms exacerbated by climate change. Projected sea level rise threatens to cause severe coastal impacts, primarily from erosion and loss of coastal habitats. Increasingly intense tropical storms pose a growing danger to coastal communities, with the potential to erode them and cause damage to infrastructure. Seventy percent of the Gulf of California coastline experiences steady erosion, while the rest is subject to accretion. Dammed river mouths and areas with high population density, as well as agricultural and aquaculture uses, experience the highest erosions. Observations showed that the northern coast of Baja California, the northern coasts of Sonora, and the coast of Sinaloa had the most prominent erosion hotspots. The coasts of BC and BCS (except the north coast) showed mainly accretion.

Coastal erosion in the Gulf of California cannot be attributed to a single factor but results from a complex interaction between natural processes that are altered by climate change such as tides, currents, and sediment dynamics, along with human activities such as infrastructure construction, urbanization, and modification of river channels. During the Anthropocene, these human activities have transformed the natural dynamics of the Gulf of California, increasing the vulnerability of its coasts. Population growth, urban expansion, and infrastructure development have altered the coast's natural resilience to erosion.

Although the erosion phenomenon in the Gulf of California has characteristics, it is part of a global problem that affects numerous coasts around the world. This underscores the need to develop specific mitigation and adaptation strategies that not only consider local conditions but also learn from international experiences. To address the challenges of coastal erosion in the Gulf of California, it is crucial to adopt sustainable management approaches that include ecosystem restoration, regulation of coastal development, and implementation of natural solutions. Future strategies must integrate climate change as a central variable to ensure long-term protection. In addition, it is critical to implement adaptation and mitigation measures, along with resilient urban planning, to protect both communities and coastal ecosystems from the increasing impacts of climate change in the region.

Acknowledgements The author gratefully acknowledges the support of the Secretaría de Ciencias, Humanidades, Tecnología e Innovación (SECITI) through the Estancias Postdoctorales por México program (CVU No. 71138), which made the development of this research at the Benemérita Universidad Autónoma de Puebla possible. The authors also wish to express their gratitude to the Universidad Autónoma de Sinaloa for its institutional and academic support, and to the LaNCTIGeSSR laboratory for its valuable its technical collaboration and assistance throughout this research process.

Authors Contributions All authors have contributed equally to this work.

Funding The research was not supported by any funding.

Data Availability Statement The datasets study is available from the corresponding author on reasonable request.

Potential Conflicts of Interest On behalf of all authors, the corresponding author states that there is no conflict of interest.

Informed Consent Informed consent was obtained from all individual participants included in the study.

References

Aedla R, Dwarakish GS, Reddy DV (2015) Automatic shoreline detection and change detection analysis of Netravati-GurpurRivermouth using histogram equalization and adaptive thresholding techniques. Aquat Proced 4:563–570. https://doi.org/10.1016/j.aqpro.2015.02.073

Ahmad SR, Lakhan VC (2012) GIS-based analysis and modeling of coastline advance and retreat along the coast of Guyana. Mar Geodesy 35(1):1–15. https://doi.org/10.1080/01490419.2011.637851

Anderson BT, Roads JO, Chen SC (2000a) Large-scale forcing of summertime monsoon surges over the Gulf of California and the southwestern United States. J Geophys Res: Atmos 105(D19):24455–24467. https://doi.org/10.1029/2000JD900337

Anderson BT, Roads JO, Chen SC, Juang HMH (2000b) Regional simulation of the low-level monsoon winds over the Gulf of California and southwestern United States. J Geophys Res Atmos 105(D14):17955–17969. https://doi.org/10.1029/2000JD900161

Ashtekar AS, Mohammed-Aslam MA, Moosvi AR (2019) Utility of normalized difference water index and GIS for mapping surface water dynamics in sub-upper Krishna Basin. J Indian Soc Remote Sens 47(8):1431–1442. https://doi.org/10.1007/s12524-019-01013-6

AVISO+ Satellite Altimetry Data (2024) Satellite altimetry. Mean Sea Level. https://www.aviso.altimetry.fr/en/home.html

Baig MRI, Ahmad IA, Shahfahad Tayyab M, Rahman A (2020) Analysis of shoreline changes in Vishakhapatnam coastal tract of Andhra Pradesh, India: an application of digital shoreline analysis system (DSAS). Ann GIS 26(4):361–376. https://doi.org/10.1080/19475683.2020.1815839

Barbaro G, Bombino G, Foti G, Borrello MM, Puntorieri P (2019) Shoreline evolution near river mouth: case study of Petrace River (Calabria, Italy). Reg Stud Mar Sci 29:100619. https://doi.org/10.1016/j.rsma.2019.100619

Bardají T, Caridad Zazo AC, Cristino JD, Goy JL, Lario J, Silva PG (2009) Impacto del cambio climático en el litoral. Enseñanza De Las Ciencias De La Tierra 17(2):141–154

Bera R, Maiti R (2019) Quantitative analysis of erosion and accretion (1975–2017) using DSAS—a study on Indian Sundarbans. Reg Stud Mar Sci 28:100583. https://doi.org/10.1016/j.rsma.2019.100583

Bernal G, Ripa P, Herguera JC (2001) Oceanographic and climatic variability in the lower Gulf of California: links with the Tropics and North Pacific. Cienc Mar 27(4):595–617. https://doi.org/10.7773/cm.v27i4.498

Boak EH, Turner IL (2005) Shoreline definition and detection: a review. J Coast Res 214:688–703. https://doi.org/10.2112/03-0071.1

Bruun P (1962) Sea-level rise as a cause of shore erosion. J Waterw Harbors Div 88(1):117–132

Carriquiry JD, Sánchez A (1999) Sedimentation in the Colorado River delta and Upper Gulf of California after nearly a century of discharge loss. Mar Geol 158(1–4):125–145. https://doi.org/10.1016/S0025-3227(98)00189-3

Carriquiry JD, Sánchez A, Camacho-Ibar VF (2001) Sedimentation in the northern Gulf of California after cessation of the Colorado River discharge. Sed Geol 144(1–2):37–62. https://doi.org/10.1016/S0037-0738(01)00134-8

Dawahidi T, Marinas DI, Gomariz-Castillo F (2019) Accuracy measurement of sea level rise due to effect of climate change through tide gauge data and NASA Topex-poseidon/Jason satellite series. The case of Valencia. Anales De Geografia De La Universidad Complutense 39(1):39–58. https://doi.org/10.5209/aguc.64676

Dorsey RJ, Umhoefer PJ, Renne PR (1995) Rapid subsidence and stacked Gilbert-type fan deltas, Pliocene Loreto basin, Baja California Sur Mexico. Sed Geol 98(1–4):181–204. https://doi.org/10.1016/0037-0738(95)00032-4

Dwivedi CS, Singh S, Pandey AC, Basheer Ahammed KK, Mitra D (2023) Indicator based approach and geospatial technology for coastal vulnerability assessment along Chennai District Coast, Tamil Nadu State, India. Thalassas: Int J Mar Sci 39(2):719–737

Ezcurra E, Barrios E, Ezcurra P, Ezcurra A, Vanderplank S, Vidal O, Villanueva-Almanza L, Aburto-Oropeza O (2019) A natural experiment reveals the impact of hydroelectric dams on the estuaries of tropical rivers. Sci Adv 5(3):9875. https://doi.org/10.1126/sciadv.aau9875

Farfán LM, D'Sa EJ, Liu KB, Rivera-Monroy VH (2014) Tropical cyclone impacts on coastal regions: the case of the Yucatán and the Baja California Peninsulas Mexico. Estuaries Coast 37(6):1388–1402. https://doi.org/10.1007/s12237-014-9797-2

Fotsi YF, Pouvreau N, Brenon I, Onguene R, Etame J (2019) Temporal (1948–2012) and dynamic evolution of the Wouri estuary coastline within the gulf of Guinea. J Mar Sci Eng 7(10):343. https://doi.org/10.3390/jmse7100343

Franco-Ochoa C, Zambrano-Medina Y, Plata-Rocha W, Monjardín-Armenta S, Rodríguez-Cueto Y, Escudero M, Mendoza E (2020) Long-term analysis of wave climate and shoreline change along the Gulf of California. Appl Sci 10(23):8719. https://doi.org/10.3390/app10238719

Gómez-Pazo A, Pérez-Alberti A, Pérez XLO (2019) Recent evolution (1956–2017) of rodas beach on the Cíes Islands, Galicia, NW Spain. J Mar Sci Eng 7(5):125. https://doi.org/10.3390/jmse7050125

Hernández Azcúnaga L (2005) Dinámica del transporte de sedimentos como carga de fondo en el Alto Golfo de California [Instituto Potosino de Investigación Científica y Tecnológica]. https://repositorio.ipicyt.edu.mx/bitstream/handle/11627/62/HernandezAzcunaga.pdf?sequence=1&isAllowed=y

Herrera-Cervantes H (2019) Sea surface temperature, ocean color and wind forcing patterns in the Bay of La Paz Gulf of California: Seasonal variability. Atmósfera 32(1):25–38. https://doi.org/10.20937/ATM.2019.32.01.03

Himmelstoss EA, Henderson RE, Kratzmann MG, Farris A (2018) Digital shoreline analysis system (DSAS) version 5.0. US Geol Surv. https://doi.org/10.3133/ofr20181179

IPCC (2022) Sexto informe de evaluación del IPCC: Cambio Climático 2022 | UNEP—UN Environment Programme. https://www.unep.org/es/resources/informe/sexto-informe-de-evaluacion-del-ipcc-cambio-climatico-2022

Jiménez-Illescas ÁR, Zayas-Esquer MM, Espinosa-Carreón TL (2019) Integral management of the coastal zone to solve the problems of erosion in Las Glorias Beach, Guasave, Sinaloa, Mexico. In: Krishnamurthy RR, Jonathan MP, Srinivasalu S, Glaeser B (eds), Coastal Management, pp 141–163. https://doi.org/10.1016/b978-0-12-810473-6.00010-8

Jonah FE, Boateng I, Osman A, Shimba MJ, Mensah EA, Adu-Boahen K, Chuku EO, Effah E (2016) Shoreline change analysis using end point rate and net shoreline movement statistics: an application to Elmina, Cape Coast and Moree section of Ghana's coast. Reg Stud Mar Sci 7:19–31. https://doi.org/10.1016/j.rsma.2016.05.003

Kallepalli A, Kakani NR, James DB (2017) Digital shoreline analysis system-based change detection along the highly eroding Krishna-Godavari delta front. J Appl Remote Sens 11(03):1. https://doi.org/10.1117/1.jrs.11.036018

Kasper-Zubillaga JJ, Carranza-Edwards A, Morales De la Garza E (2007) Textural characterization of beach sands from the Gulf of California, Mexico: implications for coastal processes and relief. Cienc Mar 33(1):83–94. https://doi.org/10.7773/cm.v33i1.1018

Kermani S, Boutiba M, Guendouz M, Guettouche MS, Khelfani D (2016) Detection and analysis of shoreline changes using geospatial tools and automatic computation: case of Jijelian Sandy Coast (East Algeria). Ocean Coast Manage 132:46–58. https://doi.org/10.1016/j.ocecoaman.2016.08.010

Laignel B, Vignudelli S, Almar R, Becker M, Bentamy A, Benveniste J, Birol F, Frappart F, Idier D, Salameh E, Passaro M, Menende M, Simard M, Turki EI, Verpoorter C (2023) Observation of the coastal areas, estuaries and deltas from space. Surv Geophys 44(5):1309–1356. https://doi.org/10.1007/S10712-022-09757-6

Ledesma-Vázquez J, Johnson ME, Gonzalez-Yajimovich O, Santamaría-del-Angel E (2009) Gulf of California geography, geological origins, oceanography, and sedimentation patterns. Atlas of Coastal Ecosystems in the Western Gulf of California. The University of Arizona Press, Tucson, pp 1–10

Lee Y, Eom J, Do J, Kim B, Ryu J (2019) Shoreline movement monitoring and geomorphologic changes of beaches using Lidar and UAVs images on the Coast of the East Sea, Korea. J Coast Res 90:409–414. https://doi.org/10.2112/SI90-052.1

Lluch-Cota SE, Parés-Sierra A, Magaña-Rueda VO, Arreguín-Sánchez F, Bazzino G, Herrera-Cervantes H, Lluch-Belda D (2010) Changing climate in the Gulf of California. Prog Oceanogr 87(1–4):114–126. https://doi.org/10.1016/j.pocean.2010.09.007

Lluch-Cota SE, Tripp-Valdez M, Lluch-Cota DB, Lluch-Belda D, Verbesselt J, Herrera-Cervantes H, Bautista-Romero JJ (2013) Recent trends in sea surface temperature off Mexico. Atmosfera 26(4):537–546. https://doi.org/10.1016/S0187-6236(13)71094-4

López Martínez J, Farach Espinoza EB, Herrera Cervantes H, García Morales R (2023) Long-term variability in sea surface temperature and chlorophyll a concentration in the Gulf of California. Remote Sens 15(16):4088. https://doi.org/10.3390/RS15164088

Mart BS (2018) Morphology and sediments of the beaches of the coastal cell of San Jose del Cabo BCS Mexico. Autonomous University of Baja California Sur La Paz. Mexico

McFeeters SK (1996) The use of the normalized difference water index (NDWI) in the delineation of open water features. Int J Remote Sens 17(7):1425–1432. https://doi.org/10.1080/014311696 08948714

Merlotto A, Bértola GR, Piccolo MC (2013) Clasificación morfodinámica estacional de playas del partido de Necochea, provincia de Buenos Aires Argentina. Cienc Mar 39(4):331–347. https://doi.org/10.7773/cm.v39i4.2180

Merlotto A, Bértola RG (2008). Evolución urbana y su influencia en la erosión costera en el balneario parque mar chiquita, argentina. Papeles de Geografía

Mondejar JP, Tongco AF (2019) Near infrared band of landsat 8 as water index: a case study around Cordova and Lapu-Lapu City, Cebu Philippines. Sustain Environ Res 1(1):1–15. https://doi.org/10.1186/s42834-019-0016-5

National Oceanic and Atmospheric Administration (2024) Historical hurricane tracks. Historical Hurricane Tracks. https://coast.noaa.gov/hurricanes/#map=4/32/-80

Nicholls RJ, Cazenave A (2010) Sea-level rise and its impact on coastal zones. Science 328(5985):1517–1520. https://doi.org/10.1126/science.1185782

Ojeda-Ruiz MÁ, Guerrero-Izquierdo CS, Petatán-Ramírez D (2020) Proyecciones de cambio climático en el Golfo de California y sus posibles implicaciones ambientales, sociales y económicas. In: Marín Monroy EA, Hernández Trejo V, Avilés Polanco G, Valdivia Alcalá R (eds), Cambio Climático Adaptación y Resilencia en el Noroeste de México. Primera ed, pp 35–57

de la Peña Olivas JM, González JF (2018). Respuesta de la costa a las variaciones del nivel del mar: efecto sobre las playas en España. Rev Digit Del Cedex 191, 74–85. http://ingenieriacivil.cedex.es/index.php/ingenieria-civil/article/view/2355

Olvera MMC, Contreras WLD, Barahona CV, Meza CGI (2024) Extractivismos en el Golfo de California: Una Expresión Colonial del Antropoceno. Historia Ambiental Latinoamericana y Caribeña (HALAC) Revista De La Solcha 14(1):77–103. https://doi.org/10.32991/2237-2717.2024V14I1.P77-103

Oyedotun TDT (2016) Shoreline evolution and metocean data behaviour in Southwest England: is there any historical link? Environ Process 3(4):939–960. https://doi.org/10.1007/s40710-016-0189-4

Oyedotun TDT, Ruiz-Luna A, Navarro-Hernández AG (2018) Contemporary shoreline changes and consequences at a tropical coastal domain. Geol Ecol Landscapes 2(2):104–114. https://doi.org/10.1080/24749508.2018.1452483

Páez-Osuna F, Sanchez-Cabeza JA, Ruiz-Fernández AC, Alonso-Rodríguez R, Piñón-Gimate A, Cardoso-Mohedano JG, Flores-Verdugo FJ, Carballo JL, Cisneros-Mata MA, Álvarez-Borrego S (2016) Environmental status of the Gulf of California: a review of responses to climate change and climate variability. Earth-Sci Rev 162:253–268. https://doi.org/10.1016/j.earscirev.2016.09.015

Passeri D, Hagen S, Irish J (2014) Comparison of shoreline change rates along the South Atlantic bight and Northern Gulf of Mexico coasts for better evaluation of future shoreline positions under sea level rise. J Coast Res 68:20–26. https://doi.org/10.2112/SI68-003.1

Pérez-Cruz L (2013) Hydrological changes and paleoproductivity in the Gulf of California during middle and late Holocene and their relationship with ITCZ and North American Monsoon variability. Quatern Res 79(2):138–151. https://doi.org/10.1016/J.YQRES.2012.11.007

Plomaritis TA, Costas S, Ferreira Ó (2018) Use of a Bayesian network for coastal hazards, impact and disaster risk reduction assessment at a coastal barrier (Ria Formosa, Portugal). Coast Eng 134:134–147. https://doi.org/10.1016/j.coastaleng.2017.07.003

Ponce Bastidas JS (2015) Dinámica costera en las playas de El Cardonal y Cabo Pulmo. Baja California Sur

Ranasinghe R (2016) Assessing climate change impacts on open sandy coasts: a review. Earth-Sci Rev 1(160):320–332. https://doi.org/10.1016/j.earscirev.2016.07.011

Rangel-Buitrago N (2009) Contribución antropogénica a los cambios geomorfológicos y evolución reciente de la costa Caribe Colombiana. Gestión y Ambiente 12(2):43–56

Reguero BG, Losada IJ, Méndez FJ (2019) A recent increase in global wave power as a consequence of oceanic warming. Nat Commun 10(1):1–14. https://doi.org/10.1038/s41467-018-08066-0

Ritchie H (2022) CO_2 emissions dataset: our sources and methods. Published Online at OurWorldinData.Org. Retrieved From: https://ourworldindata.org/co2-dataset-sources

Robles-Tamayo CM, Valdez-Holguín JE, García-Morales R, Figueroa-Preciado G, Herrera-Cervantes H, López-Martínez J, Enríquez-Ocaña LF (2018) Sea surface temperature (SST) variability of the eastern coastal zone of the Gulf of California. Remote Sens 10(9):1434. https://doi.org/10.3390/RS10091434

Sans CA (2018) The Gulf of California in its entirety as Mexican Inland or territorial waters. Revista Del Centro De Estudios Superiores Navales 39(1):95–126

Servicio Mareográfico Nacional-UNAM (2024) Servicio Mareográfico Nacional—SMN del IGEF. UNAM. http://www.mareografico.unam.mx

Soto Mardones L, Marioni S, Pares Sierra A (1999) Variabilidad espaciotemporal de la temperarura superficial del mar en el Golfo de California. Cien Mar 25(1):1–30

Thieler R, Himmelstoss E, Zichichi J, Ayhan E (2009) The digital shoreline analysis system (DSAS) version 40—an ArcGIS extension for calculating shoreline change. US Geol Surv. https://doi.org/10.3133/ofr20081278

Tseng YH, Breaker LC, Chang ETY (2010) Sea level variations in the regional seas around Taiwan. J Oceanogr 66(1):27–39. https://doi.org/10.1007/s10872-010-0003-2

Wasim M, Mohanty S, Pandey PC, Basheer AK, Pandey AC (2025) Coral reefs threat assessment and comparative analysis of Neil and Agatti Island, India, using multispectral approach. In: Aquatic ecosystems monitoring. CRC Press, pp 252–278

Zambrano-Medina YG, Plata-Rocha W, Monjardin-Armenta SA, Franco-Ochoa C (2023) Assessment and forecast of shoreline change using geo-spatial techniques in the Gulf of California

Zamora M, Lambert A, Montero G (2014) Effect of some meteorological phenomena on the wind potential of Baja California. Energy Proced 57:1327–1336. https://doi.org/10.1016/J.EGYPRO.2014.10.086

Zhang R, Chen L, Liu S, Zhang H, Gong W, Lin G (2019) Shoreline evolution in an embayed beach adjacent to tidal inlet: the impact of anthropogenic activities. Geomorphology 346:106856. https://doi.org/10.1016/j.geomorph.2019.106856

Mapping the Biogeography of Vulnerable Marine Ecosystems

Comprehensive Framework for Monitoring of Seas and Coastal Ecosystems Under Changing Environments

Uma Sankar Panda, Umakanta Pradhan, Subrat Naik, D. Kumaresan, Subhasis Pradhan, Rabindra Kumar Sahoo, and M. V. Ramanamurthy

Abstract In the face of climate change, coastal zones are among the most sensitive habitats on Earth and must be monitored to assess their health, identify changes, and develop management plans. Ineffective coordination, standardization, and integration across disciplines and stakeholders hinder coastal oceanographic monitoring activities. This study proposes systematic coastal monitoring program planning, coordination, and implementation to solve these problems. The proposed framework encompasses site selection using geospatial analysis and stakeholder input, identification of critical parameters based on environmental characteristics and management objectives, standardized protocols for data collection across physical (e.g. tides, currents, bathymetry, waves), chemical (e.g. water quality, nutrients), biological (e.g. plankton, habitat maps), and socio-economic parameters, and quality assurance measures. It leverages advanced technologies such as tide gauges, current profilers, autonomous underwater vehicles, high-frequency radars, drifters, remote sensing for bathymetry/topography, water quality sensors, sediment traps/samplers, and meteorological stations for efficient data acquisition and integration across surface, subsurface, surf zone, air-sea interface from multiple sources. Data management and statistical and predictive models help discover patterns, threats, and decision-making scenarios. Open data and stakeholder involvement promote information sharing and evidence-based policy-making. This framework addresses critical needs for a holistic understanding of complex coastal processes, climate change impact, early warnings, and science-based coastal management. Overcoming challenges related to capacity building, data accessibility, and governance requires sustained multidisciplinary collaboration among researchers, resource managers, policymakers, and coastal communities. Widespread adoption of this framework can significantly improve coastal stewardship in changing climate and promote sustainable blue economies worldwide.

U. S. Panda (✉) · U. Pradhan · S. Naik · D. Kumaresan · S. Pradhan · R. K. Sahoo · M. V. Ramanamurthy
National Centre for Coastal Research (NCCR), Ministry of Earth Sciences, Pallikaranai, Chennai, India
e-mail: uspanda@nccr.gov.in

K. K. Basheer Ahammed et al. (eds.), *Monitoring Coastal Marine Ecosystems Under Climate Change*, https://doi.org/10.1007/978-3-032-09137-6_5

Keywords Coastal monitoring · Climate change · Oceanographic measurements · Data integration · Advanced technologies · Stakeholder collaboration

1 Introduction

Life on Earth depends on the ocean., which covers 70.8% of the Earth's surface and is home to countless living species, providing benefits to human beings directly and indirectly, plays a significant role in climate change, controlling the weather system from a local to a global scale, acting as an oxygen producer and carbon sinker, and active mediator of water and other physical-nutrient cycles and many. Based on oceanographic characteristics and horizontal and vertical segregation, the ocean is divided into nearshore (continental region) and deep sea (high sea). Most human activities are undergone in the nearshore region compared to the high seas. Many environmental pressures are on coastal places because they constantly change and have significant ecological, economic, and social impacts (Caddell 2016; Ros et al. 2019). To comprehend and control these coastal regions, reliable monitoring systems are needed to record the intricate interplay of physical, chemical, and biological phenomena (Browman et al. 2004). To ensure the sustainable management of these crucial areas, extensive and ongoing coastal monitoring is necessary in light of the growing threats posed by climate change, pollution, habitat loss, and human activities (Mora et al. 2013; Landrigan et al. 2020; Gissi et al. 2021).

Figure 1 shows the dynamic processes (Panda et al. 2022; Kaiser et al. 2019) that occur in coastal and marine settings, and coastal oceanographic data are crucial for understanding these processes. Nevertheless, coastal oceanographic observations have many obstacles due to logistical issues, environmental considerations, and the complexity and changeability of coastal settings.

The primary challenges faced in coastal oceanographic measurements include environmental variability, technological limitations, logistical constraints, data integration issues and the combined issues of nature and humans (Fig. 2). The ever-changing tides, waves, and freshwater inflows that impact coastal regions are shaped by the changing seasons and the effects of storms. Consistently collecting correct data becomes challenging due to these quick changes. Sensors used in the ocean are prone to contamination from marine creatures such as barnacles and algae, which can alter readings and necessitate regular cleaning or repair. Coastal ecosystems are susceptible to temperature and chemical changes, affecting data precision and requiring more sensitive, specialized instruments.

Instruments like multi-parameter buoys, autonomous underwater vehicles (AUVs), and hyperspectral sensors are expensive to deploy and maintain, limiting their accessibility for many projects. Coastal instruments are exposed to rough conditions, currents, sediment movement, and salinity—causing wear and tear that shortens equipment lifespan and increases replacement costs. Remote sensors and AUVs often rely on battery power, which restricts operational time and requires frequent recharging or replacement, particularly in remote or deep sea applications.

Fig. 1 The coastal oceanographic processes and observations

Deploying and retrieving instruments can be challenging in many coastal places due to their extreme remoteness or extreme weather. There are data gaps because of limitations in network coverage and underwater communication concerns that prevent real-time data from offshore sensors from being transmitted to onshore stations. Some research operations, such as coastal monitoring, may be delayed or restricted due to the need to comply with local, national, or international legislation and obtain permissions.

There are various sources for coastal data, including satellites, buoys, and human sampling, and the data formats and resolutions of these sources might vary. It often takes sophisticated processing methods to integrate these data sources for analysis. It can be challenging to store, handle, and understand large amounts of high-resolution data from various equipment; doing so requires a lot of processing power and data science knowledge. Data reliability must be maintained by continuous quality control techniques to account for variations in sensor performance, data gaps, and environmental interferences.

Coastal development, dredging, and pollution are all human activities that change the seabed topography and water quality, introducing unknown variables that can affect baseline data. Rising sea levels disrupt coastal conditions, create more intense storms, and cause ocean acidification. This is making long-term monitoring operations more complicated and necessitating adaptive measurement tactics. Coastal ecosystems are interconnected; changes in one aspect, such as increased sedimentation due to construction, can cascade through the ecosystem, affecting water quality, currents, and marine life distribution. Addressing these challenges requires continued technological innovation, improved data integration techniques, and collaborative efforts to develop resilient, adaptable solutions for coastal monitoring. Coastal/ nearshore monitoring provides a detailed understanding of regional processes and

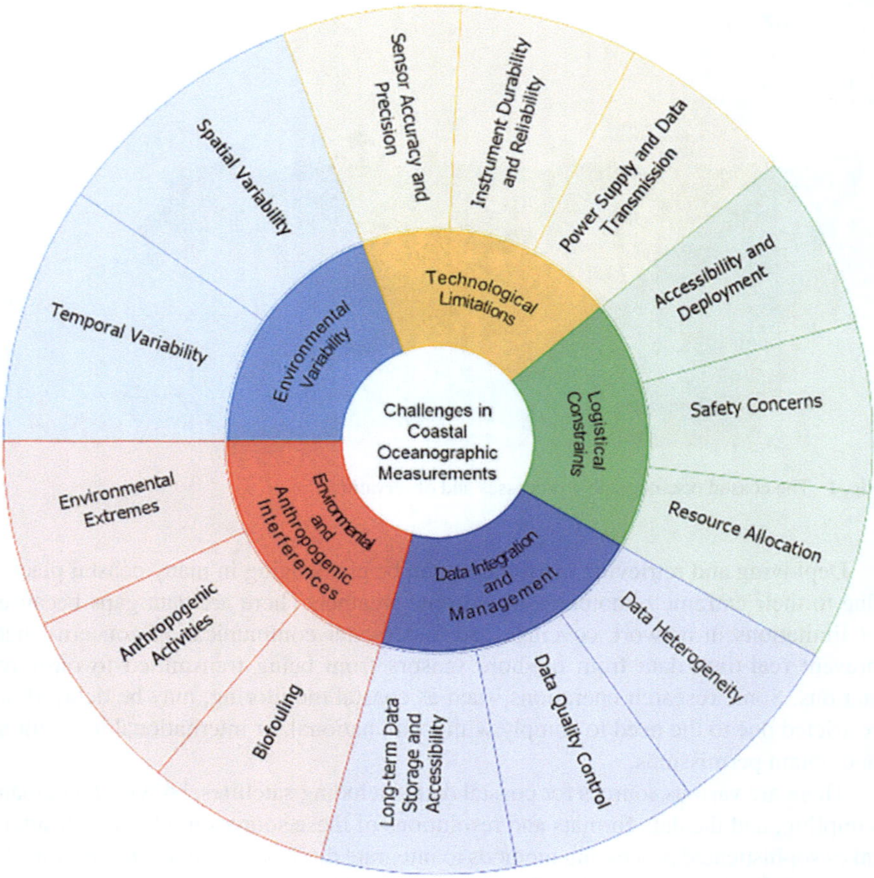

Fig. 2 Challenge in coastal oceanographic measurements

the impact of global issues on a regional scale (Levin and Bris 2015). Coastal oceanographic monitoring is paramount in preparing a better management plan and policies for users and stakeholders.

1.1 Evolution of Oceanographic Measurements

Instrumentation, remote sensing, data processing, and automation have all played significant roles in developing coastal oceanographic measuring technologies throughout the past and into the present. Through the years, notable advancements in instrumentation, remote sensing, data processing, and automation have all played significant roles in developing coastal oceanographic measuring technologies throughout the past and into the present. Through the years, there have been notable advancements in (Fig. 3):

Pre-20th Century: Early Developments

- **Navigation and Basic Measurement Tools**: Navigation was familiar throughout early coastal observations. Mariners utilized sextants, astrolabes, and compasses to determine their location and orientation, using simple instruments like lead lines to assess water depth.
- **Tidal and Current Observations**: Tidal patterns and currents were crucial for fishing and navigation for coastal people, but these findings were only approximate visual estimates and anecdotal evidence.
- **Meteorological Observations**: Early weather observations (wind speed, pressure) were taken manually at ports, providing insights into storms and seasonal changes but with low spatial coverage and precision.

Early to Mid-20th Century: Emergence of Modern Oceanographic Instruments

- **Mechanical Tide Gauges (1800s)**: The systematic recording of tidal shifts was made possible by installing mechanical tide gauges, which improved our understanding of coastal sea-level oscillations.
- **Thermometers and Hydrometers**: Initial ocean thermometers and hydrometers made measuring salinity and water temperature possible, two crucial parameters for comprehending marine characteristics. Physical collection and laboratory analysis were labour-intensive aspects of the sampling process.
- **Sounding** Techniques **and Sonar (1920s)**: Improvements in sonar technology have made it possible to study undersea topography and conduct seabed mapping with unprecedented precision.
- **Drogues and Drifters for Currents (1930s–1950s)**: Mechanical and weighted surface drifters, along with dye tracers, helped measure surface currents, although data collection was sporadic and limited to accessible areas.

Late 20th Century: Technological Advancements and Remote Sensing

- **Satellite and Remote Sensing (1970s)**: Sea surface temperature, wave heights, and surface currents were first remotely sensed on a broad scale with the launch of oceanographic satellites like SEASAT (1978), providing extensive ocean data worldwide.
- **Computational Oceanography and Numerical Models (1960s–1980s)**: Computer technologies made numerical models of coastal dynamics a reality, which aided in the forecasting and understanding of coastal processes by simulating waves, currents, and tides.
- **Moored Buoys and Drifters (1970s–1980s)**: The advent of fixed-position buoys with multi-parameter sensors shifted towards automated, continuous coastal monitoring. Drifters equipped with GPS allowed for better tracking of currents.
- **Acoustic Doppler Current Profilers (ADCPs, 1980s)**: ADCPs allowed for precise measurements of current velocities at different depths, providing a 3D view of coastal currents and sediment transport.

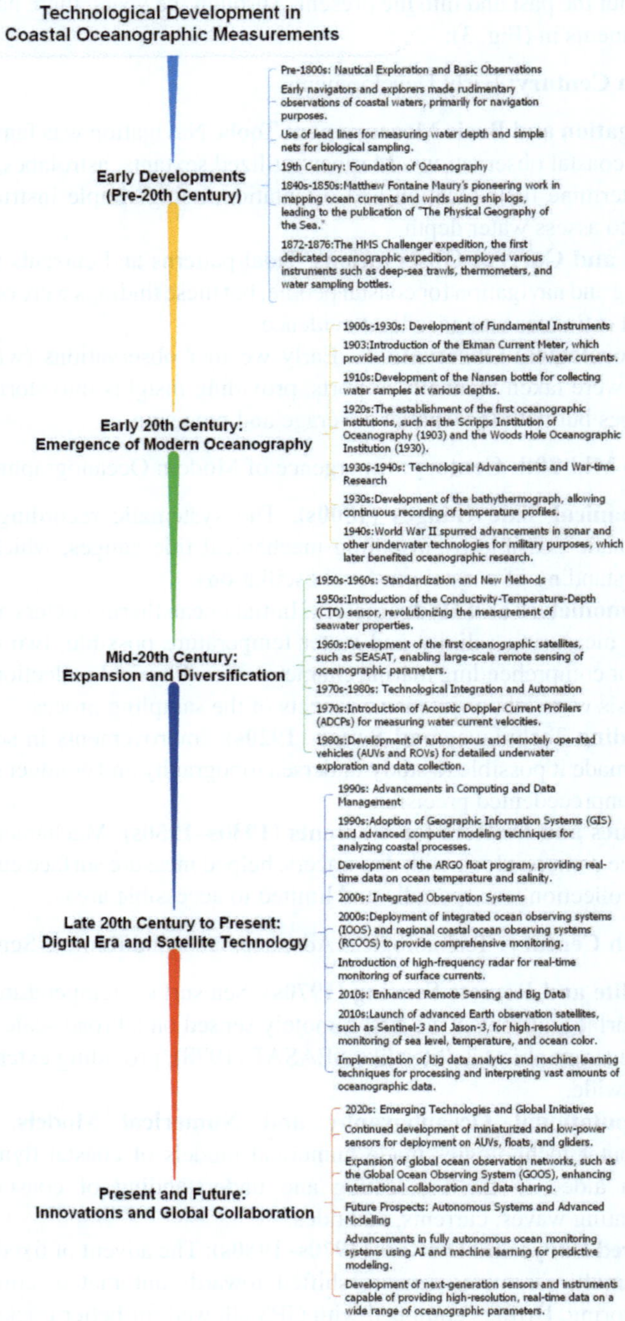

Fig. 3 Chronology of technological developments in oceanographic measurements

Early 21st Century: Digital and Autonomous Systems

- **Advanced Multi-Parameter Buoys**: Coastal buoys have advanced to the point where they can identify pollution and monitor ecosystems in real time by detecting dissolved oxygen, turbidity, pH, salinity, and temperature.
- **Autonomous Vehicles (AUVs, Gliders)**: Autonomous underwater vehicles and wave gliders emerged, conducting long-duration surveys of coastal regions and gathering high-resolution data on temperature, salinity, and currents.
- **Improved Satellite Observation (2000s)**: Enhanced satellite missions (e.g., Jason-1, Sentinel, and CryoSat) provided precise data on sea level, sea surface temperature, chlorophyll, and coastal morphology, with frequent revisit times.

Present-Day (2020s): IoT, AI, and Integrated Data Systems

- **IoT and Smart Sensor Networks**: Networks of IoT-enabled sensors collect real-time data autonomously across various coastal locations, communicating through cloud platforms and delivering instantaneous information for decision-making.
- **Artificial Intelligence (AI), Machine Learning (ML)**: AI and ML algorithms process massive datasets from buoys, satellites, and models to detect patterns, track coastal erosion, predict algal blooms, and assess pollution impacts.
- **Big Data and Cloud-Based Platforms**: Coastal conditions may now be studied and visualized in real time by policymakers, researchers, and the general public thanks to data aggregation into cloud-based systems that are accessible globally.
- **Hyperspectral and Lidar Sensors**: Drones and satellites equipped with cutting-edge sensors can map coastlines, analyse sediment, and track habitats with pinpoint accuracy.
- **Coupled Coastal Models**: Accurate coastal resilience planning and real-time data integration are made possible by state-of-the-art numerical models that simulate interactions between the ocean, the atmosphere, and land.

This evolution in coastal oceanographic technology has transformed data collection from manual, localized observations to automated, high-frequency, real-time, and large-scale monitoring, facilitating robust coastal management in the face of increasing environmental challenges.

1.2 Current Practices in Coastal Monitoring and Measurements

Coastal monitoring is an integration of pre-preparation procedures such as deciding on sites, parameters of investigations, instruments or technologies; during observation such as monitoring period with weather window, following the standard protocol, engaging of instruments and its accuracy, site accessibility, user-friendly instruments; and post-preparation on data quality, interpretation, preparation of report, and dissemination (Fig. 4). The monitoring of parameters for different issues/events

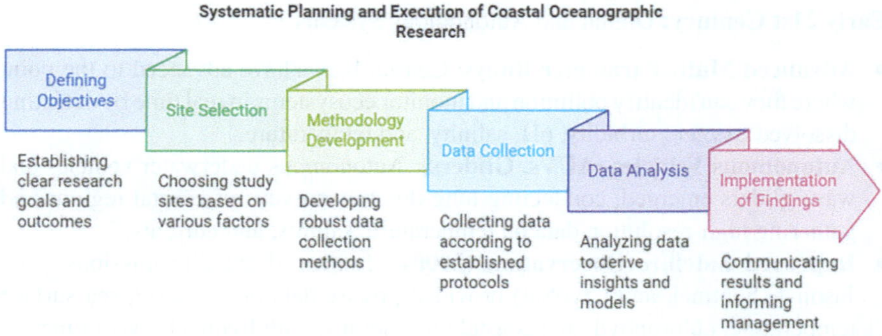

Fig. 4 Systematic planning and execution pathway for coastal oceanographic observations

are not similar; hence, care should be taken in deciding the selection of parameters, which are directly or indirectly have dependency (Miloslavich et al. 2018). Primarily, the surveyor should focus on how much a large data set can be collected within a shorter period and how cost-effective it is. Coastal monitoring and information are highly needed. Hence, several national and international agencies provide ocean data from a regional scale to a global (Crain et al. 2008). The information is available in hindcast, forecast, and actual data format. Data may be in different formats for different sources, such as point data, spatial data (two and three-dimensional), and spatiotemporal.

Coastal monitoring and ocean observation are now implemented with sophisticated measurement technologies, advanced instruments, and user-friendly tools to acquire better physical, chemical, geological and biological information on spatiotemporal scales (Lindstrom et al. 2012; Liu et al. 2017; Crise et al. 2018). The ocean observation in the present era is excellent and more accurate compared to the observations in the nineteenth century. An evolution of oceanic measurement in different periods was shown and classified into mainly five stages, which provide information from manual ocean observation to real-time data accusation from sensor-based platforms and remote sensing by establishing different national and international organizations (Aguzzi et al. 2023; Letian et al. 2024).

Worldwide, maritime observation and data networks such as EMODN and National Oceanic and Atmospheric Administration (NOAA) oversee significant coastal monitoring projects. With the help of satellites, buoys, and autonomous underwater vehicles (AUVs), NOAA's Integrated Ocean Observing System (IOOS) compiles detailed information on the ocean and its coastlines (NOAA 2020). The European Commission funds EMOD net, promoting uniform data collection and sharing among European nations by providing access to maritime datasets.

Government agencies, research organizations, and universities mainly oversee coastal monitoring programs in many nations. In India, the Ministry of Earth Sciences (MoES) houses the National Centre for Coastal Research (NCCR), crucial in studying and keeping tabs on coastal phenomena. NCCR tracks things like coastline shifts, water purity, and marine life as part of its mission. People often stick to tried-and-true techniques like in situ sampling, boat surveys, and laboratory analysis. Although these methods yield valuable data, they frequently require assistance with issues including restricted coverage in space and time, high operational expenses, and the requirement for a specialized crew. The National Centre for Medium-Range Weather Forecasting (NCMRWF), Pune, the Indian Institute of Tropical Meteorology (IITM), and the Indian Meteorology Department (IMD) all contribute to the dissemination of atmospheric data over land and sea. Regarding the Indian Ocean, you can find all you need in the INCOIS portal.

1.3 Recent Advancements in Coastal Monitoring

In recent years, significant technological and methodological advancements have been made in ocean monitoring and surveys (Tanhua et al. 2019; Aguzzi et al. 2023). The detailed survey and its type, types of data, and types of parameters about sources are discussed below.

I. **In situ observation**: In situ observation is highly essential, which is not only essential to ocean observation but also for atmospheric and land-based (Gould et al. 2013; Balan-Sarojini et al. 2024). The in situ observation helps to understand the natural processes at the site. It helps validate and calibrate remotely generated data from remote sensing and numerical modelling output. This helps in improving the forecasting system. This type of observation includes the sampling method with laboratory analysis and an automated sensor-based platform.
II. **Remote Sensing and Satellite Technology**: In the last few decades, remote sensing data has been valuable not only for ocean observation but also in various sectors of the Earth's surface, such as land cover, vegetation, and water resources (Zhang et al. 2019; Amani et al. 2022; Ho and Liu 2023). Remote sensing information is also helpful for disaster assessment, agriculture, health, and air quality. The use of high-resolution satellite imagery and aerial drones has revolutionized coastal monitoring. These technologies provide large-scale, real-time data on shoreline changes, land use and land cover, water quality and habitat conditions with improving accuracy and frequency.
III. **Autonomous Systems**: Deploying AUVs and uncrewed surface vehicles (USVs) has enhanced the ability to collect high-resolution data in hard-to-reach areas. Similarly, the automated sensor-based buoy systems such as water quality (Dadou et al. 2018; Chen et al. 2023a), waves, tsunamis and others are highly automated and provide the now casting data for public and stakeholder uses directly (Jones et al. 2019). In real time, these self-sufficient systems can monitor marine habitats, sediment movement, and water quality.
IV. **Sensor Networks and Internet of Things (IoT)**: Advanced sensor networks, equipped with IoT capabilities, enable continuous monitoring of various parameters such atmospheric parameters (air, pressure, air temperature, wind speed

and direction, humidity, rainfall, solar radiation) and water quality variables (water temperature, salinity, pH, dissolved oxygen, total dissolved solid and nutrient levels). Remote access to real-time data these sensors provide allows for more prompt decision-making.

V. **Numerical Modelling**: Computer and modelling technique advancements have made it possible to simulate physical-biochemical processes more accurately (Chassignet et al. 2007; Aremanio et al. 2019; Mishra et al. 2015). As shown in previous studies (Pradhan et al. 2020a; b; Panda et al. 2020; Pradhan et al. 2022a, b, c). Coupled models that include hydrodynamics, sediment transport, and waves can help anticipate changes in the future and plan accordingly. The linked hydro-ecological model provides similar data on ecological parameters and related processes.

VI. **Data Integration Platforms and Decision Support System (DSS)**: Developing integrated data platforms and cloud-based solutions has improved data sharing and collaboration. Coastal monitoring data is made more valuable and accessible through these platforms, which allow for the consolidation, analysis, and dissemination of data from many sources using several query-based information systems.

1.4 Gaps in Current Practices

Coastal monitoring has come a long way, yet some gaps remain. As the ocean environment is harassing and inaccessible for human beings to approach at all times, it causes failure to obtain the observed data. Similarly, catastrophic events like high waves and winds during storm periods and tsunami waves cause more devastation along nearshore regions and cause the loss of instruments and observatories, which hampers continuous ocean observation. On the other hand, remote sensing information is missing during monsoon seasons and rough weather. Several technical bogus and other unavoidable causes, such as sensor lifetime, performance, transmission, power supply, damage of sensor, and vandalism, are measured concerns for receiving continuous ocean information. Many approaches have been adopted to create a continuously updated database for a particular site and set of parameters. Various technologies and techniques are detailed here, outlining the process of making continuous data for ocean service.

Data Integration and Standardization: Integrating data from various sources and areas can be challenging due to the necessity for standardized protocols and data formats. This impedes the capacity to generate extensive datasets necessary for environmental assessments on a grand scale.

a. **Temporal and Spatial Coverage**: Due to financial and logistical limitations, traditional monitoring systems often cannot offer continuous and wide-area coverage. Data gaps that could shed light on the comprehension of dynamic coastal processes are created by this constraint.

b. **Technological Adoption**: Despite the availability of cutting-edge technology like sensor networks, AUVs, and remote sensing, its implementation varies significantly among regions and organizations. The efficiency and extent of monitoring activities are limited since many programs rely mainly on manual sampling and use antiquated technology.
c. **Stakeholder Coordination**: Governments, universities, communities, and international organizations must all work together for coastal monitoring to be effective. Unfortunately, there is usually room for improvement in the way various stakeholders communicate and work together, which might result in isolated and unnecessary initiatives.

2 Comprehensive Framework

The Framework for Ocean Observing directs establishing a cohesive and enduring ocean observation system to optimize user and society benefits (Tanhua et al. 2019). Data integration and analysis, modern technology, coordinated execution, and systematic planning are the four pillars of the suggested framework for coastal ocean monitoring, which is based on adhering to national and international norms and laws. Figure 5 shows how the various parts work together to provide a comprehensive strategy for coastal management by addressing different monitoring areas.

2.1 *Systematic Planning*

The main focus of ocean observation is systematic planning. A well-planned, rigorous timetable provides a model or demonstration of the projects. Establishing goals for

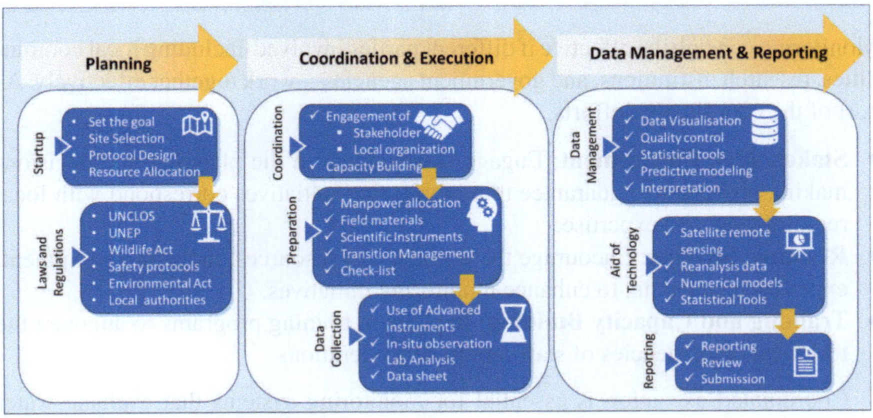

Fig. 5 Comprehensive framework for Oceanographic observation

monitoring, choosing suitable locations, and developing protocols are all parts of systematic planning.

The objectives of a project delineate the intended accomplishments by its conclusion. This may encompass deliverables and assets or more abstract goals such as enhancing productivity or motivation. The project objectives must be achievable, time-sensitive, and specified targets that can be quantified upon completion. Project objectives should be based on environmental priorities, regulatory requirements, and stakeholder needs. Every project objective's four most important aspects are what, when, where, and why. The answers to these questions determine the scope, timing, and execution of the project.

A fundamental aspect of any coastal ocean survey is choosing an appropriate study location. During the selection of sites, the major components need to be reviewed, such as ecological significance (hydrology: wind, wave, tide; bathymetry), vulnerability to impacts, accessibility, legal framework, workforce and project cost/fund. Before site selection, gathering the archive information and understanding the environmental conditions through a reconnaissance survey is a basic procedure. The site selection procedure varies according to the project objectives and the required parameters.

Standard operating procedures (SOPs) for ocean surveys govern the measurement and collection of data about the ocean's biological, chemical, and physical properties. There are many specific protocols for different aspects like instrument operation, deployment, mooring, retrieval, on board instrument operations, on board safety, laboratory analysis, sample collection, sample handling, preservation, and transportation. Similarly, the protocol is also followed by data quality control and analysis. Finally, it is a necessary protocol that follows the rules, regulations and policies furnished by local, national, and international authorities.

2.2 Coordinated Execution

Monitoring can only be effective if different parties involved-including local communities, research institutions, and government agencies—work together effectively. As part of the coordination efforts,

- **Stakeholder Engagement**: Engage stakeholders in the planning and decision-making processes to guarantee that monitoring initiatives correspond with local requirements and expertise.
- **Resource Sharing**: Encourage the exchange of resources, including equipment, experience, and data, to enhance monitoring initiatives.
- **Training and Capacity Building**: Implement training programs to augment the technical competencies of staff monitoring operations.

Coordinated execution is essential for monitoring systems that evaluate water quality and ecosystem health in coastal regions, guaranteeing efficacy and adaptability to local conditions (Jones et al. 2023). Involving stakeholders, such as local communities, government entities, and pertinent parties, is essential to this process. In coastal areas impacted by algal blooms, including local fishermen and community organizations in formulating the monitoring plan effectively tackles specific concerns, such as monitoring nutrient pollution or pinpointing contamination sources relevant to the region (Lee et al. 2022). The sharing of resources enhances the efficiency and effectiveness of monitoring initiatives by facilitating the exchange of equipment, experience, and data among diverse organizations. A university research team may collaborate with a local environmental agency by sharing sophisticated water quality sensors, offering logistical assistance and data analysis. This partnership diminishes expenses and utilizes the strengths of several collaborators, resulting in more thorough and precise data acquisition (Martínez et al. 2023). Training and capacity building are crucial to guarantee that all participants, especially local volunteers and field technicians, have the requisite skills for efficient monitoring. Conducting workshops on utilizing water quality testing apparatus and interpreting outcomes ensures elevated data precision and dependability (Nguyen et al. 2024a, b). Coordinated execution aligns monitoring programs with local requirements, maintains cost-effectiveness, and provides valuable insights into water quality and ecosystem health by incorporating stakeholder insights, sharing resources, and investing in training. This method promotes community engagement, improves data quality, and facilitates informed decision-making, aiding in effectively managing and conserving coastal and marine ecosystems (Jones et al. 2023).

2.3 *Laws and Regulation*

During the ocean survey, it needs to follow the active rules and regulations issued for the specific sites. There are several international rules for the different aspects, such as the United Nations Law of the Sea Convention (UNCLOS) for uses of sea boundaries, the Protection of the marine environment and biodiversity by UNEP, the Wildlife Act, marine litter, and pollution policy by local to international authorities, marine, and navigational regulation. Before the sampling, the survey planning and period must be circulated to the concerned authorities for better management, similar to the coastal defence authorities regarding safety points of issue.

2.4 *Integration of Advanced Technologies*

The integration of advanced technologies significantly enhances the capability and efficiency of coastal monitoring. Key technologies include.

- **Remote Sensing**: Use satellite and aerial imagery to monitor large-scale coastal changes, such as shoreline erosion and habitat loss.

- **Autonomous Systems**: Deploy autonomous underwater vehicles (AUVs) and uncrewed aerial vehicles (UAVs) for high-resolution, real-time data collection.
- **Sensor Networks**: Implement sensor networks to continuously monitor physical and chemical parameters such as temperature, salinity, and nutrient concentrations.
- **Numerical Modelling**: Apply numerical models to simulate coastal processes and predict future changes based on various scenarios.

Integrating new technology significantly improves the monitoring of coastal water quality and ecosystems, providing detailed information that was previously inaccessible (Smith et al. 2023). Remote sensing technologies, including satellite and aerial imagery, enable extensive monitoring of coastal water quality by detecting algal blooms, sediment plumes, and changes in water colour, which can indicate shifts in nutrient levels and pollution (Cloern et al. 2016; Wang et al. 2022). These tools allow for assessing water quality over large areas and extended periods, facilitating the identification of problematic trends. Autonomous underwater vehicles (AUVs) and un crewed aerial vehicles (UAVs) provide high-resolution, real-time data on various aspects of coastal ecosystems (Veeravalli et al. 2023). AUVs measure variables such as temperature, salinity, and nutrient concentrations at various depths, whereas UAVs provide aerial perspectives beneficial for monitoring shoreline conditions and the health of coastal vegetation (Lee et al. 2023).

Furthermore, sensor networks established along coastlines and in estuarine ecosystems consistently monitor essential parameters, including temperature, salinity, and pollutants, delivering continuous data that facilitates real-time detection of changes and possible concerns (Chen et al. 2023a, b). Numerical modelling utilizes data from these technologies to replicate coastal processes and forecast future water quality scenarios under diverse settings, including pollution levels and climate change effects. This holistic approach facilitates more precise water quality evaluations, prompt environmental alteration identification, and enhanced management techniques. By integrating remote sensing, autonomous systems, sensor networks, and numerical models, researchers, and managers gain a holistic view of coastal water quality and ecosystem health, leading to better-informed decisions and improved conservation outcomes (Smith et al. 2023).

2.5 *Major Ocean Monitoring Equipment*

The details of instruments used for ocean monitoring and their resulting parameters including accuracy and solution required for the instrument's extended life are provided in Table 1, and detail methodology of the instruments is presented in Table 2.

Table 1 Detail of the technical limitation involved in the coastal observation system

Sensors	Limitations	Solution Required
Conductivity-temperature-depth (CTD) sensors	Accuracy and drift: Inaccuracies due to calibration drift and biofouling	Use anti-fouling coatings; schedule regular cleaning and calibration
Turbidity sensors	Biofouling and sensitivity: Prone to fouling and affected by sediments	Install wipers; deploy multiple sensors for better data accuracy
Dissolved oxygen sensors	Durability and power: Membrane wear and high energy use	Switch to optical sensors; use energy-efficient designs and renewable power sources
pH sensors	Calibration and sensitivity: Drift over time; affected by temperature	Use multi-parameter sensors; apply real-time calibration and correction methods
Acoustic doppler current profilers (ADCPs)	Cost and durability: High cost; vulnerable in strong currents	Use durable materials; plan strategic deployment and regular maintenance
Chlorophyll fluorescence sensors	Biofouling and drift: Fouling affects readings; requires frequent recalibration	Use anti-fouling methods; employ automated cleaning systems and recalibration
Pressure sensors	Durability and power: Degrades in harsh conditions; high power use	Use corrosion-resistant materials; optimize power management
Nitrate sensors	Sensitivity and cost: Affected by temperature; expensive maintenance	Apply correction algorithms; streamline calibration to reduce costs
Optical sensors	Biofouling and data management: Obstructed optics; large data output	Use self-cleaning optics; employ data compression and intermittent transmission
Sonar sensors	Durability and cost: Wears out under pressure; high deployment costs	Use rugged materials; optimize maintenance and deployment strategies

2.6 *Mooring Preparation for Oceanographic Deployments*

In order to collect long-term data on a variety of oceanographic characteristics, moorings are essential for placing oceanographic sensors at stable locations. According to Venkatesan et al. (2018), several steps are involved in establishing a mooring, such as planning, designing, deploying, and maintaining it. This is a detailed guidance on the preparation of moorings for oceanographic instruments, as given in Fig. 6. There are several types of mooring based on the sensor types, deployment location, and safety measures.

Table 2 Methodology for measuring different oceanographic parameters

TD (conductivity, temperature, depth) sensors	
Physics/ Technology	Conductivity: Measures electrical conductivity to infer salinity using inductive or electrode-based sensors
	Temperature: Uses thermistors or platinum resistance thermometers (PRTs) to measure the seawater temperature
	Depth: Pressure sensors, typically strain-gauge or quartz crystal, measure the pressure exerted by the water column to determine depth
Methodology	Deployed from ships or autonomous platforms
	Data collection involves lowering the CTD rosette to desired depths while continuously recording data
	Provides vertical profiles of temperature, salinity, and depth
ADCP (acoustic Doppler current profiler)	
Physics/ Technology	Uses Doppler shift principle in acoustic signals
	Emits sound pulses and measures the frequency shift of the return signal scattered by particles in the water column to determine water velocity
Methodology	Deployed from fixed or moving platforms (e.g. ships, moorings)
	Provides profiles of current speed and direction at various depths
Wave buoys	
Physics/ Technology	Measures wave height, period, and direction using accelerometers, gyroscopes, and GPS
	Often include sensors for atmospheric pressure, wind speed, and direction
Methodology	Deployed at specific locations to capture real-time wave data
	Buoy motion is translated into wave parameters through on board processing
Tide gauges	
Physics/ Technology	Measures sea level relative to a fixed-point using float-operated devices, pressure sensors, or acoustic/optical distance measurement
Methodology	Installed along coastlines, harbours, and estuaries
	Continuous monitoring provides data on tidal patterns, storm surges, and long-term sea level changes
Water quality sensors (multi-parameter sondes)	
Physics/ Technology	Integrate various sensors to measure parameters such as pH, dissolved oxygen, turbidity, chlorophyll, nitrates, etc
	Use optical, electrochemical, and ion-selective electrode technologies
Methodology	Deployed in situ for continuous or periodic measurements
	Can be part of moored platforms, autonomous vehicles, or handheld units
Spectrophotometer	
Physics/ Technology	Measure the absorbance or transmittance of specific wavelengths of light to determine concentrations of dissolved substances
Methodology	Used for laboratory analysis of water samples or integrated into field-deployable devices

(continued)

Table 2 (continued)

TD (conductivity, temperature, depth) sensors	
	Analysis involves passing a light beam through a water sample and detecting the intensity of transmitted light
Benthic landers	
Physics/ Technology	Autonomous platforms equipped with various sensors to measure parameters like sediment properties, bottom currents, temperature, etc
Methodology	Deployed on the seabed for extended periods
	Collect data on benthic boundary layer dynamics and interactions
Sediment traps	
Physics/ Technology	Collect sinking particulate matter using funnel-shaped collectors
Methodology	Deployed at various depths to collect and quantify sedimentation rates and composition
HF radar systems	
Physics/ Technology	Use high-frequency radio waves to measure surface currents over large areas of the coastal ocean
	Based on the Doppler shift of backscattered radio waves from surface waves
Methodology	Installed along the coast, providing continuous real-time current data
	Useful for mapping surface currents and tracking pollutant dispersion
Remote sensing instruments	
Physics/ Technology	Use satellite or airborne sensors to capture data on sea surface temperature, chlorophyll concentration, sea level, etc
	Technologies include optical sensors, radar altimeters, and radiometers
Methodology	Data acquisition through satellite passes or airborne surveys
	Provides large-scale, synoptic views of oceanographic parameters
Gliders	
Physics/ Technology	Autonomous underwater vehicles (AUVs) that glide through the water column by adjusting buoyancy and using wings for lift
	Equipped with various sensors for measuring physical-biochemical parameters
Methodology	Programmed to follow specific transects or grid patterns
	Collects continuous profiles of water column properties over extended periods
ARGO floats	
Physics/ Technology	Autonomous floats that drift with ocean currents and periodically descend and ascend to collect profile data
	Use pressure, temperature, and conductivity sensors
Methodology	Deployed globally, real-time data on temperature, salinity, and currents
	Contribute to global ocean observing systems
Drifters	
Physics/ Technology	Surface floats equipped with GPS and sensors to measure currents, temperature, and other surface parameters

(continued)

Table 2 (continued)

TD (conductivity, temperature, depth) sensors	
Methodology	Deployed to drift with surface currents, providing data on circulation patterns and surface properties
Seismic and acoustic monitoring systems	
Physics/ Technology	Use hydrophones and seismometers to detect underwater earthquakes, volcanic activity, and other geophysical events
Methodology	Deployed on the seafloor or in the water column
	Provide data on geophysical processes and submarine hazards
Underwater cameras and video systems	
Physics/ Technology	Use high-resolution cameras and lighting systems to capture images and videos of the seabed and marine life
Methodology	Deployed on ROVs, AUVs, or fixed platforms
	Provide visual documentation for habitat mapping and biological studies
Drones (UAVs)	
Physics/ Technology	Unmanned aerial vehicles equipped with cameras and sensors for aerial surveys
Methodology	Conduct coastal and nearshore surveys, providing high-resolution imagery for mapping and monitoring coastal changes
LIDAR systems	
Physics/ Technology	Light Detection and Ranging systems use laser pulses to measure distances and create high-resolution topographic maps
Methodology	Deployed on aircraft or drones
	Useful for coastal mapping, beach erosion, and habitat characterization
Automated weather stations	
Physics/ Technology	Equipped with sensors for measuring atmospheric parameters such as wind speed, direction, temperature, humidity, and barometric pressure
Methodology	Installed on land or marine platforms
	Provide continuous meteorological data relevant to oceanographic studies
Acoustic backscatter system	
Physics/ Technology	Emit acoustic pulses and measure the intensity of the returning signal scattered by particles and organisms in the water column
	Used to infer sediment concentration, plankton distribution, and bottom topography
Methodology	Deployed on ships, moorings, or AUVs
	Provides data on sediment transport dynamics and biological distributions
Microscale temperature loggers	
Physics/ Technology	High-resolution temperature sensors that record temperature variations at fine spatial and temporal scales
	Use thermistors or microelectromechanical systems (MEMS)
Methodology	Deployed in situ on moorings or attached to other instruments

(continued)

Table 2 (continued)

TD (conductivity, temperature, depth) sensors	
	Useful for studying microhabitat conditions and fine-scale thermal dynamics
Flow cytometers	
Physics/ Technology	Use laser-based technology to analyse the physical and chemical characteristics of particles in a fluid stream
	Measure cell size, granularity, and fluorescence for identifying different plankton types
Methodology	Water samples are passed through a laser beam, and detectors capture the resulting fluorescence and light scatter signals
	Useful for detailed plankton community analysis and water quality monitoring
Nutrient analysers	
Physics/ Technology	Use chemical reactions and colorimetric or fluorometric detection to measure concentrations of nutrients like nitrates, phosphates, and silicates
Methodology	Deployed in situ or used in laboratory settings for water sample analysis
	Essential for understanding nutrient dynamics and their impact on primary productivity
DNA sequencers	
Physics/ Technology	Use next-generation sequencing (NGS) technologies to analyse genetic material from environmental samples
	Provide data on the biodiversity and genetic diversity of marine organisms
Methodology	Environmental DNA (eDNA) is extracted from water samples and sequenced
	Useful for biodiversity assessments, species monitoring, and ecological studies
Biogeochemical sensors	
Physics/ Technology	Measure various chemical and biological parameters such as dissolved oxygen, pH, chlorophyll, and organic carbon
	Utilize electrochemical, optical, and fluorometric techniques
Methodology	Deployed on buoys, moorings, or autonomous platforms
	Provide real-time monitoring of biogeochemical cycles and water quality
Hydrophone arrays	
Physics/ Technology	Use multiple hydrophones in an array to detect and localize underwater sounds
	Record acoustic signals from marine life, human activities, and natural events
Methodology	Deployed on the seafloor or suspended in the water column
	Useful for marine mammal monitoring, underwater noise pollution studies, and passive acoustic monitoring
Submersible UV–visible spectrophotometers	
Physics/ Technology	Measure the absorption of UV and visible light by dissolved substances in water
	Used for detecting and quantifying substances like nitrates, dissolved organic carbon, and pollutants
Methodology	Deployed in situ on moorings or AUVs

(continued)

Table 2 (continued)

TD (conductivity, temperature, depth) sensors	
	Provides continuous monitoring of water quality parameters
Chemosensors	
Physics/ Technology	Use chemical-sensitive materials and electrochemical methods to detect specific chemical compounds in seawater
	Can measure parameters such as heavy metals, nutrients, and organic pollutants
Methodology	Deployed on buoys, moorings, or handheld devices
	Provide real-time data on chemical pollutants and water quality
Piston corers	
Physics/ Technology	Use gravity and piston mechanisms to collect sediment cores from the seabed
	Provide undisturbed samples of sediment layers for geological and geochemical analysis
Methodology	Deployed from research vessels to penetrate and retrieve deep sediment cores
	Essential for studying sediment deposition, historical climate records, and palaeoceanography
Marine magnetometers	
Physics/ Technology	Measure variations in the Earth's magnetic field caused by seafloor geology
	Use fluxgate or proton precession magnetometers
Methodology	Towed behind research vessels or deployed on AUVs
	Useful for geological mapping, detecting underwater magnetic anomalies, and archaeological surveys
Seafloor mapping sonars	
Physics/ Technology	Use multi-beam or side scan sonar technology to create detailed maps of the seafloor
	Emit sound waves and record the time it takes for the echoes to return to the sensor
Methodology	Mounted on research vessels, AUVs, or ROVs
	Provide high-resolution bathymetric data and seafloor imagery
Isotope ratio mass spectrometers (IRMS)	
Physics/ Technology	Measure the ratios of stable isotopes in water and sediment samples
	Use mass spectrometry to analyse isotopic composition for tracing biogeochemical processes
Methodology	Samples collected and processed in laboratories
	Useful for studying nutrient cycling, carbon sources, and paleo environmental reconstructions
Resistivity profilers	
Physics/ Technology	Measure the electrical resistivity of sediments to infer properties such as porosity, grain size, and fluid content
	Use electrodes to inject electrical current and measure the resulting voltage
Methodology	Deployed on the seafloor or towed behind vessels

(continued)

Table 2 (continued)

TD (conductivity, temperature, depth) sensors	
	Provide data on sediment structure and sub-bottom features
Underwater laser scanners	
Physics/ Technology	Use laser technology to create high-resolution 3D maps of underwater structures and habitats
	Emit laser beams and capture the reflected signals to construct detailed images
Methodology	Mounted on ROVs or AUVs
	Useful for habitat mapping, structural inspections, and archaeological surveys
Autonomous surface vehicles (ASVs)	
Physics/ Technology	Unmanned surface vehicles equipped with various sensors for oceanographic measurements
	Use GPS, sonar, and other navigation systems for autonomous operation
Methodology	Programmed to follow specific routes or survey patterns
	Collect data on surface water properties, currents, water column, etc
Biofouling sensors	
Physics/ Technology	Detect and monitor the growth of biofouling organisms on marine structures and sensors
	Use optical, acoustic, and electrochemical methods to assess biofouling levels
Methodology	Deployed on moorings, buoys, and other marine installations
	Provide data on biofouling dynamics to develop antifouling strategies
Mooring systems	
Physics/ Technology	Consist of anchored buoy systems that hold various sensors in place within the water column
	Sensors can measure parameters like temperature, salinity, currents, and biological variables
Methodology	Deployed in fixed locations to provide long-term, continuous data
	Useful for monitoring time series data of oceanographic conditions
Hydrophones	
Physics/ Technology	Use piezoelectric materials to convert underwater sound into electrical signals
	Can detect a wide range of frequencies from natural and anthropogenic sources
Methodology	Deployed in arrays or singly on moorings, AUVs, or ROVs
	Used for passive acoustic monitoring of marine life, human activities, and environmental noise
Autonomous underwater vehicles (AUVs)	
Physics/ Technology	Unmanned submersibles equipped with navigation systems and a suite of oceanographic sensors
	Operate independently following pre-programmed missions
Methodology	Conduct high-resolution surveys of the water column and seafloor
	Collect data on physio-bio-chemical properties over large spatial areas

(continued)

Table 2 (continued)

TD (conductivity, temperature, depth) sensors	
Remotely operated vehicles (ROVs)	
Physics/ Technology	Tethered submersibles controlled from the surface, equipped with cameras, manipulators, and scientific instruments
	Use thrusters for manoeuvrability and various sensors for data collection
Methodology	Deployed for targeted studies of specific sites
	Useful for detailed visual inspections, sampling, and intervention tasks
Water samplers (e.g. Niskin bottles)	
Physics/ Technology	Collect water samples at specific depths using pressure-triggered bottles that seal to preserve sample integrity
Methodology	Deployed on CTD rosettes or other profiling systems
	Used for subsequent laboratory analysis of chemical and biological properties
Bathythermographs (BT and XBT)	
Physics/ Technology	Measure water temperature at various depths using thermistors or resistance thermometers
	XBTs (Expendable Bathythermographs) are disposable probes that transmit data back to the surface
Methodology	Deployed from ships to provide vertical temperature profiles
	Useful for studying thermoclines and ocean heat content
Multi-parameter probe	
Physics/ Technology	Integrated devices that combine multiple sensors (e.g., temperature, salinity, turbidity, chlorophyll) in one unit
	Can include optical, electrochemical, and pressure sensors
Methodology	Used for in situ measurements in diverse aquatic environments
	Provides comprehensive data on water quality and physical properties
Plankton nets and samplers	
Physics/ Technology	Use fine mesh nets to capture planktonic organisms from the water column
	Equipped with flow metres to measure the volume of water filtered
Methodology	Towed behind ships or deployed in fixed locations
	Samples analysed to study plankton communities, distribution, and abundance
Sediment corers	
Physics/ Technology	Extract sediment cores from the seabed using gravity, piston, or vibro-coring techniques
	Preserve stratification and structure of sediment layers
Methodology	Deployed from ships or platforms to study sediment composition, deposition rates, and historical records
	Used for geochemical, biological, and paleoclimatic analyses
Side-scan sonar	
Physics/ Technology	Uses sound waves to create detailed images of the seafloor
	Transducers emit acoustic pulses and measure the intensity of backscatter

(continued)

Table 2 (continued)

TD (conductivity, temperature, depth) sensors	
Methodology	Towed behind ships or mounted on AUVs
	Used for mapping seafloor features, habitat characterization, and identifying submerged objects
Multi beam echo sounders	
Physics/ Technology	Emit multiple beams of sound to map the seafloor topography in high resolution
	Measure the time it takes for the sound to return after bouncing off the seafloor
Methodology	Mounted on ships or underwater vehicles
	Provides detailed bathymetric maps for navigation, habitat mapping, and geological studies
Optical backscatter sensors	
Physics/ Technology	Measure the scattering of light by particles suspended in the water to estimate turbidity and suspended sediment concentration
	Use optical sensors that detect scattered light at specific angles
Methodology	Deployed on moorings, AUVs, or ROVs
	Provide real-time data on water clarity and sediment transport
Satellite altimeters	
Physics/ Technology	Measure sea surface height using radar pulses that reflect off the ocean surface
	Calculate the distance between satellite and sea surface with high precision
Methodology	Mounted on satellites in polar orbits
	Provide global sea level data, wave heights, and ocean circulation patterns
Marine magnetometers	
Physics/ Technology	Measure the Earth's magnetic field and its anomalies caused by the presence of ferromagnetic materials
	Use fluxgate, proton precession, or optical pumping technologies
Methodology	Towed behind ships or deployed on underwater vehicles
	Used for geological surveys, locating shipwrecks, and identifying mineral deposits
Current metres	
Physics/ Technology	Measure the speed and direction of water flow using mechanical (e.g., rotor) or acoustic (e.g. Doppler) sensors
Methodology	Deployed on moorings, AUVs, or ROVs
	Provide data on current profiles and help in understanding circulation patterns
Passive acoustic monitoring (PAM) systems	
Physics/ Technology	Use hydrophones to detect and record sounds from marine life, human activities, and natural processes
Methodology	Deployed on moorings, AUVs, or towed behind ships
	Used for monitoring marine mammal presence, ship noise, and environmental assessments

(continued)

Table 2 (continued)

TD (conductivity, temperature, depth) sensors	
In situ nutrient analysers	
Physics/ Technology	Use chemical reagents and optical detection to measure concentrations of nutrients like nitrates, phosphates, and silicates
Methodology	Deployed in situ for continuous or discrete sampling
	Provide real-time data on nutrient levels to study productivity and eutrophication
Wave radar systems	
Physics/ Technology	Use radar to measure sea surface roughness and wave characteristics
	Analyse the Doppler shift and return signal strength to determine wave height and direction
Methodology	Installed on coastal structures or offshore platforms
	Provide continuous monitoring of wave conditions for safety and research
Water samplers (Niskin bottles)	
Physics/ Technology	Collect water samples at specific depths using a cylindrical container that seals at both ends upon activation
Methodology	Deployed on a CTD rosette or similar sampling apparatus
	Water samples are analysed in laboratories for chemical, biological, and physical properties
Plankton nets	
Physics/ Technology	Fine mesh nets towed through the water to collect plankton samples
Methodology	Deployed from boats or fixed structures
	Plankton samples are analysed to study species composition, abundance, and distribution
Piezometers	
Physics/ Technology	Measure the hydraulic head in groundwater systems using a pressure sensor or a manual measurement device
Methodology	Installed in the sediment to measure pore water pressure
	Used to study groundwater-surface water interactions and sediment stability
Chemiluminescent sensors	
Physics/ Technology	Use chemiluminescent reactions to detect specific chemical species, such as reactive oxygen species
Methodology	Deployed in situ or used in laboratory analysis
	Useful for studying chemical processes in marine environments
Fluorometers	
Physics/ Technology	Measure the fluorescence emitted by specific substances (e.g., chlorophyll, dissolved organic matter) when excited by light
Methodology	Deployed in situ or used in laboratory settings
	Provides data on phytoplankton biomass and water quality

(continued)

Table 2 (continued)

TD (conductivity, temperature, depth) sensors	
Optical backscatter sensors (OBS)	
Physics/ Technology	Measure suspended sediment concentrations by detecting scattered light from particles in the water column
Methodology	Deployed on moorings, AUVs, or ROVs
	Provides real-time data on turbidity and sediment transport
Passive acoustic monitoring (PAM) systems	
Physics/ Technology	Use hydrophones to detect and record sounds from marine mammals, fish, and other sources
Methodology	Deployed on moorings, AUVs, or buoys
	Useful for monitoring marine life and anthropogenic noise impacts
Underwater gliders	
Physics/ Technology	Autonomous underwater vehicles that glide by adjusting buoyancy, equipped with various sensors for measuring physical and biochemical parameters
Methodology	Programmed to follow specific transects or grid patterns
	Provides continuous profiles of water column properties over extended periods
Side-scan sonar	
Physics/ Technology	Uses acoustic waves to create detailed images of the seafloor
Methodology	Towed behind a boat or mounted on a vehicle
	Used for mapping seabed features and habitat characterization
Multi-beam echo sounders	
Physics/ Technology	Emit multiple beams of sound to map the seafloor topography
Methodology	Mounted on ships or underwater vehicles
	Provides high-resolution bathymetric data for geomorphological studies
Marine magnetometers	
Physics/ Technology	Measure variations in the Earth's magnetic field caused by seafloor structures
Methodology	Towed behind ships or deployed on underwater vehicles
	Used for geological and archaeological surveys
Resistivity profilers	
Physics/ Technology	Measure electrical resistance in the seafloor sediments to infer sediment composition and pore water salinity
Methodology	Deployed on the seafloor from ships or underwater vehicles
	Used for sedimentological studies and groundwater assessments
Hydrographic survey equipment	
Physics/ Technology	Includes a variety of instruments such as GPS, sonar, and LIDAR for mapping coastal and underwater topography

(continued)

Table 2 (continued)

TD (conductivity, temperature, depth) sensors	
Methodology	Deployed from ships, aircraft, or land-based platforms
	Provides accurate maps of coastal and marine environments
Dissolved oxygen sensors	
Physics/ Technology	Electrochemical or optical sensors measure the concentration of oxygen dissolved in water
Methodology	Deployed in situ for real-time monitoring or used in laboratory analysis
	Provides critical data on water quality and biogeochemical cycles
Radiometers	
Physics/ Technology	Measure solar radiation and its interaction with the ocean surface
Methodology	Deployed on buoys, ships, or aircraft
	Used for studies of ocean–atmosphere interactions and primary productivity
Methane sensors	
Physics/ Technology	Detect methane concentrations in the water column using infrared absorption or other techniques
Methodology	Deployed on underwater vehicles or in situ monitoring stations
	Important for studying methane seeps and carbon cycling
Piston corers	
Physics/ Technology	Extract long, undisturbed sediment cores from the seabed for geological and palaeoceanographic studies
Methodology	Deployed from ships using winches and cables
	Provides sediment samples for analysing historical climate and ocean conditions
Trace metal analysers	
Physics/ Technology	Detect and quantify trace metal concentrations in water samples using techniques such as inductively coupled plasma mass spectrometry (ICP-MS)
Methodology	Used in laboratory settings to analyse water or sediment samples
	Important for studying pollution and biogeochemical cycles
Carbon dioxide (CO_2) sensors	
Physics/ Technology	Measure the concentration of CO_2 in water using non-dispersive infrared (NDIR) or other methods
Methodology	Deployed in situ or used for laboratory analysis
	Essential for studying ocean acidification and carbon cycling
Acoustic release systems	
Physics/ Technology	Use acoustic signals to release moored instruments for retrieval
Methodology	Deployed on moorings for long-term monitoring
	Allows recovery of submerged equipment without surface intervention

(continued)

Table 2 (continued)

TD (conductivity, temperature, depth) sensors	
Current metres	
Physics/ Technology	Measure the speed and direction of water flow using mechanical (e.g., rotor) or acoustic (e.g., Doppler) sensors
Methodology	Deployed on moorings, tripods, or autonomous vehicles
	Provide detailed data on water column currents
Conductivity sensors	
Physics/ Technology	Measure the electrical conductivity of seawater to determine salinity
Methodology	Deployed on CTDs, buoys, or autonomous vehicles
	Provides critical data for understanding salinity and water mass properties

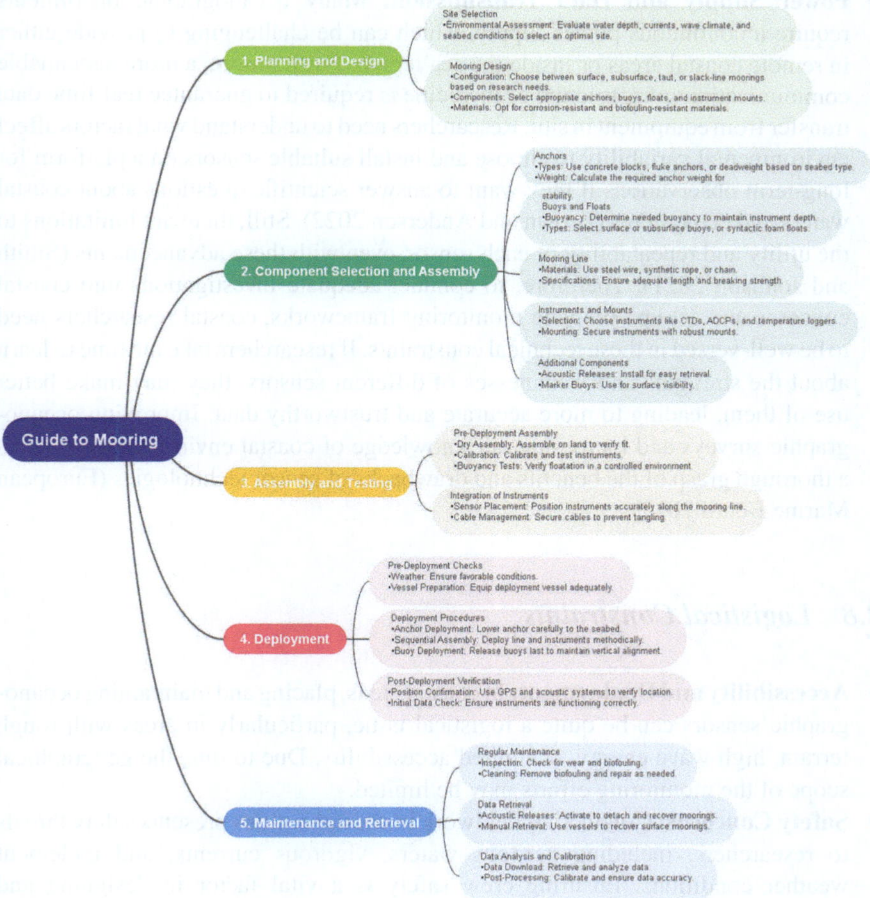

Fig. 6 SOP for mooring preparation for oceanographic deployment

2.7 Technological Limitations

- **Sensor Accuracy and Precision:** Biofouling, calibration drift, and interference from different environmental variables are some of the issues affecting oceanographic sensors' accuracy and precision. Ensuring data quality requires regular calibration and maintenance, which can be challenging in remote or unfavourable environments. Since the performance of any given sensor degrades with time, regular maintenance and calibration are necessities for any instrument or sensor that is expected to operate for a long time.
- **Instrument Durability and Reliability:** The harsh coastline conditions include strong currents, high salinity, and erosive sediments that can damage or destroy instruments. The success of long-term monitoring programs depends on the reliability and durability of the equipment used.
- **Power Supply and Data Transmission:** Many oceanographic instruments require a continuous power supply, which can be challenging to provide either in remote coastal areas or inside the sea region. Furthermore, a more dependable communication infrastructure or medicine is required to guarantee real-time data transfer from equipment in situ. Researchers need to understand what factors affect environmental variability to choose and install suitable sensors on a platform for long-term observations if they want to answer scientific questions about coastal waters and ecosystems (Brown and Anderson 2022). Still, there are limitations to the utility and repeatability of each sensor, even with these advancements (Smith and Johnson 2021). Therefore, to conduct adequate investigations into coastal concerns and set up complete monitoring frameworks, coastal researchers need to be well-versed in these technical constraints. If researchers take the time to learn about the strengths and weaknesses of different sensors, they may make better use of them, leading to more accurate and trustworthy data. Improving oceanographic surveys and increasing our knowledge of coastal environments requires a thorough grasp of the benefits and drawbacks of sensor technologies (European Marine Board 2019).

2.8 Logistical Constraints

- **Accessibility and Deployment:** In coastal areas, placing and maintaining oceanographic sensors can be quite a logistical issue, particularly in areas with rough terrain, high wave energy, or limited accessibility. Due to this, the geographical scope of the monitoring efforts may be limited.
- **Safety Concerns:** Conducting fieldwork in coastal regions presents safety threats to researchers, including perilous waters, vigorous currents, and inclement weather conditions. Ensuring crew safety is a vital factor in designing and implementing field missions.
- **Resource Allocation:** The costs of purchasing, deploying, and maintaining oceanographic instruments can be substantial. More financial and human

resources are needed to constrain the scope and duration of monitoring programs, especially in developing regions.

Deploying and maintaining oceanographic sensors in coastal regions presents challenges due to rugged terrain, elevated wave energy, and restricted access points, which can limit the spatial coverage of monitoring activities (Brown 2022). Deploying and maintaining systems efficiently is essential to collect comprehensive data in different settings (Anderson et al. 2012). Researchers must be well-prepared and adhere to strict safety protocols to protect field crews from dangers such as strong currents, rough waters, and stormy weather (Marine Conservation Society 2023). To ensure that monitoring activities are both practical and sustainable, it is vital to allocate resources optimally so that available financial and intellectual assets are maximized (European Marine Board 2019). Researchers can successfully build comprehensive monitoring frameworks that yield valuable insights into coastal ecosystems and facilitate their conservation and management by tackling these logistical challenges, ensuring appropriate equipment deployment, prioritizing safety, and optimizing resource utilization.

2.9 Data Integration, Analysis, Dissemination

Effective data integration and analysis are critical for transforming raw data into actionable insights. Different tools and technology can adapt the data to a single-user portal. Different types of ocean data have been published in different file formats by different organizations. The key parameters, which are considered for better data integration and display, are given below.

- **Data Management Systems:** Effective data integration and analysis are crucial for transforming raw data into actionable insights for managing and conserving coastal ecosystems (Garcia et al. 2024). Data management systems provide a structured framework for handling large volumes of data from various sources such as observation, satellite imagery, real-time sensor networks, and output of numerical simulations. (Martinez et al. 2023). The data management system ensures that better-organized data is easy to access, secure, and easy to process and integrate.
- **Statistical Methods:** Use advanced statistical methods to analyse data, find trends, and assess risks. In order to analyse complicated datasets, spot trends, and evaluate uncertainty, advanced statistical approaches are subsequently used. When studying how nutrient runoff impacts coastal ecosystems, for example, these methods can track changes in nutrient concentrations, link them to algal blooms, and evaluate how accurate these links are (Zhao et al. 2023). In order to formulate effective management strategies, this research clarifies the interrelationships among numerous environmental factors.
- **Visualization Tools**: Make it easy for stakeholders and the general public to understand monitoring results by creating user-friendly visualization tools. In

addition, visualization technologies play a crucial role in simplifying complex data into more approachable forms, such as interactive graphs and maps. The public, stakeholders, and lawmakers can all benefit from these tools because they make it easier to understand the data's implications, such as the contaminants' spatial distribution and water quality variations (Nguyen et al. 2024a, b). According to Garcia et al. (2024), researchers can enhance the management and sustainability of coastal ecosystems by combining robust data management systems with advanced statistical analysis and good visualization (Sujith Kumar et al. 2022). This could lead to significant discoveries that help with decision-making.

2.9.1 Environmental Variability

The coastal environment and ecosystem changes naturally due to changing the environment variables locally and globally. Therefore, the monitoring categories, such as spatial and temporal variability, must be considered while investigating the variability of environmental variables. The spatial variability of variable is mostly Lagrangian by nature, while the temporal variability is mostly Eulerian. The two situations are so different in measuring and survey procedures, each with its own benefits and drawbacks. However, it is prudent for coastal observers to use appropriate methods and instruments suitable for coastal monitoring close to marine and coastal environments. So, here are two separate ways to classify environmental variability.

Extreme weather events like hurricanes and tide cycles add much temporal variability to coastal ecosystems. Due to these temporal fluctuations, high-frequency sampling is required for precise dynamics capture; however, this approach can be resource-intensive and challenging to maintain for long periods. To understand the temporal variability of coastal water features, one must keep an eye on several dynamic processes, such as changes in sea level, circulation along the coast, currents, waves, discharge, erosion along the coast, changes in water quality, and sedimentology. Sea level variability is monitored using tide gauges and satellite altimeters, which assess changes caused by tides, meteorological circumstances, and prolonged sea level elevation (Williams et al. 2023). Current metres and drifters provide valuable information about coastal circulation dynamics and currents by measuring the speed and direction of water flows. This helps to understand how circulation patterns affect coastal regions (Brown et al. 2023). The influence of waves on coastal erosion and sediment movement can be better understood using wave height, frequency, and direction data, which can be obtained from wave buoys and radar systems (Lee et al. 2024). Discharge from rivers and streams is quantified using stream gauges and flow metres, which evaluate the volume of freshwater entering the coastal zone and its effects on salinity and nutrient concentrations (Chen et al. 2023a, b). In order to have a better understanding of the changes occurring to shorelines and erosion rates, coastal cameras, satellite imagery, and differential GPS are utilized (Nguyen et al. 2023). Monitoring parameters such as temperature, salinity, nutrient concentrations, and pollutants is done utilizing buoys fitted with sensors and autonomous sensor networks to assess variability in water quality (Zhao et al. 2024). Finally, sedimentology is examined using sediment samplers, sediment traps, and acoustic profilers, which gather and analyse sediment samples while monitoring alterations in sediment composition and deposition patterns (Garcia et al. 2024). Integrating data from various instruments enables researchers to thoroughly understand temporal variations in coastal processes (Table 3) and environmental parameters (Table 4), facilitating more effective management and conservation strategies for dynamic coastal and marine environments (Williams et al. 2023).

- **Spatial Variability**: The spatial heterogeneity of coastal zones, influenced by factors such as bathymetry, coastline morphology, and hydrodynamic conditions, necessitates dense spatial sampling. This variability can complicate the deployment and retrieval of instruments and the interpretation of collected data. There are several methods adapted in recent era to preparing spatial information such as remote sensing data, interpolation of multiple point data through GIS platform, numerical approach, artificial intelligence, and machine learning.

3 Overview of Oceanographic Variables, Observation, and Implementation

The proposed technique includes data collection, analysis, interpretation of results, and implications for improving management, policy-making, and ecological restoration. This section focuses on advanced data and analysis related to geomorphology, oceanography, biogeochemical processes, and meteorology.

3.1 Geomorphology

The coastal and oceanic geomorphology primarily consists of beaches and seabeds. The coastal geomorphology encompasses beaches (sandy, rocky, and mudflats), dunes, cliffs, bars/troughs, sand spits, atolls, and islands while the ocean floor comprises a continental shelf, slopes, seamounts, ridges, abyssal plains, hills, trenches, and more formations. To map these geomorphologies, an advanced global positioning system (GPS) is required to observe the topography and variations of the shoreline. Similarly, high-precision GPS data facilitates coastal mapping, beach profiling, and dune measurements. GIS systems such as ArcGIS (DSAS) and QGIS are employed to analyse coastal data, yielding rates of coastline deposition and erosion (Pradhan et al. 2020a). Simultaneously, satellite data is essential for identifying shorelines and their alterations.

Similarly, many instruments are utilized to assess beach width, volume, dune dimensions, and their fluctuations over spatiotemporal scales (Mohanty et al. 2011).

Table 3 Detail list and its supplementary information for understanding the temporal variability of coastal attributes

Instrument	Purpose	Working principle	Accuracy	Deployment location	Site condition
Satellite altimeters	Provide measurements of sea surface height from space, tracking changes in sea level over time	Emit radar pulses from satellites and measure the time it takes for the pulses to return, calculating sea surface height	Accuracy of about 1–2 cm for sea surface height measurements	High-altitude satellites in orbit	Covers large-scale, open ocean and coastal regions; data is not location specific
Radar systems	Track wave characteristics and surface currents, offering high-resolution time-series data	Emit electromagnetic waves and measure their reflections off the water surface to analyse wave height and movement	Accuracy of about 1–5 cm for wave height and 1–2 cm/s for surface currents	Coastal regions, often mounted on platforms or ships	Effective in clear water conditions; performance may degrade in highly turbid areas
Tide gauges	Measure sea level changes over time, including tides, storm surges, and long-term sea level rise	Use pressure sensors or float-based systems to record water level changes relative to a fixed reference point	Typically accurate to within 1–2 cm for sea level measurements	Coastal platforms, piers, and harbours	Must be deployed in stable, open-water areas with minimal wave and current interference
Current metres	Monitor water flow speed and direction, capturing variations in coastal circulation and current dynamics	Utilize acoustic Doppler technology to emit sound waves and measure the frequency shift caused by moving water, or mechanical sensors to directly measure flow speed	Accuracy of about 1–5 cm/s for current velocity and 1–2 degrees for direction	Fixed in-water locations like buoys or seabed platforms	Suitable for areas with stable, predictable current patterns; avoid high-energy zones

(continued)

Table 3 (continued)

Instrument	Purpose	Working principle	Accuracy	Deployment location	Site condition
Drifters	Track the movement of water masses by floating with currents and recording positional changes	Equipped with GPS or other positioning systems, drifters follow water currents and provide real-time location data	Accuracy of about 1–3 m in position	Deployed at sea and in coastal areas	Effective in open water and coastal zones; less suitable for areas with heavy ship traffic
Wave buoys	Measure wave height, frequency, and direction, providing time-series data on wave dynamics	Contain accelerometers and gyroscopes to detect wave movements and calculate wave parameters based on buoy motion	Accuracy of about 5–10 cm for wave height measurements and 1–2 degrees for wave direction	Offshore and coastal waters	Deployed in areas representative of wave conditions; avoid regions with extreme weather
Buoys with sensors	Continuously measure water quality parameters such as temperature, salinity, and nutrient concentrations	Equipped with various sensors to detect and record water temperature, salinity, and nutrient levels, transmitting data periodically or in real time	Accuracy of about 0.1–0.5 °C for temperature, 0.01–0.1 PSU for salinity, and 1–5% for nutrient concentrations	Coastal and open waters	Effective in regions with minimal physical interference; performance may vary with fouling
Stream gauges	Measure river and stream discharge into coastal areas, tracking changes in freshwater flow over time	Use float-based, pressure sensor, or ultrasonic technology to measure water flow volume and velocity	Accuracy of about 2–5% for discharge measurements	Rivers and streams entering coastal areas	Best placed in stable river channels with minimal obstructions; may be affected by sediment
Flow metres	Assess the volume and velocity of water flow in rivers and streams entering coastal zones	Employ various techniques such as electromagnetic, ultrasonic, or mechanical sensors to measure flow rate and velocity	Accuracy of about 1–3% for flow rate and velocity measurements	River channels, estuaries, and coastal inlets	Effective in clear water; performance can be affected by high sediment loads or debris

(continued)

Table 3 (continued)

Instrument	Purpose	Working principle	Accuracy	Deployment location	Site condition
Shoreline cameras	Capture images of coastal erosion and shoreline changes over time, providing visual time-series data	Use high-resolution digital cameras to take periodic images of the shoreline, enabling analysis of erosion and sediment movement	Accuracy depends on camera resolution; typically within 1–5 m for shoreline position	Coastal viewpoints and monitoring stations	Best used in accessible, stable locations with clear lines of sight; avoid obstructed areas
Differential GPS	Monitor precise changes in shoreline position and coastal erosion using high-accuracy positional data	Utilize multiple GPS receivers to measure precise positional changes over time with high accuracy, tracking shoreline shifts	Accuracy of about 1–10 cm for positional measurements	Coastal and estuarine areas	Suitable for areas with stable terrain; avoid locations with frequent signal interference
Autonomous sensor networks	Collect real-time data on chemical parameters like pH, dissolved oxygen, and pollutants	Deploy a network of sensors that measure chemical properties of the water autonomously, providing continuous data transmission or periodic data collection	Accuracy varies: pH (± 0.01–0.02 units), dissolved oxygen (± 0.1–0.5 mg/L), pollutants (varies)	Coastal areas and estuaries	Best in stable water conditions; may be impacted by biofouling or sediment
Sediment Samplers	Collect sediment samples from the seabed at regular intervals to analyse changes in sedimentology	Use mechanical or gravity-based devices to capture sediment from the seabed, which is then analysed for composition and deposition patterns	Accuracy in sediment volume and composition varies by sampler type; typically within 5–10%	Seabed, estuaries, and river mouths	Effective in calm areas; sediment collection can be affected by strong currents or waves

(continued)

Table 3 (continued)

Instrument	Purpose	Working principle	Accuracy	Deployment location	Site condition
Sediment traps	Capture and measure sediment accumulation over time to study deposition patterns	Place traps on the seabed that accumulate sediment over a specified period, with periodic retrieval and analysis of sediment collected	Accuracy in sediment accumulation measurement typically within 5–10%	Seabed and estuarine environments	Suitable for areas with low to moderate current; performance can be affected by high sediment loads
Acoustic profilers	Provide data on sediment thickness and distribution using sound waves, monitoring sediment dynamics	Emit sound waves through the water column and measure the time it takes for echoes to return, calculating sediment thickness and distribution	Accuracy of about 1–5 cm for sediment thickness measurements	Coastal and estuarine zones	Effective in clear waters; performance may degrade in highly turbid environments
Spectrophotometers	Measure light absorbance in water samples to determine concentrations of various chemical substances over time	Pass light through water samples at specific wavelengths and measure the amount of light absorbed, correlating it with the concentration of chemical substances	Accuracy of about 1–2% for concentration measurements	Laboratory setting (requires water sample)	Best used in controlled environments; sample handling can affect accuracy

Table 4 Environmental parameters and corresponding instruments

Parameters	Instruments
Physical parameters	
Temperature	Temperature, depth sensor (CTD), thermistor chains
Salinity	CTD, refractometer, argentometric titration methods
Pressure	CTD, bottom pressure recorder
Sea level	Tide recorder, acoustic Doppler current profiler (ADCP)
Current velocity	Aquatic Doppler current Profiler (ADCP), Recording Current Meter (RCM), Flow meter
Wave	Wave rider buoys, wave-tide recorder (WTR), ADCP, radar based wave sensor
Turbidity	Turbidity meter, RCM, optical backscatter sensor (OBS), multi parameter water quality probe
Suspended sediment concentration	Optical backscatter sensors, laser in situ scattering and transmissometer (LISST)
Bathymetry	Echo sounder (single-beam multi beam)
Chemical parameter	
Dissolved oxygen	Multi parameter water quality probe, oxygen sensor, winkler titration kits
pH	Multi parameter water quality probe, pH sensor, pH meter
Nutrients (e.g.: Nitrate, Phosphate, Silicate)	Nutrient sensors, spectrophotometer, auto analysers
Carbon dioxide (CO_2)	pCO_2 sensors, Gas chromatography
Total alkalinity	Auto titrators
Dissolved organic carbon (DOC)	Total organic carbon (TOC) analyser
Biological parameter	
Chlorophyll-a	Fluorometer, HPLC analysis, spectrophotometer
Phytoplankton	Flow cytometers, microscopic analysis
Zooplankton	Plankton nets, optical plankton counters
Benthic organism distribution	Van veen grab sampler, Underwater video systems
Microbiology	Laminar flow, Microscope
Harmful algal blooms (HABs)	Imaging flow cytobot, satellite remote sensing
Meteorological parameter	
Wind speed & Direction	Anemometers, weather stations
Air temperature	Thermometers, weather stations

(continued)

Table 4 (continued)

Parameters	Instruments
Relative humidity	Hygrometer, weather stations
Precipitation	Rain gauges, weather stations
Solar radiation	Pyranometer, solar radiation sensors
Sediment parameter	
Grain size distribution	Sieve Shaker, Laser diffraction particle size analysers
Sediment composition	X-ray fluorescence (XRF) analysers, Mass spectrometers
Sediment accumulation rates	Sediment traps, Radiometric dating (e.g., Pb-210, Cs-137)
Remote sensing parameters	
Sea surface temperature (SST)	Satellite remote sensing (e.g., MODIS, AVHRR)
Ocean colour	Satellite remote sensing (e.g., MODIS, OCM3, Sentinel)
Sea surface height	Satellite altimetry (e.g., TOPEX/Poseidon, Jason series)
Coastal landforms	Lidar, Aerial photography
Water quality parameters	
Total suspended solids (TSS)	Filtration and gravimetric analysis
Contaminants (e.g., Heavy metals, Pesticides)	Mass spectrometers, Atomic absorption spectrophotometers, ICP-OES
Microplastics	FTIR spectroscopy, Raman spectroscopy
Advanced technologies	
Autonomous underwater vehicles (AUVs)	High-resolution mapping and oceanographic monitoring
Unmanned surface vehicles (USVs)	Used for surface monitoring and data collection in difficult-to-reach areas
Drifters and gliders	Used for tracking currents and collecting long-term data over large spatial scales
Integrated ocean observing systems (IOOS)	Network of sensors and platforms providing comprehensive real-time oceanographic data

The eco-sounder and bottom scanner are utilized to investigate the seabed, which is crucial for navigating accuracy in coastal areas at ports and inlets. Human involvement along the coast primarily alters beach and shallow water bathymetry through the construction of groins, sea walls, and ports, which impede sediment transport (Rao et al. 2009; Mohanty et al. 2011). Coastal monitoring utilizing dependable data facilitates beach and wetland restoration, navigation, natural resource management, and military operations. A key challenge in the observational analysis of coastal geomorphology is the validation of remotely acquired data from satellites, drones, and various sensors.

3.1.1 Importance of Studying Coastal and Ocean Geomorphology

Monitoring coastal and ocean geomorphology is essential for comprehending and managing dynamic environments where natural habitats, human communities, and environmental processes intersect. This area examines the influence of coastal and oceanic structures, including beaches, coral reefs, and estuaries, on habitat formation for various species and their role in mitigating natural hazards such as cyclones and storm surges. Climate change is leading to rising sea levels, ocean acidification, and increased storm intensity, which are altering geomorphic features and affecting both ecosystems and human settlements. Comprehension of coastal erosion, sediment transport, and geomorphic stability is crucial for formulating policies that safeguard shorelines, promote biodiversity, and guarantee sustainable resource utilization. Coastal geomorphology has significant cultural and economic implications, as numerous communities rely on fishing, tourism, and marine resources that are closely associated with coastal stability. Comprehending these dynamics facilitates the development of more effective conservation and management strategies, thereby protecting both natural and human systems in a period of swift environmental change.

3.1.2 Implication of Geomorphology Study

Geomorphology offers critical understanding of the processes that shape landscapes, contributing to environmental science, hazard mitigation, and sustainable development. Geomorphology aids in predicting landscape responses to climate impacts, including erosion, flooding, and sea-level rise, by elucidating the formation and alteration of natural features. This knowledge underpins resilient urban planning, the conservation of critical habitats, and sustainable resource management. Geomorphological studies contribute to the preservation of cultural sites and natural landscapes, facilitating a balance between human needs and environmental resilience and sustainability.

3.2 Oceanography

Nearshore area is particularly susceptible to vulnerability due to exposure to natural risks, including flooding, storm effects, sea-level rise, coastal erosion, and anthropogenic activities such as urbanization, transportation, and industrialization. Nearshore regions are vital to the national economy, encompassing sectors such as fishing, aquaculture, recreation and tourism, water supply, wastewater management, building, harbour operations, energy generation, and ecology, including crucial habitats and tourist destinations—oceanographic measurements of wind, wave, tide, and current yield essential insights into nearshore circulation. Monitoring oceanic and coastal wind data significantly aids in assessing atmospheric weather phenomena such as storms, cyclones, and land and sea breezes and in generating energy via wind turbines. Precise and regular wave data is a crucial factor for assessing sediment transport, pollution, and its consequences in coastal areas. Wind and wave dispersion data is illustrated via rose plots and distribution tables, commonly utilized by coastal researchers, stakeholders, and communities for enhanced management (Patra et al. 2016).

Tide information is utilized for navigation, defence, and mitigating the impacts of rising sea levels attributed to climate change. An investigation of tides and currents can be conducted for various constituents to comprehend the interactions of astronomical bodies with shallow water bodies, including nearshore areas, estuaries, lagoons, and extensive waterfronts. The observed oceanic data is subsequently utilized for numerical model simulations and validations (Panda et al. 2020a; Pradhan et al. 2020b). The measured surface wind, wave, current, and current profiling, assist in validating satellite data. Precise oceanographic data facilitates the comprehension of bio-geochemical processes inside marine water masses across several dimensions.

3.2.1 Importance of Oceanography Observation

Oceanographic observation is essential for understanding both global and local changes in the ocean and coastal environments, with significant implications for climate, ecosystems, and human communities. Locally, monitoring ocean conditions such as temperature, salinity, sea level, and sediment movement allows scientists to track changes directly affecting coastal stability, shoreline erosion, and habitat health. For example, data on coastal currents and wave energy are vital for predicting sediment dynamics that impact beaches, barrier islands, and essential habitats like coral reefs and mangroves. Observations also reveal changes in water quality due to nutrient runoff, pollution, and harmful algal blooms, which can harm fisheries, tourism, and public health. By capturing these localized shifts, oceanographic observation supports early detection and targeted responses to environmental challenges, enabling informed decisions on sustainable development, disaster preparedness, and ecosystem management to enhance resilience against natural and human-induced changes in coastal areas.

3.2.2 Implication of Oceanography

Oceanography has significant implications for climate science, weather patterns, environmental sustainability, resource management, and public safety. Oceanography examines ocean currents, temperature, salinity, and biogeochemical cycles, offering essential insights into the ocean's role in regulating global climate, distributing heat, and influencing weather patterns, including extreme events such as tropical cyclones and El Niño, as well as large-scale circulation. This comprehension is crucial for forecasting climate changes, including sea-level rise and ocean acidification, which affect marine ecosystems and coastal communities. Oceanographic research contributes to sustainable resource management by providing insights into fisheries, habitat conservation, and pollution control, which are essential for biodiversity and food security. Oceanography contributes to coastal protection by monitoring erosion, assessing flooding risks, and analysing sediment dynamics that affect shoreline stability. It enhances maritime safety through the optimization of navigation and shipping routes, leading to reduced fuel consumption and minimized environmental impact. Oceanography provides a comprehensive understanding of the ocean's function within Earth's systems, facilitating societal adaptation to climate change and promoting the sustainable utilization of marine resources.

3.3 Weather Observation

Coastal weather observations are crucial for monitoring and forecasting conditions that affect marine ecosystems, coastal infrastructure, and human activities. Coastal weather is distinctive due to the interplay between land and sea, resulting in sea breezes, fog, and abrupt temperature fluctuations. This document summarizes the components of coastal weather observations, the instruments employed, and their significance.

- **Wind Speed and Direction**: Coastal areas experience distinctive wind patterns, such as sea and land breezes, driven by temperature differences between land and water. Winds impact the waves, currents, air quality, transport pollutants, and salt spray inland. Real-time monitoring is crucial for navigation, aviation, and recreational activities.
- **Temperature**: Coastal temperatures can vary significantly between day and night and are influenced by ocean temperatures. Monitoring air and sea surface temperatures helps detect warming trends and is essential for weather forecasting, tourism, and climate studies.
- **Humidity**: Coastal areas are usually more humid due to proximity to water bodies. High humidity influences thermal comfort, cloud formation, and precipitation patterns. Observing humidity levels aids in weather prediction and understanding local climate.
- **Atmospheric Pressure**: Pressure patterns affect wind and storm formation. Monitoring pressure helps identify incoming weather systems, such as tropical storms or cyclones, which is critical for coastal areas prone to extreme weather events.
- **Precipitation**: Coastal regions often receive substantial rainfall due to moisture-laden air from the sea, which leads to orographic rain and seasonal monsoon systems. Monitoring precipitation helps manage water resources, forecast floods, and analyse climate patterns.

- **Visibility**: Visibility is influenced by fog, mist, and marine aerosols. Coastal fog is common and can reduce visibility for maritime and aviation sectors. Observing visibility is crucial for safe navigation and transportation.
- **Wave Height and Direction**: Coastal wave observations are vital for maritime safety, recreational activities, and erosion monitoring. Waves are influenced by local winds and distant storms, with data helping predict erosion rates and coastal flooding.
- **Sea Level**: Tidal and storm surge measurements are critical for coastal management, especially in areas vulnerable to flooding. Observing sea levels helps us understand tides, coastal flooding, and the impacts of sea-level rise on shorelines.
- **Solar Radiation**: Solar radiation data is essential for understanding temperature variations, energy budgeting, and weather dynamics. Coastal areas can have unique radiation patterns due to the reflective properties of the ocean.

3.3.1 Importance of Weather Observation

Weather observation is crucial for public safety, agriculture, transportation, and environmental management, as it supplies vital data on atmospheric conditions including temperature, humidity, wind speed, and precipitation. Continuous monitoring utilizing advanced technologies such as satellites, radar systems, and ground-based meteorological stations facilitates precise forecasting of short-term weather events and long-term climate patterns. Real-time data is essential for providing early warnings regarding severe weather events, including cyclones and floods, enabling communities to prepare and reduce potential losses. In the agricultural sector, accurate weather information guides farmers in making informed decisions regarding planting, irrigation, and harvesting, thereby enhancing crop yields and minimizing losses. The transportation sector depends on weather observations to maintain safe navigation and improve operational efficiency by circumventing hazardous conditions. Weather observation is essential for monitoring air quality and evaluating the effects of precipitation on water resources, thereby supporting urban planning and environmental management. Reliable weather observation strengthens resilience to climate impacts and fosters the well-being and safety of societies globally.

3.3.2 Implication of Weather Observation

Weather observation is critical for public safety, agriculture, transportation, and environmental management because it provides vital information on atmospheric conditions such as temperature, humidity, wind speed, and precipitation. Continuous monitoring using advanced technologies such as satellites, radar systems, and ground-based meteorological stations allows for accurate forecasting of both short-term weather events and long-term climate trends. Real-time data is critical for providing early warnings about severe weather events such as cyclones and floods, allowing communities to prepare and mitigate potential losses. Precise weather data informs farmers' decisions on planting, irrigation, and harvesting, maximizing crop yields while reducing losses. The transportation industry relies on weather observations to ensure safe navigation and increase operational efficiency by avoiding hazardous conditions. Furthermore, weather observation is critical for monitoring air quality and assessing the effects of precipitation on water resources, which helps with urban planning and environmental management. As a result, reliable weather observation improves resilience to climate impacts while also promoting global well-being and safety.

3.4 Biogeochemical Observations

A biogeochemical study along the coastline examines the cycling and distribution of nutrients, elements, and organic materials across biological, geological, and chemical processes. Coastal regions are notably dynamic due to the interplay of terrestrial, marine, and atmospheric processes, which affect nutrient concentrations, organic carbon cycling, and the availability of diverse chemical components. This review delineates the typical subjects of examination in such investigations, along with the associated instruments and procedures employed.

- **Nutrient Cycling**: Coastal regions are essential for the cycling of nutrients like nitrogen, phosphorus, and silicon, which affect primary productivity, including phytoplankton proliferation. Nutrient inputs originate from rivers, terrestrial runoff, and anthropogenic sources, rendering nutrient balance studies crucial for comprehending the hazards of eutrophication and hypoxia.
- **Carbon Cycling**: Coastal habitats such as mangroves, salt marshes, and seagrasses function as carbon sinks, sequestering organic carbon in sediments. Examining the transfer of carbon among these systems, oceanic waters, and the atmosphere is essential for comprehending coastal impacts on the global carbon budget.
- **Oxygen Dynamics**: Oxygen concentrations in coastal waters vary as a result of biological processes and water stratification. Hypoxic zones frequently result from nutrient surpluses, adversely affecting marine organisms. Research examines the causes, magnitude, and seasonal variations in oxygen dynamics to guide the management of coastal waterways.
- **Sediment and Trace Element Transport**: The transport of sediments and trace metals, such as iron, copper, and zinc, influences coastal productivity and may act as markers of pollution. Studies on sediment movement frequently investigate the origins, trajectories, and deposits of materials that affect coastal morphology and ecology.
- **Interactions with Biological Systems**: Biogeochemical studies are intricately connected to ecosystems, investigating the roles of microorganisms, plants, and animals in contributing to or being influenced by these cycles. Phytoplankton's

nutrient absorption directly influences nutrient availability, whereas microbial communities decompose organic materials and recycle nutrients.

3.4.1 Physicochemical Observation and Principles

A. **Potential Hydrogen Ion (pH)**

The pH of marine waters is essential for sustaining the health and equilibrium of marine ecosystems. It impacts the chemistry of saltwater, affecting the availability of vital nutrients, the solubility of gases such as CO_2 (Bakker et al. 2014), and the biological processes of marine animals.

Importance of pH in Marine Waters

Marine species, such as corals and plankton, depend on constant pH levels for the formation of calcium carbonate shells and skeletons. Ocean acidification, resulting from elevated CO_2 levels, disrupts food webs, diminishes biodiversity, and adversely affects sensitive species. The pH level affects nutrient speciation, bioavailability, and the exchange of gases such as CO_2 and O_2 between the ocean and the atmosphere. Acidic environments can hinder respiration, reproduction, and growth, potentially resulting in population decreases.

Methodology for Measuring pH in Marine Waters

Accurate pH measurement in marine environments is vital for understanding ocean chemistry and the effects of anthropogenic changes. Various methodologies are used:

1. pH is quantified using diverse methods, such as glass electrodes, solid-state sensors, spectrophotometers, autonomous pH sensors, and discrete water sampling. Glass electrodes are utilized in laboratory environments and field applications. Conversely, solid-state sensors exhibit durability and precision. Spectrophotometric techniques quantify absorbance variations in saltwater samples, whereas autonomous pH sensors track pH levels across time and depth. Discrete water samples are suitable for high-precision pH analysis in laboratories.
2. **Calibration and Standardization**
   - Calibration of pH sensors is done using certified standard buffers (pH: 4.0, 7.0, 10.0), typically based on the NBS (National Bureau of Standards) scale or the total hydrogen ion concentration scale (total scale). The total scale is preferred for marine measurements since it accounts for ionic strength variations in seawater.

Continuous monitoring of ocean pH, particularly due to global climate change and rising CO_2 concentrations, is crucial for forecasting future effects on marine ecosystems and biogeochemical cycles.

B. **Dissolved Oxygen (DO)**

Dissolved oxygen (DO) in marine environments is essential for the survival and functioning of aquatic organisms and ecosystems. It is pivotal in numerous biogeochemical cycles, encompassing respiration, decomposition, and nutrient cycling. This document provides an overview of the significance of dissolved oxygen and the techniques employed for its measurement in marine ecosystems.

Importance of Dissolved Oxygen in Marine Waters

Numerous marine animals, encompassing fish, invertebrates, and plankton, depend on dissolved oxygen for respiration. Sufficient quantities of dissolved oxygen are crucial for sustaining metabolic functions, development, and reproduction. Reduced amounts of dissolved oxygen, referred to as hypoxia, can induce stress, diminish growth rates, and even result in mass mortality among marine organisms. The oxygen concentration of water significantly affects the distribution and abundance of marine life. Regions with elevated dissolved oxygen levels generally sustain abundant and varied ecosystems, but hypoxic zones can result in 'dead zones', where the majority of higher creatures are unable to survive. Oxygen levels also influence predator–prey dynamics, migratory behaviours, and habitat preferences for numerous species.

1. **Biogeochemical Processes**:
   - Dissolved oxygen (DO) is essential in biogeochemical activities, including the breakdown of organic materials by aerobic microorganisms. These mechanisms are essential for nutrient recycling and the preservation of water quality.
   - In oxygen-rich environments, aerobic microorganisms effectively decompose organic materials. In oxygen-deficient conditions, anaerobic processes, like as denitrification, predominate, modifying food availability and generating detrimental by-products such hydrogen sulphide (H_2S).
2. **Ocean Circulation and Climate**:
   - Dissolved oxygen concentrations can indicate variations in oceanic circulation and climate. Oxygen minimum zones (OMZs) arise in areas with inadequate circulation, where biological oxygen consumption surpasses replenishment by physical mixing.
   - Global climate change is anticipated to elevate the prevalence and scope of hypoxic and anoxic zones, hence exacerbating stress on marine ecosystems.
3. **Fisheries and Economic Impact**:

   Oxygen concentrations influence the distribution and prevalence of fish and other economically significant species. Hypoxic conditions can result in fish mortality, species displacement, and the degradation of fishing areas, adversely affecting the lives of those reliant on fisheries.

Methodology for Measuring Dissolved Oxygen in Marine Waters

The precise measurement of dissolved oxygen (DO) is crucial for comprehending the health of marine ecosystems and the effects of anthropogenic activities such as pollution and climate change. Diverse techniques and technologies are employed to quantify dissolved oxygen in marine ecosystems:

The Winkler method is a standard technique for measuring dissolved oxygen concentration in water samples. Optical dissolved oxygen (DO) sensors, or optodes, are contemporary, non-invasive techniques for measuring dissolved oxygen levels. These sensors are utilized on autonomous platforms to deliver continuous profiles and time-series data, enhancing worldwide monitoring networks. Consistent calibration guarantees data precision, particularly for extended deployments.

Applications of DO Monitoring

Monitoring dissolved oxygen levels helps evaluate the impact of anthropogenic activities, including nutrient runoff (eutrophication) and pollution, on marine ecosystems. Variations in dissolved oxygen levels, particularly the proliferation of oxygen minimum zones, signify ocean warming and modified circulation patterns associated with climate change. Data assist in the management of fisheries by offering insights on habitat appropriateness for species and forecasting hypoxia-induced stress episodes. Dissolved oxygen is essential for marine organisms and ecological processes, and precise monitoring is crucial for comprehending environmental alterations, directing conservation initiatives, and overseeing fisheries management.

C. **Biochemical Oxygen Demand (BOD)**

Biochemical oxygen demand (BOD) is a crucial indicator for assessing organic contamination and water quality in both marine and freshwater ecosystems. It quantifies the dissolved oxygen necessary for aerobic microbes to decompose organic materials in a water sample. Elevated BOD levels indicate a significant presence of organic contaminants, potentially resulting in oxygen deprivation and negatively affecting aquatic organisms. BOD is extensively utilized in environmental monitoring, wastewater treatment, and water quality evaluation. The significance of BOD in maritime environments is as follows:

Indicators of Organic Pollution:

BOD directly quantifies the concentration of biodegradable organic matter in water, including plant detritus, sewage, and industrial effluents. Elevated BOD levels generally signify significant contamination from organic origins, such wastewater discharge, agricultural runoff, or decomposing algal blooms. Microorganisms consume oxygen while decomposing organic material. Elevated BOD levels can diminish dissolved oxygen, resulting in hypoxic (low oxygen) or anoxic (absence of oxygen) situations. This can induce stress or mortality in marine organisms such as fish, shellfish, and other aerobic species, hence diminishing biodiversity and disrupting food webs.

Methodology for Measuring Biochemical Oxygen Demand (BOD)

BOD is generally assessed by comparing the dissolved oxygen levels in a water sample prior to and following a designated incubation period. The typical incubation period is 5 days, referred to as BOD_5; however, alternative durations may be employed based on the study's specifications. Multiple techniques exist for assessing BOD in marine environments:

The Standard BOD_5 Method (5-Day BOD Test) involves collecting and diluting water, measuring initial dissolved oxygen concentration, incubating at 20 °C for 5 days, and calculating BOD. Results indicate low BOD (1–2 mg/L), moderate BOD (3–5 mg/L), high BOD (>5 mg/L), and very high BOD (>10 mg/L), indicating water quality, organic pollution, and potential hypoxic or anoxic conditions.

Applications of BOD Monitoring

The study examines environmental standards, the assessment of marine ecosystem health, the identification of organic pollution sources, the management of agricultural runoff, nutrient enrichment, algal bloom dynamics, and the links with oxygen demand. BOD is essential for assessing water quality, regulating pollution, and safeguarding marine ecosystems from oxygen depletion and habitat deterioration.

D. **Nutrients**

Nutrients in marine environments, including nitrogen (N), phosphorus (P), and silicon (Si), are vital for the growth and production of marine organisms, particularly primary producers such as phytoplankton. These nutrients are essential to marine biogeochemical cycles and substantially influence ecosystem health, productivity, and biodiversity.

Essential Nutrients and Their Importance in Marine Waters

Nitrogen, an essential ingredient for phytoplankton proliferation, frequently serves as the limiting nutrient in marine environments. Excessive nitrogen influx from agriculture, sewage, and air deposition can lead to imbalances, eutrophication, toxic algal blooms, and oxygen depletion. Phosphorus, a vital component for ATP, DNA, and cellular membranes, can induce eutrophication and disturb ecosystems. Silicon, vital for diatoms, is crucial for modulating diatom output. Variations in silicon availability may alter phytoplankton community structure, potentially benefiting hazardous algae species.

Ecosystem Functions and Impacts of Nutrients

Nutrients are essential for phytoplankton growth, which underpins the marine food web, and their productivity is intricately linked to their balance. These nutrients experience biological and physical modifications, preserving nutritional availability and regulating distribution. Nutrient imbalances can alter phytoplankton community composition, resulting in toxic algal blooms, toxin production, fishery closures, and economic detriment.

Methodology for Measuring Nutrients in Marine Waters

Assessing nutrient concentrations in marine waters is essential for comprehending water quality, nutrient cycling, and the impacts of anthropogenic activities on marine ecosystems. Numerous nutritional analysis approaches encompass in situ sampling, laboratory analysis, and sophisticated sensor technology.

Nutrient samples are obtained using Niskin bottles affixed to a CTD rosette, thereafter filtered and preserved to avert contamination. Nutrient sensors are installed on moorings, buoys, or autonomous platforms to measure nutrient amounts in real time. Autonomous nutrient profilers deliver vertical profiles of nutrient concentrations, indicating variations in nutrient availability with depth. Nutrient auto analysers employ colorimetric techniques to quantify nutrient contents in water samples, whereas spectrophotometric methods assess nutrient levels through reactions with reagents. These approaches are vulnerable and extensively employed in oceanography. Dissolved inorganic nutrients (NO3−, NO2−, NH4 +, SiO4^4−, PO4^3−) were monitored (Strickland and Parsons 1972; Grasshoff et al. 1999).

Applications of Nutrient Monitoring

Nutrient monitoring is crucial for pinpointing sources of nutrient contamination and regulating eutrophication. By regulating nutrient inputs (e.g. via wastewater treatment and agricultural management), ecosystems can be safeguarded against nutrient over-enrichment. Nutrient monitoring evaluates the influence of alterations in ocean circulation and temperature on nutrient availability and primary productivity, offering insights into the effects of climate change on marine ecosystems. Comprehending nutrient dynamics is essential for the management of fisheries and aquaculture. The availability of nutrients affects the growth and distribution of phytoplankton, which are the foundation of the marine food web and sustain fish populations.

E. **Heavy Metals in Water and Sediment**

Heavy metals in marine waters and sediments represent a significant environmental issue owing to their potential toxicity, persistence, and bioaccumulation within marine ecosystems. In contrast to organic pollutants, heavy metals are persistent contaminants that do not disintegrate over time, hence posing dangers to marine organisms and human health. The principal heavy metals of concern in marine ecosystems comprise mercury (Hg), lead (Pb), cadmium (Cd), arsenic (As), chromium (Cr), copper (Cu), zinc (Zn), and nickel (Ni).

Critical Heavy Metals and Their Impacts

Mercury, lead, cadmium, arsenic, copper, zinc, and nickel are hazardous chemicals that infiltrate marine habitats via industrial operations, coal combustion, mining, and incorrect disposal. Mercury can be transformed into methylmercury, a very poisonous variant that can hinder neurological development in marine species and humans. Lead can impact the neurological, renal, and reproductive systems of marine species, particularly invertebrates and fish. Cadmium, originating from mining, smelting, industrial waste, and fertilizers, can additionally infiltrate marine ecosystems via air deposition and runoff. Arsenic contamination originates from industrial activities, mining, pesticide use, and naturally occurring deposits in geological formations and sediments. Copper, zinc, and nickel are vital trace elements for marine species; nevertheless, at increased concentrations, they become hazardous. Chromium, emitted from industrial activities, can induce cancer, harm respiratory and digestive systems, and bioaccumulate in fish, presenting hazards to predators and humans.

Ecological and Biological Effects of Heavy Metals

Heavy metals accumulate in marine creatures, especially in filter feeders such as mussels and clams, resulting in biomagnification in apex predators like fish, marine mammals, and seabirds. Humans face the risk of exposure to polluted seafood, specifically methylmercury and cadmium. Heavy metals can impede phytoplankton proliferation, disturb benthic species, and induce developmental anomalies in fish and marine mammals. Sediments function as reservoirs for metals, polluting benthic habitats and impacting the ecology.

Methodology for Measuring Heavy Metals in Marine Waters and Sediments

The analysis of heavy metals in marine environments involves sampling water and sediments, followed by laboratory analysis using various chemical techniques. Below are the commonly used methodologies:

Water samples are collected with Niskin bottles, filtered to eliminate particle matter, and preserved in acid-washed containers. Sediment samples are obtained using grabs or corers, thereafter dried, crushed, and sieved prior to chemical analysis. Laboratory analysis include atomic absorption spectroscopy (AAS), inductively coupled plasma mass spectrometry (ICP-MS), inductively coupled plasma optical emission spectrometry (ICP-OES), and cold vapour atomic absorption (CVAA) for mercury detection. Atomic absorption spectroscopy (AAS) quantifies metal concentrations by examining the light absorption of metal atoms in a sample at certain wavelengths. ICP-MS exhibits exceptional sensitivity and can concurrently identify trace concentrations of several metals. CVAA is optimal for assessing mercury pollution in marine ecosystems. Analysis of heavy metals adheres to established protocols (Persaud et al. 1993; MacDonald et al. 2000; Simpson et al. 2013).

Applications of Heavy Metal Monitoring

Monitoring heavy metals in marine ecosystems is essential for detecting sources of contamination, evaluating pollution control strategies, and informing regulatory frameworks. It assesses ecological hazards, dredging initiatives, and the disposal of contaminated substances. Surveillance is crucial for seafood safety, particularly for hazardous metals such as mercury and cadmium. Precise metal monitoring is essential for preserving marine ecosystems, controlling pollution, and ensuring human health.

3.4.2 Geology

The analysis of sediment texture and geochemistry is crucial in identifying this extrinsic characteristic in ancient systems, utilizing element ratios to ascertain tectonic settings and examining the correlations between grain size in fluvial systems, subsidence, and sediment supply (Roser and Korsch 1986; Duller et al. 2010). Textural analyses of surface and core sediments yield critical insights into deposition, erosion, and sedimentation throughout the marine ecosystem.

The textural analysis yields metrics such as sediment mean, median, sorting, skewness, kurtosis, and D50, which are widely utilized in marine engineering for the development of structures like ports, sea barriers, dykes, and offshore platforms. Various instruments are employed to analyse the textural characteristics of marine sediment, including GRADISAT and grain size trend analysis within the GIS platform. Numerous sediment textural investigations have been conducted to elucidate beach dynamics (Dora et al. 2014; Pradhan et al. 2020a, b), inlet stabilization (Sahoo et al. 2023), and nearshore sediment transport (Pradhan et al. 2022b; Pradhan et al. 2025). The textural analysis can be further integrated for modelling and environmental impact assessment studies to assess water nutrients and biological and microbiological interactions.

A. **Sediment Organic Carbon (SOC)**

In marine ecosystems, sediment organic carbon (SOC) is essential because it affects several ecological and biogeochemical processes. This is a comprehensive overview of its significance and the methods by which it is evaluated.

Importance of Sediment Organic Carbon

Sediment organic carbon is an essential component of the global carbon cycle, functioning as both a carbon sink and a nutrition supply for benthic animals and phytoplankton. It enhances sediments, sustains varied ecosystems, and affects species dispersion. Soil organic carbon can sequester contaminants and stabilize sediments, so avoiding erosion and preserving water purity. Variations in SOC levels may signify modifications in sedimentation rates, organic matter influx, and human-induced effects.

Methodology for Assessing Sediment Organic Carbon

Sampling Techniques

Sediment sampling encompasses grab sampling and core sampling for vertical profiling. Laboratory analysis offers loss on ignition, elemental combustion analysis, and wet chemistry techniques. Geochemical analysis encompasses stable isotope analysis and molecular techniques. Field measurements encompass sediment profile imaging and in situ sensors. These techniques facilitate the identification of organic carbon sources, evaluation of biogeochemical processes, and visualization of sedimentary layers and structures. The sediment organic carbon (SOC %) was quantified using dried sediment samples via the chromic acid oxidation method (Gaudette et al. 1974; Walkley and Black 1934).

Effective ecosystem management and climate change mitigation rely on comprehending the significance of sediment organic carbon in marine environments. Accurate evaluation methods are crucial to assess SOC's function and consequences, hence facilitating informed decision-making in resource management and marine conservation. Researchers can enhance the monitoring and protection of maritime environments by integrating ecological expertise with technology advancements.

3.4.3 Biological Parameters

A. Phytoplankton

Phytoplankton, the marine food chain's foundation, is a primary energy producer through photosynthesis, sustaining various marine organisms. They play a crucial role in the global carbon cycle, absorbing atmospheric CO_2 and regulating greenhouse gases. They produce 50% of the world's oxygen, and their health and abundance are critical for maintaining oxygen levels. Monitoring phytoplankton can reveal early signs of ocean stress. Phytoplankton communities respond rapidly to environmental changes, such as shifts in temperature, nutrient availability, and pH. Monitoring phytoplankton can reveal early signs of ocean stress, including harmful algal blooms (HABs), eutrophication, and shifts in oceanic conditions linked to climate change. Fisheries and aquaculture depend on the productivity of phytoplankton, which supports higher trophic levels. Changes in phytoplankton communities can affect fish stocks and marine resources vital to human populations.

Methodologies for Studying Phytoplankton

1. Microscopy:
    - Light Microscopy: A traditional method for identifying and quantifying phytoplankton species. It allows for direct observation of morphology and taxonomy.
    - Electron Microscopy: Offers higher resolution to study detailed cell structures, especially for small or cryptic phytoplankton species.

2. Flow Cytometry:

- Flow cytometers measure individual phytoplankton cells' optical properties (e.g., fluorescence and light scattering) in suspension. This technique allows for rapid quantification and sorting of different types of phytoplankton based on size and pigment content.

Phytoplankton species are identified under the microscope, and cells were enumerated using the Sedgwick rafter cell and following the standard manuals (Sournia 1970; Subrahmanyan 1971; Desikachary 1987).

B. **Chlorophylls**

Chlorophyll-a is an essential pigment in marine phytoplankton that facilitates photosynthesis, rendering it a fundamental component of marine ecosystem. Research on chlorophyll is essential for comprehending marine production, biogeochemical cycles, and environmental alterations. This document outlines the significance and procedures associated with chlorophyll research in marine ecosystems.

Methodologies for Studying Chlorophyll in Marine Environments

Chlorophyll sampling involves the collection of seawater and the extraction of chlorophyll-a utilizing solvents like acetone or methanol. The quantification of chlorophyll-a is conducted via spectrophotometry or fluorometry. Satellite remote sensing employs ocean colour sensors to assess chlorophyll concentrations in surface waters. Fluorometers and profiling instruments, including in situ fluorometers and continuous flow-through systems, quantify chlorophyll concentration at many depths in real time. High-performance liquid chromatography (HPLC) isolates and measures chlorophyll and other photosynthetic pigments in phytoplankton. Optical sensors on autonomous platforms, including gliders, autonomous underwater vehicles, and aircraft, quantify chlorophyll fluorescence, yielding essential data on chlorophyll distribution and primary productivity. Spectral fluorometry distinguishes phytoplankton taxa by distinct light wavelengths, facilitating the comprehension of phytoplankton diversity and the photosynthetic processes of microbial communities.

Applications of Chlorophyll Studies

Chlorophyll studies help understand marine ecosystem dynamics, predict fisheries, monitor environmental changes, and predict climate change impacts. They reveal patterns in ocean productivity, aid in managing resources, and are crucial in global biogeochemical models. By studying chlorophyll, researchers can better understand marine primary production, ecosystem health, and phytoplankton's influence on global processes like the carbon cycle.

C. **Zooplankton**

Zooplankton are essential in marine ecosystems, serving as intermediates between primary producers and higher trophic levels. Their research is crucial for comprehending marine food webs, nitrogen cycling, and ecosystem dynamics. Zooplankton plays a role in carbon sequestration, nutrient cycling, and vertical migrations. Their quantity and composition affect the success of fish larvae and young fish, and their dynamics can aid in forecasting fish population variations, evaluating overfishing effects, and informing sustainable fisheries management. Comprehending zooplankton dynamics is essential for forecasting fish stock variations and managing sustainable fisheries.

Methodologies for Studying Zooplankton

Zooplankton are typically collected by many techniques, including as conventional nets, vertical tows, oblique tows, and multiple net systems. Light microscopy is employed to identify zooplankton species and assess their abundance. Long-term monitoring programs and automated samplers gather data to examine variations in abundance, variety, and community structure throughout time. Experimental methodologies encompass laboratory experiments to examine zooplankton feeding behaviour and reproduction under many settings, as well as mesocosm experiments to analyse zooplankton populations in semi-natural environments. These methodologies assist researchers in comprehending zooplankton reactions to environmental alterations and their significance in oceanic health. The taxonomic content of the zooplankton samples was assessed in accordance with identification guidelines by Kasturirangan (1963), Conway et al. (2003), and Yamani and Prusova (2003).

Applications of Zooplankton Studies

Zooplankton research aids in understanding the energy and nutrient transfer through marine food webs, predicting fish species recruitment and survival rates, and assessing marine ecosystems' health and productivity. It helps resource managers make informed decisions about sustainable fishing practices and ecosystem management strategies, enhancing our understanding of the ocean's capacity to support life on Earth.

D. **Benthos**

Benthos, or benthic organisms, inhabit the seafloor (the benthic zone). Their composition encompasses a variety of organisms, ranging from minute bacteria and meiofauna to bigger species like as corals, sponges, and burrowing invertebrates. Investigating benthic populations is essential for comprehending the architecture and dynamics of marine ecosystems and for monitoring environmental alterations. This document provides an overview of the significance of benthos studies and the approaches employed in benthic research.

Importance of Benthos Studies in the Marine Environment

Benthic organisms are essential for nutrient cycling and ecosystem functioning, as they recycle organic debris that descends to the ocean floor. They engage in bioturbation, sediment mixing, and oxygenation, which facilitates the decomposition of organic matter and affects sediment characteristics. Benthic habitats, including coral reefs, seagrass meadows, and hydrothermal vents, are biodiversity hotspots that sustain a diverse array of species and are essential for preserving marine biodiversity and ecosystem services. Benthic ecosystems significantly contribute to carbon sequestration, functioning as a carbon sink and alleviating climate change. Comprehending the carbon dynamics of benthic systems is crucial for forecasting future environmental alterations and assessing the ocean's contribution to global carbon cycle. Benthic studies are essential for the management and conservation of fisheries, as deterioration may result in diminished fish populations and threaten food security.

Methodologies for Studying Benthos in Marine Environments

Benthic community research employs diverse techniques, including Grab Samplers, Box Corers, and Multi-Corers. These instruments gather surface and subsurface sediments, encompassing deeper-dwelling creatures, for examination. Dredging and trawling harvest larger animals from the seabed, whereas benthic chambers assess respiration and nutrient fluxes. In situ incubations yield insights into carbon remineralization, oxygen utilization, and nutrient exchange. Benthic chambers quantify sediment oxygen demand, yielding real-time data on benthic respiration rates. These strategies assist researchers in comprehending the composition, abundance, and features of benthic communities. Macrobenthic taxa are classified according to established manuals (Fauvel 1953; Fauchald and Jumars 1979; Day 1967; Lang 1950; Kasturirangan 1963).

Applications of Benthos Studies:

- Ecosystem Assessment and Management: Benthic studies provide the foundation for understanding ecosystem health and functionality. They are crucial for developing conservation strategies, managing fisheries, and protecting sensitive marine habitats like coral reefs and seagrass meadows.
- Pollution Monitoring: Benthic organisms are often used as bioindicators for assessing the impacts of pollutants (e.g. heavy metals, hydrocarbons, microplastics) on marine environments. Changes in benthic community structure and function provide valuable information on ecosystem stress.
- Climate Change Research: Studying how benthic ecosystems respond to ocean acidification, warming, and sea-level rise is essential for predicting the broader effects of climate change on marine ecosystems.

E. **Microbiology**

Marine microbiology examines the varied microorganisms (bacteria, archaea, viruses, protists, and fungi) residing in oceanic environments. These bacteria are essential in marine ecosystems, facilitating biogeochemical cycles, sustaining food webs, and impacting global climate. Marine microbiology is crucial for comprehending the operational dynamics of the ocean and its responses to environmental alterations.

Here is an overview of the importance of microbiology studies in the marine environment and the methodologies used to study marine microorganisms:

Importance

Marine microorganisms, such as bacteria and phytoplankton, are crucial for marine food webs, transforming sunlight and inorganic substances into organic matter via photosynthesis and chemosynthesis. Heterotrophic bacteria decompose organic matter, facilitating nutrient recycling for energy transfer. Marine microorganisms are integral to the cycling of vital elements, including carbon, nitrogen, sulphur, and phosphorus. They serve as sensitive sensors of oceanic health and can rapidly adjust to environmental alterations. Certain bacteria are essential in bioremediation, decomposing contaminants such as oil and plastic. The diversity of marine microorganisms is crucial for comprehending ecological resilience and evolutionary mechanisms. Certain marine microorganisms engage in symbiotic partnerships with other species, facilitating essential tasks such as nutrient cycling and disease defence. Marine microorganisms possess potential applications in pharmacological research, biotechnology, and environmental stewardship.

Methodologies for Studying Marine Microorganisms

The microbiology of marine organisms entails the collection of water samples, filtration, and sediment sampling to examine their variety. Epifluorescence and electron microscopy techniques are employed to visualize and enumerate microorganisms, whereas culture-based methods such as selective culture and enrichment cultures are utilized to investigate specific groups. Molecular methodologies such as DNA/RNA extraction, PCR, metagenomics, metatranscriptomics, and metaproteomics are employed to examine microbial community composition and functionality. Environmental DNA (eDNA) is progressively utilized to assess microbial biodiversity and identify diseases or unusual species. These methodologies assist researchers in comprehending the diversity of marine microorganisms and their reactions to environmental factors. Characterization of each bacterial group according to Bergey's handbook and the American Public Health Association (APHA 1998).

Applications of Marine Microbiology Studies

- Marine microbes are crucial in understanding global biogeochemical cycles, monitoring ocean health, and developing bioremediation strategies. They are also valuable resources for discovering new drugs, enzymes, and bioactive compounds.
- Marine biotechnology is essential for discovering new drugs, enzymes, and bioactive compounds, and their sensitivity to environmental shifts makes them crucial for studying climate change impacts.

4 Conclusion

Accurate measurement of coastal and ocean environment and generating reliable data are always a challenging task for the surveyor and requires through understanding of the survey method. To supplement that information for a resiliency society in the era of Climate change, this article provides a comprehensive framework for coastal oceanographic monitoring, offering a systematic methodology for planning, coordination, and implementation. The framework will improves the ability to monitor dynamic coastal processes and tackle environmental concerns through the integration of new technology and the promotion of collaboration. Ocean monitoring is a crucial instrument for managing and mitigating the marine environment's reaction to anthropogenic activities and natural threats, various application of environment variables are described for its best use. Further, considering the significant of UN sustainable development goals (SDGs) urged governments globally to allocate additional resources for continuous ocean and coastal observation; therefore, standard operating procedure for enhancing knowledge of marine environment observation is framed.

Advancements in oceanographic technologies, including remote sensing, autonomous underwater vehicles (AUVs), and satellite-based monitoring systems, alongside innovative biological tools and ecological modelling, provide exceptional capabilities for real-time data collection and analysis. These technologies enable comprehensive evaluations of environmental alterations, biodiversity transitions, and ecosystem vitality, offering essential insights for adaptive management approaches. With the intensification of global environmental pressures, the integration of interdisciplinary approaches is essential for the preservation of coastal ecosystems, the enhancement of sustainable resource management, and the mitigation of climate change impacts on marine life. A comprehensive and data-informed strategy will be essential for maintaining the resilience of coastal and marine ecosystems amid continuous environmental changes.

Consequently, enhanced marine environment monitoring supports the Sustainable Development Goals (SDGs) and the blue economy, resulting in an increased number of blue-flag beaches and climate research. The improved understanding of the ocean, resulting from decades of scientific effort and the recognized need for continuous collaboration in observing, analysing, comprehending, and predicting oceanic phenomena, will aid in grasping coastal oceanographic processes in the context of climate change and in fostering resilient adaptations for sustainable resource management.

References

Aguzzi J, Thomsen L, Flögel S, Robinson NJ, Picardi G, Chatzievangelou D, Bahamon N, Stefanni S, Grinyo J, Fanelli E, Corinaldesi C, Fernandez JDR, Calisti M, Mienis F, Chatzidouros E, Costa C, Violino S, Tangherlini M, Danovaro R (2023) New technologies for monitoring and upscaling marine ecosystem restoration in deep-sea environments. Engineering 34:195–211. https://doi.org/10.1016/j.eng.2023.10.012

Amani M, Moghimi A, Mirmazloumi SM, Ranjgar B, Ghorbanian A, Ojaghi S, Ebrahimy H, Naboureh A, Nazari ME, Mahdavi S et al (2022) Ocean remote sensing techniques and applications: a review (Part I). Water 14:3400. https://doi.org/10.3390/w14213400

Anderson DM et al (2012) Harmful algal blooms and eutrophication: examining linkages from selected coastal regions of the United States. Estuaries Coasts 35(4):856–877
APHA (1998) Standards methods for the examination of water and wastewater (20th ed.)
Armenio E, Ben Meftah M, De Padova D, De Serio F, Mossa M (2019) Monitoring systems and numerical models to study coastal sites. Sensors 19:1552. https://doi.org/10.3390/s19071552
Bakker DCE et al (2014) An update to the surface ocean CO_2 Atlas (SOCAT version 2). Earth Syst Sci Data 6(1):69–90
Balan-Sarojini B, Balmaseda MA, Vitart F, Roberts CD, Zuo H, Tietsche S, Mayer M (2024) Impact of ocean in-situ observations on ECMWF sub-seasonal forecasts. Front Mar Sci 11:1396491. https://doi.org/10.3389/fmars.2024.1396491
Browman H, Stergiou K, Cury P, Hilborn R, Jennings S, Lotze H, Zeller D (2004) Perspectives on ecosystem-based approaches to the management of marine resources. Mar Ecol Prog Ser 274:269–303. https://doi.org/10.3354/meps274269
Brown T, Lee J, Zhang Y (2023) Monitoring coastal circulation and currents with drifters and current meters. J Coastal Res 40(2):145–160
Brown AB, Anderson KE (2022) Comprehensive survey strategies for coastal ecosystems. Mar Data Rev
Brown AB, Anderson KE, Smith J (2022) Data quality control in oceanographic research. J Mar Syst
Caddell R (2016) International environmental governance and the final frontier: the protection of vulnerable marine ecosystems in deep-sea areas beyond national jurisdiction. Yearbook Int Environ Law 27:28–63
Chassignet EP et al (2007) The HYCOM (hybrid coordinate ocean model) data assimilative system. J Mar Syst 65(1–4):60–83
Chen X, Zhang Y, Wang J (2023a) Real-time monitoring of coastal water quality using sensor networks. J Coast Res 39(2):115–128
Chen X, Zhao Y, Li Q (2023b) Discharge measurements and their impact on coastal salinity and nutrients. Mar Technol Soc J 58(3):92–106
Cloern JE et al (2016) Perspectives on the causes of harmful algal blooms. Environ Microbiol 18(5):1307–1318
Conway DVP, White RG, Hugues-Dit-Ciles J, Gallienne CP, Robins DB (2003) Guide to the coastal and surface zooplankton of the South-Western Indian Ocean. Occasional Publication of the Marine Biological Association of the United Kingdom, No. 15. Plymouth, UK, p 354
Crain CM, Kroeker K, Halpern BS (2008) Interactive and cumulative effects of multiple human stressors in marine systems. Ecol Lett 11:1304–1315. https://doi.org/10.1111/j.1461-0248.2008.01253.x
Crise AM, Ribera D, Alcala M, Patrizio M, George P, Julie R et al (2018) A conceptual framework for developing the next generation of Marine OBservatories (MOBs) for science and society. Front Mar Sci 5:318. https://doi.org/10.3389/fmars.2018.00318
Da Ros Z, Dell'Anno A, Morato T, Sweetman AK, Carreiro-Silva M, Smith CJ et al (2019) The deep sea: the new frontier for ecological restoration. Mar Policy 108:103642
Dadou I et al (2018) Coastal ocean monitoring with autonomous systems. J Mar Sci Eng 6(1):23–34
Day JH (1967) A monograph on the polychaeta of Southern Africa. Sedentaria Published by the Trustees of the British Museum, Natural History, pp 1–786
Desikachary TV (1987) Atlas of diatoms, fascicle III and IV. Madras Science Foundation. Madras, pp 222–400
Dora GU, Kumar VS, Vinayaraj P, Philip CS, Johnson G (2014) Quantitative estimation of sediment erosion and accretion processes in a micro-tidal coast. Int J Sediment Res 29(2):218–231. https://doi.org/10.1016/S1001-6279(14)60038-X
Duller RA et al (2010) From grain size to tectonics. J Geophys Res: Earth Surf 3:115
European Marine Board (2019) Managing resources for coastal and marine observing systems
Fauchald J, Jumars PA (1979) The diet of worms: a study of polychaete feeding guilds. Oceanograph Mar Biol: Annu Rev 17:193–284. https://doi.org/10.1146/annurev-marine-010814-020007

Fauvel P (1953) The Fauna of India Including Pakistan, Ceylon, Burma and Malaya. Annelida Polychaeta, Indian Press, Allahabad, vol 12, p 507

Garcia M, Brown T, Patel A (2024) Advancements in coastal data management systems for ecosystem monitoring. J Coastal Res 40(1):22–35

Garcia M, Patel A, Smith R (2024) Sedimentology studies using samplers and acoustic profilers. J Sediment Res 94(1):23–37

Gaudette HE, Wilson RF, Toner L, David WFG (1974) An inexpensive titration method for determination of organic carbon in recent sediments. J Sediment Petrol 44:249–253

Gissi E, Manea E, Mazaris AD, Fraschetti S, Almpanidou V, Bevilacqua S et al (2021) A review of the combined effects of climate change and other local human stressors on the marine environment. Sci Total Environ 755:142564

Gould J, Sloyan B, Visbeck M (2013) In situ ocean observations: a brief history, present status, and future directions. Int Geophys 1(103):59–81. https://doi.org/10.1016/B978-0-12-391851-2.00003-9

Grasshoff K, Kremling K, Ehrhardt M (Eds.) (1999) Methods of Seawater Analysis (3rd ed.). Weinheim: Wiley-VCH. https://doi.org/10.1002/9783527613984

Ho C-R, Liu AK (2023) Preface: remote sensing applications in ocean observation. Remote Sens 15:415. https://doi.org/10.3390/rs15020415

Jones DOB, Gates AR, Huvenne VAI, Phillips AB, Bett BJ (2019) Autonomous marine environmental monitoring: application in decommissioned oil fields. Sci Total Environ 668:835–853

Jones R, Smith P, Williams T (2023) Effective stakeholder engagement in coastal monitoring programs. Mar Policy 136:104321

Kaiser BA, Hoeberechts M, Maxwell KH, Eerkes-Medrano L, Hilmi N, Safa A, Horbel C, Juniper SK, Roughan M, Theux Lowen N, Short K, Paruru D (2019) The importance of connected ocean monitoring knowledge systems and communities. Front Mar Sci 6:309. https://doi.org/10.3389/fmars.2019.00309

Kasturirangan LR (1963) A key for the identification of the more common planktonic Copepoda of Indian coastal waters. Publication No. 2. In: Panikkar NK (ed), Indian National Committee on Oceanic Research. CSIR, New Delhi, p 87

Landrigan PJ, Stegeman JJ, Fleming LE, Allemand D, Anderson DM, Backer LC, Rampal P (2020) Human health and ocean pollution. Ann Glob Health 86(1):151. https://doi.org/10.5334/aogh.2831

Lang K (1950) Contribution to the systematics and synonymics of the tanaidacea. Arkiv För Zool 42:1–14

Lee J, Kim H, Chang S (2022) Assessing algal blooms: local knowledge and monitoring strategies. J Environ Manage 304:114135

Lee J, Park H, Kim S (2023) Advances in autonomous underwater vehicles for coastal ecosystem monitoring. Mar Technol Soc J 57(1):34–46

Lee H, Nguyen D, Kim S (2024) Wave dynamics and coastal erosion: insights from wave buoys and radar systems. Remote Sens Environ 275:113212

Letian M, Jiabiao L, Xuwen F, Tianyao H, Fangzhou N (2024) Marine exploration and its environmental impact assessment: insights from international standards. Front Collect. https://doi.org/10.3389/fmars.2024.1389906

Levin LA, Le Bris N (2015) The deep ocean under climate change. Science 350(6262):766–768

Lindstrom E, Gunn J, Fischer A, McCurdy A, Glover L, Alverson K et al (2012) A framework for ocean observing. By the task team for an integrated framework for sustained ocean observing. Paris: UNESCO.https://doi.org/10.5270/OceanObs09-FOO

Liu Y et al (2017) The role of advanced technologies in monitoring and managing coastal environments. Oceanography 30(3):42–51

MacDonald D, Ingersoll C, Berger T (2000) Development and evaluation of consensus-based sediment quality guidelines for freshwater ecosystems. Arch Environ Contam Toxicol 39:20–31

Marine Conservation Society (2023) Citizen science projects. Retrieved October 28, 2025, from https://www.mcsuk.org/what-you-can-do/citizen-science/

Martinez A, Li Y, Rodriguez M (2023) Data integration techniques for large-scale coastal monitoring projects. Mar Technol Soc J 58(2):78–91

Martínez A, García L, Rodriguez M (2023) Collaborative resource sharing for enhanced coastal monitoring. Ocean Sci J 12(1):45–58

Miloslavich P, Bax NJ, Simmons SE, Klein E, Appeltans W, AburtoOropeza O et al (2018) Essential ocean variables for global sustained observations of biodiversity and ecosystem changes. Glob Change Biol 24:2416–2433. https://doi.org/10.1111/gcb.14108

Mishra P, Panda US, Pradhan UK, Saravana Kumar C, Naik S, Begum M, Ishwarya J (2015) Coastal water quality monitoring and modelling off Chennai city. Proced Eng 116:955–962. https://doi.org/10.1016/j.proeng.2015.08.386

Mohanty PK, Patra SK, Bramha S, Seth B, Pradhan UK, Behera B, Mishra P, Panda US (2011) Impact of groins on beach morphology: a case study near Gopalpur Port, east coast of India. J Coastal Res. https://doi.org/10.2112/JCOASTRES-D-10-00045.1

Mora C, Wei CL, Rollo A, Amaro T, Baco AR, Billett D et al (2013) Biotic and human vulnerability to projected changes in ocean biogeochemistry over the 21st century. PLoSBiol 11(10):e1001682

National Oceanic and Atmospheric Administration (NOAA) (2020) Integrated ocean observing system (IOOS). Retrieved from https://ioos.noaa.gov

Nguyen T, Brown A, Rodriguez M (2023) Evaluating coastal erosion with satellite imagery and differential GPS. Coast Manage 51(4):267–283

Nguyen D, Kim H, Zhao X (2024a) Effective visualization tools for coastal ecosystem data analysis. Environ Monit Assess 196:134

Nguyen D, Liu X, Zhao Y (2024b) Training and capacity building for effective water quality monitoring. Environ Monit Assess 196:215

Panda US, Pradhan UK, Sujith S, Naik S, Begam M, Mishra P, Ramana Murthy MV (2020) Bathing water quality forecast for Chennai coastal waters. J Coast Res 89(sp1):139–144. https://doi.org/10.2112/SI89-023.1

Panda US, Usha T, Ramana Murthy MV (2022) Perspectives on coastal research. Geogr You 22(148):8–15

Patra SK, Mishra P, Mohanty PK, Pradhan UK, Panda US, Murthy MVR, Sanil Kumar V, Balakrishnam Nair TM (2016) Cyclone and monsoonal wave characteristics of northwestern Bay of Bengal: long-term observations and modeling. Nat Hazards. https://doi.org/10.1007/s11069-016-2233-0

Persaud D, Jaaguagi R, Hayton A (1993) Guidelines for the protection and management of aquatic sediment quality in Ontario (Toronto: Ontario Ministry of the Environment), p 27

Pradhan UK, Mishra P, Mohanty PK, Panda US, Ramanamurty MV (2020a) Modeling of tidal circulation and sediment transport near tropical estuary, East Coast of India. Reg Stud Mar Sci 37:10135. https://doi.org/10.1016/j.rsma.2020.101351

Pradhan UK, Sahoo RK, Pradhan S, Mohanty PK, Mishra P (2020b) Textural analysis of coastal sediments along east coast of India. J Geol Soc India 95:67–74. https://doi.org/10.1007/s12594-020-1387-2

Pradhan UK, Mishra P, Mohanty PK (2022a) Beach and nearshore sediment textural characteristics of a monsoonal wave dominated micro-tidal, human perturbed environment, Central east coast of India. Arab J Geosci 15:667. https://doi.org/10.1007/s12517-022-09850-5

Pradhan UK, Panda US, Naik S, Mishra P, Satish DK, Begum M, Ramana Murthy MV (2022c) Coupled hydrodynamic and water quality modelling in the coastal waters off Chennai, East coast of India. Front Mar Sci 9:987067. https://doi.org/10.3389/fmars.2022.987067

Pradhan UK, Mohanty PK, Mishra P (2022a) Coastal erosion: a threat to sea turtle nesting habitat, east coast of India, RendicontiLincei. ScienzeFisiche e Naturali. https://doi.org/10.1007/s12210-022-01046-z

Pradhan S, Mohanty PK, Samal RN, Mahanty MM, Pradhan U, Panda US, Mishra P (2025) Nearshore hydrodynamics and sediment transport modeling along the shorefront of a tropical lagoon. Reg Stud Mar Sci 81:103993. https://doi.org/10.1016/j.rsma.2024.103993

Rao VR, Murthy MVR, Bhat M, Reddy NT (2009) Littoral sediment transport and shoreline changes along Ennore on the southeast coast of India: Field observations and numerical modeling. Geomorphology, 112(1-2):158–166. https://doi.org/10.1016/j.geomorph.2009.05.015

Roser BP, Korsch RJ (1986) Determination of tectonic setting of sandstone-mudstone suites using SiO_2 content and K_2O/Na_2O ratio. J Geol 94(5):635–650

Sahoo RK, Mohanty PK, Pradhan S, Barik S, Pradhan UK (2023) Assessing stability of tidal inlet in a tropical coastal lagoon. Arab J Geosci 16:319. https://doi.org/10.1007/s12517-023-11394-1

Simpson SL, Batley GB, Chariton AA (2013) Revision of the ANZECC/ARMCANZ sediment quality guidelines. CSIRO Land Water Sci Rep 8(07):128

Smith R, Brown A, Johnson L (2023) Integration of remote sensing and numerical modeling in coastal monitoring. Environ Monit Assess 195:278

Smith J, Johnson L (2021) Advanced techniques for data integration in coastal surveys. Mar Data Rev

Sournia A (1970) Les Cyanophytesdans le plancton. Marin L Anne Biol. 9:63–76

Strickland JDH, Parsons TR (1972) A practical handbook of seawater analysis, The Fisheries Research Board of Canada. (2nd Ed.), Vol 310, p 167

Subramanyan R (1971) The dinophycaes of Indian Seas Part—II. Peridiniaceae. J Mar Biol Assoc India 334

Sujith Kumar S, Panda US, Pradhan UK, Mishra P, Ramana Murthy MV (2020) Web-based decision support system for coastal water quality. J Coast Res 89:139–144. ISSN 0749-0208

Tanhua T, McCurdy A, Fischer A, Appeltans W, Bax N, Currie K, DeYoung B, Dunn D, Heslop E, Glover LK, Gunn J, Hill K, Ishii M, Legler D, Lindstrom E, Miloslavich P, MoltmannT NG, Palacz A, Simmons S, Sloyan B, Smith LM, Smith N, Telszewski M, Visbeck M, Wilkin J (2019) What we have learned from the framework for ocean observing: evolution of the global ocean observing system. Front Mar Sci 6:471. https://doi.org/10.3389/fmars.2019.00471

Venkatesan R, Ramesh K, Kishor A, Vedachalam N, Atmanand MA (2018) Best practices for the ocean moored observatories. Front Mar Sci 5:469

Veeravalli SG, Balaganesh S, Silamban D, Alluri SKR, Ramanathan V, Panda, US (2023) UAV based Topographic Survey of Inaccessible Remote Terrains, 2023 IEEE India Geoscience and Remote Sensing Symposium (InGARSS 2023), pp 1–4. https://doi.org/10.1109/InGARSS59135.2023.10490411

Walkley A, Black IA (1934) An examination of the Degtjareff method for determining soil organic matter, and a proposed modification of the chromic acid titration method. Soil Sci 37(1):29–38

Wang M, Liu Q, Zhang H (2022) Satellite and aerial remote sensing for coastal water quality assessment. Remote Sens Environ 268:112869

Williams J, Anderson K, Brown T (2023) Sea level variability: tide gauges and satellite altimeters. Ocean Sci J 15(1):101–115

Yamani FY, Prusova I (2003) Common copepods of the Northwestern Arabian Gulf: identification guide. Kuwait Institute of Scientific Research, p 162

Zhang J et al (2019) Advances in remote sensing technology for coastal monitoring. Remote Sens 11(12):1437

Zhao Y, Wang L, Chen X (2023) Statistical methods for analyzing coastal water quality and ecosystem data. Remote Sens Environ 270:113032

Zhao X, Wang J, Liu Q (2024) Continuous monitoring of coastal water quality using autonomous sensor networks. Environ Monit Assess 196:218

Spatiotemporal Dynamics of Seagrass Habitats Under Climate Stress: A Case Study of Pramuka Island, Indonesia

Rinaldy Terra Pratama, I Wayan Gita Krishna Dananjaya, and I Wayan Gede Astaw Karang

Abstract Coastal ecosystems are declining due to natural and human influences. This study analyzes the temporal patterns of seagrass meadows around Pramuka Island, Jakarta, using Sentinel-2 imagery from 2016 to 2023. The data were processed with the dark spectrum fitting (DSF) algorithm for atmospheric correction, followed by sun glint correction. The Depth-Invariant Index (DII) and various spectral indices were used as inputs for classification with a random forest model. The annual variation in detected dense seagrass meadows showed a decreasing trend at a rate of 1.64 ha per year, while sparse and dense seagrasses together accounted for approximately 20% of the shallow water areas. Seagrass, particularly denser types, could be distinguished by its reflectance in the red edge (704 nm) band. Despite fluctuations of both gains and losses, the net result over the observed period was a total loss of 52.9 ha. The relationship between sea surface temperature (SST) and seagrass extent had a correlation of − 0.43, generally showing an inverse pattern. However, this relationship varied over time, with SST and seagrass extent occasionally fluctuating together in certain years.

Keywords Seagrass habitat change · Remote sensing · Random forest · Multi-temporal analysis · Climate change

R. T. Pratama (✉) · I. W. G. K. Dananjaya · I. W. G. A. Karang
Department of Marine Science, Faculty of Marine Science and Fisheries, Udayana University, Badung, Bali 80361, Indonesia
e-mail: rinaldyterra.16@gmail.com

I. W. G. K. Dananjaya
e-mail: krishnadananjayagita@gmail.com

I. W. G. A. Karang
e-mail: gedekarang@unud.ac.id

K. K. Basheer Ahammed et al. (eds.), *Monitoring Coastal Marine Ecosystems Under Climate Change*, https://doi.org/10.1007/978-3-032-09137-6_6

1 Introduction

Seagrasses are foundational components of marine ecosystems, with their productivity playing a vital role in supporting diverse habitats in shallow waters and coastal areas. They are commonly found in coastal areas, where they often form meadows in shallow waters that can reach depths of up to 60 m (Unsworth et al. 2019). Their productivity is vital for carbon sequestration, coastal protection, and the preservation of biodiversity, establishing interdependent relationships with various organisms (Orth and Heck, 2023). The abundance of seagrasses is influenced by these intricate interactions. For instance, restoration efforts can enhance their abundances, whereas disturbances can lead to declines. Such fluctuations in abundance ultimately impact the ecosystem services that seagrasses provide to marine environments. The effectiveness of seagrasses in providing ecosystem services depends on the health and density of the natural communities that support them, as well as the spatial extent of their influence on the surrounding environment (James et al., 2023). Due to their sensitivity to fluctuating environmental conditions, seagrasses serve as indicators of the ecological status and health of coastal ecosystems (Frederiksen et al. 2004; Madden et al., 2009).

Climate change is a global phenomenon that reshapes life on Earth, with far-reaching consequences for both terrestrial and marine ecosystems. Rising sea surface temperatures (SST), ocean acidification, and increasing sea levels are integral aspects of climate change, profoundly impacting marine ecosystems (Basheer Ahammed et al., 2024). The significant increase in SST may trigger reductions in growth, reproduction, and fitness of various marine organisms, from highly mobile animals to sessile marine plants (Williamson and Guinder 2021). A study by Carlson et al. (2018) reported high temperature anomalies in the seagrass meadows of Florida Bay, which altered large areas, leading to seagrass mortality. Regarding ocean acidification, it is still unclear how increases in CO_2 may affect the seagrass photosynthetic mechanism, but this phenomenon may inhibit the absorption of dissolved inorganic carbon (Koch et al. 2013). It is important to learn more about the effects of climate change on vegetation health including seagrass, considering that seagrass helps sustain the ecosystems around it (Kumar et al. 2024). Seagrass alleviates CO_2 concentration, which supports the calcification process of corals and many other calcifying organisms (Hendriks et al. 2014). This example illustrates the crucial role of seagrass meadows and highlights how their degradation may disrupt the balance of marine ecosystems.

On a global scale, seagrass meadows show changes that tend to indicate declines, with losses outweighing gains, although reversed conditions could be found in a few regions. Dunic et al. (2021) analyzed seagrass changes across four bioregions, showing that the Indo-Pacific regions have experienced a slow but continuous decline over the past 70 years. Unfortunately, the absence of continuous data results in gaps that complicate the study of seagrass changes. There are many approaches to monitor seagrass from conducting field works (in situ) by using transects and sampling points to estimating seagrass distribution and the latter remote sensing implementations (Veettil et al. 2020). Remote sensing can be applied directly, such as in detecting seagrass presence, mapping spatial extent, species differentiation, estimating biomass, and monitoring growth patterns and degradation. It can also be used indirectly to observe environmental factors that influence seagrass, such as sea surface temperature (SST), salinity, sea level rise, pollution, and epiphyte detection. This technology offers broad coverage at a relatively low cost compared to in situ methods.

Remote sensing, however, has its flaws when capturing seagrass information, primarily because seagrasses are often submerged (Duffy et al. 2018). This introduces the penetration limits of visible and infrared wavelengths, which are commonly used for distinguishing plants on terrestrial land. Veettil et al. (2020) thoroughly explained the current advantages and disadvantages of the remote sensing approach for seagrass mapping. A recent approach involves utilizing machine learning algorithms to classify benthic habitats in seagrass mapping studies. Machine learning algorithms are continuously developing and improving to enhance accuracy in seagrass mapping, with deep learning being a prominent advancement (Misiuk and Brown 2024). In terms of satellite imagery, Sentinel-2 is presumably the most widely chosen for short-term seagrass mapping due to its 10 m spatial resolution. For longer observation periods, the Landsat series is an alternative, though it has a coarser 30 m spatial resolution (Leblanc et al. 2021). Time-series analysis with satellite imagery, particularly over decadal periods, can be challenging due to the limited availability of ground truth datasets and some acquisition times during the observation period that may not be accounted for. Dalby et al. (2023) presented an approach to combine remotely sensed data with historical maps to observe changes over 20 years, aiding future spatially targeted restoration efforts. However, access to decadal spatial information on seagrass remains limited, especially in areas considered remote. Nonetheless, remote sensing remains the better option, as shown by Setiawati et al. (2024), who conducted a time-series analysis of coral and seagrass mapping in Western Papua.

This paper presented our attempt to observe the spatial and temporal patterns of seagrass around Pramuka Island as inexpensively as possible. We used freely available Sentinel-2 imagery, and the processes for conducting this study were accessible through open-source tools. We encourage replication of our study, with detailed procedures provided in the supplementary materials. However, uncertainties persist, which are discussed in this paper along with future directions for further studies. To better map seagrass changes, we explored the spectral reflectance responses for each benthic class. Furthermore, to explore potential correlations between seagrass changes and climate variables, we incorporated sea surface temperature (SST) data, as well as indices for the El Niño–Southern Oscillation (ENSO) and the Indian Ocean Dipole (IOD).

2 Study Area

Kepulauan Seribu is an archipelago and the only regency within Jakarta Province, characterized by diverse natural and geological features. The islands, primarily coral reefs or atolls, have been shaped by the interaction of coral and associated biota. The archipelago covers a total area of 8.69 km^2 and consists of 110 islands. The average temperature ranges from 28.1 °C to 30.1 °C, with the wet season typically occurring from November to April, influenced by the West Monsoon (Setiawati et al. 2023; Badan Pusat Statistik (BPS) 2024). The water depth varies between 0 and 60 m, with the Java Sea being the only connection to the open ocean, reaching depths of up to 50 m (Farhan and Lim 2011; Utami et al. 2018). Kepulauan Seribu has been designated as Taman Nasional Laut Kepulauan Seribu, with zoning established across four areas: the Core Zone, Protection Zone, Tourism Utilization Zone, and Settlement Zone. Administratively, the archipelago is divided into two districts: Kepulauan Seribu Utara and Kepulauan Seribu Selatan. In Pulau Pramuka, located within the Settlement Zone, a variety of seagrass species can be found, including *Cymodocea rotundata*, *Cymodocea serrulata*, *Enhalus acoroides*, *Halodule uninervis*, *Halodule ovalis*, *Thalassia hemprichii*, and *Syringodium isoetifolium* (Ningsih et al. 2021).

This study focuses on Pulau Pramuka and its surrounding islands within Pulau Panggang village which is dedicated as the Settlement Zone, located in the Kepulauan Seribu Utara sub-district of Kepulauan Seribu Regency, Jakarta. The study area is approximately 45 km northwest of the Jakarta coastline, extending from 5°46′30.1″S to 5°42′26.24″S and 106°32′59.6″E to 106°37′57.58″E (Fig. 1). Pulau Panggang village comprises 17 islands, with Pulau Panggang and Pulau Pramuka being the main islands where most settlements are concentrated. While specific population data for Pulau Panggang village are not publicly available for recent census, it is estimated that Kepulauan Seribu Utara, the administrative sub-district, has a projected population of 16,575 people, based on several census data from 1961 to 2012 (Badan Pusat Statistik (BPS) 2024). Pulau Panggang has undergone significant land reclamation by local residents to accommodate rapid population growth (Ningsih et al. 2021; Setiawati et al. 2023). On the other hand, situated to the south of Pulau Panggang and separated by a shallow inter-reef channel, Pulau Pramuka has a lower population density and functions as the administrative center of Kepulauan Seribu Regency. Additionally, Pulau Semak Daun is a lagoon characterized by a shallow reef flat and a deep central lagoon, which features a rich marine ecosystem (Utami et al. 2018). Fisheries are the primary source of income for the local community, involving both capture fishing and aquaculture with additional revenue generated from providing accommodation for tourists. An example of this is the floating marine aquaculture in Pulau Panggang, situated on the surface of the lagoon located to the west of the island (Farhan and Lim 2011). Historically, the islands also supported seagrass cultivation. However, after the 2000s, this activity was eventually abandoned due to environmental degradation that affected production rates (Bahri et al. 2017).

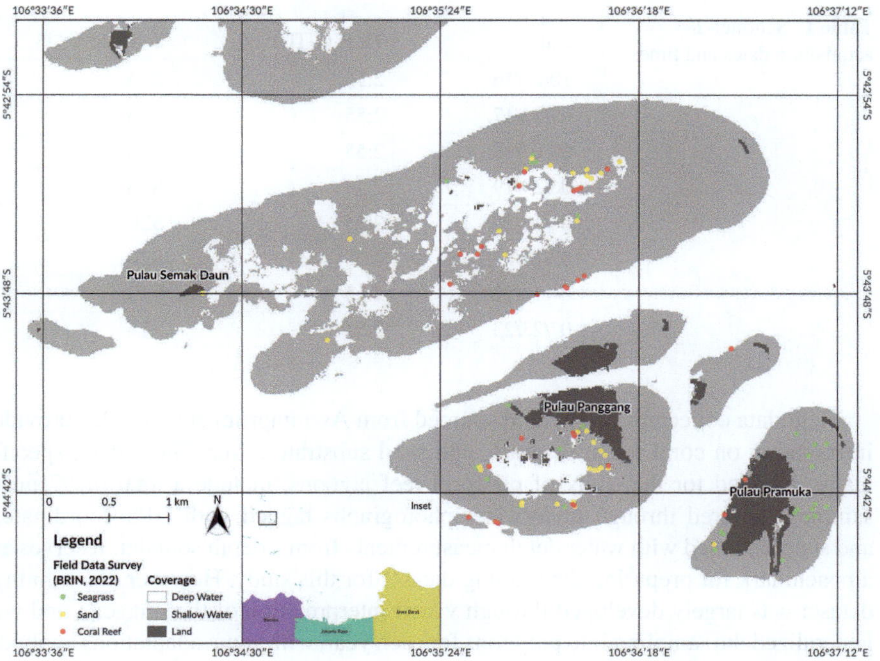

Fig. 1 Study area

3 Materials and Methods

3.1 Data

The primary data source for the benthic mapping is Sentinel-2 MultiSpectral Imager (MSI) Level-1C imagery, which we collected annually from 2016 to 2023, resulting in a total of eight scenes (Table 1). The Level-1C product is orthorectified and provides Top-of-Atmosphere (TOA) reflectance. We specifically selected images with minimal cloud cover from The Copernicus Data Space Ecosystem, accessible at https://browser.dataspace.copernicus.eu/. For this study, the spectral bands (central wavelengths) used include blue (492 nm), green (560 nm for S2A; 559 nm for S2B), red (665 nm), and NIR (833 nm), with a spatial resolution of 10 m. Additionally, Red Edge 1 (704 nm), SWIR1 (1614 nm for S2A; 1610 nm for S2B), and SWIR2 (2202 nm for S2A; 2186 nm for S2B) are used, with a spatial resolution of 20 m (Drusch et al. 2012). The imagery underwent atmospheric correction using the ACOLITE processor, followed by sun glint correction, and the development of Depth-Invariant Index (DII) as well as several water and vegetation spectral indices. The classification procedure includes model training and validation, where the imageries were first processed to distinguish between water and land and then further classified into specific benthic habitats using a random forest classifier.

Table 1 Sentinel-2 acquisition dates and times

Date	Time (UTC)	Satellite
10/07/16	2:56	S2A
09/22/17	2:55	S2A
09/07/18	2:55	S2A
06/29/19	2:55	S2B
06/13/20	2:55	S2B
07/28/21	2:55	S2B
04/14/22	2:55	S2B
07/23/23	2:55	S2A

Field data collected in 2022 and sourced from Asriningrum et al. (2022) provide information on coral reefs, seagrass, and sand substrate cover. These data, specifically intended for the study of platform reef lagoons, include a total of 97 field samples captured through underwater photographs tagged with GPS coordinates and supplemented with water depth measurements from a depth sounder. It serves as a benchmark for preparing the training dataset for this study. However, the training dataset was largely developed through visual interpretation of the imagery, and we generalized the sampling into polygons for each year, with some adaptations applied, particularly for the 2016 scene. Additionally, this field data are used to validate the 2022 classification results.

To assess the influence of climate, we utilized the Sea Surface Temperature (SST) dataset from the Operational Sea Surface Temperature and Sea Ice Analysis (OSTIA), obtained from the Copernicus Marine and Environment Monitoring Service (CMEMS), available at https://data.marine.copernicus.eu/product/SST_GLO_SST_L4_NRT_OBSERVATIONS_010_001/. This dataset provides gap-free SST data derived from satellite and in situ observations, available on both daily and monthly timescales, with a spatial resolution of 0.5° (Donlon et al. 2012). The analysis also incorporated irregular oscillation phenomena occurring in the Pacific and Indian Oceans, including the El Niño–Southern Oscillation (ENSO) and the Indian Ocean Dipole (IOD). We used the Oceanic Niño Index (ONI) and the Dipole Mode Index (DMI) to assess ENSO and IOD, respectively. Both indices are distributed by the National Oceanic and Atmospheric Administration (NOAA) Climate Prediction Center and Physical Sciences Laboratory, accessible at https://origin.cpc.ncep.noaa.gov/products/analysis_monitoring/ensostuff/ONI_v5.php and https://psl.noaa.gov/gcos_wgsp/Timeseries/DMI/.

3.2 Methodology

The process for generating benthic thematic maps (Fig. 2) includes image correction, spectral indices, and classification. Atmospheric correction was performed using the

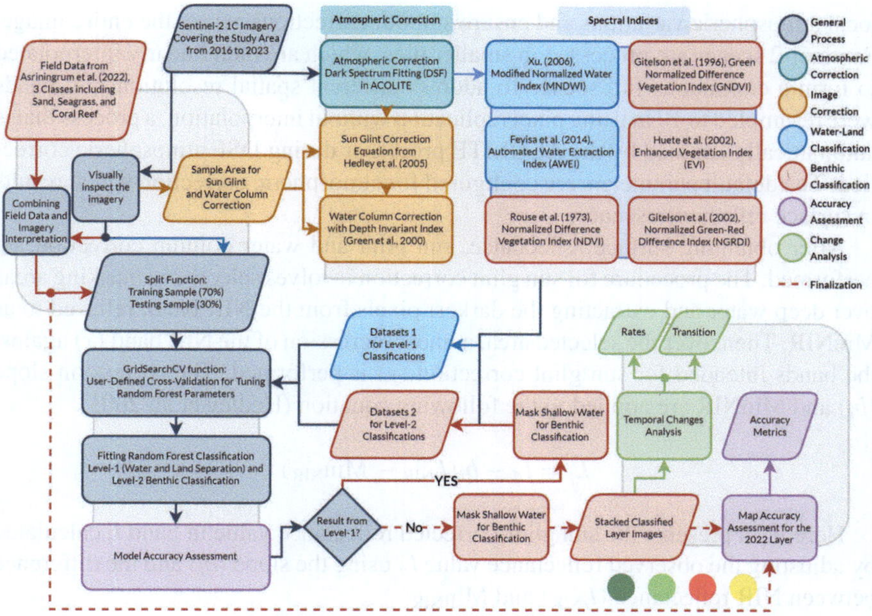

Fig. 2 Workflow for benthic habitat classification using Sentinel-2 imagery (2016–2023)

ACOLITE processor v.20231023.0, followed by analysis and classification in Jupyter Lab v.4.2.1 using Python v.3.9. Several Python libraries are utilized, including xarray, numpy, geopandas, rasterio, matplotlib, plotly, scipy, and sklearn. Map layouting is completed with QGIS v.3.34. The full code and workflow for this process are available on GitHub at https://github.com/rtp07/study_of_seagrass_dynamics.git.

3.2.1 Image Correction

Royal Belgian Institute of Natural Sciences' (RBINS) ACOLITE processor was developed as a comprehensive tool for atmospheric correction processing of imagery for coastal and inland waters application. The processor supports multiple satellite sensors and features two algorithms for atmospheric correction: exponential extrapolation (EXP) and dark spectrum fitting (DSF) with the DSF as a default algorithm (Vanhellemont and Ruddick 2014, 2018). In this study, atmospheric correction was performed using the DSF algorithm, which is developed to optimize Sentinel-2 optical imagery for aquatic applications (Vanhellemont 2019). The DSF algorithm estimates atmospheric path reflectance by constructing a dark spectrum from the satellite image scene. This dark spectrum is derived by identifying dark pixels, where water reflectance is low in certain bands. The DSF estimates the atmospheric path reflectance ρ_{path} using a lookup table (LUT) created with the 6SV radiative transfer models (RTMs) and selecting the most suitable aerosol type. To account for local atmospheric variations and ensure smooth corrections across the entire image, Sentinel-2 scenes are processed in smaller tiles, which are then linearly interpolated to form a continuous full scene. To address different spatial resolutions, all bands were resampled to 10 m using pixel replication without interpolation, a process that is automatically handled by the ACOLITE processor during DSF atmospheric correction. The default parameters are configured for atmospheric correction which results in surface reflectance scenes.

After obtaining surface reflectance, sun glint and water column correction are performed. The procedure for sun glint correction involves selectively masking areas over deep water and extracting the darkest pixels from the NIR band, referred to as MinNIR. Then, over the selected area, a linear regression of the NIR band (x) against the bands intended for sun glint correction (y) is performed. The regression slope (b_i) and MinNIR are applied to the following equation (Hedley et al. 2005):

$$L_i' = L_i - b_i(L_{\mathrm{NIR}} - \mathrm{Min}_{\mathrm{NIR}})$$

Here, L_i' represents the sun glint-corrected reflectance value in band i, calculated by adjusting the observed reflectance value L_i using the slope (b_i) and the difference between NIR reflectance (L_{NIR}) and $\mathrm{Min}_{\mathrm{NIR}}$.

The Depth-Invariant Index (DII) model from Green et al. (2000) was adopted for water column correction. Pairs of spectral bands were selected from the blue, green, red, and red edge bands for this correction. Multiple subsets of sand objects in relatively shallow water at different depths were used to calculate the attenuation ratios $\left(k_i/k_j\right)$ for each band pair (i and j), and attenuation coefficient (a) was determined. The following equation is used to calculate the attenuation ratios:

$$k_i/k_j = a + \sqrt{\left(a^2 + 1\right)}$$

Here,

$$a = \frac{\sigma_{ii} - \sigma_{jj}}{2\sigma_{ij}}$$

Here, σ_{ii} and σ_{jj}, are the variances of the reflectance values for the band i and j, respectively. And σ_{ij} is the covariance between the reflectance values of these band pairs. The calculated attenuation ratios were then implemented in the DII calculation for the scenes:

$$L_{ij} = \ln(L_i) - \left[\left(k_i/k_j\right)\ln\left(L_j\right)\right]$$

Here, a logarithmic transformation (denoted as ln) is applied to the reflectance value on each band pair (L_i and L_j) to linearize the attenuation of light with depth.

Table 2 Spectral indices, formulas, and references used for water–land classification and benthic habitat mapping

Index	Formula	Reference
MNDWI	$\text{MNDWI} = \frac{L_{\text{Green}} - L_{\text{SWIR1}}}{L_{\text{Green}} + L_{\text{SWIR1}}}$	Xu (2006)
AWEI	$\text{AWEI} =$ $4 \times (L_{\text{Green}} - L_{\text{SWIR2}}) - (0.25 \times L_{\text{NIR}} + 2.75 \times L_{\text{SWIR1}})$	Feyisa et al. (2014)
NDVI	$\text{NDVI} = \frac{L_{\text{NIR}} - L_{\text{Red}}}{L_{\text{NIR}} + L_{\text{Red}}}$	Rouse et al. (1974)
GNDVI	$\text{GNDVI} = \frac{L_{\text{NIR}} - L_{\text{Green}}}{L_{\text{NIR}} + L_{\text{Green}}}$	Gitelson et al. (1996)
EVI	$\text{EVI} = \frac{L_{\text{NIR}} - L_{\text{Red}}}{L_{\text{NIR}} + 6 \times L_{\text{Red}} - 7.5 \times L_{\text{Blue}} + 1}$	Huete et al. (2002)
NGRDI	$\text{NGRDI} = \frac{L_{\text{Green}} - L_{\text{Red}}}{L_{\text{Green}} + L_{\text{Red}}}$	Gitelson et al. (2002)
NGRDI	$\text{NGRDI} = \frac{L_{\text{Green}} - L_{\text{Red Edge}}}{L_{\text{Green}} + L_{\text{Red Edge}}}$	Gitelson et al. (2002)

3.2.2 Spectral Indices

Reflectance from the atmospherically corrected scene was further used to calculate spectral indices. Specific band combinations that respond to variations in surface conditions, water depths, and bottom types were generated for water–land and benthic habitat classification, as shown in Table 2. For water separation, the Modified Normalized Water Index (MNDWI) and Automated Water Extraction Index (AWEI) were calculated. For benthic habitat classification, the Normalized Difference Vegetation Index (NDVI), Green Normalized Difference Vegetation Index (GNDVI), Enhanced Vegetation Index (EVI), and Normalized Green–Red Difference Index (NGRDI) were used.

3.2.3 Benthic Habitat Classification and Validation

We applied a random forest classifier to categorize image features by distinguishing pixels based on their unique reflectance signatures or pixel values. This supervised classification approach used training inputs corresponding to each recognized object. The random forest algorithm was implemented using the scikit-learn ensemble module (Buitinck et al. 2013). Generally, the random forest builds multiple decision trees which then aggregate their votes to assign the class that receives the majority (Breiman 2001). We randomly split the samples into 70% training and 30% testing to validate the trained classifier. A cross-validation process was used to determine the optimal number of trees, resulting in the selection of 80 trees and 200 trees for water–land and benthic habitats classification, respectively. The classes for both classifications are described in Table 3, where the shallow water class from the land–water classification is used to mask the data, limiting the benthic habitat classification to shallow water areas. The accuracy of the models was assessed using the confusion matrix (CM)-based method, with resulting metrics including overall accuracy

Table 3 Classification categories with descriptions. Asterisks (*) indicate classes used for water–land separation

Class	Description
Land*	Terrestrial areas that not submerged by water
Shallow water*	Shallow coastal or nearshore waters where the depth is minimal
Deep water*	Regions of the ocean or sea where the depth exceeds the limits of sunlight penetration
Dense seagrass	This area is identified by a homogeneous brown to black color, which is prevalent along the coastline
Sparse seagrass	This area typically features a brown to black color with brighter spots scattered throughout
Coral or rubble	This area is indicated by a light brown to dark brown color, with a coarse texture and uneven edges
Sand	This area features a bright color with a sandy hue and texture

(OA), producer's accuracy (PA), user's accuracy (UA), and Cohen's kappa coefficient (Maxwell and Warner 2020; Tassi et al. 2021). This assessment also included a comparison between the ground truth and the predicted maps for 2022 using these metrics. Unfortunately, the ground truth sample did not provide information about the density of the seagrass. As a result, we combined the dense and sparse seagrass into a single category for the accuracy assessment, which focused on its relationship with each benthic type.

3.2.4 Change Analysis

Benthic types change analysis adopted the approach of Dalby et al. (2023) to understand seagrass transitions. We focused on the predicted sparse seagrass class to identify these changes. Thus, several classes were developed to compare the maps pairwise between years in chronological order. First, we examined areas that did not change, noting them as consistent in either the presence or absence of sparse seagrass. Subsequently, we classified the areas based on observed changes, identifying gains (transitions from absence to presence) and losses (transitions from presence to absence). Lastly, we compare the sparse seagrass that change into dense seagrass which assigned as density gain. While several classes, such as changes in dense seagrass, were ignored, focusing on sparse seagrass allowed us to identify both gains in density and presence and losses in specific areas. Dense seagrass, on the other hand, was compared with climate-driven factors such as SST, IOD, and ENSO to understand its relationship with these variables.

4 Results

4.1 Satellite Image Analysis

Benthic classification involves retrieving spectral characteristics from the image through several preprocessing steps. A statistical summary comparing variations in corrections from the blue band channel is provided (Table 4). Reflectance values resulting from the sun glint calculation that are associated with the land surface were automatically assigned negative values based on the NIR band channel. Consequently, in the DII calculation, all negative values from the sun glint calculation were treated as null values. After this calculation, an offset was applied to ensure that all values remained positive. In the classification scheme, all imagery resulting from the preprocessing steps, including several spectral indices, was used as features for fitting into the model, whereas the original TOA imagery (before correction) was excluded. Once the model was trained, we extracted the feature importances to understand which spectral characteristics contributed most significantly to the benthic classification.

Among the model features, the red band (7.5%) contributed the most overall, based on the average across all years, followed by the red edge (7.2%) and sun glint-corrected red edge (7.1%), SWIR1 (5.2%), green (5.2%), and blue (5.2%) bands. However, features from atmospheric and sun glint corrections varied—where in some years, the sun glint-corrected features were the most significant contributors, and in other years, the atmospheric corrections were more influential. When we combined both contributions, the red edge (14%) was identified as the most significant, followed by the red (11%), blue (10%), green (10%), and SWIR1 (5.2%) bands. The model assigned low importance to water column-corrected images, preferring band pairs between the green and red bands, which contributed 3.2%. Water column

Table 4 Summary statistics for reflectance values before and after atmospheric and sun glint correction, and for Depth-Invariant Indices (DII) between different band pairs

Variables	Mean	Median	Std dev	Min	Max	Range
Before atmospheric correction (TOA)	0.117	0.114	0.011	0.106	0.348	0.242
After atmospheric correction	0.044	0.041	0.014	0.031	0.321	0.29
After sun glint correction	0.04	0.038	0.016	– 0.243	0.178	0.421
DII between blue and green	8.352	8.362	0.074	0	11.28	11.28
DII between blue and red	8.288	8.293	0.095	0	13.031	13.031
DII between blue and red edge	8.576	8.576	0.108	0	11.795	11.795

correction has little to no effect on improving performance in distinguishing benthic type coverage, especially in calm, clear, and shallow waters (Mederos-Barrera et al. 2022; Widya et al. 2023). However, it moderately contributes when classifying land and water body separations and can occasionally outperform water indices. Among spectral indices for benthic types, the highest contribution was from the Normalized Green–Red Difference Index (NGRDI) using the red band (4.9%), while the variant with the red edge band provided a lower contribution of 2.4%.

We also extracted spectral reflectance from atmospherically corrected images relative to each predicted class. By calculating the average and standard deviation, we analyzed the spectral response (Fig. 3). The dense seagrass class could notably be distinguished by the average reflectance in the red edge band and by the standard deviation in the SWIR1 band. Feature importance analysis indicated that both atmospheric and sun glint-corrected red edge and SWIR1 were among the top contributors, while the red band showed more dispersion. Several spectral indices to distinguish seagrass have been proposed, including combinations of the red, red edge, and SWIR1 bands. Li et al. (2023) proposed the Submerged Seagrasses Identification Index (SSI-I), which is calculated as the ratio between the red edge (704 nm) and red (665 nm) bands, plus an additional denominator. In the Landsat series, two spectral indices were developed: Seagrass Index I (SSI-I), which combines the red, NIR, and SWIR1 bands, and Seagrass Index II (SSI-II), which uses a combination of the green and red bands to distinguish seagrass from sand (Liang et al. 2023). The SSI-II has the same combination as the NGRDI, with the only difference being that NGRDI is the Normalized Difference Index between green and red, while SSI-II uses the ratio of the difference between the green and red bands to the green band.

The ability of remote sensing approach to accurately map seagrass distribution is influenced by the condition of surrounding benthic habitat. In shallow water

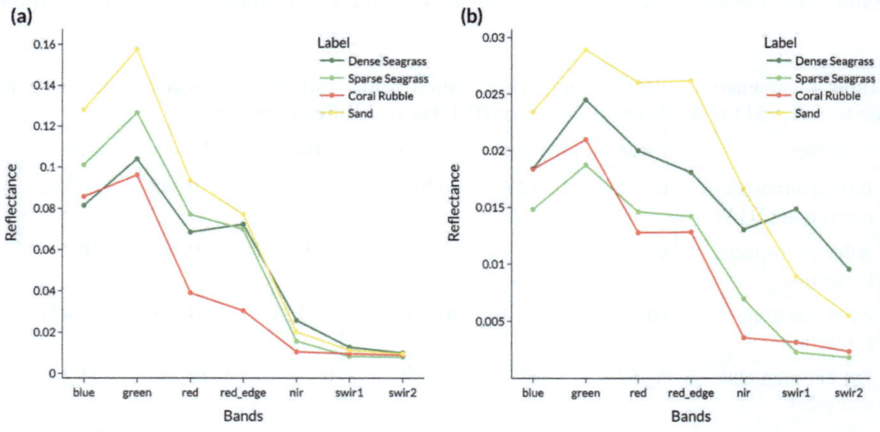

Fig. 3 Reflectance profiles for benthic habitats: **a** mean reflectance, **b** standard deviation across spectral bands for dense seagrass, sparse seagrass, coral rubble, and sand

ecosystem, object such as sand, algae, coral and rubble, and seagrass may live adjacent to one another and often create community in an area. For remote sensing, this is known as mixed benthic class, which may often cause confusion for the model to produce high-quality distribution map. As observed in the seagrass class, the pixel values of dense and sparse seagrass may be affected by the underlying substrate, such as sand. In sparse seagrass areas, the overall reflectance may be dominated by the sandy background (Lazuardi et al., 2021). This may explain the similarity of reflectance trend of sparse seagrass and sand class.

4.2 Distribution of Benthic Habitats

Through the classification scheme, we produced eight images, each representing a different period of our study (Fig. 4). Pulau Semak Daun with its smaller land area is characterized by extensive coral rubble and sand surrounding the inner atoll and deep water, while seagrass occupies the east side, extending within a shallow area up to the reef crest, often in areas where sand is present. A similar pattern is found on Panggang Island, where seagrass is evenly distributed around corals within the atoll. Pramuka Island, which is not an atoll, is characterized by seagrass encircling the island, with corals and rubble occupying the outer reef next to deep water. Seagrass is predominantly distributed near deep water, with sand cover occasionally interrupting this pattern. Dense seagrass is mainly found closer to the shore on Pramuka Island, which has a proportionally larger distribution of dense seagrass compared to the other islands.

Built-up structures were constructed during the observed years, resulting in changes to the landscape of Pramuka Island between 2016 and 2017. The south-eastern part of the island was developed into a ferry port with breakwater structures. This development led to notable changes in the local landscape (Saraswati et al. 2021). In 2018, a pier was constructed just north of the port, extending across shallow waters into deeper areas, impacting the surrounding sand habitat. This construction led to the separation of the sparse seagrass area into two distinct segments. Such constructions can modify the surrounding coastal environment by altering sediment dynamics, water quality, and hydrodynamics. These modifications are quantified either as a distance from the structure, with effects extending outward, or as the area affected by the structure itself (Bugnot et al. 2021).

The series of maps illustrate these temporal changes in benthic habitat composition from 2016 to 2023 (Fig. 5). The maps depict regions experiencing both gains (in pale green) and losses (in red) over the years. In some areas, seagrass coverage remained consistent (in light green), while other areas showed density gains (in dark green) across different locations on the islands. A comparison of sparse seagrass areas over the years shows fluctuations between gains and losses in the same locations. On the northern shore tip of Pramuka Island, two significant losses occurred during the transitions from 2018 to 2019 and from 2022 to 2023, while a major gain occurred from 2020 to 2021. In other years, both gains and losses occurred but were more

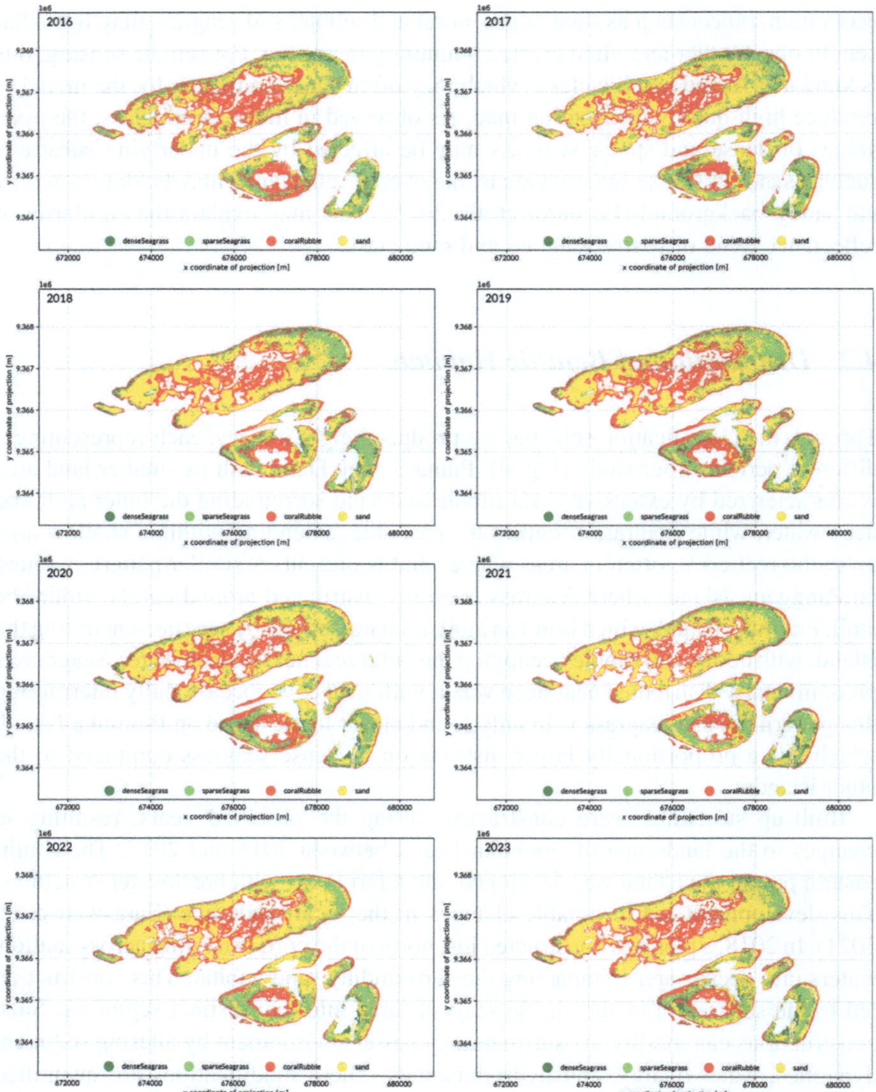

Fig. 4 Benthic habitat classification maps from 2016 to 2023, showing the spatial distribution of dense seagrass, sparse seagrass, coral rubble, and sand across the study area

partial and less consistent. Fluctuations also occurred on the southeastern part of Panggang Island and the western part adjacent to Panggang Island's atoll. These spatial and temporal variations indicate that the seagrass habitat has experienced both disturbances and recovery in different segments, with patterns varying across the observed years and locations. In Pramuka Island, benthic habitat mapping has been conducted by Rahman et al. (2020), and the distribution is similar to what has been described above. Furthermore, the distribution of sparse seagrass in our study matches with distribution of mix substrate class in Rahman et al. (2020), suggesting the existence of other benthic objects such in area of low-density seagrass.

Changes in benthic habitat composition over time show that the overall distribution of seagrass cover has consistently remained around 20% across the years (Fig. 6). The benthic areas fluctuate within a range of about 900 hectares. In both 2016 and 2022, dense seagrass cover underwent its most significant decline before increasing the following year. However, in 2016, the combined share of dense and sparse seagrass reached its highest overall proportion. In 2016, the combined share of dense and sparse seagrass reached its highest overall proportion. Following this peak, dense seagrass saw an increase in the subsequent year, but sparse seagrass lost about 50 ha by 2018 (Table 5). Gains in sparse seagrass presence occurred only in 2019 and 2021, with both years followed by a loss of the same amount in the subsequent year. The net change in sparse seagrass presence over the observed period is -52.9 hectares, indicating an overall decline despite temporary gains in specific years.

The reduction of seagrass has been an issue for seagrass meadows in Indonesia and around the globe. In the Bay of Palu, seagrass meadow extent was modeled using Landsat imagery, indicating extensive habitat loss from 127.08 ha to 43.29 ha from 2012 to 2022 (Tis'in et al., 2023). Adjacent to the seagrass meadows, the area is primarily utilized for mining and reclamation, which is suspected to cause degradation of seagrass cover. This loss is far greater compared to the model produced in the Pramuka Islands, due to highly destructive anthropogenic disturbances in the area. The loss of seagrass meadows on small islands is apparent. A recent study by Fadhlurahman et al. (2024) reveals a reduction in seagrass meadows on Derawan Island by 23.47 ha from 2003 to 2021, utilizing medium-resolution Landsat 8 imagery. The study suggested that the loss was caused by extensive watersport activities on the island, which reduced water clarity, eventually preventing seagrass from acquiring enough sunlight for photosynthesis. Both studies serve as evidence of seagrass habitat loss, which is highly associated with anthropogenic stress. However, it is also worth noting that variation in seagrass area is influenced by climate-related factors, namely sea surface temperature, which is controlled by ocean anomalies such as the Indian Ocean Dipole (IOD).

4.3 Accuracy Assessment

The accuracy assessment shows that our model provides benthic habitat distribution information with an overall accuracy (OA) of 70% and a kappa coefficient of 0.55. This result exceeds the 60% accuracy standard for benthic mapping set by the Indonesia Geospatial Agency (Badan Informasi Geospasial (BIG) 2014). The confusion matrix is included to illustrate the model's performance in assigning pixels to the designated classes. As summarized in Table 6, the model effectively distinguishes coral but struggles with sand, likely due to similarities in spectral or spatial features between these classes during classification. Our model's accuracy is comparable to

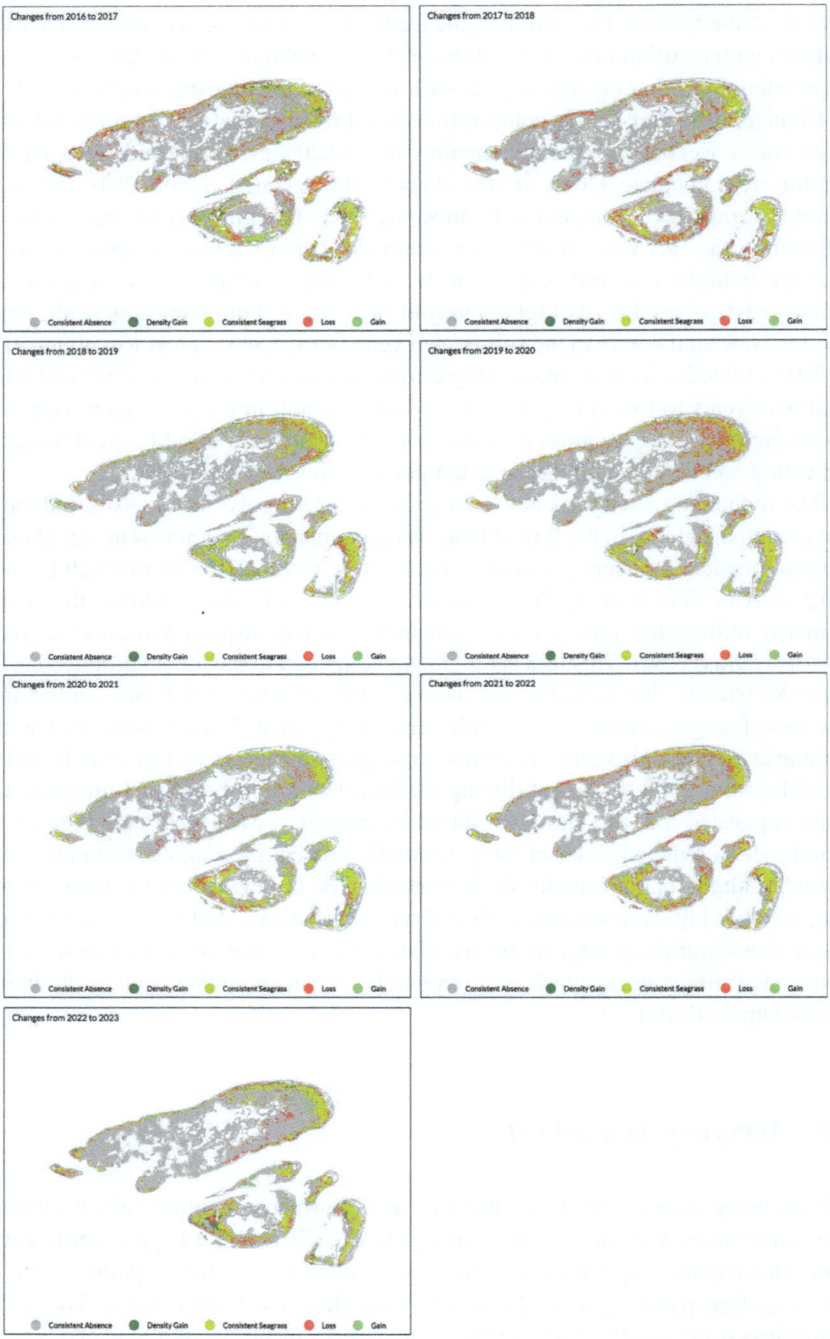

Fig. 5 Temporal change maps showing transitions in benthic habitats between consecutive years from 2016 to 2023

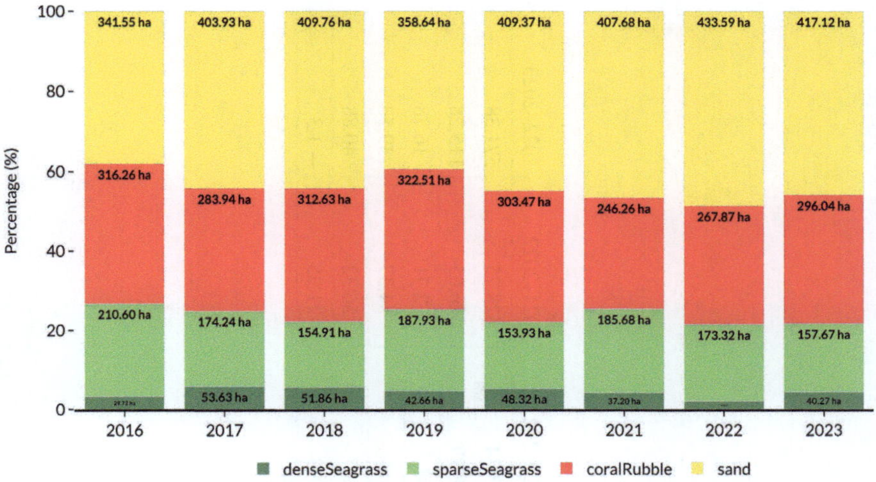

Fig. 6 Percentage distribution of benthic habitats from 2016 to 2023

other studies focused on benthic habitat mapping, particularly regarding seagrass distribution. For instance, Rahman et al. (2020) classified coral reefs, sand, seagrass, and mixed substrates, achieving a slightly higher accuracy of 73.33%. In contrast, our study outperformed the findings of Hamuna et al. (2023), who reported an OA of 65.31% using the random forest algorithm. Although these studies utilized different temporal resolutions, they highlight how variations in classification approaches can significantly affect OA. A recent study by Fadhlurahman et al. (2024) examined the multi-temporal distribution of seagrass and other benthic habitats around Derawan, reporting a similar OA of 71%, though with only 59.83% accuracy for seagrass percent cover prediction. This suggests that more complex classification schemes may lead to lower accuracy. As noted by Green et al. (2000), factors such as the number of classes, types of satellite imagery, classification methods, and the quantity of ground truth data used can all influence classification results.

4.4 Climate Pressures

This section describes the linkage between seagrass cover and climate factor pressures. We examined the annual variability in sea surface temperature (SST) and compared it with the coverage of dense seagrass classes (Fig. 7). In 2016, seagrass coverage was significantly lower than in the following years, coinciding with the peak in SST. During this period, an intense El Niño event was indicated by a high positive Oceanic Niño Index (ONI) in the first quarter of 2016, which coincide with the observed low coverage of seagrass (Fig. 8a). This event was also discussed in Setiawati et al. (2024), where the behavior of ENSO was observed to be directly related to

Table 5 Yearly changes in benthic habitat categories from 2016 to 2023

Category	2016–2017	2017–2018	2018–2019	2019–2020	2020–2021	2021–2022	2022–2023
Consistent absence	483.9	534.96	554.29	571.18	533.34	528.75	571.38
Consistent seagrass	114.04	94.34	112.97	120.07	118.29	126.65	115.28
Density gain	22.42	7.88	5.17	6.86	4.57	2.71	16.78
Gain	51.3	41.15	68.87	27.33	53	34.37	39.58
Loss	71.75	71.66	36.76	60.06	31.07	56.32	40.88
Net change	− 20.45	− 30.51	32.11	− 32.73	21.93	− 21.95	− 1.3

Table 6 Accuracy assessment for the 2022 benthic habitat classification

2022					
Overall accuracy—OA (60/86)				0.7	
Kappa coefficient				0.55	
Class	Seagrass	Coral or rubble	Sand	Total	User accuracy (%)
Seagrass	22	6	1	29	88
Coral or rubble	1	25	1	27	54
Sand	2	15	13	30	87
Total	25	46	15	86	
Producer accuracy (%)	76	93	43		

SST around Pramuka Island. Their reported seagrass coverage significantly dropped from the previous year, with an increase in SST until 2017, which differs from our SST observations that showed a gradual decrease in the following year. Unfortunately, we did not cover the 2015 period and could not determine whether a similar drop occurred in our observation location.

In 2017, SST decreased by 0.4 °C, and seagrass extent was observed to increase. The ENSO event in this period saw a La Niña, while the Indian Ocean Dipole (IOD) was positive, with both events behaving oppositely (Fig. 8b). We observed similar behaviors from the ENSO and IOD circulations in 2019, as both the El Niño and positive IOD events occurred, resulting in an increase in SST from 2018 to 2020, which coincided with a decrease in seagrass coverage in 2019. However, in 2020, when SST reached its highest peak compared to previous or following years, seagrass coverage increased, a condition we could not clarify. In the subsequent years, despite the gradual decrease in SST, seagrass coverage continued to decline, with a

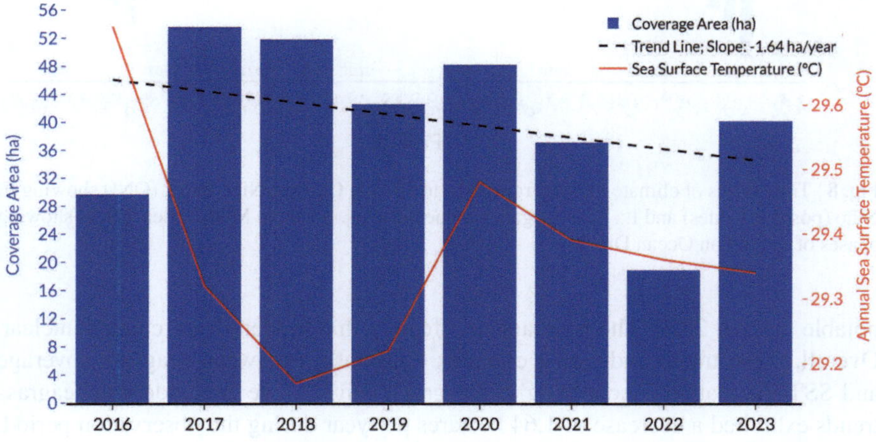

Fig. 7 Relationship between coverage area (ha) and annual sea surface temperature (°C) from 2016 to 2023

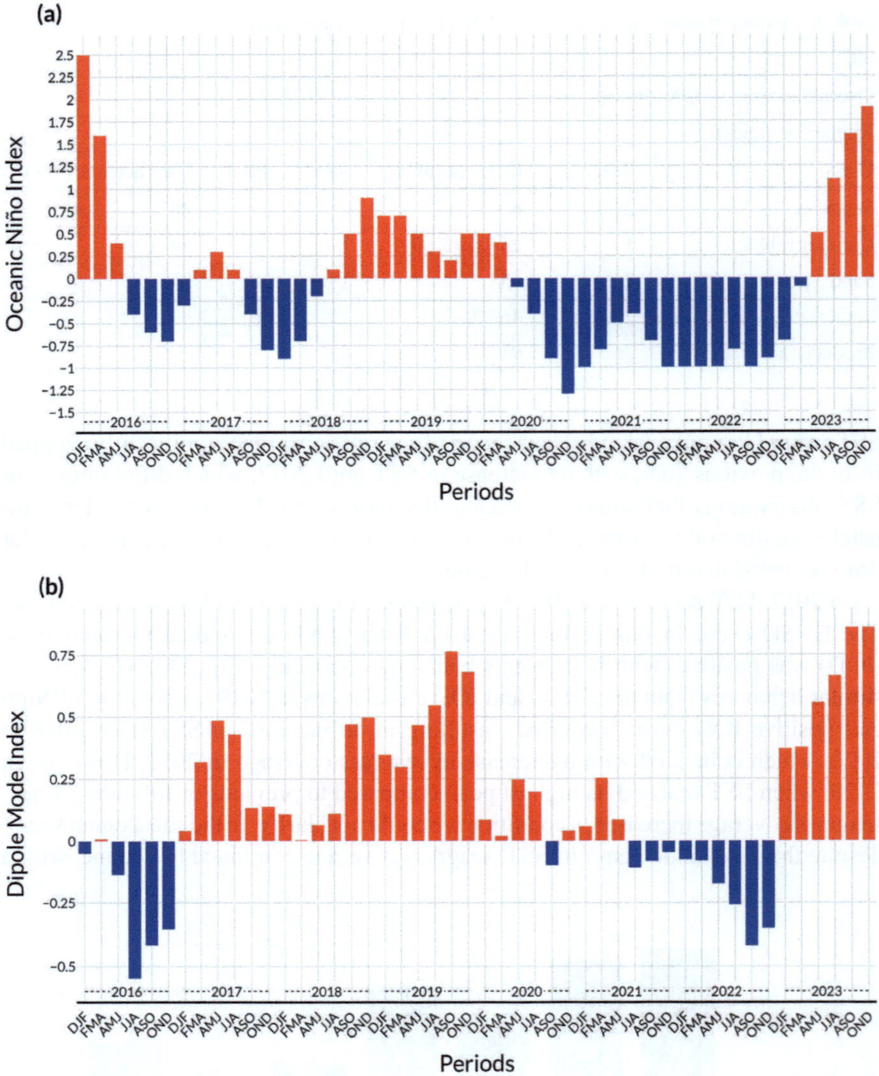

Fig. 8 Time series of climate indices from 2016 to 2023. **a** Oceanic Niño Index (ONI) showing El Niño (positive values) and La Niña (negative values) events. **b** Dipole Mode Index (DMI) showing phases of the Indian Ocean Dipole

notable drop in 2022. The mechanism affecting this sudden drop remains unclear. Overall, these trends indicate a complex relationship between seagrass coverage and SST fluctuations. According to the graph (Fig. 7), we conclude that seagrass trends exhibited a decrease of 1.64 hectares per year during the observation period. Both seagrass coverage and SST demonstrated a moderate inverse correlation, with a correlation value of – 0.432 observed in the first four years of our study period.

In this study, the observed SST is presented as a yearly average, showing only a 0.55 °C difference over the observation period. Therefore, the actual conditions at the time of image acquisition are not captured, and we present only a descriptive analysis. We observed an instantaneous association in 2017, where a significant decrease in SST from the previous year coincided with an increase in seagrass coverage. Additionally, we identified a potential delayed response in 2022, where seagrass coverage reached its lowest point, which may be related to the increasing SST in 2020 and its gradual decrease in the following years. However, we do not have sufficient evidence to determine whether SST or other factors influenced the changes in seagrass coverage between 2020 and 2022. To some extent, climate factors like sea surface temperature may not directly cause seagrass mortality, and local conditions likely contribute additional stressors (Carlson et al. 2018; Clemente et al. 2023). Another factor that could be included is the resilience of each seagrass species to SST, which is not to be discussed in this study. When combined with local disturbances such as eutrophication, overfishing, sedimentation, and physical damage from boating activities, these stressors may overwhelm the resilience of seagrass populations. For example, studies by Karang et al. (2024) identified multiple propeller scars at Samuh Beach, Bali. With intensifying tourism activities, these pressures are becoming more pronounced. Moreover, with the increasing intensity of climate change, extreme conditions like heatwaves and more frequent, intense storms are expected to further strain seagrass resilience, leading to immediate and drastic losses of seagrass meadows (Dunic et al. 2021). Ultimately, habitat loss is driven by a combination of human-induced disturbances and climate-related stressors.

4.5 *Uncertainty and Future Directions*

This study serves as a preliminary effort to demonstrate the potential of remote sensing for observing seagrass dynamics. Temporal remote sensing analysis conducted over the past eight years has shown a decline in seagrass coverage, with loss rates ranging from 1.3 to 32.73 hectares per year. However, challenges remain in providing an accurate representation due to limited knowledge of the environmental conditions at the time the imageries were acquired. First, our field data only cover 2022, and we only validated the maps' performance at that given year. Much of the training data in this study were generalized from that reference year, with a significant portion derived from visual interpretation of the imagery. This introduces potential uncertainties, particularly when applied to different times. To address these limitations, automatic adaptive signature generalization (AASG) could be considered, where one scene with high confidence in accuracy is used to train the model and then applied to other scenes, assuming that the conditions between observation periods have not undergone significant changes that could hinder the model's ability to predict (Leblanc et al. 2021). Therefore, performing exploratory data analysis (EDA) to understand the statistical summary of each scene could help in the generalization process. This analysis could also consider other key features, such as spectral signatures. In this study, dense seagrass can be distinguished in the red edge band (704 nm) due to the chlorophyll present in the seagrass meadows. Furthermore, the declining coverage rate of seagrass corresponds to a weaker signal in the red region, which cannot distinguish seagrass presence due to water absorption (Bannari et al. 2022).

Handling multiple satellite images with varying temporal resolutions requires extensive consideration to ensure consistency in the study. For instance, water column conditions can fluctuate due to tidal variations, causing depth changes (Phinn et al., 2018). Although we did not test the performance of Sentinel-2 for depth accuracy at the time of image acquisition due to tidal fluctuations, field data indicate that seagrass points are located at depths of around 3 m. Kuhwald et al. (2022) suggest that the optimal depth for seagrass observation is less than 5 m, even in turbid waters. However, turbidity should still be considered, as it can reduce the accuracy of the produced maps (Roelfsema et al. 2013). Benthic objects such as seagrass, algae, and coral often overlap, creating spectral inconsistencies that complicate the accurate classification of habitat types in marine environments. Wicaksono et al. (2021) observed misclassification between macroalgae and seagrass, both of which were incorrectly identified as coral. This problem was especially common in transitional areas where seagrass, sand, and coral co-occur, causing pixel values to be dominated by sand reflectance. As a result, accurate classification relies on the presence of dense seagrass or coral, which offer clearer spectral signatures. Similarly, Joyce et al. (2013) emphasized that spectral mixing and habitat variability can reduce classification accuracy.

To better understand seagrass coverage dynamics, future studies should incorporate climatic, local oceanographic parameters, and anthropogenic factors. Given that Pramuka Island and its surrounding islands are not remote, these factors may act as stressors on seagrass ecosystems. Seagrasses exhibit seasonal growth patterns, with different species reaching maximum coverage at different times (Metz et al. 2020). Failing to account for this variability may lead to misinterpretations in annual classification data, potentially suggesting a decline when it is simply part of the natural growth cycle. Moreover, local disturbances, such as physical damage or eutrophication, can drive a transition from late-successional seagrass species to those more tolerant of dynamic conditions, such as earlier successional or invasive species that grow rapidly (James et al. 2023). Given the impact of these disturbances, establishing permanent station for continuous monitoring would provide a clearer understanding of seagrass dynamics year-round. Such surveys could offer a more comprehensive view of the relationship between seagrass meadows and external variables, while also helping assess the effectiveness of current conservation efforts or guiding new strategies. Furthermore, remote sensing has the potential to offer valuable, broad-scale insights into seagrass ecosystems, complementing ground-based surveys for more comprehensive monitoring.

5 Conclusion

We analyzed seagrass dynamics over eight years using cost-effective methods. While there is still room for improvement in the procedure, we have discussed these areas and made our study replicable by developing a Python module and providing supplementary notebooks for the analysis. With limited access to ground truth data, we observed changes in seagrass extent, both gains and losses. Therefore, future studies with more sophisticated field data and further development of our approach could help address the limitations of this study. We know that the spectral signature of seagrass is strong in the red region, but we have not fully understood the moderate contribution of SWIR1 based on the feature importance in our model. Understanding coastal ecosystems, including seagrass dynamics, will be crucial in future as the need for mitigation and restoration efforts becomes imminent. Our observation of potential correlation between sea surface temperature (SST) and seagrass dynamics requires further assessment. In general, we observed an inverse correlation between SST and seagrass dynamics, but this finding does not yet fully explain the complexities of these interactions. Additionally, incorporating anthropogenic factors will be essential for a more comprehensive understanding of the drivers behind changes in seagrass.

Acknowledgements We thank Abd. Rahman As-syakur for valuable advice and insights throughout this study. We also would like to thanks to Wikanti Asriningrum, the first author, and Azura Ulfa for providing the ground truth data in their research, and to their team for their contributions.

References

Asriningrum W, Ulfa A, Aziz K et al (2022) Platform Reef Lagoon Detection from Sentinel-2 in Panggang Island and Semak Daun Island, Seribu Archipelago, Jakarta. Int J Remote Sens Earth Sci 19:101–118

Badan Informasi Geospasial (BIG) (2014) PERATURAN KEPALA BADAN INFORMASI GEOSPASIAL NOMOR 8 TAHUN 2014 TENTANG PEDOMAN TEKNIS PENGUMPULAN DAN PENGOLAHAN DATA GEOSPASIAL HABITAT DASAR PERAIRAN LAUT DANGKAL. Indonesia

Badan Pusat Statistik (BPS) (2024) Kepulauan Seribu dalam Angka 2024. Badan Pusat Statistik (BPS), Jakarta

Bahri AD, Hamdani A, Wibowo A (2017) Di Balik Krisis Agraria dan Ekosistem Kepulauan Seribu: Apakah Wisata Bahari adalah Jawabannya? Bogor

Bannari A, Ali TS, Abahussain A (2022) The capabilities of Sentinel-MSI (2A/2B) and Landsat-OLI (8/9) in seagrass and algae species differentiation using spectral reflectance. Ocean Sci 18:361–388. https://doi.org/10.5194/os-18-361-2022

Basheer Ahammed KK, Karang IWGA, Nuarsa IW, Indrawan IGS, Hendrawan IG, Dewi NMNBS, Pandey AC (2024) Mapping of the spatio-spectral dynamics of mangrove chlorophyll concentrations via sentinel-2 satellite imagery. In: Forum Geografi, vol 38, no 2, pp 244–256

Breiman L (2001) Random forests. Mach Learn 45.https://doi.org/10.1023/A:1010933404324

Bugnot AB, Mayer-Pinto M, Airoldi L et al (2021) Current and projected global extent of marine built structures. Nat Sustain 4:33–41. https://doi.org/10.1038/s41893-020-00595-1

Buitinck L, Louppe G, Blondel M et al (2013) API design for machine learning software: experiences from the scikit-learn project. arXiv preprint arXiv:13090238

Carlson DF, Yarbro LA, Scolaro S et al (2018) Sea surface temperatures and seagrass mortality in Florida Bay: spatial and temporal patterns discerned from MODIS and AVHRR data. Remote Sens Environ 208:171–188. https://doi.org/10.1016/j.rse.2018.02.014

Clemente KJE, Thomsen MS, Zimmerman RC (2023) The vulnerability and resilience of seagrass ecosystems to marine heatwaves in New Zealand: a remote sensing analysis of seascape metrics using PlanetScope imagery. Remote Sens Ecol Conserv 9:803–819. https://doi.org/10.1002/rse2.343

Dalby O, Pucino N, Tan YM et al (2023) Identifying spatio-temporal trends in seagrass meadows to inform future restoration. Restor Ecol 31:e13787. https://doi.org/10.1111/rec.13787

Donlon CJ, Martin M, Stark J et al (2012) The operational sea surface temperature and sea ice analysis (OSTIA) system. Remote Sens Environ 116:140–158. https://doi.org/10.1016/j.rse.2010.10.017

Drusch M, Del Bello U, Carlier S et al (2012) Sentinel-2: ESA's optical high-resolution mission for GMES operational services. Remote Sens Environ 120:25–36. https://doi.org/10.1016/j.rse.2011.11.026

Duffy JP, Pratt L, Anderson K et al (2018) Spatial assessment of intertidal seagrass meadows using optical imaging systems and a lightweight drone. Estuar Coast Shelf Sci 200:169–180. https://doi.org/10.1016/j.ecss.2017.11.001

Dunic JC, Brown CJ, Connolly RM et al (2021) Long-term declines and recovery of meadow area across the world's seagrass bioregions. Glob Chang Biol 27:4096–4109. https://doi.org/10.1111/gcb.15684

Fadhlurahman YN, Manessa MDM, Semedi JM et al (2024) Multitemporal analysis of seagrass dynamics on Derawan Island (2003–2021) using remote sensing techniques. Ilmu Kelaut 29:1–14. https://doi.org/10.14710/ik.ijms.29.1.1-14

Farhan AR, Lim S (2011) Resilience assessment on coastline changes and urban settlements: a case study in Seribu Islands, Indonesia. Ocean Coast Manage 54:391–400. https://doi.org/10.1016/j.ocecoaman.2010.12.003

Feyisa GL, Meilby H, Fensholt R, Proud SR (2014) Automated water extraction index: a new technique for surface water mapping using Landsat imagery. Remote Sens Environ 140:23–35. https://doi.org/10.1016/j.rse.2013.08.029

Frederiksen M, Krause-Jensen D, Holmer M, Laursen JS (2004) Spatial and temporal variation in eelgrass (Zostera marina) landscapes: influence of physical setting. Aquat Bot 78:147–165. https://doi.org/10.1016/j.aquabot.2003.10.003

Gitelson AA, Kaufman YJ, Merzlyak MN (1996) Use of a green channel in remote sensing of global vegetation from EOS—MODIS. Remote Sens Environ 58:289–298. https://doi.org/10.1016/S0034-4257(96)00072-7

Gitelson AA, Kaufman YJ, Stark R, Rundquist D (2002) Novel algorithms for remote estimation of vegetation fraction. Remote Sens Environ 80:76–87. https://doi.org/10.1016/S0034-4257(01)00289-9

Green EP, Mumby PJ, Edwards AJ, Clark CD (2000) Remote sensing handbook for tropical coastal management. UNESCO

Hamuna B, Pujiyati S, Gaol JL, Hestirianoto T (2023) Spatial distribution of benthic habitats in Kapota Atoll (Wakatobi National Park, Indonesia) using remote sensing imagery. Biodiversitas 24. https://doi.org/10.13057/biodiv/d240706

Hedley JD, Harborne AR, Mumby PJ (2005) Technical note: simple and robust removal of sun glint for mapping shallow-water benthos. Int J Remote Sens 26:2107–2112. https://doi.org/10.1080/01431160500034086

Hendriks IE, Olsen YS, Ramajo L et al (2014) Photosynthetic activity buffers ocean acidification in seagrass meadows. Biogeosciences 11:333–346. https://doi.org/10.5194/bg-11-333-2014

Huete A, Didan K, Miura T et al (2002) Overview of the radiometric and biophysical performance of the MODIS vegetation indices. Remote Sens Environ 83:195–213. https://doi.org/10.1016/S0034-4257(02)00096-2

James RK, Keyzer LM, van de Velde SJ et al (2023) Climate change mitigation by coral reefs and seagrass beds at risk: how global change compromises coastal ecosystem services. Sci Total Environ 857:159576. https://doi.org/10.1016/j.scitotenv.2022.159576

Joyce K, Phinn S, Roelfsema C (2013) Live coral cover index testing and application with hyper-spectral airborne image data. Remote Sens (Basel) 5:6116–6137. https://doi.org/10.3390/rs5116116

Karang IWGA, Pravitha NLPR, Nuarsa IW et al (2024) High-resolution seagrass species mapping and propeller scars detection in Tanjung Benoa, Bali through UAV imagery. J Ecol Eng 25:161–174. https://doi.org/10.12911/22998993/174943

Koch M, Bowes G, Ross C, Zhang X (2013) Climate change and ocean acidification effects on seagrasses and marine macroalgae. Glob Chang Biol 19:103–132. https://doi.org/10.1111/j.1365-2486.2012.02791.x

Kuhwald K, Schneider von Deimling J, Schubert P, Oppelt N (2022) How can sentinel-2 contribute to seagrass mapping in shallow, turbid Baltic Sea waters? Remote Sens Ecol Conserv 8:328–346. https://doi.org/10.1002/rse2.246

Kumar KCVN, Ahammed KB, Lele N, Nimmi N, Kubo S, Rao KN, Harikumar PS, Gopinath G, Bhattacharya BK (2024) Assessing the significance of leaf chlorophyll content for measuring the vegetation health in mangrove species along the Kerala coast, India–a multi-proxy approach. Regional Stud Mar Sci 77:103702

Lazuardi W, Wicaksono P, Marfai MA (2021) Remote sensing for coral reef and seagrass cover mapping to support coastal management of small islands. IOP Conf Ser Earth Environ Sci 686:012031. https://doi.org/10.1088/1755-1315/686/1/012031

Leblanc M-L, LaRocque A, Leblon B et al (2021) Using landsat time-series to monitor and inform seagrass dynamics: a case study in the Tabusintac Estuary, New Brunswick, Canada. Can J Remote Sens 47:65–82. https://doi.org/10.1080/07038992.2021.1893672

Li Y, Bai J, Chen S et al (2023) Mapping seagrasses on the basis of Sentinel-2 images under tidal change. Mar Environ Res 185:105880. https://doi.org/10.1016/j.marenvres.2023.105880

Liang H, Wang L, Wang S et al (2023) Remote sensing detection of seagrass distribution in a marine lagoon (Swan Lake), China. Opt Express 31:27677. https://doi.org/10.1364/OE.498901

Madden CJ, Rudnick DT, McDonald AA et al (2009) Ecological indicators for assessing and communicating seagrass status and trends in Florida Bay. Ecol Indic 9:S68–S82. https://doi.org/10.1016/j.ecolind.2009.02.004

Maxwell AE, Warner TA (2020) Thematic classification accuracy assessment with inherently uncertain boundaries: An argument for center-weighted accuracy assessment metrics. Remote Sens (Basel) 12:1905. https://doi.org/10.3390/rs12121905

Mederos-Barrera A, Marcello J, Eugenio F, Hernández E (2022) Seagrass mapping using high resolution multispectral satellite imagery: a comparison of water column correction models. Int J Appl Earth Obs Geoinf 113:102990. https://doi.org/10.1016/j.jag.2022.102990

Metz JL, Harris RJ, Arrington DA (2020) Seasonal occurrence patterns of seagrass should influence resource assessment and management decisions: a case study in the Indian River Lagoon and Loxahatchee River Estuary. Florida. Reg Stud Mar Sci 34:101093. https://doi.org/10.1016/j.rsma.2020.101093

Misiuk B, Brown CJ (2024) Benthic habitat mapping: a review of three decades of mapping biological patterns on the seafloor. Estuar Coast Shelf Sci 296:108599. https://doi.org/10.1016/j.ecss.2023.108599

Ningsih EN, Setiawan A, Hartoni, Fauziyah (2021) Perubahan luasan Pulau Pramuka, Pulau Panggang dan Pulau Karya Kepulauan Seribu DKI Jakarta menggunakan data citra satelit penginderaan jauh

Orth RJ, Heck KL (2023) The dynamics of seagrass ecosystems: history, past accomplishments, and future prospects. Estuaries Coasts 46:1653–1676. https://doi.org/10.1007/s12237-023-01252-4

Phinn S, Roelfsema C, Kovacs E et al (2018) Mapping, monitoring and modelling seagrass using remote sensing techniques. Seagrasses of Australia. Springer International Publishing, Cham, pp 445–487

Rahman TA, Supriatna S, Saraswati R (2020) Coral reefs distribution using Sentinel-2A imagery in Belanda Island, Bira Besar Island and Pramuka Island at Kepulauan Seribu National Park. IOP Conf Ser Earth Environ Sci 538:012004. https://doi.org/10.1088/1755-1315/538/1/012004

Roelfsema C, Kovacs EM, Saunders MI et al (2013) Challenges of remote sensing for quantifying changes in large complex seagrass environments. Estuar Coast Shelf Sci 133:161–171. https://doi.org/10.1016/j.ecss.2013.08.026

Rouse JW, Haas RH, Schell JA, Deering DW (1974) Monitoring vegetation systems in the Great Plains with ERTS. NASA special publication, NASA special publication, p 24

Saraswati M, Yulianda F, Samosir AM (2021) Kondisi Ekosistem Lamun Pasca Pembangunan Pelabuhan Baru di Pulau Pramuka. Skripsi, IPB University, Taman Nasional Kepulauan Seribu

Setiawati MD, Chatterjee U, Djamil YS et al (2023) Seribu islands in the megacities of Jakarta on the frontlines of the climate crisis. Front Environ Sci 11:1280268. https://doi.org/10.3389/fenvs.2023.1280268

Setiawati MD, Ismayanti GA, Hafizt M et al (2024) Changes in Benthic habitat under climate pressure in Western Papua, Indonesia: remote sensing-based approach. J Indian Soc Remote Sens 52:291–304. https://doi.org/10.1007/s12524-024-01813-5

Tassi A, Gigante D, Modica G et al (2021) Pixel- vs. Object-based Landsat 8 data classification in google earth engine using random forest: the case study of Maiella National Park. Remote Sens (Basel) 13:2299. https://doi.org/10.3390/rs13122299

Tis'in M, Ambo-Rappe R, Supriadi S, Faizal A (2023) Decadal remote sensing analysis of seagrass changes in Palu Bay, Central Sulawesi. Indones J Geogr 55:291. https://doi.org/10.22146/ijg.78701

Unsworth RKF, McKenzie LJ, Collier CJ et al (2019) Global challenges for seagrass conservation. Ambio 48:801–815. https://doi.org/10.1007/s13280-018-1115-y

Utami RT, Zamani NP, Maddupa HH (2018) Molecular identification, abundance and distribution of the coral-killing sponge Terpios hoshinota in Bengkulu and Seribu Islands, Indonesia. Biodiversitas 19:2238–2246. https://doi.org/10.13057/biodiv/d190632

Vanhellemont Q (2019) Adaptation of the dark spectrum fitting atmospheric correction for aquatic applications of the Landsat and Sentinel-2 archives. Remote Sens Environ 225:175–192. https://doi.org/10.1016/j.rse.2019.03.010

Vanhellemont Q, Ruddick K (2014) Turbid wakes associated with offshore wind turbines observed with Landsat 8. Remote Sens Environ 145:105–115. https://doi.org/10.1016/j.rse.2014.01.009

Vanhellemont Q, Ruddick K (2018) Atmospheric correction of metre-scale optical satellite data for inland and coastal water applications. Remote Sens Environ 216:586–597. https://doi.org/10.1016/j.rse.2018.07.015

Veettil BK, Ward RD, Lima MDAC et al (2020) Opportunities for seagrass research derived from remote sensing: a review of current methods. Ecol Indic 117:106560. https://doi.org/10.1016/j.ecolind.2020.106560

Wicaksono P, Wulandari SA, Lazuardi W, Munir M (2021) Sentinel-2 images deliver possibilities for accurate and consistent multi-temporal benthic habitat maps in optically shallow water. Remote Sens Appl 23:100572. https://doi.org/10.1016/j.rsase.2021.100572

Widya LK, Kim C-H, Do J-D et al (2023) Comparison of satellite imagery for identifying seagrass distribution using a machine learning algorithm on the eastern coast of South Korea. J Mar Sci Eng 11:701. https://doi.org/10.3390/jmse11040701

Williamson P, Guinder VA (2021) Effect of climate change on marine ecosystems. In: The impacts of climate change. Elsevier, pp 115–176

Xu H (2006) Modification of normalised difference water index (NDWI) to enhance open water features in remotely sensed imagery. Int J Remote Sens 27:3025–3033. https://doi.org/10.1080/01431160600589179

Establishing Baseline Data and Monitoring Strategies for Coral Reefs in the Grand Island, Goa, and Palk Bay, Tamil Nadu, Through 3D Photogrammetry

Satya Swaroop Nanda, Ramesh Chatragadda, Narayan Dalvi, Samson Rokkarukala, and Ashish Aswin Kumar

Abstract Coral reef ecosystems play a vital role in sustaining the health of oceanic life and the livelihoods of coastal population. Despite their importance, these reefs face severe threats from climate change, pollution, and human activities. Hence, continuous monitoring of coral reefs around the world has become a vital research to address reef health, reef restoration, reef resources, reef services, and coastal livelihood. Existing reef monitoring methods are limited in their capacity to deliver cost-effective, noninvasive, and objective evaluations of coral reef structure and complexity and fail to provide a permanent record of reef images, making photogrammetry a more effective alternative. This is the first ever study to employ 3D photogrammetry method and establish the baseline 3D photogrammetry-based coral reef maps in India. By establishing permanent monitoring stations and utilizing advanced photogrammetry, this study aims to provide a comprehensive assessment of the coral cover status, reef structure, and complexity of coral reefs at Grand Island, Goa, and Palk Bay, Tamil Nadu. A detailed transect mosaic reconstructions were generated through in situ underwater photogrammetry to evaluate coral cover, health indicators, and substrate composition. In the Grande Island region, Navy Bay and Shelter Cove were surveyed, revealing *Turbinaria* sp. as a dominant coral with a total live coral cover of 41.16 ± 18.77%. Dead coral with algae (DCA) represents 13.3 ± 5.77% of the surveyed area. Significantly, the coral killing invasive encrusting sponge, *Cliona*, impacting *Turbinaria* sp. and *Favites* sp. was observed. Coral substrate was primarily composed of sand, gravel, and coral rubble, covering 44.65 ± 14.33% of the transects. In the Palk Bay region, surveys were conducted at Mandapam Camp and Rameswaram, where *Acropora* sp. emerged as the dominant coral genus, with a total live coral cover of 32.17 ± 13.86%. However, the reef ecosystem in Palk Bay

S. S. Nanda · R. Chatragadda (✉) · N. Dalvi · S. Rokkarukala · A. A. Kumar
Biological Oceanography Division, CSIR-National Institute of Oceanography, Goa, India
e-mail: chramesh.nio@csir.res.in

R. Chatragadda
Academy of Scientific and Innovative Research (AcSIR), Ghaziabad, Uttar Pradesh, India

K. K. Basheer Ahammed et al. (eds.), *Monitoring Coastal Marine Ecosystems Under Climate Change*, https://doi.org/10.1007/978-3-032-09137-6_7

is evidently under degradation, with apparent evidences of broken and dead fragments of *Acropora* sp. constituting a major component of the reef substrate (64.27 $\pm$ 7.03%). This degradation is likely attributed to heavy boat traffic and fishing, which has caused extensive fragmentation of the coral colonies. Additionally, the infestation of Serpulidae worms and prevalence of pink spot on *Porites* sp. represented 2.02% cover of one transect. Coral-infesting sponges were also observed, with diseased and infected corals covering 0.6 $\pm$ 0.05% of the surveyed area. The anthropogenic impact in Palk Bay was more pronounced than in Grande Island, as evidenced by toppled *Acropora* sp. colonies, likely caused by boat anchor dragging, and the alarming discovery of plastic cement bags covering coral colonies. These bags smother and weaken corals, promoting the growth of turf algae and sponges, further compromising reef health. The absence of reef-associated fishes across survey sites highlights the severe impact of overfishing and other human activities. Compared to the relatively well-maintained reefs of Grande Island, the proximity of Palk Bay to the shore has exacerbated the degradation. Continued long-term monitoring of Indian reefs is in progress, aiming to develop a permanent repository of photo mosaics that will provide a valuable baseline dataset for future studies and research endeavors. The findings from this assessment underscore the urgent need for ongoing monitoring, expanded surveys, and targeted conservation efforts to mitigate further damage and restore the health of coral reef ecosystems in these regions.

Keywords Scleractinia · Coral reef monitoring · 3D photogrammetry · Reef resilience · Coral-associated biodiversity · Coral reef invasive species

1 Introduction

Corals are invertebrates from the phylum Cnidaria, which also contains Portuguese men-of-war, anemones, jellyfishes, and hydroids. These organisms are identified by their radial symmetry, lack of a head, and often a crown of tentacles enclosing the mouth. Corals reside within calcium carbonate skeletons produced by their tissue (Birkeland 2015). A coral's living tissue is made up of two layers: an outer layer called the epidermis, which secretes the coral skeleton, and an inner layer called the gastrodermis. Between these layers is a thin, fibrous matrix known as the mesoglea, which provides buoyancy and support (Branes, 1987). Within the gastrodermal layer of most corals are zooxanthellae—microalgae capable of photosynthesis, which produce energy-rich sugars by harnessing sunlight. This relationship between corals and zooxanthellae is a mutualistic symbiotic association, where both parties benefit. The zooxanthellae find shelter, sunlight, and a stable environment within the coral tissue, while also gaining nutrients from the coral's waste products. In return, the coral benefits from the energy-rich compounds produced by the zooxanthellae and the removal of waste from its cells. Coral polyps typically retract into their protective skeletons diurnally to avoid predation. Nocturnally, the polyps extend and deploy a network of tentacles equipped with stinging cells, enabling the coral

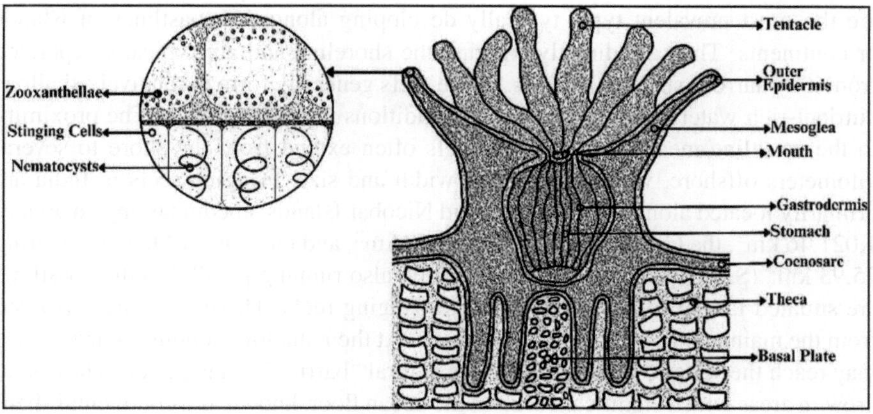

Fig. 1 Anatomy of coral polyp. *Illustration* ©Satya Swaroop Nanda

to capture and consume plankton and other small organisms from the surrounding water column (Fig. 1) (Barnes, 1987; Barnes and Hughes 1999; Lalli and Parsons 1995; Levinton 1995).

There are two main categories of corals: Hermatypic corals, also known as Scleractinians (hard corals), build unique and essential ecosystems, harboring extraordinary biodiversity and providing critical ecological services, despite their limited coverage; a recent study by Lyons et al. (2024) estimated the global shallow coral reef cover at 348,361 km^2, with coral reef habitat specifically accounting for 80,213 km^2. These corals have a remarkable ability to secrete calcium carbonate, which forms a solid, rigid skeleton that supports the entire coral colony. As new coral generations grow on the skeletal remains of older ones, they contribute to the expansion and further development of the reef. The structure of hermatypic corals is composed of individual polyps, each residing within a calcareous, cup-like exoskeleton called a corallite. Inside the corallite, the polyp extends its tentacles to capture food and interact with its surroundings. In contrast, Ahermatypic corals, commonly referred to as soft corals or non-reef-building corals, differ significantly from their hard coral counterparts. Unlike hermatypic corals, soft corals do not produce a rigid calcium carbonate skeleton; instead, they have a more flexible and pliable structure, often giving them a fleshy appearance. Though they do not play a major role in the formation of large coral reefs, soft corals are nonetheless essential to the reef ecosystem. They coexist with hard corals, enhancing the overall biodiversity and complexity of the reef environment. Soft corals exhibit a wide range of shapes, colors, and sizes, adding to the beauty of the underwater world (Veron and Stafford-Smith 2000).

India's coral reefs encompass around 5790 km^2 and contribute significantly to the country's marine biodiversity (Raghuraman et al., 2013). Coral reefs in Indian waters are classified into three types: fringing reefs, barrier reefs, and atolls. They are primarily found in the Andaman and Nicobar Islands, Gulf of Kutch, Gulf of Mannar, and Lakshadweep Islands (Raghuraman et al., 2013). Fringing reefs are the most prevalent type, typically developing along the coastlines of islands or continents. They are directly fringing the shoreline and are typically separated from it by narrow, shallow lagoons. These reefs generally form in relatively shallow, nutrient-rich waters, providing optimal conditions for coral growth. The proximity to the coastline means that fringing reefs often extend from the shore to several kilometers offshore, with variations in width and size. Fringing reefs in India are primarily located along the Andaman and Nicobar Islands, encompassing an area of 1,021.46 km^2, the Gulf of Kutch with 352.50 km^2, and the Gulf of Mannar covering 75.93 km^2 (SAC, 2010). Barrier reefs, while also running parallel to the coastline, are situated farther offshore compared to fringing reefs. These reefs are separated from the mainland by deeper, wider lagoons. At their shallowest points, barrier reefs may reach the water's surface, forming a natural "barrier" to navigation. They often grow in areas with a significant drop in the ocean floor, known as a continental shelf. Barrier reefs in India are notably present in the Andaman and Nicobar Islands (SAC 2010). These reefs play a crucial role in protecting coastlines and supporting marine biodiversity. Atolls are distinct from other types of reefs due to their circular or oval shape, which encloses a central lagoon. Typically found in the middle of the ocean, far from any landmass, atolls often originate from the remnants of volcanic islands surrounded by fringing reefs. Over time, as these volcanic islands subside or erode, the fringing reefs continue to grow, eventually forming a ring-shaped reef structure enclosing a lagoon. In India, atolls are primarily located in the Lakshadweep Islands, covering an area of 933.7 km^2 (SAC 2010).

In addition to the major reef types, additional formations such as patch reefs and bank reefs are distributed within the Indian Exclusive Economic Zone (EEZ). Patch reefs, generally smaller and isolated, are commonly situated within the lagoons of larger reef systems and are distinguished by their relatively low generic diversity. Coral reefs in India's Exclusive Economic Zone make up 2.48% of the global overall tropical coral reef areas (http://www.seaaroundus.org). Significant discoveries in Indian waters include the identification of a living coral bank approximately 100 km off the coast of Malpe, north of Mangalore, known as the Gaveshani bank (Nair and Qasim 1978) and Angria bank located 130–135 km off the Indian mainland along the central west coast in the Arabian Sea (Nair et al. 1966). Hermatypic corals have also been documented along the shoreline, stretching from Quilon in Kerala to Enayam in Tamil Nadu (Pillai 1996). Qasim and Wafar (1979) documented the occurrence of corals in the intertidal zones along India's central west coast, particularly in Malvan, Redi, and Ratnagiri. On the east coast, Ramaiyan and Adiyapatham (1985) reported the presence of corals between Parangipettai (Porto Novo), located south of Cuddalore (10°50'N, 79°80'E), and Pondicherry. Further, Vora and Almeida (1990) identified a subaqueous reef system along the continental shelf of the same coastal region, extending between Vengurla and Vijaydurga at depths ranging from 60 to 110 m. Collectively, these studies underscore the diversity and ecological significance of coral reef ecosystems in Indian waters, emphasizing their vital role in supporting marine biodiversity and maintaining regional ecosystem health.

Although the combined area of all continents, islands, and freshwater habitats is more than 460 times larger than that of coral reefs, these regions contain only

19 animal phyla, while coral reefs support at least 30 (Birkeland, 2015). Structural complexity is one of the most striking characteristics of coral reef ecosystems. Coral reefs act as ecosystem engineers or foundation species by forming intricate three-dimensional structures through the precipitation of calcium carbonate (Samson et al., 1996). These structures create a variety of microhabitats that support breeding, feeding, and refuge for many marine species (Darling et al., 2017; Friedlander et al., 2003; Graham and Nash, 2013; Quadros et al., 2019). Although they cover only a tiny fraction of the Earth's surface—about 0.00063%—coral reefs have a profound influence on global systems, affecting atmospheric dynamics, ocean chemistry, and the shape of coastlines (Birkeland, 2015). They contribute significantly to climate regulation by functioning as carbon sinks, sequestering large amounts of atmospheric carbon dioxide (Birkeland, 2015). Additionally, coral reefs offer natural coastal defense against sea-level rise and storm impacts, thereby helping to protect shorelines and sustain marine biodiversity (Moberg and Folke, 1999). Beyond their ecological roles, coral reefs also underpin the livelihoods of around 500 million people globally, providing vital services such as fisheries, shoreline protection, medicinal resources, and tourism opportunities (Moberg and Folke, 1999).

Coral reef ecosystems around the world have declined significantly and are at high risk of extinction as a result of climate change and anthropogenic disruptions. Recent years have seen an increase in elevated temperature-driven coral bleaching events, leading to worldwide coral loss and leading to extensive environmental consequences (Hughes et al., 2017) (Fig. 2). Anthropogenic processes have increased atmospheric carbon dioxide concentrations by 36%, which has led to a decline in the pH of ocean surface waters by about 0.1 units, from 8.2 to 8.1, since the dawn of the age of industrialization (Caldeira and Wickett, 2003). This uptake of CO_2 by the ocean leads to a reduction in pH and a decrease in carbonate ion (CO_32^-) concentration, leading to a series of chemical alterations commonly known as ocean acidification (Orr et al., 2005). Ocean acidification presents a severe threat to coral reefs, with predictions indicating that by 2050, these reefs might shift from net calcium carbonate deposition to net dissolution (Eyre et al., 2018). The emerging environmental stressors have led to a significant decline in coral reproduction and recruitment capacity, gradually impairing the resilience of coral ecosystems (Done et al., 2010; Hughes et al., 2003, 2010). The sensitivity of corals to climate change and their limited resilience can be traced back to the global decline in coral cover that began in the late 1970s to early 1980s (Gardner et al., 2003; Bruno and Selig, 2007; Schutte et al., 2010). Coral cover in the Indo-Pacific region, for example, dropped from 42.5% in the early 1980s to 22.1% by 2003, with a projected decline rate of 1–2% annually (equivalent to 3168 km^2 per year) across the region (Bruno and Selig, 2007). The situation is similarly dire in other regions. According to a recent study, the cumulative impact of local and global stressors has led to a 50% reduction in reef-building corals across the Great Barrier Reef, one of the world's most biologically diverse marine ecosystems (Lough et al., 2018). If the ongoing trend of coral degradation persists, coral reef ecosystems are likely to evolve into environments dominated by organisms such as cyanobacteria, algae, sponges, and invertebrates (Bell et al., 2013; Bellwood et al., 2004). Future projections suggest that coral cover could potentially decrease by

70–90% compared to current levels by 2050 if carbon dioxide emissions persist at their current rate, even if the Paris Climate Agreement's targets are achieved (Beyer et al., 2018; Hoegh-Guldberg et al., 2018). A recent study by Wasim et al. (2024) examined coral reefs on selected Pacific islands, including Heron Island in the Great Barrier Reef, Namena Island in Fiji, and Aitutaki Island in the Cook Islands. The researchers used remote sensing datasets and the Depth Invariant Index algorithm to map and classify coral features and identify areas affected by coral bleaching. The results showed decline of coral reefs significantly across the islands due to bleaching, although some adapted corals showed resilience. El Niño was identified as a major contributor to the 2015–2016 bleaching events, while rising sea levels and increased tourism in 2020 led to higher turbidity, exacerbating coral bleaching. Additionally, high sea surface temperatures (SST) of over 30 °C and elevated photosynthetically active radiation (PAR) levels of 47 E/m^2/day caused photoinhibition, negatively affecting coral health. Fluctuating ocean salinity from 2002 to 2020 also harmed coral tissues.

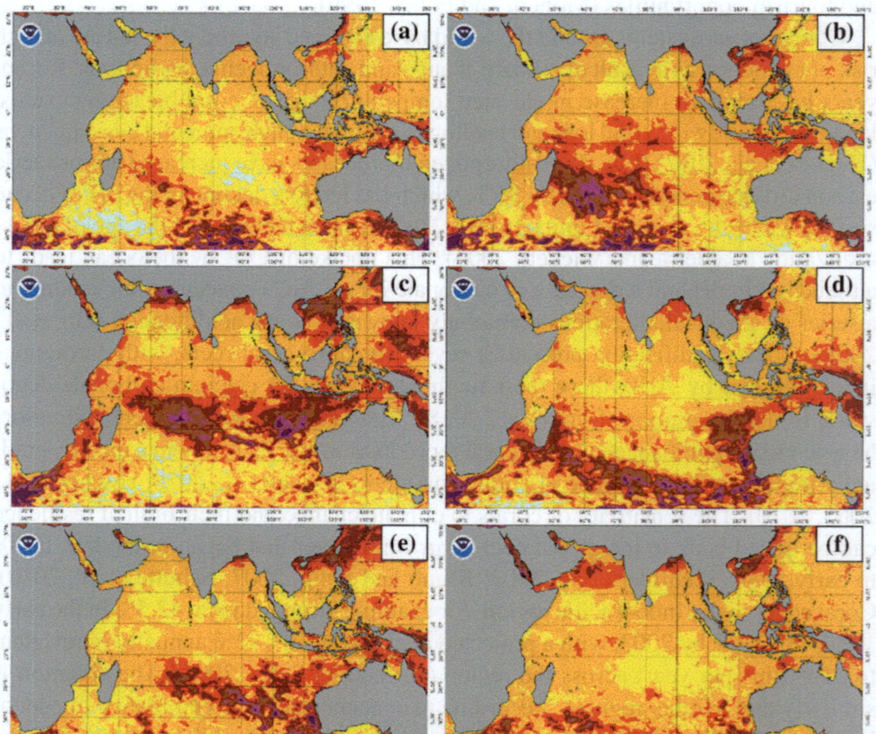

Fig. 2 Coral bleaching advisory for Indian ocean reef ecosystems **a** 2018, **b** 2019, **c** 2020, **d** 2021, **e** 2022, **f** 2023. *Data source* NOAA-CRW 5 km SST data; data were visualized using NOAA-CRW platform version 3.1, https://coralreefwatch.noaa.gov/

In India, mass coral bleaching events have caused extensive coral mortality. Arthur (2000) reported varying levels of coral bleaching and mortality in three different regions of India during the period from April to July 1998. In the Gulf of Kutch, 11% of the corals experienced bleaching, but there was no noticeable death of corals due to this bleaching. In Lakshadweep, the situation was more severe, with 82% of the corals bleached, and 26% of them died as a result. Similarly, in the Gulf of Mannar, 89% of the corals were bleached, and 23% of them succumbed to the bleaching. Harithsa et al. (2005) observed that in April and May 2002, the seawater temperatures in the Kavaratti Atoll, Lakshadweep, exceeded the bleaching threshold, leading to widespread coral bleaching. While the typical monthly mean seawater temperature in the Lakshadweep Archipelago ranges between 28 and 29 °C throughout the rest of the year, during April and May 2002, the region experienced unusually high temperatures, reaching 31–32 °C. More recently, Ravindran et al. (2012) studied the alarming rate of coral reef degradation in Palk Bay, driven by a combination of coral bleaching and other stressors. They reported that elevated sea surface temperatures, combined with increased irradiance, led to coral bleaching in Palk Bay during April 2010. Similarly, Marimuthu et al. (2013) documented extensive coral bleaching of several coral species in the South Andaman Islands during the 2010 bleaching event, with minimal signs of recovery. Ramesh (2020) reported extensive coral bleaching in the Gulf of Mannar and Palk Bay regions, particularly in shallower waters. Corals such as *Porites solida*, *Porites lutea*, *Montipora digitata*, and *Acropora hyacinthus* were completely bleached at depths of zero to two meters. At slightly greater depths of two to four meters, partial bleaching was observed in species including *Acropora formosa*, *A. hyacinthus*, *Montipora digitata*, *M. foliosa*, *Pocillopora damicornis*, *Goniastrea retiformis*, *Platygyra sinensis*, *Dipsastraea favus*, *Turbinaria* sp., and *Dipsastraea speciosa*. However, in the Palk Bay region, all *Porites* species observed at these depths were entirely bleached. Interestingly, corals located at depths greater than five meters showed no signs of bleaching, suggesting that deeper waters might provide some protection against rising sea temperatures. The sea surface temperature in these areas has reached 36 °C, far exceeding the bleaching threshold of 30.60 °C identified for the Gulf of Mannar (Arora et al., 2021). Notably, in some Gulf of Mannar sites, massive *Porites* species were completely bleached, while nearby colonies of branching corals such as *Montipora digitata* and *Acropora* species remained unaffected (Ramesh et al., 2020). The situation in the Palk Bay region is particularly severe, with 98% of *Porites* colonies showing signs of bleaching (Ramesh et al., 2020). Hussain and Ingole (2020) conducted a series of surveys around Grande Island, Goa, examining coral bleaching and reef health at two sites with varying depths. Site 1, the shallow site (2–4 m), and site 2, the deeper site (4–8 m), showed mild bleaching during October 2014, with 6% and 4% affected corals at sites 1 and 2, respectively. By April 2015, bleaching increased slightly to 7% and 5.34%. However, a significant rise in bleaching was recorded in November 2015, with 51% of corals at the first site and 50% at the second site being affected. By April 2016, recovery signs were observed in *Porites* colonies, with bleaching rates decreasing to 7% at site 1 and 8% at site 2. Despite this recovery, the study also reported an increase in turf algae, macroalgae, and rubble, with their composition rising from a mean of 17.88% in November 2015 to 35.83% in April 2016. This suggests that although coral recovery was underway, other ecological changes, such as increased algal growth, were impacting the reef's overall health.

Coral reefs hold vital ecological and socio-economic significance, but they are facing increasing degradation. Effective monitoring and protection of these vital ecosystems are essential, yet the process can be costly and require specialized technical expertise. Due to the remote locations of many coral reefs and the need for scuba diving, long-term monitoring often results in fragmented or spatially limited datasets (Hughes et al., 2010). To reduce costs and broaden monitoring efforts, the use of underwater digital photography has expanded across modest spatial scales (Aronson et al., 1994; Ninio et al., 2003; Ninio and Meekan, 2002). However, analyzing the digital data in these images—such as RGB intensity and texture—requires a significant amount of time from experts, delaying the availability of information that is critical for conservation decisions. This creates a bottleneck in the flow of data from monitoring programs to conservation practitioners (Gonzalez-Rivero et al., 2020). A promising solution to this bottleneck is the application of artificial intelligence (AI), specifically through machine learning. AI, in this context, refers to the ability of computers to simulate aspects of human cognition, such as learning and decision-making, to autonomously complete specific tasks (Russell and Norvig, 2016). Machine learning, a branch of computer science, allows computers to learn tasks without being explicitly programmed for them (Samuel, 1959). Increasingly, these applications have leveraged deep learning techniques (LeCun et al., 2015), which enable the automatic discovery of features within input images that are necessary for classification. This capability makes deep learning highly effective at resolving complex patterns in high-dimensional data (LeCun et al., 2015).

In the present study, we employed 3D photogrammetry, a highly accurate and noninvasive technology for creating comprehensive photomosaics of reef structures (Urbina-Barreto et al., 2021). These photomosaics produce detailed visual records, capturing the reefs' current state as a benchmark for future research. Our primary objective is to establish permanent monitoring stations at key sites, enabling systematic assessments of coral reef health on a seasonal and annual basis. By employing photogrammetry, we created baseline data on coral cover and evaluated the present condition of reefs at Grande Island, Goa, and Palk Bay, Tamil Nadu. The establishment of permanent transects ensures that future surveys and data collection are consistent and accurate. Continuous monitoring of these transects, combined with regular photomosaic updates, is key for tracking changes in coral cover, species composition, and structural complexity—all of which are important indications of reef resilience and deterioration. These insights will drive conservation efforts and inform management practices aimed at mitigating the effects of climate change, pollution, and unsustainable fishing, ensuring the long-term protection of these ecosystems.

2 Materials and Methods

2.1 Study Sites

A series of systematic SCUBA surveys were conducted across two coral reef sites in India (Fig. 3) from November 2023 to January 2024.

(i) Grande Island, Goa: Reef assessments were carried out at two sites on Grande Island: Navy Bay (15°21′10.6128″N, 73°47′6.684″E) and Shelter Cove (15°21′11.4912″N, 73°46′9.4332″E). Located approximately 18 km off the

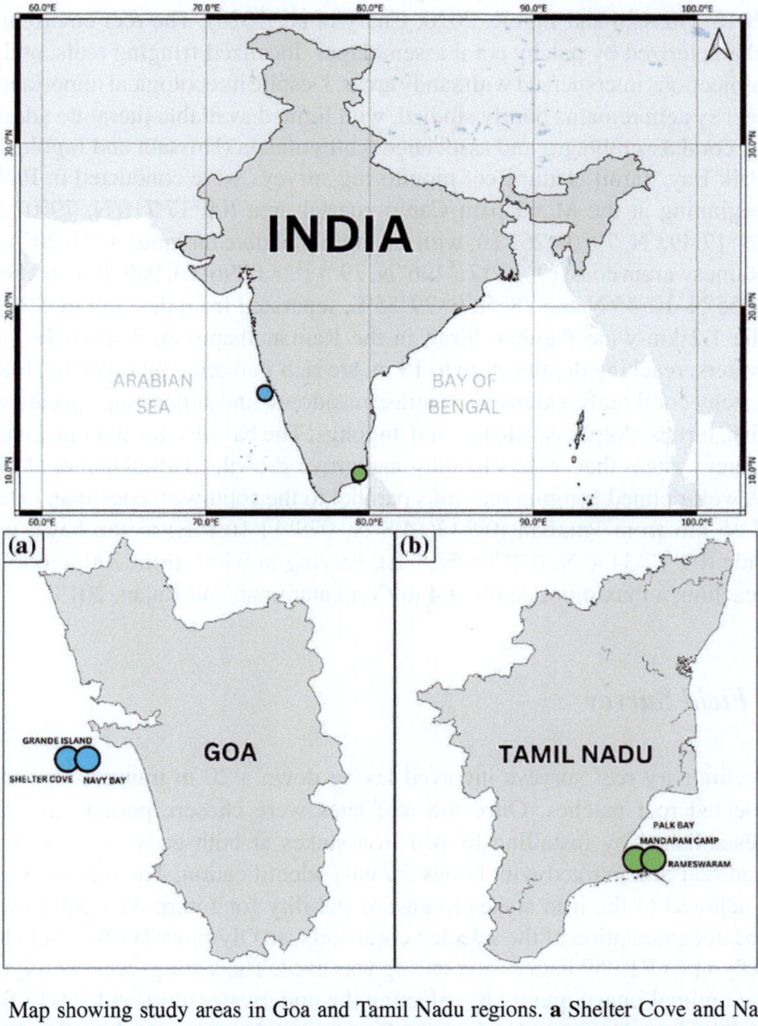

Fig. 3 Map showing study areas in Goa and Tamil Nadu regions. **a** Shelter Cove and Navy Bay sites in Grande Island, Goa. **b** Mandapam Camp and Rameswaram in Palk Bay, Tamil Nadu

coast of Goa near Mormugao Port, Grande Island lies along India's Konkan coastline. The marine environment surrounding the island is influenced by freshwater inputs from the Mandovi and Zuari rivers (Chary et al., 2020). Grande Island consists of two distinct landmasses—the smaller São Jorge Island and the larger Ilha Grande, the latter commonly referred to simply as Grande Island (Hussain and Ingole, 2020). The islands are predominantly composed of granite and laterite substrates. Extending from the island, rocky cliffs continue underwater, providing hard substrates conducive to coral colonization. Notably, the southern coast hosts Suzy's Wreck, an old shipwreck that now functions as an artificial reef, with its structure reaching depths of up to 14 m (Sreekanth et al., 2019). Coral communities have been recorded to this depth, with greater coral cover and diversity typically found between 6 and 8 m (Manikandan et al., 2016; Hussain and Ingole, 2020; Chary et al., 2020). The reef environment is characterized by patchy coral assemblages, localized fringing reefs, and rocky projections interspersed with sandy areas. Despite its ecological importance, this reef system remains poorly studied, with limited available literature addressing its coral assemblages and associated ichthyofauna (Hussain and Ingole, 2020).

(ii) Palk Bay, Tamil Nadu: Reef monitoring surveys were conducted in Palk Bay, beginning at the Mandapam Camp coastal area (09°17.716′N, 79°07.385′E; 09°17.493′N, 79°09.262′E), with an additional dive on January 23, 2024, at the Rameswaram coast (9°19′17.3496"N, 79°17′28.0896"E). Palk Bay lies between 9°55′N–10°45′N and 78°58′E–79°55′E, separated from the Gulf of Mannar by the 1.2 km-wide Pamban Strait in the Ramanathapuram district. Its shallow waters, reaching depths of up to 13 m, are rich in marine biodiversity, featuring patchy coral reefs, extensive seagrass meadows, and supporting species such as fish, turtles, dolphins, whales, and dugongs. The bay also hosts a brackish backwater system that fosters healthy mangrove growth (Sadhukhan et al., 2024). A well-defined fringing reef runs parallel to the southwest coastline, extending 7.46 km from Pamban (09°17′14.8"N, 079°11′16.4"E) to the bay's western side (09°17′34.4"N, 079°06′57.8"E), varying in width from 200 to 600 m and reaching a maximum depth of 4 m (Venkataraman and Rajan, 2013).

2.2 *Field Survey*

The preliminary reef surveys involved laying down a 20 m transect tape to mark the selected reef patches. Once the reef sites were chosen, permanent transects were established by installing looped iron stakes at both ends, securely fixed to the dead reef and marked with buoys for easy identification. The transect tape was then anchored to the iron stakes to ensure stability for future data collection. For detailed documentation of the selected coral reefs, an Olympus TG-6 camera housed in an Olympus PT-059 underwater casing was used. The settings were configured to achieve optimal image quality by selecting the underwater ultra-wide 60 FPS mode to record high-resolution video footage. Camera height was carefully adjusted based on light penetration and water turbidity, ensuring optimal image quality. During data acquisition, the camera was held steady at a distance of 1.5–2 m from the transect line, depending on water movement, while capturing continuous video footage as the diver moved along the entire transect. This approach offered a detailed visual documentation of the coral reef habitat, enabling in-depth analysis of both structural and biological characteristics. In total, two transects were surveyed at Grande Island, Goa, and three transects were covered in the Palk Bay, Tamil Nadu region, offering valuable baseline data for future monitoring efforts.

2.3 Visual Data Acquisition and Processing

The acquired underwater video footage underwent thorough post-processing to ensure their suitability for detailed analysis. Editing was carried out using DaVinci Resolve, focusing on enhancing critical visual elements and optimizing overall video quality. Subsequently, VLC Media Player was used to extract around 90% of the frames from the processed videos for further analysis. To ensure a structured and coherent dataset for subsequent analysis, the extracted frames were systematically organized with clear links to the corresponding transect metadata. The processed images were later imported into Agisoft Metashape Professional, a platform renowned for its advanced photogrammetric processing capabilities. Sophisticated algorithms within the software aligned the image series by identifying invariant features, known as "keypoints," common to multiple images. These keypoints facilitated the construction of geometric projective matrices, which precisely determined the spatial position and orientation of each camera. This alignment process resulted in the generation of a cohesive three-dimensional model of the surveyed site. In parallel, the software utilized camera positions, identified feature points, and intrinsic parameters (e.g., focal length from image metadata) to reconstruct the 3D geometry on the 2D image plane. This resulted in the formation of a sparse 3D point cloud, offering a preliminary spatial model of the surveyed transect. The sparse cloud provided the structural framework for generating a dense point cloud; it represented a higher-resolution, detailed reconstruction that underpinned the development of a seamless and accurate transect mosaic.

Mosaic reconstructions of each transect were carefully conducted using datasets consisting of approximately 2500–2800 images. After a stringent selection process, 2000 high-resolution images were retained for detailed analysis. Through Agisoft Metashape Professional, photomosaic reconstructions were successfully developed, offering an extensive visual record of the surveyed habitats and facilitating accurate assessments of coral cover and associated health metrics.

2.4 Coral Cover Assessment and Data Processing

ImageJ software was employed to analyze coral cover on reef transect mosaics. The mosaics were initially imported into ImageJ, where a calibrated scale was used to ensure measurement accuracy. In the field, coral colonies were marked with stainless steel tags and identified to the genus level following Corals of the World (https://www.coralsoftheworld.org/). The software methodically traced the tagged colonies, calculating the associated surface area for each colony. This methodology generated extensive quantitative data on the spatial distribution and extent of coral assemblages along the transects. Further data processing and regional visualization of coral cover were carried out using Python.

3 Results

A thorough coral reef assessment was conducted across two distinct regions of India: the Grande Island region of Goa and the Palk Bay area along the Tamil Nadu coast (Fig. 4). The study involved careful transect mosaic reconstructions and detailed analyses of coral cover, health indicators, and substrate composition.

At Navy Bay and Shelter Cove within the Grande Island region, two transects were surveyed to evaluate coral cover and associated health indicators. Transect 1 (Fig. 5a), surveyed from a height of 1 m, covered a substantial area of 38.3 m^2, while Transect

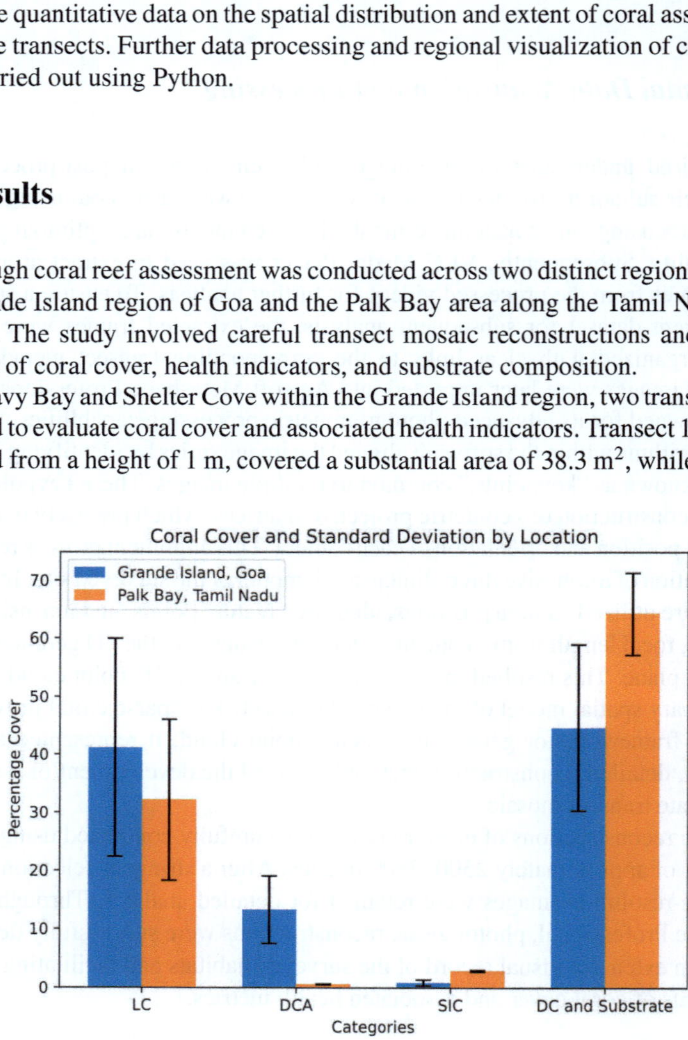

Fig. 4 Coral cover percentages at Grande Island, Goa, and Palk Bay, Tamil Nadu. Bars show mean cover with error bars representing standard deviation across categories: live coral (LC), dead coral with algae (DCA), sponge-infested coral (SIC), and dead coral (DC) and substrate

2 (Fig. 5b), located at the same dive site, encompassed 20 m^2. Coral identification, conducted up to the genus level, revealed *Turbinaria* sp. as the dominant genus in both transects, covering 38.44 ± 9.24% of the surveyed area. *Porites* sp. contributed 1.14 ± 0.98% to the coral cover, while *Favites* sp. and *Favia* sp. collectively accounted for 1.54 ± 0.16%. *Goniopora* sp. was recorded with minimal presence, covering 0.04% in Transect 1 and was absent in Transect 2. The total live coral cover across the two transects was calculated at 41.16 ± 18.77%, indicating a moderate level of coral presence. In terms of coral health indicators, the presence of dead coral with algae (DCA) was significant, covering 13.3 ± 5.77% of the surveyed area. Brown and orange encrusting sponges from the genus *Cliona*: *Cliona orientalis* and *Cliona celata*, were consistently observed (Fig. 12). These sponges primarily encrusted *Turbinaria* sp. and *Favites* sp., covering 0.87 ± 0.50% of the substrate. The substrate itself, composed of a mix of sand, gravel, and coral rubble, covered 44.65 ± 14.33% of the two transects (Table 1 and Fig. 4). Crustose coralline algae were also observed encrusting dead corals, particularly those of the genus *Turbinaria*. The study's findings underscore the current status of coral reefs in the Grande Island region and highlight the importance of ongoing monitoring to track changes in coral health and cover.

In the Palk Bay area along the Tamil Nadu coast, three transects were surveyed to assess coral cover and habitat conditions. Transect 1 (Fig. 6a), located at Mandapam Camp, covered 19.181 m^2, Transect 2 (Fig. 6b), also at Mandapam Camp, spanned 21.658 m^2, and Transect 3 (Fig. 6c), situated along the Rameswaram coast, covered 12.658 m^2. Coral identification up to the genus level revealed *Acropora* sp. as the dominant coral species across all three transects, covering 28.72 ± 22.77% of the surveyed area. *Porites* sp. was identified, contributing 3.13 ± 2.66% to the coral cover, while *Favites* sp. and *Favia* sp. collectively accounted for 0.02 ± 0.01%. *Dipsastraea* sp. was also present, covering 0.28 ± 0.37%. The total live coral cover across the three transects was calculated at 32.17 ± 13.86%, reflecting a moderate presence of live corals in the area. The assessment revealed significant habitat degradation, particularly the prevalence of broken and dead fragments of *Acropora* sp., which constituted a major component of the reef substratum. This fragmentation, presumably driven by intense fishing vessel activity, represents a significant threat to the structural integrity and ecological resilience of the reef ecosystem. The degradation not only signifies a loss of biodiversity but also raises concerns about the disruption of essential reef-building processes critical for the ecosystem's stability. Further observations of *Porites* sp. revealed extensive worm burrows, likely caused by members of the Family Serpulidae, along with manifestations of pink spots and pink lines, indicative of disease, across all surveyed sites (Fig. 12). Coral-infesting sponges were also observed, with diseased and infected corals covering 0.6 ± 0.05% of the surveyed area (Fig. 12). Notably, the pink spots on *Porites* sp. were most prevalent in Transect 2, where they covered 2.02% of the transect. The substrate in the Palk Bay area was heavily composed of fragments of dead *Acropora* sp., sand, and gravel, covering 64.27 ± 7.03% of the surveyed area, further indicating significant habitat degradation. These findings emphasize the urgent need for conservation

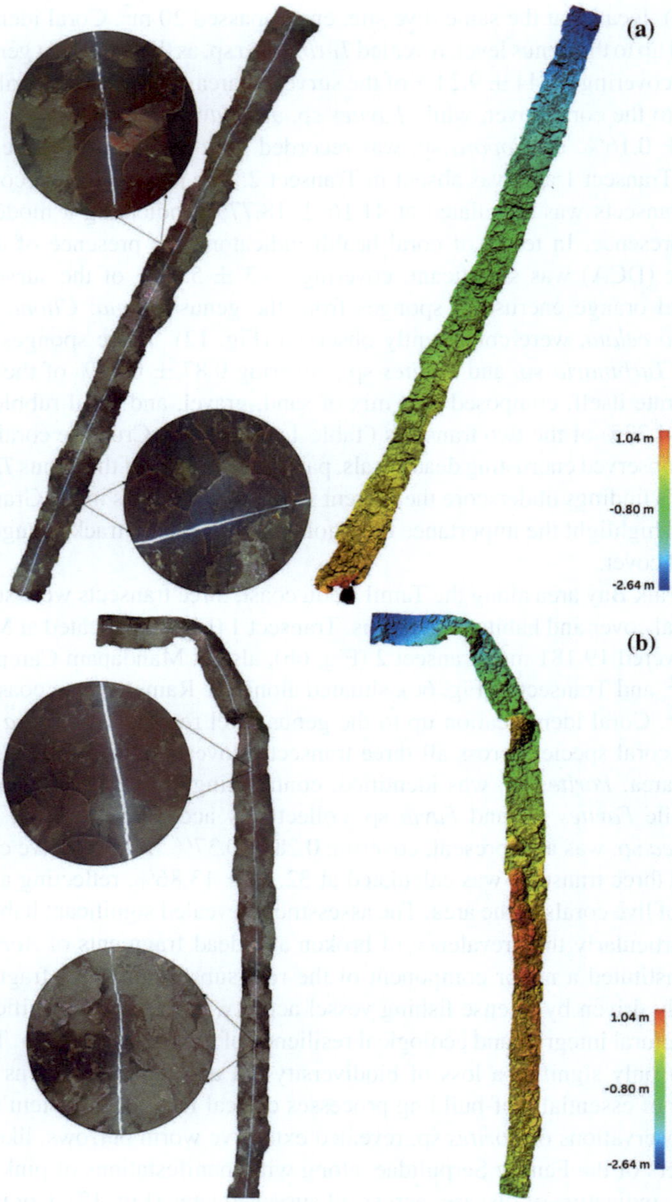

Fig. 5 Photo mosaic and depth elevation model (DEM) of **a** Transect 1: Shelter Cove, **b** Transect 2: Navy Bay, Grande Island, Goa

Table 1 Coral cover analysis of various study sites (LC: live coral, DC: dead coral, DCA: dead coral with algae, SIC: sponge-infested corals)

S. No.	Coral genera/Species	No. of colonies/Patches	Area	Percentage cover (%)
Life Form Analysis for Grande Island, Goa				
1	LC	61	24.002	41.16 ± 18.77
2	DCA	18	7.755	13.3 ± 5.77
3	SIC	11	0.511	0.87 ± 0.50
4	DC and substrate	–	26.032	44.65 ± 14.33
Life Form Analysis for Palk Bay, Tamil Nadu				
1	LC	125	17.213	32.17 ± 13.86
2	SIC	8	0.325	0.6 ± 0.05
3	DCA	5	1.56	2.91 ± 0.12
4	DC and substrate	–	34.387	64.27 ± 7.03

efforts to mitigate further damage and restore the health of the coral reefs in the Palk Bay region.

The coral reef at Grande Island, Goa, hosts a diverse array of marine life (Figs. 8 and 9). Among the fish species observed are the honeycomb moray eel (*Gymnothorax favagineus*), twinspot damselfish (*Chrysiptera biocellata*), sergeant major (*Abudefduf vaigiensis*), red-tailed butterflyfish (*Chaetodon collare*), whitecheek monocle bream (*Scolopsis vosmeri*), goldband goatfish (*Karalla daura*), blackstriped sweeper (*Pempheris mangula*), yellowtail dottyback (*Pseudochromis caudalis*), as well as species of *Amblygobius*, black-spotted puffer (*Chelonodontops patoca*), common lionfish (*Pterois miles*), redtooth triggerfish (*Odonus niger*), and whitetail triggerfish (*Sufflamen verres*). The site is also rich in algae and seaweeds, including *Caulerpa scalpelliformis*, *Padina* sp., *Sargassum* sp., *Caulerpa* sp., and *Amphiroa anceps*. The invertebrate diversity is equally impressive, with species such as the *Tenguella granulate*, *Nassarius* sp., Balanidae, *Elysia marginata*, *Stomopneustes variolaris*, *Trochus* sp., *Stomopneustes toreumaticus*, *Holothuria leucospilota*, *Phidiana militaris*, and *Baeolidia palythoae*. In contrast, the coral reefs of Palk Bay exhibited limited fish diversity, but the invertebrate diversity at the site was more pronounced during the survey (Figs. 10, 11 and 12). Among the species observed were the Crescent grunter (*Terapon jarbua*), Whipfin Silver-Biddy (*Gerres filamentosus*), White-Spotted Puffer (*Arothron hispidus*), Immaculate Puffer (*Arothron immaculatus*), Bluefin Trevally (*Cephalopholis formosa*), the Indopacific Tropical Sand Goby (*Favonigobius reichei*), *Pentaceraster mammillatus*, *Sabellastarte* sp., *Linckia multifora*, *Stichodactyla haddoni*, *Elysia marginata*, *Peronia verruculata*, *Plakobranchus ocellatus*, *Periclimenes brevicarpalis*, *Amphioctopus aegina*, and *Laevistrombus canarium*. The site was rich in algae and seaweeds, including *Amphiroa anceps*, *Halymenia dilatata*, *Caulerpa racemosa*, and *Boergesenia forbesii*.

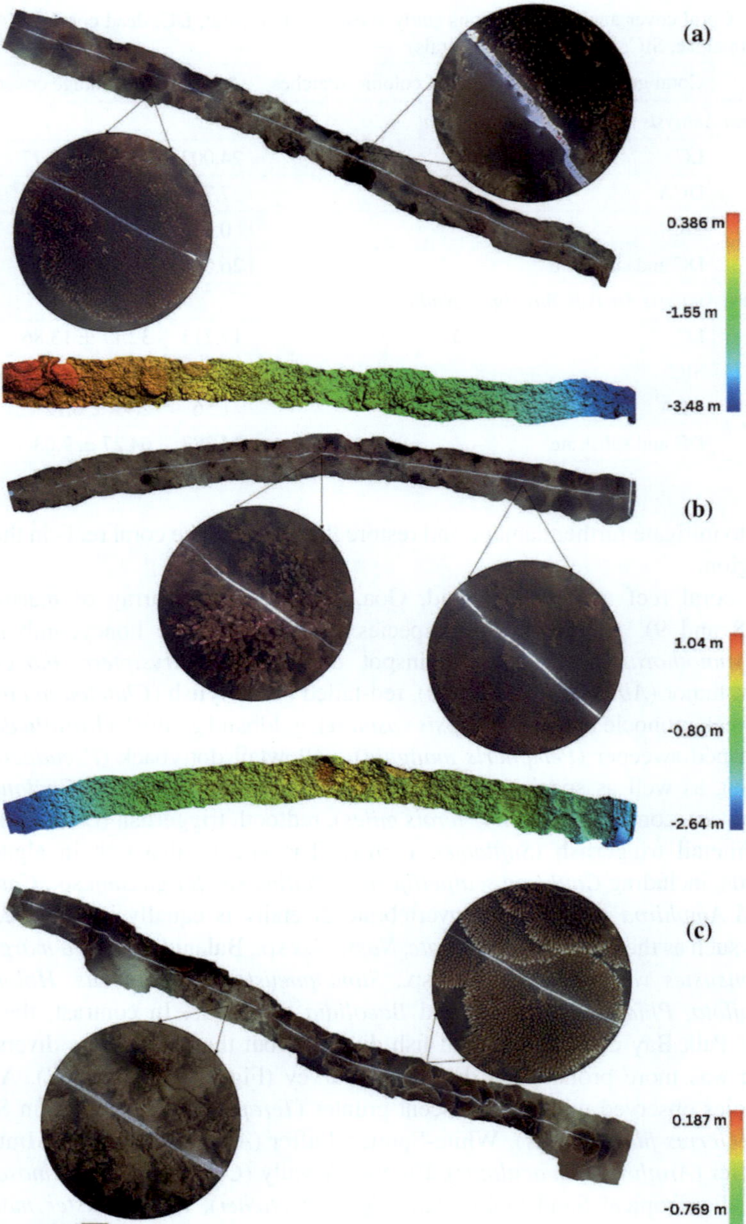

Fig. 6 Photo mosaic and depth elevation model (DEM) of **a** Transect 1: Palk Bay, Mandapam Camp, **b** Transect 2: Palk Bay, Mandapam Camp, **c** Transect 3: Palk Bay, Rameswaram, Tamil Nadu

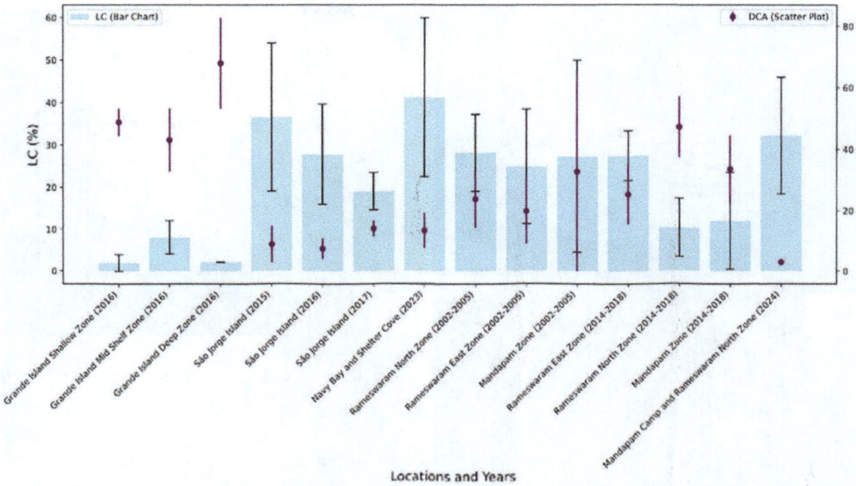

Fig. 7 Reported LC and DCA data for the study locations. The bar chart represents the LC values, while the scatter plot shows the DCA values

4 Discussion

The assessment of coral reefs within Grande Island, Goa, and Palk Bay, Tamil Nadu, has provided valuable insights into the structural dynamics and health of these critical marine ecosystems. By employing transect mosaic reconstructions, the study achieved a comprehensive examination of coral communities within the surveyed regions. This methodology facilitated the acquisition of high-resolution imagery, allowing for a detailed analysis of coral structures and their surrounding environments. The current study conducted at Navy Bay and Shelter Cove involved two transects, with Transect 1 covering 38.3 m^2 and Transect 2 covering 20 m^2. It found that *Turbinaria* sp. was the dominant genus, covering $38.44 \pm 9.24\%$ of the surveyed area, while *Porites* sp. contributed $1.14 \pm 0.98\%$, and *Favites* sp. and *Favia* sp. together accounted for $1.54 \pm 0.16\%$. Goniopora sp. was present only minimally in Transect 1 and absent in Transect 2. The total live coral cover across the two transects was $41.16 \pm 18.77\%$, reflecting a moderate level of coral presence. Additionally, Dead coral with algae (DCA) covered $13.3 \pm 5.77\%$ of the area, and encrusting sponges from the genus *Cliona* covered $0.87 \pm 0.50\%$ of the substrate, primarily encrusting Turbinaria and Favites. The substrate itself, composed of sand, gravel, and coral rubble, covered $44.65 \pm 14.33\%$. In contrast, the earlier study by Manikandan et al. (2016) examined coral community structure and health across various depths at Grande Island. Their findings revealed significant differences in coral cover across depth zones: $1.92 \pm 2.01\%$ in the shallow zone (< 5 m), $8.05 \pm 3.98\%$ in the mid-shelf zone (5–8 m), and $2.12 \pm 0.05\%$ in the deep zone (> 8 m) (Table 2 and Fig. 7). Additionally, a study by Hussain (2022) conducted surveys at São Jorge Island of Grande Island, reporting $27.78 \pm 11.82\%$ live coral cover (LC) and $7.40 \pm 3.30\%$ dead coral with

Fig. 8 Some corals of Grande Islands, Goa: **a** *Turbinaria mesenterina,* **b** *Goniopora columna,* **c** *Favites pentagona,* **d** *Pseudosiderastrea tayamai,* **e** *Dendrophyllia* sp., **f** White *Acanthogorgia* sp., **g** Purple *Acanthogorgia* sp., **h** *Zoanthus sansibaricus*

Fig. 9 Coral reef biodiversity of Grande Island, Goa: **a** Chrysaora sp., **b** *Gymnothorax favagineus*, **c** *Baeolidia palythoae*, **d** *Elysia hirasei*, **e** *Stomopneustes variolaris*, **f** *Chaetodon collare*, **g** *Parablennius thysanius*, **h** *Hyastenus* sp

Fig. 10 Some corals of Palk Bay, Tamil Nadu: **a** *Acropora* sp., **b** *Porites* sp., **c** *Dipsastraea* sp., **d** *Goniopora* sp., **e** *Favia pallida*, **f** *Favites pentagona*, **g** *Platygyra* sp., **h** *Dipsastraea* sp

Fig. 11 Coral reef biodiversity of Palk Bay, Tamil Nadu: **a** *Pentaceraster mammillatus,* **b** *Arothron hispidus,* **c** *Sabellastarte* sp., **d** *Linckia multifora,* **e** *Stichodactyla haddoni,* **f** *Elysia marginata,* **g** *Periclimenes brevicarpalis,* **h** *Amphioctopus aegina*

Fig. 12 Prevalence of coral-infesting organisms and disease syndromes across the two study sites: **a** orange boring sponge *Cliona celata* on *Turbinaria* sp., **b** turf algae growing on *Acropora* sp., **c** brown *Cliona orientalis* infesting *Favites* sp., **d** *Acropora* sp. growing on *Cliona sp.* infested coral boulder, **e** *Haliclona coerulea* growing on *Dipsastraea* sp., **f** white plaque disease observed on *Porites* sp., **g** pink pigmentation on *Porites* sp., **h** vermetid snails and rhodoliths infestation on *Porites* sp.

algae (DCA) in 2016. In 2017, these figures shifted to 19.03 ± 4.43% LC and 13.97 ± 2.63% DCA (Table 2 and Fig. 7).

Meanwhile, the coral reef ecosystem in Palk Bay has experienced considerable degradation due to both natural disturbances and anthropogenic pressures (Wilson 2009). Marimuthu et al. (2020) conducted surveys at various study sites along the Mandapam and Rameswaram coast and observed that the overall mean live coral cover, which was estimated at 26.59 ± 13.14% during the years 2002–2005, declined

Table 2 Previously recorded LC and DCA data across Grande Island, Goa, and Palk Bay, Tamil Nadu

S. No.	Location	Year	LC (%)	DCA (%)	Reference
Grande Island, Goa					
01	Grande Island Shallow Zone	2016	1.92 ± 2.01	48.8 ± 4.48	Manikandan et al. (2016)
02	Grande Island Mid Shelf Zone	2016	8.05 ± 3.98	43.03 ± 10.25	Manikandan et al. (2016)
03	Grande Island Deep Zone	2016	2.12 ± 0.05	68.00 ± 14.72	Manikandan et al. (2016)
04	São Jorge Island	2015	36.56 ± 17.49	8.79 ± 6.01	Hussain (2022)
05	São Jorge Island	2016	27.78 ± 11.82	7.40 ± 3.30	Hussain (2022)
06	São Jorge Island	2017	19.03 ± 4.43	13.97 ± 2.63	Hussain (2022)
07	Navy Bay and Shelter Cove	2023	41.16 ± 18.77	13.3 ± 5.77	Present Study
Palk Bay, Tamil Nadu					
07	Rameswaram North zone	2002–2005	28.13 ± 9.08	23.80 ± 9.45	Marimuthu et al. (2020)
08	Rameswaram East zone	2002–2005	24.81 ± 13.62	19.71 ± 10.73	Marimuthu et al. (2020)
09	Mandapam Zone	2002–2005	27.22 ± 22.71	32.71 ± 32.80	Marimuthu et al. (2020)
10	Rameswaram East zone	2014–2018	27.35 ± 5.89	25.16 ± 9.70	Marimuthu et al. (2020)
11	Rameswaram North zone	2014–2018	10.36 ± 7.0	47.19 ± 9.82	Marimuthu et al. (2020)
12	Mandapam Zone	2014–2018	11.96 ± 11.54	33.59 ± 10.94	Marimuthu et al. (2020)
13	Mandapam Camp and Rameswaram North	2024	32.17 ± 13.86	2.91 ± 0.12	Present study

to 16.98 ± 11.79% between 2014 and 2018, indicating a decade-long degradation of approximately 36.14% (Table 2 and Fig. 7). In the present study, transects 1, 2, and 3 surveyed from Mandapam Camp and Rameswaram coast of Palk Bay region of Tamil Nadu, displayed distinct coral compositions, with *Acropora* species emerging as the dominant taxa across all transects. Notably high coral cover (29.23%) in Transect 3 underscores the ecological significance of the Rameswaram coast as a thriving coral habitat. Additionally, the presence of *Porites, Favites, Favia*, and *Dipsastraea* species highlights the taxonomic diversity within this region. However, alongside these observations of coral richness, the assessment also uncovered concerning indicators of habitat degradation and coral distress.

Encrusting sponges, particularly those from the genus *Cliona*, were consistently observed across sites, indicating detrimental interactions within the coral habitat. The ability to excavate calcium carbonate substrata, such as Scleractinian corals, mollusc shells, and coralline algae, makes the Clionaidae family a critical group of excavating sponges (Rützler, 2002). While the rates of bioerosion and calcification in healthy reef ecosystems are generally comparable (Hein and Risk, 1975), increased populations of these sponges can potentially hinder coral reef accretion. Bioerosion by this genus of sponge is a deteriorating process that adds to reef sediment generation and compromises coral skeletons' integrity (Rützler, 1975). Excavating sponges of the genus *Cliona* have been dominating reef structure in the Caribbean Sea for decades, causing skeletal erosion and tissue loss in coral colonies. Coral tissue loss induced by these sponges, which in tandem with excavation and encroachment, results in long-term alterations in reef community structure and physical stability (López-Victoria et al., 2006; Williams et al., 1999). Additionally, manifestations of disease and infestation, particularly noted in *Porites* species, raise further concerns regarding coral health and ecosystem resilience. *Porites* species have been reported to produce pink pigmentation as an active strategy to combat epibionts, such as vermetid snails and rhodoliths, and limit their expansion on the coral surface, thereby defending themselves and minimizing further surface tissue damage (Ramesh et al., 2023).

In Palk Bay, the assessment revealed significant anthropogenic influences, including several toppled *Acropora* sp. colonies likely caused by boat anchor dragging. The discovery of plastic cement bags covering coral colonies (Fig. 13) is particularly concerning, as these bags smother and weaken the corals, encouraging the growth of turf algae and sponges, which further deteriorates reef health. According to Ramesh et al. (2022), these plastic bags likely originated from fishermen who use them filled with beach sand as a stable holdfast for attaching date palm (*Phoenix sylvestris*) leaves, which serve as a substrate for reef squids to lay their eggs. Fishermen tie bundles of *P. sylvestris* leaves to each plastic sack and leave them for a day to attract squids, particularly the big fin reef squid (*Sepioteuthis lessoniana*), a common species in the Gulf of Mannar and Palk Bay region. According to the fishermen, this setup remains intact on reef flats for up to one or two weeks, after which the *P. sylvestris* leaves begin to decompose, and the plastic bags lose their sand content due to wave action and currents. Eventually, these empty bags drift away, contributing to the pollution of coral reefs. The damage here is more severe when

Fig. 13 Effects of anthropogenic influence of corals at the Palk Bay, Tamil Nadu, study sites: **a** a toppled *Acropora sp.* colony caused by anchor dragging, **b** plastic cement bag entangled on *Acropora* sp. and *Porites* sp.

compared to Goa, exacerbated by its proximity to the shore and its role as a major hub for fisheries, which further contributes to the loss of biodiversity.

Grande Island in Goa has relatively well-maintained reefs, partly due to their distance from the mainland. The primary threats to reef health are linked to poorly managed and unprofessional diving activities by unlicensed adventure companies, although the overall anthropogenic impact remains lower than in Palk Bay.

To mitigate the damage to coral reefs in Palk Bay and Goa, a comprehensive approach that includes regulatory measures, community engagement, and habitat restoration is essential. Strengthening regulations and enforcement is a key strategy. In Palk Bay, implementing designated anchoring zones with permanent mooring buoys in sensitive coral areas can prevent anchor dragging and subsequent coral damage. Stricter waste disposal policies should also be enforced to prevent the entry of harmful materials like plastic cement bags in reef environment. Regular patrolling and imposing fines on offenders can serve as deterrents. In Goa, stricter licensing and monitoring of diving activities should be implemented, ensuring only licensed and trained operators conduct tours. Conducting awareness programs and monitoring adventure companies will ensure eco-friendly practices, reducing the impact of poorly managed diving activities. Additionally, fisheries management should be enhanced by introducing no-fishing zones or seasonal restrictions in critical reef areas, particularly in Palk Bay, to protect coral ecosystems during recovery periods and peak spawning seasons. A crucial aspect of fisheries management is the regulation of destructive fishing practices. Sri Lanka, which shares the waters of Palk Bay with India, has imposed a ban on bottom trawling to protect marine ecosystems. Similar regulations should be enforced in India, especially in these sensitive zones, to minimize damage caused by this unsustainable fishing method, which often devastates coral reefs and seafloor habitats. A ban on bottom trawling in Palk Bay would not only protect the coral reefs but also help restore biodiversity in this ecologically important area.

Community engagement and education are also critical components of a successful conservation strategy. In Palk Bay, Tamil Nadu, fishermen education programs can inform local communities about the ecological value of coral reefs and the long-term economic benefits of reef conservation. Introducing biodegradable alternatives to plastic cement bags for squid egg-laying substrates would reduce the damage caused by current fishing practices. Public awareness campaigns in Goa, aimed at tourists and locals, can raise awareness about the importance of coral reef ecosystems and promote responsible diving and waste disposal practices. By collaborating with diving companies to foster eco-friendly tourism, the long-term health of the reefs can be protected.

Finally, the restoration of adjacent coastal habitats, such as mangroves and seagrass beds, is critical in both Palk Bay, Tamil Nadu, and Grande Island, Goa. These ecosystems act as natural buffers against pollution and sedimentation, which negatively impact coral reefs by smothering corals and reducing water quality. Healthy coastal habitats, like mangroves and seagrass beds, enhance the resilience of coral reefs to environmental stresses, playing a key role in the long-term sustainability of these ecosystems. In Tamil Nadu and Goa, mangrove restoration has been undertaken on a considerable scale, helping to stabilize coastlines and improve water quality. However, very little attention has been given to seagrass bed restoration, despite the vital role seagrasses play in supporting marine biodiversity and carbon sequestration.

Photogrammetry offers several advantages over traditional methods such as video line transect (VLIT). First, it enables high-resolution, three-dimensional mapping of coral reefs, capturing intricate details that are often missed with VLIT and other methodologies. Photogrammetry also allows for comprehensive coverage, enabling large sections of the reef to be surveyed in a single effort, whereas VLIT focuses on linear sections, often leaving out important areas. In terms of accuracy, photogrammetry produces precise 3D models that can quantify coral cover, complexity, and volume, minimizing human error. VLIT, on the other hand, often relies on subjective visual estimations, which can introduce variability and inter-observer bias. Additionally, photogrammetry creates permanent datasets, offering a long-term record for future comparisons and trend studies. VLIT's qualitative data may not always be as useful for ongoing monitoring efforts beyond initial visual interpretation. Photogrammetry can also be optimized with the integration of artificial intelligence (AI) and machine learning, automating processes like frame extraction, filtering out blurry images, and color correction—making data analysis faster and more efficient compared to the manual, time-consuming frame-by-frame analysis required by VLIT. This methodology can be applied across various coral reef habitats in India, recorded by divers or remotely operated vehicles (ROVs). At deeper depths, underwater video lights can be used to ensure clear footage. The flexibility and adaptability of photogrammetry further enhance its utility, making it a superior tool for long-term coral reef monitoring and conservation when compared to traditional methods.

Water clarity and turbidity are critical factors influencing the success of photogrammetry in assessing coral reef health. High levels of sediment or plankton in the water can reduce visibility, making it difficult to capture high-quality images or videos. Additionally, underwater lighting conditions can vary significantly due to natural fluctuations in sunlight and water depth. Low light can result in grainy images, while excessive sunlight may lead to overexposure, both of which compromise the quality of the collected data. Furthermore, the complexity of reef structures presents another challenge; coral reefs often feature intricate formations that can be difficult to capture comprehensively. Ensuring that all relevant features are documented requires careful attention. However, many of these challenges can be mitigated through effective planning and preparation. By assessing water conditions, such as clarity and turbidity, as well as evaluating lighting parameters before conducting the survey, researchers can optimize their data collection efforts. This proactive approach allows for better timing and equipment adjustments, ensuring higher-quality imagery and more accurate representations of the reef ecosystem. With thoughtful planning, the difficulties associated with photogrammetry can be significantly reduced, leading to more reliable results.

This comprehensive assessment highlights the urgent need to expand surveys across India by incorporating more survey sites and increasing the number of transects at each location, potentially conducted at various depths. Such efforts would provide a more complete understanding of coral reef health and inform more effective conservation strategies, safeguarding these invaluable marine ecosystems for future generations (Table 3).

5 Conclusions

In the face of global warming and climate change, long-term monitoring of coral reefs has become imperative due to the alarming impact of environmental and anthropogenic stressors on these vital habitats. The degradation or loss of coral reefs directly affects the livelihoods and economies of both local and global communities. Therefore, monitoring coral reefs using photogrammetry offers numerous benefits: It provides permanent 3D datasets for future comparisons, uncovers changes in reef spatial patterns and temporal dynamics, reveals invasive species distribution, and assesses associated biodiversity metrics. Furthermore, photogrammetry facilitates a more efficient and accurate analysis of reef health, minimizing the time and expertise required for traditional monitoring methods. Our current study is pioneering in its provision of such comprehensive data from Indian reefs, laying the groundwork for future initiatives to generate photogrammetry-based datasets from other coral reef regions in India. Notably, our surveys have revealed that the reefs in Goa and Palk Bay exhibit geographically distinct coral diversity and associated biodiversity patterns. Moreover, our findings indicate that anthropogenic stress is significantly higher in Palk Bay compared to Grande Island in Goa, underscoring the urgent need for awareness programs targeting fisherfolk and tourists to protect the deteriorating coral reefs in both regions. Our current study, conducted along both the West and East coasts of India, incorporates variable environmental parameters and demonstrates that this methodology can be effectively implemented in coral reefs along the entire Indian

Table 3 Checklist of coral species of Grande Island, Goa, and Palk Bay, Tamil Nadu

S. No.	Species	Location		Reference
		Grande Island	Palk Bay	
Family Acroporidae				
Genus *Acropora*				
01	*Acropora humilis* (Dana, 1876)	–	+	Venkataraman et al. (2007)
02	*Acropora digitifera* (Dana, 1846)	–	+	Venkataraman et al. (2007)
03	*Acropora formosa* (Dana, 1846)	–	+	Venkataraman et al. (2007)
04	*Acropora cytherea* (Dana, 1846)	–	+	Venkataraman et al. (2007)
05	*Acropora hyacinthus* (Dana, 1846)	–	+	Venkataraman et al. (2007)
06	*Acropora indica* (Brook, 1893)	–	+	Venkataraman et al. (2007)
Genus *Montipora*				
07	*Montipora spumosa* (Lamarck, 1816)	–	+	Venkataraman et al. (2007)
08	*Montipora digitata* (Dana, 1846)	–	+	Venkataraman et al. (2007)
09	*Montipora hispida* (Dada, 1846)	–	+	Venkataraman et al. (2007)
10	*Montipora foliosa* (Pallas, 1746)	–	+	Venkataraman et al. (2007)
11	*Montipora exesa* (Quelch, 1886)	–	+	Venkataraman et al. (2007)
12	*Montipora monasteriata* (Forskal, 1775)	–	+	Venkataraman et al. (2007)
Genus *Astreopora*				
13	*Astreopora myriophthalma* (Lamarck, 1860)	–	+	Venkataraman et al. (2007)
Family Astrocoeniidae				
Genus *Stylocoeniella*				
14	*Stylocoeniella guentheri* (Bassett-Smith, 1890)	+	–	Hussain (2022)
Family Agarciidae				
Genus *Pavona*				
15	*Pavona varians* (Venill, 1864)	–	+	Venkataraman et al. (2007)

(continued)

Table 3 (continued)

S. No.	Species	Location		Reference
		Grande Island	Palk Bay	
Family Psammocoridae				
Genus *Psammocora*				
16	*Psammocora contigua* (Esper, 1797)	–	+	Venkataraman et al. (2007)
Family Rhizangiidae				
Genus *Pseudosiderastrea*				
17	*Pseudosiderastrea tayami* (Yabe Sugiyama, 1935)	+	+	Venkataraman et al. (2007), Hussain (2022)
Genus *Siderastrea*				
18	*Siderastrea savignyana* (Milne and Haime, 1849)	+	–	Hussain (2022)
Family Coscinaraeidae				
Genus *Coscinaraea*				
19	*Coscinaraea monile* (Forskål, 1775)	+	–	Hussain (2022)
Family Poritidae				
Genus *Porites*				
20	*Porites solida* (Forksal, 1775)	–	+	Venkataraman et al. (2007)
21	*Porites lutea* (Milne Edwards Heime, 1860)	+	+	Venkataraman et al. (2007), Hussain (2022)
22	*Porites lichen* (Dana, 1846)	+	+	Venkataraman et al. (2007), Hussain (2022)
23	*Porites lobata* (Dana, 1846)	–	+	Venkataraman et al. (2007)
24	*Porites annae* (Crossland, 1952)	–	+	Manikandan et al. (2014)
Genus *Goniopora*				
25	*Goniopora stokesi* (Milne and Heime, 1851)	+	+	Venkataraman et al. (2007)
26	*Goniopora planulata* (Ehrenberg, 1834)	–	+	Venkataraman et al. (2007)
27	*Goniopora nigra* (Pillai, 1967)	–	+	Venkataraman et al. (2007)
28	*Goniopora columna* (Dana, 1846)	+	–	Hussain (2022)
29	*Goniopora djiboutiensis* (Vaughan, 1907)	+	–	Hussain (2022)

(continued)

Table 3 (continued)

S. No.	Species	Location		Reference
		Grande Island	Palk Bay	
30	*Goniopora pedunculata* (Quoy and Gaimard, 1833)	+	–	Hussain (2022)
Family Faviidae				
Genus *Favia*				
31	*Favia fragum* (Esper, 1797)	–	+	Venkataraman et al. (2007)
Family Merulinidae				
Genus *Dipsastraea*				
32	*Dipsastraea favus* (Forskål, 1775)	–	+	Venkataraman et al. (2007)
33	*Dipsastraea pallida* (Dana, 1846)	–	+	Venkataraman et al. (2007)
34	*Dipsastraea albida* (Veron, 2000)	–	+	Venkataraman and Rajan (2013)
35	*Dipsastraea lizardensis* (Veron, Pichon and Wijsman, 1977)	+	+	Venkataraman and Rajan (2013) Hussain (2022)
36	*Dipsastraea rotumana* (Gardiner, 1899)	–	+	Venkataraman and Rajan (2013)
37	*Dipsastraea matthaii* (Vaughan, 1918)	+	–	Hussain (2022)
38	*Dipsastraea veroni* (Moll and Best, 1984)	+	–	Hussain (2022)
39	*Dipsastraea speciose* (Dana, 1846)	–	+	Manikandan et al. (2014)
Genus *Favites*				
40	*Favites abdita* (Ellis and Solander, 1876)	–	+	Venkataraman et al. (2007)
41	*Favites complanata* (Ehrenberg, 1834)	+	+	Venkataraman et al. (2007) Hussain (2022)
42	*Favites halicora* (Ehrenberg, 1834)	+	+	Venkataraman and Rajan (2013) Hussain (2022)
43	*Favites valenciennesii* (Milne & Haime, 1849)	–	+	Venkataraman and Rajan (2013)
44	*Favites flexuosa* (Dana, 1846)	+	–	Hussain (2022)
45	*Favites pentagona* (Esper, 1795)	+	+	Hussain (2022)
46	*Favites spinosa* (Klunzinger, 1879)	+	–	Hussain (2022)

(continued)

Table 3 (continued)

S. No.	Species	Location		Reference
		Grande Island	Palk Bay	
Genus *Goniastrea*				
47	*Goniastrea retiformis* (Lamarck, 1816)	+	+	Venkataraman et al. (2007) Hussain (2022)
48	*Goniastrea pectinata* (Ehrenberg, 1834)	−	+	Venkataraman et al. (2007)
Genus *Platygyra*				
49	*Platygyra daedalea* (Ellis and Solander, 1786)	−	+	Venkataraman et al. (2007)
50	*Platygyra sinensis* (Milne & Haime, 1849)	−	+	Venkataraman and Rajan (2013)
Genus *Cyphastrea*				
51	*Cyphastrea serailia* (Forskal, 1775)	+	+	Venkataraman et al. (2007)
52	*Cyphastrea microphthalma* (Lamarck, 1.816)	−	+	Venkataraman et al. (2007)
Genus *Echinopora*				
53	*Echinopora lamellosa* (Esper, 1795)	−	+	Venkataraman et al. (2007)
Genus *Merulina*				
54	*Merulina ampliata* (Ellis and Solander, 1876)	−	+	Venkataraman et al. (2007)
Genus *Hydnophora*				
55	*Hydnophora exesa* (Pallas, 1866)	−	+	Venkataraman et al. (2007)
56	*Hydnophora microconos* (Lamarck, 1816)	−	+	Venkataraman et al. (2007)
Genus *Paragoniastrea*				
57	*Paragoniastrea russelli* (Wells, 1954)	−	+	Venkataraman and Rajan (2013)
Genus *Coelastrea*				
58	*Coelastrea aspera* (Verrill, 1866)	−	+	Venkataraman and Rajan (2013)
Family Lobophylliidae				
Genus *Lobophyllia*				
59	*Lobophyllia radians* (Milne and Haime, 1849)	−	+	Manikandan et al. (2014)

(continued)

Table 3 (continued)

S. No.	Species	Location		Reference
		Grande Island	Palk Bay	
60	*Lobophyllia recta* (Dana, 1846)	–	+	Venkataraman et al. (2007)
Family Leptastreidae				
Genus *Leptastrea*				
61	*Leptastrea purpurea* (Dana, 1846)	–	+	Venkataraman et al. (2007)
62	*Leptastrea transversa* (Klunzinger, 1879)	–	+	Venkataraman et al. (2007)
Family Euphylliidae				
Genus *Galaxea*				
63	*Galaxea fascicularis* (Linnaeus, 1767)	–	+	Venkataraman et al. (2007)
Family Caryophyllidae				
Genus *Paracyathus*				
64	*Paracyathus profundus* (Duncan, 1889)	–	+	Venkataraman et al. (2007)
Genus *Polycyathus*				
65	*Polycyathus verrilli* (Duncan, 1889)	–	+	Venkataraman et al. (2007)
Genus *Heterocyathus*				
66	*Heterocyathus aequicostatus* (Milne and Haime, 1848)	–	+	Venkataraman et al. (2007)
Family Dendrophyllidae				
Genus *Duncanopsammia*				
67	*Duncanopsammia peltata* (Esper, 1790)	–	+	Venkataraman and Rajan (2013)
Genus *Turbinaria*				
68	*Turbinaria frondens* (Dana, 1846)	+	+	Venkataraman and Rajan (2013)
69	*Turbinaria mesenterina* (Lamarck, 1816)	+	–	Hussain (2022)
Family Fungiidae				
Genus *Podabacia*				
70	*Podabacia crustacea* (Pallas, 1766)	+	–	Hussain (2022)
Family Plesiastreidae				
Genus *Plesiastrea*				
71	*Plesiastrea versipora* (Lamarck, 1816)	+	–	Hussain (2022)

coastline. This adaptability underscores the potential of photogrammetry as a valuable tool for ongoing coral reef monitoring efforts in diverse marine environments. We strongly recommend that Indian researchers adopt photogrammetry techniques to generate permanent 3D datasets from reef sites across the country. This approach will provide a valuable resource for researchers, conservators, policymakers, and stakeholders, enabling them to take informed actions to conserve and protect coral reefs. By harnessing the capabilities of photogrammetry, we can enhance our understanding of coral reef ecosystems and better equip ourselves to address the challenges they face in a rapidly changing environment.

Acknowledgements Authors thank the National Centre for Coastal Research, Ministry of Earth Sciences, Chennai, for funding support under the Grant-in-aid project: MoES/NCCR/CRAM/Proposal/3/2019 dated 13/01/2023 and constant support to enhance the knowledge of Indian reefs. This is the CSIR-NIO contribution: xxxx.

Author Contribution Statement The authors confirm their contributions as follows: Ramesh Chatragadda was responsible for funding acquisition, project administration, supervision, resources, investigation, and validation. Satya Swaroop Nanda and Ramesh Chatragadda were involved in the conceptualization, data analysis, methodology, visualization, and the writing of the original draft. Field surveys and data collection were conducted by Satya Swaroop Nanda, Ramesh Chatragadda, Narayan Dalvi, Samson Rokkarukala, and Ashish Aswin Kumar. All authors contributed to the writing, review, and editing of the manuscript. All authors reviewed the results and approved the final version of the manuscript.

Competing Interest Authors do not have any competing interest to declare.

References

Aronson RB, Edmunds PJ, Precht WF, Swanson DW, Levitan DR (1994) Large-scale, long-term monitoring of Caribbean coral reefs: simple, quick, inexpensive techniques. Atoll Research Bulletin

Arora M, Gujrati A, Chaudhury NR, Chauhan P, Patel RC (2021) Assessment of coral reef thermal stress over India based on remotely sensed sea surface temperature. Geocarto Int 36(7):740–757

Arthur R (2000) Coral bleaching and mortality in three Indian reef regions during an El Niño southern oscillation event. Current Science, pp 1723–1729

Barnes RSK, Hughes RN (1999) An introduction to marine ecology, 3rd edn. Blackwell Science Ltd., Oxford, UK, pp 117–141

Barnes RD (1987) Invertebrate zoology; fifth edition. Fort Worth. Harcourt Brace Jovanovich College Publishers, TX, pp 92–96, 127–134, 149–162

Bell JJ, Davy SK, Jones T, Taylor MW, Webster NS (2013) Could some coral reefs become sponge reefs as our climate changes? Glob Change Biol 19(9):2613–2624

Bellwood DR, Hughes TP, Folke C, Nyström M (2004) Confronting the coral reef crisis. Nature 429(6994):827–833

Beyer HL, Kennedy EV, Beger M, Chen CA, Cinner JE, Darling ES, Eakin CM, Gates RD, Heron SF, Knowlton N, Obura DO (2018) Risk-sensitive planning for conserving coral reefs under rapid climate change. Conserv Lett 11(6):e12587

Birkeland C (2015) Coral reefs in the anthropocene. Springer, Netherlands, pp 1–15

Bruno JF, Selig ER (2007) Regional decline of coral cover in the indo-pacific: timing, extent, and subregional comparisons. PLoS ONE 2(8):e711

Caldeira K, Wickett ME (2003) Anthropogenic carbon and ocean pH. Nature 425(6956):365–365

Darling ES, Graham NA, Januchowski-Hartley FA, Nash KL, Pratchett MS, Wilson SK (2017) Relationships between structural complexity, coral traits, and reef fish assemblages. Coral Reefs 36:561–575

Done TJ, DeVantier LM, Turak E, Fisk DA, Wakeford M, Van Woesik R (2010) Coral growth on three reefs: development of recovery benchmarks using a space for time approach. Coral Reefs 29:815–833

Eyre BD, Cyronak T, Drupp P, De Carlo EH, Sachs JP, Andersson AJ (2018) Coral reefs will transition to net dissolving before end of century. Science 359(6378):908–911

Friedlander AM, Brown EK, Jokiel PL, Smith WR, Rodgers KS (2003) Effects of habitat, wave exposure, and marine protected area status on coral reef fish assemblages in the Hawaiian archipelago. Coral Reefs 22:291–305

Gardner TA, Côté IM, Gill JA, Grant A, Watkinson AR (2003) Long-term region-wide declines in Caribbean corals. Science 301(5635):958–960

Gonzalez-Rivero M, Beijbom O, Rodriguez-Ramirez A, Bryant DE, Ganase A, Gonzalez-Marrero Y, Herrera-Reveles A, Kennedy EV, Kim CJ, Lopez-Marcano S, Markey K (2020) Monitoring of coral reefs using artificial intelligence: a feasible and cost-effective approach. Remote Sensing 12(3):489

Graham NA, Nash KL (2013) The importance of structural complexity in coral reef ecosystems. Coral Reefs 32:315–326

Harithsa S, Raghukumar C, Dalal SG (2005) Stress response of two coral species in the Kavaratti atoll of the Lakshadweep Archipelago, India. Coral Reefs 24:463–474

Hein FJ, Risk MJ (1975) Bioerosion of coral heads: inner patch reefs, Florida reef tract. Bull Mar Sci 25(1):133–138

Hoegh-Guldberg O, Kennedy EV, Beyer HL, McClennen C, Possingham HP (2018) Securing a long-term future for coral reefs. Trends Ecol Evol 33(12):936–944

Hughes TP, Graham NA, Jackson JB, Mumby PJ, Steneck RS (2010) Rising to the challenge of sustaining coral reef resilience. Trends Ecol Evol 25(11):633–642

Hughes TP, Kerry JT, Álvarez-Noriega M, Álvarez-Romero JG, Anderson KD, Baird AH, Babcock RC, Beger M, Bellwood DR, Berkelmans R, Bridge TC (2017) Global warming and recurrent mass bleaching of corals. Nature 543(7645):373–377

Hughes TP, Baird AH, Bellwood DR, Card M, Connolly SR, Folke C, Grosberg R, Hoegh-Guldberg O, Jackson JB, Kleypas J, Lough JM (2003). Climate change, human impacts, and the resilience of coral reefs. Science 301(5635):929–933

Hussain A, Ingole B (2020) Massive coral bleaching in the patchy reef of Grande Island, along the eastern Arabian Sea during the 2015/16 global bleaching event. Regional Stud Marine Sci 39:101410

Hussain A (2022) Studies on coral patches along the central west coast of India. Doctoral dissertation, Goa University

Indo-German Biodiversity Programme (IGBP), New Delhi, 1–82 pp

Lalli CM, Parsons TR (1995) Biological oceanography: an introduction. Butterworth-Heinemann Ltd., Oxford, UK, pp 220–233

LeCun Y, Bengio Y, Hinton G (2015) Deep Learning. Nature 521(7553):436–444

Levinton JS (1995) Marine biology: function, biodiversity, ecology. Oxford University Press, Inc., New York, pp 306–319

López-Victoria M, Zea S, Weil E (2006) Competition for space between encrusting excavating Caribbean sponges and other coral reef organisms. Mar Ecol Prog Ser 312:113–121

Lough JM, Anderson KD, Hughes TP (2018) Increasing thermal stress for tropical coral reefs: 1871–2017. Sci Rep 8:6079

Lyons MB, Murray NJ, Kennedy EV, Kovacs EM, Castro-Sanguino C, Phinn SR, Acevedo RB, Alvarez AO, Say C, Tudman P, Markey K (2024) New global area estimates for coral reefs from high-resolution mapping. Cell Reports Sustain 1(2)

Manikandan B, Ravindran J, Shrinivaasu S, Marimuthu N, Paramasivam K (2014) Community structure and coral status across reef fishing intensity gradients in Palk Bay reef, southeast coast of India. Environ Monit Assess 186:5989–6002

Manikandan B, Ravindran J, Mohan H, Periasamy R, Murali RM, Ingole BS (2016) Kk form health status of corals across the spatial scales at different depths in Grande Island. Such descriptions. Regional Stud Mar Sci 7:150–158

Marimuthu N, Jerald Wilson J, Vinithkumar NV, Kirubagaran R (2013) Coral reef recovery status in South Andaman Islands after the bleaching event 2010. J Ocean Univ China 12:91–96

Marimuthu N, Wilson JJ, Kumaraguru AK (2020) Tracking decadal changes in Palk Bay coral community structure: a multivariate approach. J Coast Conserv 24:1–10

Moberg F, Folke C (1999) Ecological goods and services of coral reef ecosystems. Ecol Econ 29(2):215–233

Nair RR, Qasim SZ (1978) Occurrence of a bank with living corals off the south-west coast of India

Nair KV, Bhattathiri PM, Chapgar BF (1996) Angria bank expedition. Report No. AEET/HP/PM3, Atomic Energy Establishment, Trombay

Ninio R, Meekan M (2002) Spatial patterns in benthic communities and the dynamics of a mosaic ecosystem on the Great Barrier Reef, Australia. Coral Reefs 21:95–104

Ninio R, Delean S, Osborne K, Sweatman H (2003) Estimating cover of benthic organisms from underwater video images: variability associated with multiple observers. Mar Ecol Prog Ser 265:107–116

Orr JC, Fabry VJ, Aumont O, Bopp L, Doney SC, Feely RA, Gnanadesikan A, Gruber N, Ishida A, Joos F, Key RM (2005) Anthropogenic ocean acidification over the twenty-first century and its impact on calcifying organisms. Nature 437(7059):681–686

Pillai CSG (1996) Coral reefs of India, their conservation and management

Qasim SZ, Wafar MVM (1979) Occurrence of living corals at several places along the west coast of India. Mahasagar

Quadros AL, Barros F, Blumstein DT, Meira VH, Nunes JAC (2019) Structural complexity but not territory sizes influences flight initiation distance in a damselfish. Mar Biol 166:1–6

Raghuraman R, Raghunathan C, Venkataraman K (2013) Present status of coral reefs in India. In: Ecology and conservation of tropical marine faunal communities, pp 351–379

Ramaiyan V, Adiyapatham S (1985). Food and feeding adaptations of some coral fishes of PortoNovo (Southeast coast of India). In: Proceedings of the fifth international coral reefs congress, Tahiti, vol 2, no 4

Ramesh CH, Koushik S, Shunmugaraj T, Murthy MR (2022) Tracking the origin of polypropylene and polyethylene bags in coral reefs of the Gulf of Mannar. J Mar Biol Assoc India 64:100–103

Ramesh CH, Koushik S, Shunmugaraj T, Murthy MR (2020). Exhibition of bleaching resistance via adaptive bleaching pattern by coral reefs of the Gulf of Mannar during massive bleaching event 2019. J Mar Sci 2(2)

Ramesh C, Shunmugaraj T, Koushik S, Murthy MVR (2023). Porites sp. deposits pink pigmentation to limit the propagation of vermitid gastropods and rhodoliths at coral surface interfaces. Discov Environ 1(1):20

Ravindran J, Kannapiran E, Manikandan B, Mani Murali R, Joseph A (2012). Bleaching and secondary threats on the corals of Palk Bay: a survey and proactive conservation needs

Russell SJ, Norvig P (2016). Artificial intelligence: a modern approach. Pearson

Rützler K (1975) The role of burrowing sponges in bioerosion. Oecologia 19(3):203–216

Rützler K (2002) Family Clionaidae D'Orbigny, 1851. Systema Porifera: a guide to the classification of sponges, pp 173–185

SAC (2010). Coral reef atlas of the world. Vol. I Central Indian Ocean. Space Applications Centre, Indian Space Research Organization, Ahmedabad, India, pp 281

Sadhukhan K, Shanmuagaraj T, Murthy R (2024) Coral Restoration in Palk Bay on the Southeast Coast of India

Samson FB, Knopf FL, Jones CG, Lawton JH, Shachak M (1996). Organisms as ecosystem engineers. In: Ecosystem management: selected readings, pp 130–147

Samuel AL (1959) Some studies in machine learning using the game of checkers. IBM J Res Dev 3(3):210–229

Schutte V, Selig E, Bruno J (2010) Regional spatio-temporal trends in Caribbean coral reef benthic communities. Mar Ecol Prog Ser 402:115–122

Sreekanth GB, Lekshmi NM, Patil A (2019) Performance of a shipwreck as an artificial fish habitat along Goa, west coast of India. J Environ Biol 40(2):170–176

Urbina-Barreto I, Garnier R, Elise S, Pinel R, Dumas P, Mahamadaly V, Facon M, Bureau S, Peignon C, Quod JP, Dutrieux E (2021) Which method for which purpose? A comparison of line intercept transect and underwater photogrammetry methods for coral reef surveys. Front Mar Sci 8:636902

Venkataraman K, Rajan R (2013) Status of coral reefs in Palk Bay. Records of the Zoological Survey of India, pp 1–11

Venkataraman K, Krishnamoorthy P, Satyanarayana C, Sivaleela G (2007). Studies on faunal diversity and coral reef ecosystems of Palk Bay, India. Rec Zool Surv India, Occ Paper No 279(1):47

Veron JEN, Stafford-Smith M (2000). Corals of the World: Australian Institute of Marine Science Townsville MC. Qld, Australia

Vora KH, Almeida F (1990) Submerged reef systems on the central western continental shelf of India. Mar Geol 91(3):255–262

Wasim M, Pandey AC, Ahammed KB, Dwivedi CS (2024) Two decades of coral bleaching in selected islands of Pacific Ocean: a holistic impact assessment. Regional Stud Mar Sci 77:103648

Williams EH, Bartels PJ, Bunkley-Williams L (1999) Predicted disappearance of coral-reef ramparts: a direct result of major ecological disturbances. Glob Change Biol 5:839–845

Wilson JJ (2009). Natural and anthropogenic impact on Palk Bay corals, southeast coast of India. Doctoral dissertation, Ph.D thesis submitted to Madurai Kamaraj University, Madurai, Tamil Nadu, India

The Changing Landscape of Coral Reefs: A Biogeographic and AI-Integrated Perspective

Atanu Kumar Panja and Soumya Haldar

Abstract Coral reefs represent one of Earth's most diverse and sensitive marine ecosystems, yet their global distribution and ecological stability are increasingly threatened by climate change and human pressures. A biogeographic perspective provides insight into how environmental parameters such as temperature, salinity, water clarity, depth, and nutrient regimes shape the spatial distribution, diversity, and resilience of reef systems across regions including the Indo-Pacific, Red Sea, Caribbean, and Great Barrier Reef. Recent advances in artificial intelligence (AI) are transforming coral reef science by enabling high-resolution mapping, automated species identification, ecosystem monitoring, and predictive modelling. Using satellite, drone, acoustic, and in-situ sensor datasets, AI methods—including convolutional neural networks, recurrent neural networks, Random Forests, and habitat suitability models—enable classification of coral types, detection of bleaching, evaluation of reef health, forecasting of future distribution shifts under climate scenarios, and design of optimized marine protected areas. Sensitivity and scenario-based analyses further identify key environmental drivers of resilience. By integrating multi-source data streams and enabling adaptive early-warning systems for bleaching, invasive species, and pollution, AI offers a powerful framework for future-focused coral conservation. This chapter highlights how AI-assisted biogeography can accelerate early diagnostics, improve decision support, and strengthen restoration and management strategies to safeguard coral reef biodiversity in a rapidly changing ocean.

Keywords Coral biogeography · Coral reef conservation · Climate change · Artificial intelligence · Machine learning · Predictive modelling · Remote sensing · Habitat suitability · Marine protected areas · Ecosystem resilience

---

A. K. Panja (✉) · S. Haldar
CSIR-CSMCRI, Bhavnagar, Gujarat, India
e-mail: atanubappa7@gmail.com

AcSIR, Ghaziabad, Uttar Pradesh, India

K. K. Basheer Ahammed et al. (eds.), *Monitoring Coastal Marine Ecosystems Under Climate Change*, https://doi.org/10.1007/978-3-032-09137-6_8

1 Introduction

Coral reefs are among the most biodiverse ecosystems on the planet, hosting thousands of marine species, including fish, invertebrates, and algae. The biogeographic perspective on coral reefs focuses on the distribution patterns of coral species and the factors that influence these patterns across different regions of the world. Coral reefs are predominantly found in tropical and subtropical waters, between 30°N and 30°S latitude (Saravanan et al. 2017). It has been estimated that littoral coral reefs, which are approximately half the size of Madagascar, cover approximately 284,300 square kilometres worldwide. This is less than 1.2% of the global continental shelf area and only 0.09% of the total oceanic area (Spalding et al. 2001) Major regions consist of the Red Sea, the Caribbean, and the Indo-Pacific. The Coral Triangle, situated in the western Pacific Ocean, is renowned for its exceptional marine biodiversity, in which it is home to approximately 76% of the world's coral species (WWF 2024). The global centre of marine biodiversity, the Indo-Pacific region, is home to the highest diversity of coral species and includes Indonesia, Malaysia, the Philippines, Papua New Guinea, the Solomon Islands, and Timor-Leste. Great Barrier Reef (GBR) of Australia located off the coast of Queensland, Australia; it is the world's largest coral reef system and a UNESCO World Heritage Site (NOAA 2024a; b). GBR is the world's largest coral reef system and a UNESCO World Heritage Site. Red Sea is known for its unique and diverse marine life, and the Red Sea reefs are resilient to high temperatures and salinity levels. While Caribbean region includes Mesoamerican Barrier Reef System, which stretching from the Yucatan Peninsula down to Honduras, this is the largest barrier reef in the Western Hemisphere. Another one is Florida Reef tract, extending from the Florida Keys, and this reef system is the only barrier reef in the continental United States.

The distribution of coral reefs is influenced by a combination of environmental, biological, and human factors. Corals thrive in warm water temperatures, typically between 23 °C and 29 °C. (NOAA 2024a; b) Extreme temperatures outside this range can cause thermal stress and lead to coral bleaching. The distribution of corals is being affected by the widespread bleaching and mortality that are a result of the increasing sea temperatures caused by global warming. Corals rely on sunlight for photosynthesis through their symbiotic algae (zooxanthellae). This restricts most coral reefs to shallow waters (up to 60 m deep) where light penetration is sufficient (US-EPA 2024). Clear waters with low turbidity allow better light penetration, supporting coral growth (Henkel 2010). Corals require stable salinity levels, usually between 32 and 42 parts per thousand. Significant changes in salinity, often caused by freshwater runoff or changes in evaporation rates, can stress corals (NOAA 2024a; b). Corals prefer oligotrophic (nutrient-poor) waters, and high nutrient levels can promote algal blooms that compete with corals for space and light (Rädecker et al 2015). While runoff containing sediments, nutrients, and pollutants can damage coral reefs by reducing water quality and light availability, ocean currents play a crucial role in the dispersal of coral larvae, (Thompson et al. 2018) affecting the distribution of coral species. Currents can carry larvae over long distances, facilitating gene flow and colonisation of new areas.

Examining the changing landscape of coral reefs from a biogeographic perspective involves understanding how the distribution and diversity of coral reefs are influenced by both natural and anthropogenic factors over time and across different geographic regions (Reverter et al. 2022). This perspective helps to identify patterns, processes, and drivers of change in coral reef ecosystems. Investigating where coral reefs are currently found and how their distribution has changed historically is essential, as it enables researchers to identify regions that have experienced reef expansion or decline over time (Wasim et al. 2025). Studying the biogeographic patterns of different coral species across various regions is important because it highlights areas of high biodiversity, endemism, and the presence of unique coral assemblages. Assessing the climate change impact on coral reefs includes rising sea temperatures, ocean acidification, and sea level rise on coral reefs, understanding how these factors affect coral distribution and health (Hoegh-Guldberg et al. 2017). Analysing how changes in water quality, driven by pollution and runoff, influence coral reef health and distribution is critical for understanding the degree to which anthropogenic stressors alter reef resilience. Examining how light availability, salinity, and ocean currents affect where coral reefs can thrive is essential for understanding the environmental limits that control coral biogeography.

With its cutting-edge tools and solutions that improve data collection, analysis, and decision-making, artificial intelligence (AI) is making a big impact on environmental research and conservation (Bibri et al. 2024). Camera and drone footage is analysed using AI algorithms, specifically in the areas of computer vision and machine learning. These technologies are efficient and capable of tracking wildlife populations, species identification, and automated monitoring of ecosystem changes. AI listens to audio recordings made in natural settings to track species diversity, find changes in the soundscape that signal threats to the environment or ecological shifts, and monitor biodiversity. Artificial intelligence (AI) models are utilised to forecast the consequences of climate change on ecosystems, including biodiversity shifts, species migration, and habitat loss (Amiri et al. 2024). These models aid in planning conservation strategies and comprehending future scenarios. Researchers can investigate how different factors, such as human activities and natural disturbances, affect ecosystems by using AI simulations to model complex ecosystem interactions. AI prioritises areas for protection according to ecological value, biodiversity, and threat levels, which maximises conservation efforts. By doing this, conservation results are maximised and resources are distributed more wisely (Eiswerth and Haney 2001). In order to preserve genetic diversity and population health, AI is used in the design of wildlife corridors, which allow animals to move safely between habitats. AI-driven surveillance systems, such as drones and sensors, are able to quickly respond to and enforce law enforcement by detecting illicit activities such as poaching and logging in real time (Nandutu et al. 2023). AI helps create focused management strategies by detecting and tracking invasive species, forecasting their impact, and estimating their spread. AI guides habitat restoration efforts by analysing data on climate, vegetation, and soil to make sure that restored areas are resilient to future changes and suitable for target species. Artificial intelligence (AI) helps restore coral reefs by identifying hardy coral species and refining transplantation techniques to increase retention rates. By offering instruments and techniques for evaluating sizable datasets, keeping an eye on coral ecosystems, and creating predictive models, artificial intelligence (AI) significantly advances the study of coral biogeography. Drone and satellite imagery are processed by AI algorithms to map coral reefs across broad regions. This aids in monitoring the health of the reefs, determining their distribution and extent, and spotting changes over time. Using underwater photographs and videos, AI-driven computer vision algorithms identify different types of coral and evaluate the health of the reef. This makes it possible to track species composition and biodiversity in great detail across various biogeographic regions. Based on environmental parameters like temperature, availability of light, and water quality, artificial intelligence models, such as machine learning algorithms, forecast coral species' ideal habitats. These models aid in comprehending distribution patterns both now and potentially as a result of climate change. Artificial intelligence (AI) is used to model how coral distribution will change in the future under various climate change scenarios (Leal Filho et al. 2022). This helps identify regions at risk of coral loss and anticipate possible range shifts. Artificial intelligence models analyse intricate relationships like symbiotic partnerships, competition, and predator–prey dynamics that occur within coral reef ecosystems. Understanding how biogeographic factors affect ecosystem resilience and stability is aided by this. AI evaluates coral reef biodiversity patterns, pinpointing hotspots for endemism and species richness. Planning for conservation and setting protection priorities depend heavily on this data. By examining data on coral distribution, connectivity, and biodiversity, artificial intelligence (AI) improves the design of marine protected areas (MPAs). This guarantees the efficient conservation of vital habitats and the maintenance of coral resilience in protected areas. By locating transplantable sites and choosing coral species with the highest resistance to environmental stressors, artificial intelligence (AI) aids in the restoration of coral reefs. Artificial intelligence (AI) systems evaluate temperature data along with other environmental factors to give early alerts for coral bleaching events, allowing for prompt mitigation and intervention (Sultana et al. 2024). AI helps create focused management plans to counter these threats by spotting trends in disease outbreaks and pollution that affect coral reefs. An all-encompassing understanding of coral biogeography is made possible by artificial intelligence (AI), which makes it easier to integrate diverse datasets from a variety of sources, including citizen science contributions, in-situ observations, and remote sensing.

Coral biogeography uses a number of contemporary AI tools and technologies to analyse data, track ecosystems, and aid in conservation efforts. These open-source machine learning libraries, such as PyTorch and TensorFlow, are used to create and implement models for the analysis of sizable coral reef datasets. Examples of these analyses include the classification of images to identify species and the prediction of environmental impacts. The Python machine learning library Scikit-learn is widely used for modelling and data analysis tasks, such as classifying and clustering coral reef data. Through image processing and analysis, researchers can identify and categorise coral species from underwater photographs and videos with the aid of OpenCV, an open-source computer vision library. In order to support real-time monitoring of coral health and biodiversity, YOLO (You Only Look Once) is an object detection system that can be trained to quickly and accurately identify and locate coral species in images. The cloud-based platform Google Earth Engine is used to map coral reefs, track changes over time, and examine environmental variables that impact coral distribution. It is also used for processing geospatial data and satellite imagery (Zapata-Ramírez 2023). However, researchers can examine biogeographic patterns and use this information to guide conservation planning by using Geographic Information System (GIS) software such as ArcGIS and QGIS for spatial analysis and visualisation of coral reef data. With the aid of NetLogo, a multi-agent simulation environment, researchers can better understand coral population dynamics and forecast future changes by modelling intricate coral reef ecosystems and interactions. Simulation modelling software such as AnyLogic can be used to construct dynamic coral reef ecosystem models, allowing researchers to investigate the effects of both natural drivers and anthropogenic stressors. IBM Watson is an artificial intelligence platform that supports decision-making in conservation efforts by analysing massive datasets and offering insights into the health, distribution, and environmental threats facing coral reefs. The Google Cloud AI Platform makes it easier to integrate and analyse a variety of coral reef datasets from different sources by providing tools for creating, implementing, and maintaining machine learning models (Gonzalez-Rivero et al. 2020). Zooniverse is a platform for citizen science initiatives that makes use of artificial intelligence (AI) to help analyse massive datasets, like volunteer-submitted photographs of coral reefs, to improve research and public participation (Morales 2024). Researchers can use the AI-based web application 'CoralNet' to automatically annotate images of coral reefs, giving them the tools to monitor and analyse changes in coral ecosystems over time. While Wildbook is a platform designed for tracking coral species and evaluating the health of reefs, it uses artificial intelligence (AI) and computer vision to identify specific animals and track populations.

2 Material Method

2.1 Mapping of the Coral Biogeography

It is important to understand how coral reefs are changing from a biogeographic point of view in order to protect biodiversity, study how climate change affects coastal ecosystems, plan marine space, and deal with the global challenges of coral conservation. From a biogeographic point of view, studying how coral reefs are changing means looking at different patterns of where corals are found, their diversity, and their health over time. Field research, remote sensing, laboratory studies, and data analysis are all things that can be used together to do this.

- **The Satellite Imagery**

The effectiveness of direct mapping of coral reefs through satellite imagery is restricted by a number of challenges. Complex and intricate structures, coral reefs frequently necessitate imagery with a spatial resolution of sub-metres. The majority of satellites do not offer such high-resolution imagery, which complicates the identification of individual coral colonies and small-scale reef features. Satellite sensors require clear water to acquire images of the underwater environment. The visibility of coral reefs from space can be significantly obscured by turbidity, which can be caused by sediments, algae, or pollution. This can result in a reduction in water clarity and an obstruction of light penetration. The efficacy of satellite imagery diminishes as the depth of the water increases. The reduced light levels make it particularly challenging to observe corals that are located at depths exceeding 20–30 m (Kahng et al. 2019). Standard satellite sensors may lack the necessary spectral bands to accurately distinguish between a variety of benthic habitats, including coral, algae, and sand. It is difficult to accurately identify and map corals due to the fact that the spectral signatures of coral reefs can overlap with those of other underwater features. Satellites may not pass over the same location frequently enough to monitor rapid changes in coral health, such as bleaching events that can occur in days or weeks, due to their fixed orbits. The majority of coral reefs are situated in tropical regions, which frequently experience cloud cover. This can obstruct satellite views, resulting in incomplete or unusable images. The reflection of sunlight off the ocean surface, referred to as sun glint, can result in the formation of bright spots in satellite images, which can obscure the view of the reefs below. The image can be distorted by surface turbulence and waves, which can make it challenging to see the underwater features clearly. The intricate three-dimensional structures of coral reefs are challenging to accurately capture using two-dimensional satellite images. The interpretation of images can be complicated by the shadows cast by coral structures and other features, which can obscure portions of the reef.

Drones and manned aircraft can capture high-resolution imagery, providing detailed views of coral reefs. These platforms can be deployed as needed, avoiding some of the limitations of satellite orbits and cloud cover.

- **AI Integration into Drone Imagery**

AI plays a pivotal role in processing satellite and drone images to map coral reefs over large areas, providing real-time and accurate information on reef distribution and health. High-resolution satellite images from sources like Landsat, Sentinel, and commercial satellites provide extensive coverage of coral reefs worldwide. Drones equipped with cameras capture detailed images of specific reef areas, offering higher resolution and more localised data compared to satellites. AI algorithms preprocess raw images by correcting for distortions, removing noise, and adjusting for variations in light and water conditions. This step is crucial for ensuring the quality and consistency of the data. The images are georeferenced to align them with real-world coordinates, allowing accurate mapping and spatial analysis. AI uses machine learning algorithms, such as convolutional neural networks (CNNs), to analyse the images (Anwar et al. 2018). These models are trained to recognise different coral types, other marine life, and non-living components like sand or algae. AI techniques perform semantic segmentation to divide images into regions representing different categories (e.g. healthy coral, bleached coral, and seagrass). This provides detailed maps of coral cover and condition.

A convolutional neural network (CNN) is a deep learning model that is specifically designed to handle structured grid data, such as images. In the context of coral reef image classification, a CNN can automatically learn and extract relevant features from images to differentiate between various coral types (Gómez-Ríos et al. 2019).

A basic example of how convolutional neural network (CNN) can be used with Python and TensorFlow/Keras to perform image classification for coral reef detection. This example will demonstrate how to set up a simple CNN to classify images into different categories, such as different coral types and non-living components like sand or algae.

```python
import tensorflow as tf
from tensorflow.keras.models import Sequential
from tensorflow.keras.layers import Conv2D, MaxPooling2D, Flatten, Dense, Dropout
from tensorflow.keras.preprocessing.image import ImageDataGenerator
import matplotlib.pyplot as plt

# Define image dimensions and path to dataset
img_width, img_height = 150, 150
train_data_dir = 'data/train'
validation_data_dir = 'data/validation'
batch_size = 32

# Image data augmentation
train_datagen = ImageDataGenerator(
    rescale=1.0/255,
    shear_range=0.2,
    zoom_range=0.2,
    horizontal_flip=True
)

validation_datagen = ImageDataGenerator(rescale=1.0/255)

# Training data generator
```

```
train_generator = train_datagen.flow_from_directory(
    train_data_dir,
    target_size=(img_width, img_height),
    batch_size=batch_size,
    class_mode='categorical'
)

# Validation data generator
validation_generator = validation_datagen.flow_from_directory(
    validation_data_dir,
    target_size=(img_width, img_height),
    batch_size=batch_size,
    class_mode='categorical'
)

# Build the CNN model
model = Sequential([
    Conv2D(32, (3, 3), activation='relu', input_shape=(img_width, img_height, 3)),
    MaxPooling2D(pool_size=(2, 2)),
    Conv2D(64, (3, 3), activation='relu'),
    MaxPooling2D(pool_size=(2, 2)),
    Conv2D(128, (3, 3), activation='relu'),
    MaxPooling2D(pool_size=(2, 2)),
    Flatten(),
    Dense(256, activation='relu'),
    Dropout(0.5),
    Dense(train_generator.num_classes, activation='softmax')
])

# Compile the model
model.compile(optimizer='adam', loss='categorical_crossentropy', metrics=['accuracy'])

# Train the model
history = model.fit(
    train_generator,
    steps_per_epoch=train_generator.samples // batch_size,
    epochs=10,
    validation_data=validation_generator,
    validation_steps=validation_generator.samples // batch_size
)

# Plot training and validation accuracy
plt.plot(history.history['accuracy'], label='Training Accuracy')
plt.plot(history.history['val_accuracy'], label='Validation Accuracy')
plt.xlabel('Epochs')
plt.ylabel('Accuracy')
plt.legend()
plt.show()

# Save the model
model.save('coral_reef_model.h5')
```

- **Explanation of This Code**

3 Data Preparation

'ImageDataGenerator' used to preprocess and augment the images. This includes rescaling pixel values, applying transformations like zoom and flip, and creating batches of images for training and validation.

- **CNN Architecture:**

Three convolutional layers (Conv2D) comprise the model, each of which is succeeded by a maximum pooling layer (MaxPooling2D). To prevent overfitting, the output is flattened and sent through a dense layer with a dropout layer. A 'softmax' activation function is employed in the final dense layer to facilitate multi-class classification, with units that correspond to the number of classes (e.g. various coral types and other components). Using the training and validation data generators, the model is trained for a specified number of epochs after being compiled with the 'Adam' optimiser and categorical 'crossentropy' loss. To monitor the model's performance over the epochs, a plot has been generated to illustrate the accuracy of the training and validation.

Image files in JPEG, PNG, or other common formats are contained in each subdirectory under train and validation, which represents a class label (e.g. various coral types, sand, and algae). This structure enables the 'flow_from_directory' method to automatically label the images according to the names of their directories (Fig. 1).

3.1 Key Components of a CNN

- Convolutional Layers:

Fig. 1 Directory structure for a dataset organised for a machine learning task

```
data/
    train/
        coral_type_1/
            image1.jpg
            image2.jpg
            ...
        coral_type_2/
            image1.jpg
            image2.jpg
            ...
    validation/
        coral_type_1/
            image1.jpg
            image2.jpg
            ...
        coral_type_2/
            image1.jpg
            image2.jpg
            ...
```

Convolutional layers apply a set of filters (also called kernels) to the input image to create feature maps. These filters slide over the image and perform element-wise multiplication and summation to extract features such as edges, textures, and patterns. In the case of coral reef images, convolutional layers help identify distinctive features of different coral species, colours, shapes, and textures.

- Activation Function:

Rectified Linear Unit (ReLU): The most commonly used activation function in CNNs is ReLU, which introduces nonlinearity into the model by converting negative values to zero and keeping positive values unchanged. ReLU activation enables the CNN to model complex relationships and interactions between image pixels.

- Pooling Layers:

Pooling layers reduce the spatial dimensions (width and height) of the feature maps while retaining the most important information. The most common pooling operation is max pooling, which selects the maximum value in a specified window. Pooling layers help reduce the computational load and control overfitting by down sampling the feature maps, thus focusing on the most salient features.

- Flattening Layers:

The flattening layer transforms the 2D feature maps into a 1D vector, which can be fed into fully connected layers. Flattening prepares the extracted features for classification by connecting them to a dense network of neurons.

- Fully Connected Layers:

Fully connected layers (dense layers) process the 1D vector of features to perform classification. Each neuron in a dense layer is connected to every neuron in the previous layer. In the final dense layer, a softmax activation function is typically used to output probabilities for each class, such as different coral types, algae, or sand.

- Dropout Layers:

Dropout is a regularisation technique that randomly sets a portion of the neurons in a layer to zero during training, preventing the model from becoming too dependent on specific features. Dropout helps prevent overfitting, especially when the training dataset is small.

3.2 Procedure for CNN Classification of Coral Reef Images

1. The CNN takes an input image (e.g. a photograph of a coral reef) with three colour channels (RGB) and a specific resolution.
2. Convolutional and pooling layers work together to extract hierarchical features from the input image. Early layers capture low-level features like edges and

corners, while deeper layers capture high-level features like specific coral patterns.
3. The extracted features are flattened and passed through fully connected layers, where the model learns to associate specific feature patterns with the target classes (e.g. coral type 1, coral type 2, and sand).
4. The final output layer uses the 'softmax' function to generate probabilities for each class, indicating the model's confidence that the input image belongs to each category.

3.3 Outcome of the Image Analysis of Coral Reef Using CNN

The result for the coral reef image study using convolutional neural network can be the following.

1. The coral species in an image are identified and classified by the CNN. The output is typically a list of species labels, along with the corresponding confidence scores or probabilities that indicate the probability of each species' presence in the image.
2. CNN assesses the coral's health by examination of visual indicators in the images, including texture and colour. This could be a binary output (healthy vs. unhealthy) or a more fine-grained rating, such as a bleaching severity scale.
3. The CNN calculates the proportion of the area that is occupied by live coral in comparison with non-living components, such as sand, rocks, or dead coral. This could take the form of continuous values or categories that denote varying degrees of coverage.
4. Sequential images of the same location are fed to the CNN, which detects changes in coral reefs over time. The output would indicate regions of growth, decline, or substantial changes in the coral structure, which are essential for the monitoring of reef dynamics.
5. The CNN detects anomalies or unusual features in coral reef images, including the presence of invasive species, unusual bleaching patterns, or pollution effects. These anomalies are typically identified in the output for further investigation by researchers.
6. The CNN can assist in the quantification of biodiversity by identifying a diverse array of species and their abundance in coral reef ecosystems. The output may consist of a list or a count of the various species that were identified, thereby contributing to the field of biodiversity research.

3.4 Artificial-Intelligence-Driven Predictive Modelling for Coral Reef

AI-powered predictive modelling is an effective tool for assessing the effects of climate change and human activity on coral reefs. Using large datasets and advanced machine learning techniques, researchers can forecast future changes in coral reef ecosystems, assess the resilience of various coral species, and guide conservation efforts. AI models can forecast future sea temperature and pH levels based on historical and current data from various climate change scenarios. These models forecast coral bleaching events and calcification rates, which are critical for understanding coral health and survival. As ocean conditions change, AI models can predict shifts in coral distributions, identifying potential new habitats for coral growth as well as areas where corals may no longer exist. This is useful for planning marine protected areas and conservation strategies. AI models use satellite imagery, water quality sensors, and historical pollution records to predict how coastal development, agricultural runoff, and industrial discharges will affect coral reefs. These predictions can help guide mitigation measures like land-use regulation and stricter pollution controls. AI can predict the stress placed on coral reefs by analysing data on fishing activities, tourism, and local economic actions. Predictive models aid in the development of sustainable practices and policies that strike a balance between human activity and coral health. AI models can simulate complex interactions in coral reef ecosystems, predicting how changes in one species will affect others. This is critical in understanding the cascading effects of biodiversity loss caused by climate change and human activities. Machine learning algorithms can identify patterns and factors that influence coral resilience. Predictive models use this data to predict how well various coral species will recover from bleaching events or adapt to new environmental conditions.

AI-powered tools combine multiple data sources to generate risk maps that highlight areas most vulnerable to climate change and human impacts. These maps play an important role in conservation and resource allocation decisions. AI models can simulate multiple scenarios to evaluate the efficacy of various conservation strategies under changing climatic and anthropogenic conditions. This enables policymakers and conservationists to create adaptable and robust management plans. AI algorithms use real-time data from underwater sensors and satellites to detect early signs of stress in coral reefs, such as temperature anomalies or unusual sediment levels. This allows for quick responses to mitigate adverse effects. Predictive models can send automated alerts to local managers and policymakers as coral health conditions approach critical thresholds, allowing for timely interventions. Collect comprehensive and diverse data from a variety of sources, including satellite imagery, underwater sensor networks, historical climate data, and ecological surveys. Use machine learning techniques to create predictive models. Techniques such as neural networks, decision trees, and ensemble methods may be used depending on the complexity of the data and the specificity of the predictions required. To improve accuracy and reliability, models should be validated and refined on a regular basis using new data and ecological assessment feedback.

3.5 Tools that Are Used for Predictive Modelling Using AI

AI tools and frameworks have become essential for modelling complex ecological systems such as coral reefs. These tools provide powerful capabilities for data analysis, predictive modelling, and simulation, all of which are critical for understanding and managing environmental changes and their effects on ecosystems.

TensorFlow and Keras are widely used for developing complex models, including deep learning networks that can process vast datasets (Vasilev et al. 2019), TensorFlow paired with its high-level API Keras is ideal for building, training, and deploying models involved in image recognition, predictive analytics, and time series forecasting. In research environments, PyTorch is well-known for its adaptability and simplicity; among academics and researchers, (Molinari 2020) it is especially well-liked for creating deep learning models needing dynamic computation graphs. Projects like those in ecological research that change quickly will find it especially helpful. For conventional machine learning tasks, Scikit-learn is ideal; it provides a broad spectrum of algorithms for classification, regression, clustering, and dimensionality reduction, so fitting for predictive modelling where deep learning may be overkill. H_2O.ai is an open-source data science and machine learning platform that enables the development of highly scalable machine learning models. Most conventional machine learning models are supported by H_2O.ai, which also performs especially well for businesses wishing to rapidly and effectively apply models. Although Apache Spark is great for managing vast amounts of data, it can rapidly execute complicated algorithms, which qualifies for ecological data analysis requiring streaming data or processing big datasets. Designed especially for geospatial data analysis and environmental monitoring, Google Earth Engine lets researchers run intricate simulations and analyses on an extensive collection of environmental datasets kept on Google's systems (Brooke et al. 2020).

Several kinds of artificial intelligence models can be used to examine, forecast, and interpret intricate ecological data for the modelling of coral reefs. These models differ in their method and the particular chores they are meant for—classification, regression, or clustering.

Sequential data, such temporal variations in coral reef conditions, can be modelled using recurrent neural networks (RNNs) and long short-term memory networks (LSTMs) (Li et al. 2020). Past observations allow RNNs and LSTMs to forecast future reef states. Sequential data, such temporal fluctuations in coral reef conditions, can be modelled with RNN and LSTMs. Past observations allow RNNs and LSTMs to forecast future reef states. These models help to forecast changes in coral reefs brought about by environmental pressures including temperature rises, acidification, or human activity, so affecting policy and conservation plans.

Robust to overfitting and able of managing nonlinear data with interactions, random forests (RF) is an ensemble learning method for classification and regression (Kiangala and wang 2021). These help to forecast coral reef resilience by combining several biotic and abiotic elements and guide decisions on the probability of reef survival under several conditions. By combining many biotic and abiotic elements and guiding decisions on the probability of reef survival under different conditions, RF can help to forecast coral reef resilience. Effective in high-dimensional spaces, support vector machines (SVM) are applied for classification and regression tasks including differentiating between healthy and unhealthy coral reefs from multi-spectral images or other sensor data (Wicaksono et al. 2021). Based on their spectral fingerprints in remote sensing data, SVM is categorising forms of coral and other sessile organisms. Particularly suited for handling heterogeneous data, gradient boosting machines (GBMs) are strong machine learning method that creates models in a stage-wise fashion. They examine environmental impact studies in which several kinds of data must be combined to forecast changes in ecosystems and coral condition. Unsupervised learning of efficient coding uses autoencoders; these also enable dimensionality reduction of ecological data, noise reduction from sensor data, and anomaly detection. In coral reef environments, autoencoders assist in identifying odd trends or anomalies that would point to emerging problems including pollution or disease outbreak. Natural groupings in data are found using K-means clustering and other clustering algorithms; this can help to identify different ecological zones inside a bigger coral reef ecosystem. (Barve et al. 2023) They help with focused conservation by allowing coral reefs to be segmented into zones depending on similar traits or stresses. Useful for making sequence of decisions, deep reinforcement learning (DRL) can replicate management strategies to ascertain the best course of action under uncertain circumstances (Zai and Brown 2020). Through simulating various intervention scenarios and their long-term effects on coral reef health, DRL helps in optimising conservation strategies and resource management. General purpose models for classification and prediction, artificial neural networks (ANNs) can be taught to forecast results depending on intricate input patterns. Forecasting the growth rates of various coral species under diverse environmental conditions—including nutrient availability and water quality—can benefit from ANNs.

4 Habitat Suitability Model Using Artificial Neural Network

Habitat suitability modelling using artificial neural networks (ANNs) is the prediction of appropriate habitats for species dependent on environmental circumstances (Olden and Jackson 2001). Using the neuralnet package, I will walk you through building a simple ANN model in R to forecast habitat suitability for coral reefs depending on a hypothetical dataset of environmental variables.

```
> library(neuralnet)
> library(caret)
> # Example data creation
> set.seed(123)
> coral_data <- data.frame(
> Temperature = runif(100, 24, 30),  # Temperature in degrees Celsius
> Salinity = runif(100, 34, 38),     # Salinity in PSU
> Depth = runif(100, 5, 20),         # Depth in meters
> Suitable = sample(0:1, 100, replace = TRUE)  # Habitat suitability
> )

> # Normalize data: Scale the predictors
> preproc <- preProcess(coral_data[, -4], method = c("center", "scale"))
> normalized_data <- predict(preproc, coral_data[, -4])
> normalized_data$Suitable <- coral_data$Suitable
> # Set the formula
> formula <- Suitable ~ Temperature + Salinity + Depth

> # Train the neural network
> set.seed(123)
> nn_model <- neuralnet(formula, data = normalized_data, hidden = c(5),
  linear.output = FALSE)
> plot(nn_model)
> # Assuming new_data is available and preprocessed similarly
> new_data <- data.frame(
> Temperature = runif(10, 24, 30),
> Salinity = runif(10, 34, 38),
> Depth = runif(10, 5, 20)
```

```
> )
> new_data_normalized <- predict(preproc, new_data)

> # Predict using the neural network
> predictions <- compute(nn_model, new_data_normalized)
> predicted_classes <- ifelse(predictions$net.result > 0.5, 1, 0)

> # Print predictions
> print(predicted_classes)
> # Example: Calculating accuracy if true labels are known
> true_labels <- sample(0:1, 10, replace = TRUE)  # hypothetical true labels
> accuracy <- sum(predicted_classes == true_labels) / length(true_labels)
> print(paste("Accuracy:", accuracy))
```

5 Outcome of This Model

The outcome for this model will be as following. The 'predicted_classes' output from the artificial neural network (ANN) model represents the model's predictions about the suitability of habitats for coral reefs based on the environmental conditions provided in the 'new_data' (Fig. 2).

> print(predicted_classes)
[,1]
[1,] 1
[2,] 0
[3,] 0
[4,] 0
[5,] 1
[6,] 0
[7,] 1
[8,] 0
[9,] 1
[10,] 0

For 'predicted_classes', column [1] shows the predicted suitability class for each of the 10 observations (habitat sites) in the new_data. Value1 indicates that the model predicts the habitat which is suitable for coral reefs. 0 indicates that the model predicts the habitat which is not suitable for coral reefs.

The Outcome and Their Interpretation:

[1,] 1: The model predicts that the first habitat site is suitable.

[2,] 0: The model predicts that the second habitat site is not suitable.

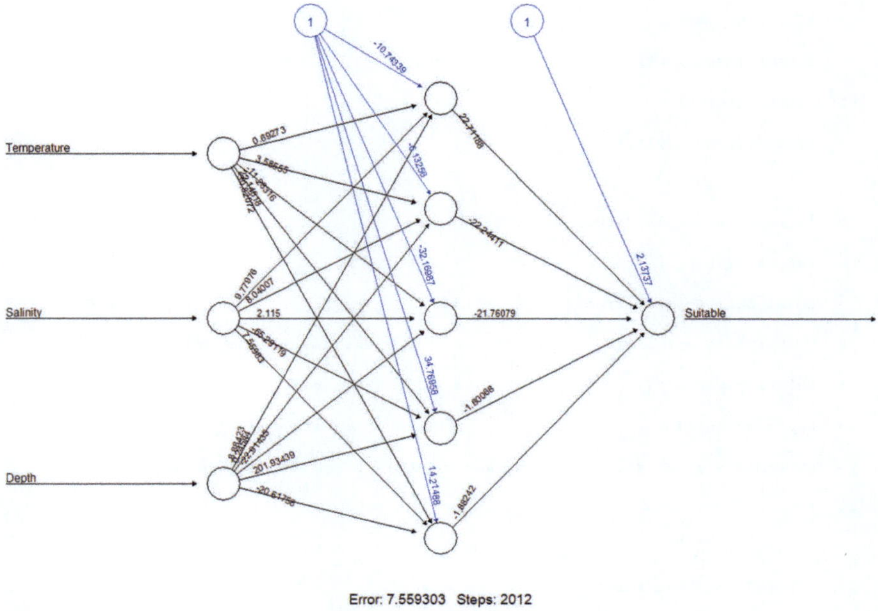

Fig. 2 Neural network architecture for the habitat suitability modelling

[3,] 0: The model predicts that the third habitat site is not suitable.
[4,] 0: The model predicts that the fourth habitat site is not suitable.
[5,] 1: The model predicts that the fifth habitat site is suitable.
[6,] 0: The model predicts that the sixth habitat site is not suitable.
[7,] 1: The model predicts that the seventh habitat site is suitable.
[8,] 0: The model predicts that the eighth habitat site is not suitable.
[9,] 1: The model predicts that the ninth habitat site is suitable.
[10,] 0: The model predicts that the tenth habitat site is not suitable.

5.1 Sensitivity Analysis for This Model and the Possible Outcome

Understanding how changes in input variables affect the output of a model depends on first sensitive analysis. It facilitates the identification of the factors with most influence on the forecasts of the model. Sensitivity analysis can help one understand how various environmental factors such as temperature, salinity, and depth affect the expected suitability in the framework of the ANN model for coral reef habitat suitability.

```
> # Load necessary libraries
> library(neuralnet)
> library(caret)
> library(sensitivity)

> # Create a sample dataset
> set.seed(123)
> coral_data <- data.frame(
> Temperature = runif(100, 24, 30),  # Temperature in degrees Celsius
> Salinity = runif(100, 34, 38),     # Salinity in PSU
> Depth = runif(100, 5, 20),         # Depth in meters
> Suitable = sample(0:1, 100, replace = TRUE)  # Habitat suitability
> )

> # Normalize the data
> preproc <- preProcess(coral_data[, -4], method = c("center", "scale"))
> normalized_data <- predict(preproc, coral_data[, -4])
```

```
> normalized_data$Suitable <- coral_data$Suitable

> # Define the model
> formula <- Suitable ~ Temperature + Salinity + Depth

> # Train the neural network
> set.seed(123)
> nn_model <- neuralnet(formula, data = normalized_data, hidden = c(5),
  linear.output = FALSE)

> # Define a function to predict using the neural network
> predict_nn <- function(model, newdata) {
> # Predict using the neural network model
> predictions <- compute(model, newdata)
> return(predictions$net.result)
> }

> # Generate two sets of input samples (X1 and X2) for sensitivity analysis
> set.seed(123)
> X1 <- as.data.frame(apply(normalized_data[, -4], 2, function(x)
  runif(1000, min(x), max(x))))
> X2 <- as.data.frame(apply(normalized_data[, -4], 2, function(x)
  runif(1000, min(x), max(x))))

> # Perform sensitivity analysis using variance-based sensitivity
> sensitivity_analysis <- sobol2007(
> model = function(X) predict_nn(nn_model, X),
> X1 = X1,
> X2 = X2,
> nboot = 100
```

```
> )

> # Print sensitivity indices
> print(sensitivity_analysis)

> # Plot sensitivity analysis results
> plot(sensitivity_analysis)
```

5.2 The Result of Sensitivity Analysis

```
Call:
sobol2007(model = function(X) predict_nn(nn_model, X), X1 = X1, X2 = X2,nboot =
100)
Model runs: 5000
First order indices:
               original          bias std. error   min. c.i.  max. c.i.
Temperature  0.06983743 -0.000136500 0.05520678 -0.04897244 0.16540718
Salinity    -0.01019335  0.010128951 0.04805855 -0.12307223 0.07367023
Depth        0.23326076  0.002997735 0.07103269  0.07322790 0.35945908
Total indices:
             original         bias std. error min. c.i. max. c.i.
Temperature 0.5473356  0.004827802 0.05584640 0.4403414 0.6638009
Salinity    0.5369107 -0.008539925 0.03458884 0.4663877 0.6148887
Depth       0.6804626 -0.005969786 0.05840903 0.5551984 0.7999871
```

The Sobol method's sensitivity analysis (Ganji et al. 2016) outcomes reveal how each input variable influences the output of your artificial neural network (ANN) model for coral reef habitat suitability. The Sobol method breaks out the variance of the model output into contributions from every input variable and their interactions. Considering that variable independently, first-order indices (S1) indicate the contribution of every input to the variance of the model output, so measuring the main effect of every variable. Indicating how much each variable directly and via interactions influences the output variance, total indices (ST) evaluate the total effect including the main effect and all interaction effects with other variables. The model was evaluated 5000 times to estimate the sensitivity indices, so guaranteeing statistical dependability in the results. First order indices are temperature, salinity, and depth; one can understand them as follows.

Temperature: 0.0698 suggests that the temperature by itself explains almost 7% of the variance in the model output. This indicates that while temperature is not the most important factor by itself, it has a modest direct effect on habitat suitability.

Depth: 0.2333 is a significant independent factor influencing habitat suitability since it indicates that roughly 23% of the variance is ascribed to depth. Among the three variables, depth has the biggest independent influence and implies it is a fundamental factor influencing the habitat suitability for coral reefs.

Salinity: − 0.0102 implies that, given negative values usually indicate minimal effects, salinity by itself does not significantly influence the variance in the model output. When taken without interactions, salinity has little effect on the suitability in any meaningful sense.

Examining the overall indices (ST), one finds temperature value 0.5493 indicates that, considering interactions with other variables, temperature considerably influences the model output variance. A key variable in determining habitat suitability since temperature interacts strongly with other elements. Although the first-order effect of salinity is small, the total Sobol index of 0.5369 indicates that, through interactions with other variables, salinity significantly influences habitat suitability. Although salinity has a small independent impact, its interactions make it a crucial component. The habitat suitability for coral depends much on depth. When interactions are taken into account, value 0.6805 indicates that depth is the most important variable; it accounts for almost 68% of the variance. With a given level of confidence—usually 95%—confidence intervals (min. c.i. and max. c.i.) offer the range within which the true sensitivity indices are expected to fall. A smaller interval implies more exact estimations.

Ultimately, from the model depth, both independently and through interactions, the most important variable shows to be though less important on their own, temperature and salinity greatly affect the model output in view of interactions (Fig. 3).

6 Discussion

6.1 *Deep Neural Network for Predicting Climate Change Impact on Coral*

By means of a structured approach comprising data collecting, data preprocessing, building a deep learning model, and training of it, using a deep neural network (DNN) in Python forecasts the effect of climate change on coral reefs. Data on coral reefs could comprise sea surface temperature, salinity, pH levels, depth, and other environmental factors affecting coral state. Once acquired, the data must be preprocessed and cleaned, so it is suitable for training. This addresses data standardising, data missing management, and data organisation fit for DNN input.

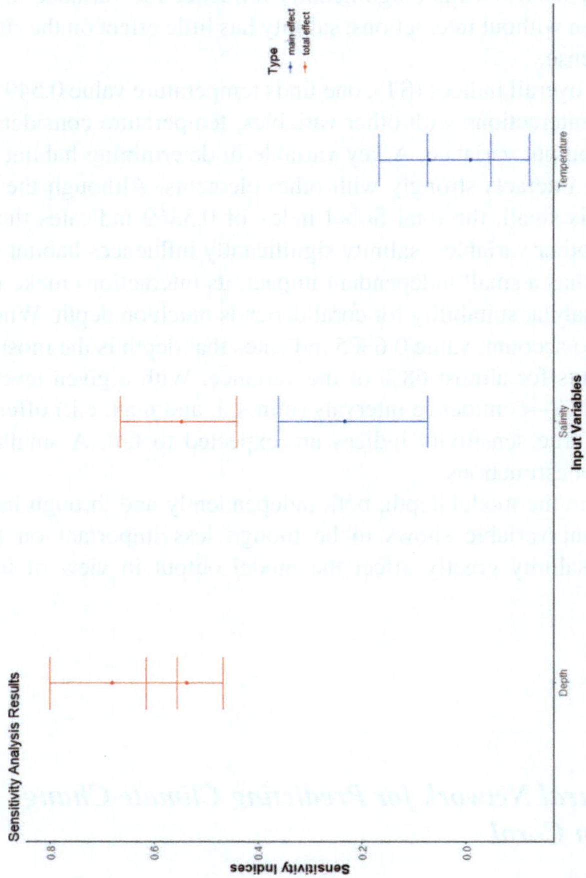

Fig. 3 Sensitivity analysis for the coral reef habitat suitability

```
import pandas as pd
from sklearn.model_selection import train_test_split
from sklearn.preprocessing import StandardScaler

# Load data
data = pd.read_csv('coral_reef_data.csv')

# Handle missing values
data.fillna(data.mean(), inplace=True)

# Normalize data
scaler = StandardScaler()
features = ['Temperature', 'Salinity', 'pH', 'Depth']
data[features] = scaler.fit_transform(data[features])

# Split data
X = data[features]
y = data['Coral_Health']  # Assuming Coral_Health is the target variable
X_train, X_test, y_train, y_test = train_test_split(X, y, test_size=0.2,
random_state=42)
import tensorflow as tf
from tensorflow.keras.models import Sequential
from tensorflow.keras.layers import Dense, Dropout

# Build the model
model = Sequential([
    Dense(128, activation='relu', input_shape=(X_train.shape[1],)),
    Dropout(0.2),
```

```
    Dense(64, activation='relu'),
    Dense(1, activation='sigmoid')  # Change to 'softmax' if it's a
classification problem
])

# Compile the model
model.compile(
    optimizer='adam',
    loss='binary_crossentropy',  # Or 'categorical_crossentropy' for multi-class
    metrics=['accuracy']
)
# Train the model
history = model.fit(X_train, y_train, epochs=50, batch_size=32,
validation_split=0.2)
# Evaluate the model
model.evaluate(X_test, y_test)

# Make predictions
predictions = model.predict(X_test)
# Evaluate the model
model.evaluate(X_test, y_test)

# Make predictions
predictions = model.predict(X_test)
import matplotlib.pyplot as plt

# Plot training history
plt.plot(history.history['accuracy'], label='accuracy')
plt.plot(history.history['val_accuracy'], label='val_accuracy')
plt.title('Model Accuracy')
plt.ylabel('Accuracy')
plt.xlabel('Epoch')
plt.legend()
plt.show()
```

The 'sequential' model in Keras is a linear stack of layers. It enables you to construct models incrementally, following a systematic approach. In the sequential model, multiple layers are defined. The 'dense' layer in a neural network is fully connected, with each neuron in this layer being connected to all neurons in the previous layer.

'128'. This layer contains a specific number of neurons or units. Choosing 128 neurons can have a significant impact on the model's ability to learn from intricate data. The activation function utilised in this case is the rectified linear unit (ReLU). It brings in nonlinearity to the model, enabling it to grasp more intricate patterns in the data. ReLU is typically defined as

$$f(x) = \max(0, x)$$

Understanding the structure of the input data that the model will process. The value 'X_train.shape[1]' represents the number of features present in the training dataset. Input shape is only specified for the initial layer in Keras models, as it requires knowledge of the input dimensionality. During training, a fraction of input units (in this case, 20%) are randomly set to 0 by the dropout layer at each update. This approach helps mitigate overfitting by decreasing the model's dependence on individual neurons, encouraging independent learning of representations across neurons. There is an additional dense layer with 64 neurons that utilises the ReLU activation function. It helps to further process the transformed features from the previous layers. When dealing with binary classification problems, the sigmoid activation function is commonly employed. The output is a value ranging from 0 to 1, which indicates the likelihood of a particular class, usually referred to as the 'positive' class. In the case of a multi-class classification problem, an alternative approach would be to utilise a 'softmax' activation function. This function would generate a probability distribution across various classes.

This model architecture is designed to process input features through a sequence of dense layers, while also incorporating a dropout layer and specific activation functions. The goal is to create increasingly abstract representations of the data, which helps to reduce overfitting. Ultimately, the model makes a final prediction in a binary classification format. This structure is commonly used for straightforward yet powerful deep learning models in a range of applications, such as predicting environmental data like coral reef health.

7 AI Models in Ecosystem Resilience

AI models can analyse complex interactions within coral reef ecosystems by combining different types of data and applying advanced analytical techniques (Salcedo 2016). Creating an AI model entails several step-by-step methods for understanding ecosystem resilience. Developing an AI model entails creating a predictive model that assesses how various factors contribute to the resilience of an ecosystem, such as a coral reef. This model will use environmental, biological, and human impact data to forecast resilience and identify key drivers. The first step is to define the objectives, which requires three basic procedures.

Step 1: Forecast resilience scores for ecosystems under various scenarios.

Step 2: Determine which factors most significantly impact ecosystem resilience.
Step 3: Offer recommendations for conservation and management.

The detailed procedure consists of data collection, data preprocessing, feature engineering, model selection and training, model construction, model evaluation, deployment, and visualisation.

An example of how to create an AI model using the Python language is presented. In this example, the random forest regressor was used to create the model.

```
import numpy as np

import pandas as pd

from sklearn.model_selection import train_test_split

from sklearn.ensemble import RandomForestRegressor

from sklearn.metrics import mean_absolute_error, mean_squared_error, r2_score

import shap

import matplotlib.pyplot as plt

import os


# Set random seed for reproducibility

np.random.seed(42)
```

```
# Number of samples
n_samples = 1000

# Generate synthetic data
environmental_data = pd.DataFrame({
    'common_key': np.arange(n_samples),
    'temperature': np.random.uniform(20, 30, n_samples),
    'salinity': np.random.uniform(30, 40, n_samples),
    'nutrient_levels': np.random.uniform(1, 10, n_samples),
    'water_quality': np.random.uniform(1, 10, n_samples)
})

biological_data = pd.DataFrame({
    'common_key': np.arange(n_samples),
    'species_count': np.random.randint(10, 100, n_samples),
    'population_size': np.random.uniform(1000, 10000, n_samples),
    'coral_cover': np.random.uniform(10, 90, n_samples)
})

human_impact_data = pd.DataFrame({
    'common_key': np.arange(n_samples),
    'pollution_level': np.random.uniform(1, 10, n_samples),
    'fishing_pressure': np.random.uniform(1, 10, n_samples),
    'tourism_intensity': np.random.uniform(1, 10, n_samples)
})

# Merge the data into a single dataframe
data = pd.merge(environmental_data, biological_data, on='common_key')
data = pd.merge(data, human_impact_data, on='common_key')

# Generate a synthetic resilience score (target variable)
```

```
data['resilience_score'] = np.random.uniform(0, 1, n_samples)


# Feature Engineering
def feature_engineering(data):
    data['temperature_anomaly'] = (data['temperature'] -
data['temperature'].mean()) / data['temperature'].std()
    data['salinity_level'] = data['salinity'] / data['salinity'].max()
    data['species_richness'] = data['species_count'] / data['population_size']
    data['human_impact_index'] = (data['fishing_pressure'] +
data['tourism_intensity'] + data['pollution_level']) / 3
    return data


data = feature_engineering(data)


# Split data into features and target
features = data.drop(['resilience_score', 'common_key'], axis=1)
target = data['resilience_score']


# Split into training and test sets
X_train, X_test, y_train, y_test = train_test_split(features, target,
test_size=0.2, random_state=42)


# Train the Random Forest model
model = RandomForestRegressor(n_estimators=100, random_state=42)
model.fit(X_train, y_train)


# Make predictions
y_pred = model.predict(X_test)


# Evaluate the model
mae = mean_absolute_error(y_test, y_pred)
rmse = np.sqrt(mean_squared_error(y_test, y_pred))
```

```
r2 = r2_score(y_test, y_pred)

print("MAE:", mae)
print("RMSE:", rmse)
print("R²:", r2)

# Save evaluation metrics to a text file
results_dir = 'results'
os.makedirs(results_dir, exist_ok=True)
with open(os.path.join(results_dir, 'model_evaluation.txt'), 'w') as f:
    f.write(f"MAE: {mae}\n")
    f.write(f"RMSE: {rmse}\n")
    f.write(f"R²: {r2}\n")

# Interpret the model using SHAP
explainer = shap.TreeExplainer(model)
shap_values = explainer.shap_values(X_test)

# Summary plot of SHAP values
summary_plot_path = os.path.join(results_dir, 'G:/Book chapter/T1-
book/shap_summary_plot.tif')
shap.summary_plot(shap_values, X_test, show=False)
plt.savefig(summary_plot_path)
plt.close()

# Dependence plot for a specific feature
dependence_plot_path = os.path.join(results_dir,
'shap_dependence_plot_temperature_anomaly.png')
shap.dependence_plot("temperature_anomaly", shap_values, X_test, show=False)
plt.savefig(dependence_plot_path)
plt.close()
```

The model in the provided example is a random forest regressor. Random forest is an ensemble learning technique that combines the predictions of multiple decision trees in order to improve predictive accuracy and prevent overfitting (Seni and Elder 2010). This option is appropriate for the initial implementation due to its robustness and ability to handle a combination of numerical and categorical features without extensive preprocessing. Furthermore, random forest models strike an appropriate balance between performance and interpretability, which is critical for understanding the factors that drive ecosystem resilience.

Random forests are less prone to overfitting than other models, such as decision trees. They provide feature importance scores to assist in identifying key drivers of resilience. They can detect complex, nonlinear interactions between features. They require less parameter tuning than other models, such as support vector machines or neural networks.

8 The Random Forest Architecture

Gini impurity is used in classification to measure the impurity of a node (Muchai and Odongo 2014). The formula for Gini impurity for a node is

$$Gini(m) = 1 - \sum_{i=1}^{C} p_i^2$$

where p_i is the proportion of samples belonging to class in node m, and C is the number of classes.

Information gain is used in classification to measure the reduction in entropy from a split. For a node m with N samples, entropy H is,

$$H(m) = -\sum_{i=1}^{C} p_i \log_2(p_i)$$

Information gain (IG) from a split into nodes m_1 and m_2 is

$$IG = H(m) - \left(\frac{N_1}{N} H(m_1) + \frac{N_2}{N} H(m_2)\right)$$

where N_1 and N_2 are the number of samples in nodes m_1 and m_2, respectively.

Random forest algorithm is for dataset with N samples.

$$D = \{(x_i, y_i)\}_{i=1}^{N}$$

where create B bootstrap samples D_b from D.

Tree construction for each bootstrap sample D_b:

- At each node, randomly select m features from the total M features ($m \ll M$).
- Split the node using the best feature among the mmm selected features.
- Repeat until the stopping criteria are met (e.g. maximum depth, minimum number of samples in a node).

Final prediction of model:

1. Classification: For a new input x, the final prediction $\hat{y}$ is the mode of the predictions from all trees T_b

$$\hat{y} = mode\left(\{T_b(x)\}_{b=1}^{B}\right)$$

2. Regression: For a new input x, the final prediction $\hat{y}$ is the mode of the predictions from all trees T_b

$$\hat{y} = \frac{1}{B}\sum_{b=1}^{B} T_b(x)$$

Random forests can also provide estimates of feature importance, which helps in understanding which features contribute most to the model's predictions.

- **Mean Decrease Impurity**: For each feature, calculate the total decrease in impurity (e.g. Gini impurity); it causes across all trees. Features that result in larger decrease in impurity are considered more important.
- **Mean Decrease Accuracy**: Measure the decrease in model accuracy when the values of a particular feature are permuted. A significant drop in accuracy indicates that the feature is important.

Ecological resilience refers to an ecosystem's ability to withstand disturbances while maintaining its basic structure and function. In this case, a random forest model can help identify key factors that contribute to resilience and predict how changes in these factors will affect ecosystem stability. SHapley Additive Explanations (SHAP) values provide a detailed breakdown of each feature's contribution to individual predictions (García and Aznarte 2020). This helps us understand how each factor affects resilience scores. The SHAP summary plot depicts the total impact of each feature on the model's predictions. Features with larger spreads have more influence (Fig. 4).

Higher coral cover (pink) improves ecosystem resilience, as demonstrated by the clustering of pink dots on the positive side of the SHAP values. Higher human impact (pink) generally reduces ecosystem resilience, as evidenced by the pink dots on the negative side of the SHAP values. Better water quality (pink) promotes ecosystem resilience. Large temperature anomalies (pink) appear to have a negative impact on ecosystem resilience, indicating that extreme temperature changes can be harmful to the ecosystem. Increased fishing pressure (pink) is associated with a decrease in resilience, implying that overfishing could harm the ecosystem. Increased species richness (pink) contributes positively, demonstrating that diverse ecosystems are more resilient.

Except for random forest regressor, gradient model boosting machines (GBM), and neural networks such as convolutional neural networks (CNN) and recurrent neural networks (RNNs), XGBoost, or LightGBM, these models frequently outperform random forests, particularly with extensive hyperparameter tuning.

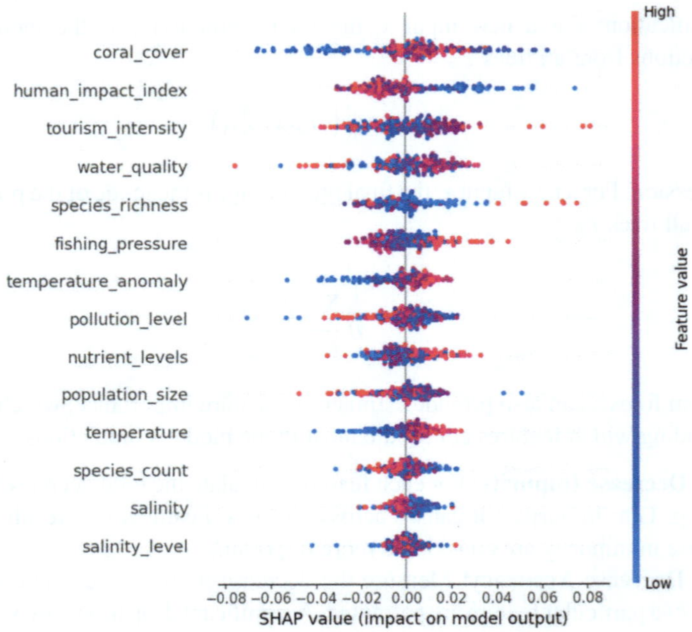

Fig. 4 Parameters importance for random forest regressor

9 Optimising Conservation Efforts

AI can help advance marine conservation efforts by leveraging data-driven insights to identify priority areas, optimise marine-protected area (MPA) design, and support coral restoration (Trice et al. 2021). AI can analyse massive amounts of data, such as satellite imagery, oceanographic data, and biodiversity records, to identify regions of high ecological value or areas at risk from human activity. Machine learning algorithms can predict the distribution of marine species, assisting in the identification of critical habitats that require protection. These models can take into account climate change, pollution, and habitat degradation. AI can assess and predict the impact of various threats to marine ecosystems, such as overfishing, pollution, and climate change, allowing conservationists to prioritise high-risk areas.

AI can optimise MPA design by analysing spatial data and balancing ecological needs with socioeconomic considerations. This ensures that MPAs are effective at protecting biodiversity while also reducing conflicts with human activities like fishing and tourism. AI can evaluate the connectivity of various marine habitats, ensuring that MPAs are designed to sustain ecological networks that support species migration, breeding, and feeding. AI can help with adaptive management of MPAs by continuously monitoring environmental conditions and human behaviour. This allows for dynamic adjustments to MPA boundaries and regulations in response to changing conditions. Artificial intelligence can analyse genetic, environmental, and ecological data to identify coral species that are more resistant to stressors such as rising sea temperatures and acidification. This information is critical for selecting species that are more likely to survive and thrive in changing ocean environments. AI can identify the best locations for coral transplantation by analysing water quality, temperature, light availability, and existing biodiversity. This increases the likelihood that transplanted corals will survive and grow successfully. AI-powered monitoring tools can track the health and growth of restored coral reefs, providing feedback on the effectiveness of restoration efforts and allowing for real-time strategy adjustments.

```
import geopandas as gpd

import numpy as np

from scipy.optimize import minimize

from sklearn.ensemble import RandomForestClassifier

from shapely.geometry import Polygon

import matplotlib.pyplot as plt

import rasterio

from rasterio.plot import show

from pyproj import Proj, transform
```

```
# Load shapefile of the area of interest
area_of_interest = gpd.read_file('path_to_shapefile.shp')

# Load raster data (e.g., sea surface temperature, chlorophyll, etc.)
with rasterio.open('path_to_raster.tif') as src:
    environmental_data = src.read(1)
    transform = src.transform
    crs = src.crs

# Load biodiversity data as a GeoDataFrame
biodiversity_data = gpd.read_file('path_to_biodiversity_data.shp')

# Load human activity data (e.g., fishing zones)
human_activity_data = gpd.read_file('path_to_human_activity.shp')
# Extract environmental features from raster data
def extract_raster_values(geometry, raster_data, transform):
    # Transform the polygon to match the raster's coordinate system
    coords = [(p.x, p.y) for p in geometry.exterior.coords]
    px = [int((x - transform[2]) / transform[0]) for x, y in coords]
    py = [int((y - transform[5]) / transform[4]) for x, y in coords]

    values = raster_data[py, px]
    return values.mean()

biodiversity_data['env_value'] = biodiversity_data.geometry.apply(lambda x:
extract_raster_values(x, environmental_data, transform))
def objective_function(x, biodiversity_data, human_activity_data):
    # Calculate biodiversity score
    biodiversity_score = np.sum(biodiversity_data['env_value'] * x)

    # Calculate human activity cost
    human_activity_cost = np.sum(human_activity_data['impact'] * x)
```

```
    # Objective is to maximize biodiversity score and minimize human activity
cost
    return -(biodiversity_score - human_activity_cost)
def area_constraint(x, total_area):
    protected_area = np.sum(x * biodiversity_data['area'])
    return total_area - protected_area
# Initial guess (all areas unprotected)
x0 = np.zeros(len(biodiversity_data))

# Constraints
constraints = [{'type': 'ineq', 'fun': area_constraint, 'args': (total_area,)}]

# Bounds (each area can either be protected (1) or not (0))

bounds = [(0, 1) for _ in range(len(biodiversity_data))]

# Run optimization
result = minimize(objective_function, x0, args=(biodiversity_data,
human_activity_data),
                  constraints=constraints, bounds=bounds, method='SLSQP')

# Extract the optimal design
optimal_design = result.x
# Add the optimization results to the GeoDataFrame
biodiversity_data['protection_status'] = optimal_design

# Plot the results
fig, ax = plt.subplots(figsize=(10, 10))
area_of_interest.plot(ax=ax, color='white', edgecolor='black')
biodiversity_data[biodiversity_data['protection_status'] > 0.5].plot(ax=ax,
color='green', alpha=0.5)
plt.title('Optimized Marine Protected Areas')
```

This model is intended to optimise the design of marine-protected areas (MPAs) by balancing the need to protect biodiversity against the impact of human activities such as fishing and tourism.

Data collection and preprocessing include environmental, biodiversity, and human activity data. The model computes environmental values (such as average sea surface temperature) for various areas within the region of interest. These features contribute to the ecological value of each area. Similarly, the model assesses the effects of human activity on these areas. This could include calculating the intensity of fishing or the proximity to shipping lanes. The model assigns a score to each area based on its environmental characteristics. This score reflects the area's ecological value, such as its importance in biodiversity. The model also assigns a cost to each area based on the intensity of human activity. This cost reflects the potential conflict between conservation and human use. The objective function is intended to maximise biodiversity protection while minimising the costs associated with limiting human activity. This balance is critical in MPA design. One common constraint is the total area that can be designated as an MPA. For example, if only a portion of the total area can be protected, the model must ensure that it does not exceed that limit.

The model starts with a guess, typically assuming that no areas are protected. The model iteratively adjusts the protection status of each area using an optimisation algorithm (for example, sequential least squares programming, or SLSQP). The algorithm aims to improve the objective function with each iteration. After running the optimisation, the model returns the 'optimal' configuration, which indicates which areas should be protected in order to strike the best balance between conservation and human activity. The results are displayed on a map, with areas designated for protection highlighted. This visualisation assists stakeholders in understanding the proposed MPA boundaries and their implications. The initial design is compared to additional criteria, such as connectivity between protected areas or specific species protection objectives. Based on this evaluation, the model may be rerun with different parameters or additional constraints to improve the MPA design. The model can be rerun on a regular basis with updated data, allowing it to adjust to new environmental conditions or changes in human activity. This ensures that the MPA network is still effective over time, even as external factors change.

Sequential least squares programming (SLSQP) is an optimisation algorithm. It is specifically designed to solve nonlinear optimisation problems involving equality and inequality constraints (Ma et al. 2024). SLSQP can handle nonlinear objective functions, making it ideal for complex optimisation problems. It can handle both equality and inequality constraints, which is critical in many real-world optimisation problems. SLSQP allows for bound constraints on variables, which means you can limit the values that the decision variables can have. SLSQP employs gradient information (i.e. derivatives) from the objective function to guide the search for the best solution. This makes it more efficient for problems in which derivatives can be calculated. For design optimisation, where there are numerous constraints due to physical laws and performance criteria, multi-objective optimisation frameworks are required to balance trade-offs and achieve feasible solutions. Portfolio optimisation requires consideration of risk, return, and other factors. To optimise supply chains, resource allocation, and scheduling under complex constraints, advanced optimisation models are required to identify feasible and cost-effective outcomes. SLSQP is based on the sequential quadratic programming method, which involves approximating the nonlinear problem with a quadratic programming problem in each iteration. After solving the quadratic approximation, the method employs a line search to ensure that each step moves towards a lower value of the objective function. The process continues until the changes in the objective function and solution variables are sufficiently small, indicating convergence to an optimal solution.

10 AI System in the Early Warning System

Early warning systems about environmental problems like pollution, invasive species invasions, and coral bleaching are made possible by artificial intelligence (AI) systems, which enables prompt management and intervention measures (Ning et al. 2024). Artificial intelligence systems monitor sea surface temperatures through satellite data, which is a crucial signpost for possible instances of coral bleaching. In order to enable real-time monitoring of conditions that may cause coral bleaching, artificial intelligence (AI) systems also process data from underwater sensors. These sensors measure variables like pH, light levels, and water temperature (Noor et al. 2022). Machine learning models are employed by artificial intelligence systems to forecast the probability of coral bleaching incidents. To find patterns and correlations, these models are trained on historical data, such as past bleaching events and related environmental conditions. AI is capable of analysing time series data to predict temperature trends in the future and pinpoint coral bleaching high-risk times. AI models can send out early warning alerts when they identify circumstances that are favourable for coral bleaching. In order to enable them to take preventative measures like shading corals, lowering local stressors, and organising restoration activities, these alerts are sent to marine biologists, conservationists, and local communities. In order to identify the existence of invasive species, artificial intelligence systems can analyse photographs and videos taken by underwater drones or cameras. Based on visual traits, computer vision techniques like convolutional neural networks (CNNs) are used to identify particular invasive species (Rauf et al. 2019). Artificial intelligence (AI) systems can also identify invasive species using acoustic data, especially those that have distinctive sounds, like some fish or marine invertebrates. Artificial intelligence models can forecast the distribution of invasive species by evaluating environmental parameters like salinity, temperature, and type of habitat. This aids in prioritising monitoring efforts and identifying ecosystems that are vulnerable. With the use of historical trends and current data, artificial intelligence systems can track the movement and population growth of invasive species and forecast their spread (Martinez et al. 2020). Targeted interventions, such as physical removal, the application of biological control agents, or the construction of barriers to stop future spread, are made possible by early detection of invasive species. By pinpointing the regions' most vulnerable invasive species, artificial intelligence can aid in more efficient resource allocation and enable more focused management initiatives. Artificial intelligence systems track pollution levels, including chemical pollutants, oil spills, and plastic waste, by utilising data from networks of sensors placed in maritime environments (Fazri et al. 2023). These sensors produce real-time data streams by detecting pollutants in the air and water. AI-analysed satellite imagery can identify changes in water colour, turbidity, or surface reflectance to identify large-scale pollution events, like oil spills or algal blooms. AI is able to identify patterns in pollution and follow pollutants back to their origin. AI, for instance, can assist in locating pollution sources by recognising potential routes like sewage outfalls, agricultural runoff, and industrial discharge (Subramaniam et al. 2022). AI models enable timely pollution containment and mitigation measures by predicting the movement of pollutants based on currents, wind patterns, and other environmental factors. AI systems have the ability to alert relevant authorities in advance when they identify pollution levels that surpass safe thresholds. These alerts enable prompt action, such as prohibiting swimming or fishing in the impacted areas, sending out clean up teams, or enforcing laws to stop pollution before it starts. By continuously updating pollution models with real-time data, artificial intelligence (AI) can support adaptive management by enabling quicker responses to changing conditions and more successful long-term management plans.

Artificial intelligence (AI) systems enhance the ability to identify environmental threats, such as pollution, invasive species, and coral bleaching, early and respond appropriately. By combining data from many sources, utilising sophisticated predictive models, and automating detection procedures, artificial intelligence (AI) enables more proactive and successful management strategies. These systems play a vital role in conservation efforts, the preservation of marine ecosystems, and the reduction of environmental stressors.

11 AI Use in Monitoring Activities

AI can help monitor human activities like fishing and coastal development to assess their impact on coral reefs and promote sustainable practices. AI algorithms analyse data from vessel monitoring systems (VMS), which track fishing vessels' positions, speeds, and courses. This enables real-time monitoring of fishing activity in and around coral reef areas. Similar to VMS, AI uses automatic identification system (AIS) data to track commercial vessels (Birchenough et al. 2021). Machine learning models can identify fishing patterns like trawling and longlining, as well as detect illegal, unreported, and unregulated (IUU) fishing activities. AI uses spatial data from satellite imagery, AIS, and VMS to map fishing pressure on coral reefs. AI can determine the level of stress on reef ecosystems by analysing the frequency and intensity of fishing activities in specific areas. AI models predict the potential impact of fishing activities on coral reefs by comparing fishing intensity to changes in reef health indicators like coral cover, fish biomass, and species diversity. AI can recommend the establishment or modification of MPAs based on real-time data on fishing activity and reef health. This dynamic approach ensures that protective measures adapt to changing conditions. AI systems can improve fisheries management by making recommendations on sustainable catch limits, seasonal closures, and gear restrictions, ensuring that fishing activities do not exceed the reef ecosystems' ability to recover. AI examines satellite and aerial imagery to identify changes in land use along coastlines, such as urbanisation, agriculture, and infrastructure development. This aids in identifying areas where coastal development may be negatively impacting coral reefs due to increased runoff, sedimentation, or pollution. AI-powered image recognition algorithms can monitor construction activities such as resorts, ports, and seawalls that may have an impact on coral reefs, either directly or indirectly. This includes monitoring the growth of coastal facilities and their proximity to sensitive reef areas.

AI systems use data from water quality sensors to evaluate the impact of coastal development on coral reefs. For example, AI can detect increases in nutrients, pollutants, or sediments caused by construction or land-use changes and correlate them with reef health. AI models can predict how coastal development will affect erosion and sedimentation patterns, both of which are important factors affecting coral reefs. By simulating various development scenarios, AI can assist planners in reducing the impact on reefs. AI tools can streamline the EIA process by quickly assessing the potential impact of proposed developments on coral reefs and recommending mitigation measures. This ensures that development projects prioritise reef conservation from the start. AI can help design and place green infrastructure, such as mangrove restoration or artificial reefs, to reduce the impact of coastal development on coral reefs by stabilising shorelines, reducing runoff, and increasing habitat complexity. AI systems generate reports and visualisations that highlight the effects of human activity on coral reefs. These tools are critical for educating stakeholders, such as policymakers, developers, and the general public, on the value of sustainable practices. AI-powered platforms can engage communities by providing readily available information about the health of nearby reefs and the impact of local activities. This can encourage community-led conservation efforts and promote sustainable practices at the grassroots level. AI can help enforce coral reef conservation regulations by monitoring compliance with fishing quotas, coastal development guidelines, and MPA boundaries. This ensures that human activities remain within sustainable limits. By analysing trends and impacts over time, AI can make data-driven recommendations to policymakers about how to change regulations and policies to better protect coral reefs from human activity.

Artificial intelligence systems enable adaptive management by continuously monitoring human activity and reef health. This allows for timely adjustments to management strategies to address emerging threats and ensure long-term viability. AI can simulate various scenarios based on different levels of human activity and environmental conditions, allowing stakeholders to plan for the future and make informed decisions that balance development needs and reef conservation. AI is useful for monitoring and managing the effects of human activities on coral reefs. By tracking fishing activities and coastal development, assessing their effects on reef ecosystems, and promoting sustainable practices, AI contributes to the protection and survival of coral reefs in the face of increasing human pressures. These AI-powered solutions are critical for balancing human development needs with the imperative to protect one of the planet's most important ecosystems.

12 AI Application on the Coral Biogeography

Artificial intelligence (AI) is poised to transform coral reef research and conservation, providing new ways to understand and manage these vital ecosystems from a biogeographic perspective (Clampitt 2023). Traditionally, coral reef research relied heavily on manual data collection, field surveys, and limited technological tools, which frequently provided only a snapshot of reef conditions. However, artificial intelligence introduces a number of advanced capabilities that can significantly improve the depth, accuracy, and timeliness of coral reef biogeographic analyses.

Enhanced Monitoring and Mapping: Artificial intelligence can process massive amounts of data from satellite imagery, underwater sensors, and other remote sensing technologies to generate detailed maps of coral reef distributions around the world. These AI-powered maps are not only more accurate, but also more frequently updated than traditional methods, allowing researchers to track changes in coral reef health in real time. For example, AI can detect early signs of coral bleaching by analysing temperature data and visual indicators in images, allowing for timely interventions to protect vulnerable reefs.

Predictive Modelling: One of AI's most transformative capabilities is its ability to simulate various environmental scenarios and predict their effects on coral reefs. AI models can predict how coral reefs will shift geographically as a result of climate change by combining data on sea temperature, salinity, ocean currents, and other factors. These predictions aid in identifying future areas of coral reef decline, as well as potential new habitats where corals could thrive as conditions change. This foresight is critical for developing effective conservation strategies and marine protected areas (MPAs) that can withstand future environmental changes. AI can identify which coral species are most likely to survive changing environmental conditions, directing conservation and restoration efforts towards the most resilient species. AI can also determine the best locations for coral transplantation by analysing environmental factors such as water quality, light availability, and biodiversity. This targeted approach improves the success rate of restoration projects and ensures that resources are used efficiently to protect and restore coral reefs.

Data Integration and Ecosystem Modelling: Artificial intelligence excels at integrating diverse datasets, such as ecological surveys and citizen science contributions, to create a comprehensive picture of coral reef ecosystems. This comprehensive understanding enables more accurate modelling of complex interactions within coral reef ecosystems, such as predator–prey dynamics, species competition, and the effects of human activity. AI can simulate how these interactions might evolve under different scenarios, providing insights into coral reef resilience and the efficacy of various conservation strategies.

One of AI's most promising applications in coral reef conservation is its ability to facilitate adaptive management. AI systems can detect changes in environmental conditions or human activities that endanger coral reefs by constantly analysing real-time data (Gramer et al. 2023). These systems can then recommend changes to management strategies, such as changing MPA boundaries or enforcing new regulations to reduce local stressors. This dynamic approach ensures that conservation efforts are effective even when conditions change.

Artificial intelligence has the potential to revolutionise coral reef biogeography and conservation. AI can provide the tools required to protect these ecosystems in the face of unprecedented environmental challenges by improving data collection, enabling predictive modelling, and facilitating adaptive management (Williams et al. 2024). This technological advancement represents a critical step towards preserving coral reef biodiversity and ecological function for future generations.

References

Amiri Z, Heidari A, Navimipour NJ (2024) Comprehensive survey of artificial intelligence techniques and strategies for climate change mitigation. Energy 132827

Anwar SM, Majid M, Qayyum A, Awais M, Alnowami M, Khan MK (2018) Medical image analysis using convolutional neural networks: a review. J Med Syst 42:1–13

Barve S, Webster JM, Chandra R (2023) Reef-insight: a framework for reef habitat mapping with clustering methods using remote sensing. Information 14(7):373

Bibri SE, Krogstie J, Kaboli A, Alahi A (2024) Smarter eco-cities and their leading-edge artificial intelligence of things solutions for environmental sustainability: a comprehensive systematic review. Environ Sci Ecotechnology 19:100330

Birchenough SE, Cooper PA, Jensen AC (2021) Vessel monitoring systems as a tool for mapping fishing effort for a small inshore fishery operating within a marine protected area. Mar Policy 124:104325

Brooke SA, D'Arcy M, Mason PJ, Whittaker AC (2020) Rapid multispectral data sampling using google earth engine. Comput Geosci 135:104366

Clampitt M (2023) Indexing coral health status by an artificial intelligence approach, Doctoral dissertation, Université Côte d'Azur

Eiswerth ME, Haney JC (2001) Maximizing conserved biodiversity: why ecosystem indicators and thresholds matter. Ecol Econ 38(2):259–274

Fazri MF, Kusuma LB, Rahmawan RB, Fauji HN, Camille C (2023) Implementing artificial intelligence to reduce marine ecosystem pollution. IAIC Trans Sustain Digit Innov (ITSDI) 4(2):101–108

Ganji A, Maier HR, Dandy GC (2016) A modified Sobol′ sensitivity analysis method for decision-making in environmental problems. Environ Model Softw 75:15–27

García MV, Aznarte JL (2020) Shapley additive explanations for NO_2 forecasting. Ecol Inform 56:101039

Gómez-Ríos A, Tabik S, Luengo J, Shihavuddin ASM, Herrera F (2019) Coral species identification with texture or structure images using a two-level classifier based on convolutional neural networks. Knowl-Based Syst 184:104891

Gonzalez-Rivero M, Beijbom O, Rodriguez-Ramirez A, Bryant DE, Ganase A, Gonzalez-Marrero Y, Hoegh-Guldberg O (2020) Monitoring of coral reefs using artificial intelligence: a feasible and cost-effective approach. Remote Sens 12(3):489

Gramer LJ, Soden M, Hendee JC (2023) Operational ecoforecasting for coral reefs using artificial intelligence and integrated near real-time environmental data. Bull Mar Sci 99(3):379–394

Henkel TP (2010) Coral Reefs. Nat Educ Knowl 3(10):12

Hoegh-Guldberg O, Poloczanska ES, Skirving W, Dove S (2017) Coral reef ecosystems under climate change and ocean acidification. Front Mar Sci 4:158

Kahng SE, Akkaynak D, Shlesinger T, Hochberg EJ, Wiedenmann J, Tamir R, Tchernov D (2019) Light, temperature, photosynthesis, heterotrophy, and the lower depth limits of mesophotic coral ecosystems. Mesophotic Coral Ecosyst 801–828

Kiangala SK, Wang Z (2021) An effective adaptive customization framework for small manufacturing plants using extreme gradient boosting-XGBoost and random forest ensemble learning algorithms in an Industry 4.0 environment. Mach Learn Appl 4:100024

Leal Filho W, Wall T, Mucova SAR, Nagy GJ, Balogun AL, Luetz JM, Gandhi O (2022) Deploying artificial intelligence for climate change adaptation. Technol Forecast Soc Chang 180:121662

Li XM, Ma Y, Leng ZH, Zhang J, Lu XX (2020) High-accuracy remote sensing water depth retrieval for coral islands and reefs based on LSTM neural network. J Coast Res 102(SI):21–32

Ma Y, Gao X, Liu C, Li J (2024) Improved SQP and SLSQP algorithms for feasible path-based process optimisation. Comput Chem Eng 108751

Martinez B, Reaser JK, Dehgan A, Zamft B, Baisch D, McCormick C, Giordano AJ, Aicher R, Selbe S (2020) Technology innovation: advancing capacities for the early detection of and rapid response to invasive species. Biol Invasions 22(1):75–100

Molinari A (2020) Designing a performant ablation study framework for PyTorch. TRITA-EECS-EX 2020:790

Morales J, Haywood E, Krebs L, Safsten M (2024) Crowdsourcing conservation: how to use community science to advance public and private conservation. Cent Growth Oppor

Muchai E, Odongo L (2014) Comparison of crisp and fuzzy classification trees using gini index impurity measure on simulated data. Eur Sci J 10(18)

Nandutu I, Atemkeng M, Okouma P (2023) Integrating AI ethics in wildlife conservation AI systems in South Africa: a review, challenges, and future research agenda. AI Soc 1–13

Ning J, Pang S, Arifin Z, Zhang Y, Epa UPK, Qu M, Zhao J, Zhen F, Chowdhury A, Guo R, Deng Y (2024) The diversity of artificial intelligence applications in marine pollution: a systematic literature review. J Mar Sci Eng 12(7):1181

NOAA (2024) In what types of water do corals live? 06/16/2024, https://oceanservice.noaa.gov/facts/coralwaters.Html

NOAA (2024) What is the great barrier reef? National Ocean Service website. https://oceanservice.noaa.gov/facts/gbrlargeststructure.html. © UNESCO World Heritage Centre 1992–2024

Noor NM, Maulud KNA, Harun SN (2022) A smart sensing technology for monitoring marine environment conditions. In: Intelligent systems and machine learning for industry. CRC Press, pp 163–184

Olden JD, Jackson DA (2001) Fish–habitat relationships in lakes: gaining predictive and explanatory insight by using artificial neural networks. Trans Am Fish Soc 130(5):878–897

Rädecker N, Pogoreutz C, Voolstra CR, Wiedenmann J, Wild C (2015) Nitrogen cycling in corals: the key to understanding holobiont functioning? Trends Microbiol 23(8):490–497

Rauf HT, Lali MIU, Zahoor S, Shah SZH, Rehman AU, Bukhari SAC (2019) Visual features based automated identification of fish species using deep convolutional neural networks. Comput Electron Agric 167:105075

Reverter M, Helber SB, Rohde S, de Goeij JM, Schupp PJ (2022) Coral reef benthic community changes in the anthropocene: biogeographic heterogeneity, overlooked configurations, and methodology. Glob Chang Biol 28(6):1956–1971

Salcedo-Sanz S, Cuadra L, Vermeij MJ (2016) A review of computational intelligence techniques in coral reef-related applications. Ecol Inform 32:107–123

Saravanan R, Ranjith L, Jasmine S, Joshi KK (2017) Coral bleaching: causes, consequences and mitigation. Mar Fish Inf Serv Tech Ext Ser (231):3–9

Seni G, Elder J (2010) Ensemble methods in data mining: improving accuracy through combining predictions. Morgan & Claypool Publishers

Spalding M, Ravilious C, Green EP (2001) World atlas of coral reefs. Univ of California Press

Subramaniam S, Raju N, Ganesan A, Rajavel N, Chenniappan M, Prakash C, Pramanik A, Basak AK, Dixit S (2022) Artificial intelligence technologies for forecasting air pollution and human health: a narrative review. Sustainability 14(16):9951

Sultana M, Bhattacharya S, Adhikari N, Sengupta D, De D (2024) Artificial intelligence for sustainable. Artif Intell Edge Comput Sustain Ocean Health 3

Thompson DM, Kleypas J, Castruccio F, Curchitser EN, Pinsky ML, Jönsson B, Watson JR (2018) Variability in oceanographic barriers to coral larval dispersal: do currents shape biodiversity? Prog Ocean 165:110–122

Trice A, Robbins C, Philip N, Rumsey M (2021) Challenges and opportunities for ocean data to advance conservation and management. Ocean Conservancy, Washington, DC, USA

USEPA (2024) Basic information about coral reefs. https://www.epa.gov/coral-reefs/basic-information-about-coral-reefs

Vasilev I, Slater D, Spacagna G, Roelants P, Zocca V (2019) Python deep learning: exploring deep learning techniques and neural network architectures with Pytorch, Keras, and TensorFlow. Packt Publishing Ltd.

Wasim M, Mohanty S, Pandey PC, Basheer AK, Pandey AC (2025) Coral reefs threat assessment and comparative analysis of Neil and Agatti Island, India, using multispectral approach. In: Aquatic ecosystems monitoring. CRC Press, pp 252–278

Wicaksono P, Wulandari SA, Lazuardi W, Munir M (2021) Sentinel-2 images deliver possibilities for accurate and consistent multi-temporal benthic habitat maps in optically shallow water. Remote Sens Appl Soc Environ 23:100572

Williams B, Balvanera SM, Sethi SS, Lamont TAC, Jompa J, Prasetya M, Richardson L, Chapuis L, Weschke E, Hoey A, Beldade R, Mills SC, Haguenauer A, Zuberer F, Simpson SD, Curnick D, Jones KE (2024) Unlocking the soundscape of coral reefs with artificial intelligence: pretrained networks and unsupervised learning win out. BioRxiv 2024.02.02.578582. https://doi.org/10.1101/2024.02.02.578582

WWF (2024) Coral triangle. https://www.worldwildlife.org/places/coral-triangle

Zai A, Brown B (2020) Deep reinforcement learning in action. Manning Publications

Zapata-Ramírez PA, Hernández-Hamón H, Fitzsimmons C, Cano M, García J, Zuluaga CA, Vásquez RE (2023) Development of a google earth engine-based application for the management of shallow coral reefs using drone imagery. Remote Sens 15(14):3504

Natural Hazards in Coastal Zones

Natural Hazards in Coastal Zones

Annisa Utami Rauf, Ratna Dwi Puji Astuti, and K. Sarinah Basri

Abstract Coastal zones are dynamic environments highly susceptible to natural hazards such as tropical cyclones, tsunamis, storm surges, coastal erosion, and sea-level rise. These hazards pose significant threats to both human populations and environmental health. Climate change exacerbates the frequency and severity of these events, leading to public health challenges, including water contamination, habitat loss, and the spread of vector-borne and waterborne diseases. Pollution from industrial, agricultural, and urban sources compounds these risks, causing further degradation of coastal ecosystems and human health. This chapter provides a comprehensive analysis of natural hazards in coastal zones, which includes the deployment of early warning systems, sustainable coastal development practices, and the restoration of natural barriers such as mangroves and coral reefs. These measures enhance coastal resilience and protect environmental health. The environmental health implications of coastal hazards are multifaceted. Contamination of potable water sources due to saltwater intrusion and runoff from polluted land can lead to gastrointestinal diseases and other health issues. The rise in vector-borne diseases, such as malaria and dengue fever, is particularly concerning in coastal areas where stagnant water from flooding and poor drainage creates breeding grounds for disease-carrying mosquitoes. Community-based adaptation strategies are essential for addressing these health risks. Engaging local stakeholders in decision-making processes ensures that mitigation efforts are context-specific and more effective. Policymakers, planners, and community leaders must collaborate to develop sustainable and resilient coastal management practices.

A. U. Rauf (✉)
Department of Health Behavior, Environment and Social Medicine, Faculty of Medicine, Public Health and Nursing, Universitas Gadjah Mada, Yogyakarta, Indonesia
e-mail: annisaur@ugm.ac.id

R. D. P. Astuti
Department of Environmental Health, Faculty of Public Health, Universitas Airlangga, Surabaya, Indonesia

K. Sarinah Basri
Faculty of Sports and Health, Universitas Negeri Gorontalo, Gorontalo, Indonesia

K. K. Basheer Ahammed et al. (eds.), *Monitoring Coastal Marine Ecosystems Under Climate Change*, https://doi.org/10.1007/978-3-032-09137-6_9

Keywords Coastal resilience · Climate adaptation · Environmental pollution · Water contamination · Mitigation strategies

1 Introduction

Coastal zones are characterised by their ever-changing nature, where the land interfaces with the sea, creating environments that are both unique and continuously evolving (Agarwal and Deb 2023; Cabral et al. 2019). These regions, with their low-lying landscapes and extensive coastlines, are highly vulnerable to a range of natural hazards (Basheer Ahammed and Pandey 2019). For instance, tropical cyclones can unleash powerful winds, heavy rains, and storm surges, leading to widespread flooding and infrastructure damage. Coastal zones are subject to gradual environmental changes such as erosion and rising sea levels, which pose significant long-term risks (Astuti et al. 2022; Hecht and Oguchi 2017; Torresan et al. 2012). Coastal erosion, driven by the combined effects of wave action and sediment movement, progressively alters shorelines, resulting in loss of land and natural habitats (Kroeger et al. 2017; Lu et al. 2018; Nicolodi and Pettermann 2012; Torresan et al. 2012).

Human activities, including coastal development and alterations to natural landscapes, can exacerbate these effects by undermining shoreline stability (Nunn et al. 2021). Sea-level rise, driven by global climate change and the melting of polar ice, intensifies these risks by increasing the frequency and severity of coastal flooding (Griggs and Reguero 2021; Kisacik et al. 2022; Wu 2021). This rise accelerates erosion and causes flooding in areas that were once stable, presenting ongoing challenges for managing and adapting coastal environments. The significant impact of these threats on coastal communities necessitates their preparedness and awareness of the potential adverse effects (Benazir et al. 2023; Catane et al. 2024; Cisternas et al. 2024). Community adaptation can be enhanced through targeted education and preparedness training programs, which enable residents to comprehend the conditions and impacts they may encounter. These initiatives can also provide them with the necessary skills and knowledge to recover and rebuild effectively following disasters (Gaborit 2022; Wu 2021).

Environmental degradation in coastal areas due to natural hazards presents significant scientific and management challenges, particularly concerning water pollution. Coastal ecosystems, such as mangroves, coral reefs, and wetlands, are critical for shoreline protection against storm surges and erosion (Basheer Ahammed et al. 2024; Hecht and Oguchi 2017). Post-disaster scenarios frequently lead to heightened pollution levels, introducing various contaminants such as heavy metals, organic pollutants, and other hazardous substances into coastal waters (Cuaton 2019). Flooding and storm surges can mobilise industrial waste, agricultural runoff, and untreated sewage, severely compromising water quality. Heavy metals like mercury, lead, and cadmium can have toxicological impacts on marine biota and pose significant health risks to humans through bioaccumulation and biomagnification in the food web (Al-Kahtany et al. 2023; Buzzi et al. 2022; Marchellina et al. 2024). Organic pollutants, including pesticides and hydrocarbons, disrupt marine ecosystems, affecting species diversity and ecosystem functioning (Camacho-Jiménez et al. 2023; Sanganyado et al. 2021).

Addressing the challenges posed by rising sea levels and coastal storm surges necessitates comprehensive efforts to enhance the resilience of coastal communities. This requires robust intersectoral collaboration, involving local leaders, the enhancement of disaster alarm systems, and both natural and artificial environmental engineering along coastlines (Lu et al. 2018; Nicolodi and Pettermann 2012). Coastal regions, which are critical for their economic resources from fisheries and tourism, must be meticulously protected. The implications of natural hazards in these areas are profound, leading to severe socio-economic disruptions such as infrastructure damage, loss of biodiversity, and displacement of populations. Strengthening coastal resilience is imperative for mitigating these impacts, ensuring economic stability, and safeguarding communities against the escalating threats of climate change (Chen et al. 2021; Tracey 2022).

2 Impact of Climate Change on Coastal Hazards

Although projections of future sea levels usually only extend to 2100 due to growing uncertainties, sea-level rise will persist for decades and potentially centuries beyond that. Even without additional greenhouse gas emissions, the inertia of sea-level rise will continue to drive increases. The ice sheets and glaciers of Antarctica, Greenland, and the world's mountain glaciers contain roughly 66 m (about 216 feet) of potential sea-level rise (Griggs and Reguero 2021). Salinisation occurs as rising sea levels cause seawater to intrude into coastal aquifers, surface waters, and soils. It is further exacerbated by land-based droughts, reduced river flows, and increased water extraction (Chandía et al. 2024; Astuti et al. 2022; Tracey 2022). In regions like the Mekong Delta in Southeast Asia, seawater intrusion is already leading to the conversion of land and freshwater ponds into brackish or saline aquaculture. Sea-level rise impacts agriculture primarily through land submersion, salinisation of soil and freshwater resources, and land loss due to permanent flooding and erosion (Renaud et al. 2015). Oversaturation can lead to the shifting, corrosion, and cracking of underground pipes, impairing wastewater infrastructure's functionality and capacity. This can result in the unintended release of untreated wastewater, potentially contaminating groundwater wells, surface water sources, coastal recreational areas, and shellfish harvesting sites (Chandía et al. 2024; Tracey 2022).

Vector-borne infectious diseases are major contributors to death and illness in both humans and animals. Predictive models indicate that the climate changes associated with global warming will modify the transmission risk of numerous vector-borne diseases in different global regions (Balasubramanian et al. 2021; Lyth et al. 2005; Ramasamy and Surendran 2011). Furthermore, global warming will lead to higher sea levels, increasing the presence of saline and brackish water bodies in coastal regions (Alencar et al. 2023; Farrell et al. 2024). The genome of *An. aquasalis*, the primary coastal malaria vector in Central and South America and the Caribbean Islands, holds significant value. Apart from its role as a vector for human malaria parasites, *An. aquasalis* is ecologically unique because its larvae develop in the brackish waters of mangroves. Consequently, its genomic sequences may uncover genes associated with *Plasmodium sp.* interactions and distinctive adaptations for surviving in saline environments, unlike other anophelines. Encephalitis outbreaks in coastal areas have been reported in many countries, spread by saline-tolerant mosquitoes that bite birds and mammals. Coastal wetlands, which support juvenile fish and invertebrates, create habitats for these mosquitoes. The rise of brackish water habitats promotes saline-tolerant mosquitoes and allows freshwater vectors to adapt to brackish breeding grounds (Balasubramanian et al. 2021; Freitas et al. 2012).

In Brazil, the vulnerability of coastal zones has been notably assessed in the Macrodiagnóstico da Zona Costeira e Marinha, a 2008 report by the Ministry of the Environment. This assessment revealed that areas characterised by low elevation, high population density, social disadvantage, and complex technological infrastructure are especially vulnerable. Metropolitan regions such as Belém, capitals in the Northeast, Rio de Janeiro and Santos were identified as particularly high-risk areas (Nicolodi and Pettermann 2012). Research in the Mekong Delta of Vietnam (MDV) indicates that drought and saline intrusion frequently occur during dry seasons, causing severe socio-economic impacts, including water shortages, crop failures, and social unrest (Nghia et al. 2024). Climate change has intensified these hazards, making them more complex and unpredictable, especially in the coastal provinces of the MDV. Both Brazil and Vietnam contend with substantial challenges in their coastal zones due to environmental vulnerabilities, which are amplified by climate change, highlighting the need for targeted adaptation and mitigation strategies.

In addition to environmental degradation, the damage to coastlines due to natural hazards has profound social and economic repercussions (Chen et al. 2021; Mattah et al. 2023). Socially, the destruction of coastal areas can lead to the displacement of communities, disruption of social structures, and increased vulnerability of populations to subsequent hazards. Economically, the impact extends to declines in tourism revenue, losses in fisheries and aquaculture, and decreased property values, which collectively contribute to financial instability and reduced economic resilience in affected regions (Levi et al. 2015). Previous studies have examined the effects of climate change, rising sea levels, and pluvial flooding on tourism. By integrating assessments of hazards, exposure, and physical and environmental vulnerabilities, researchers have estimated the pluvial flooding risk for the Venice municipality. The resulting regional risk map indicates that 41% of commercial zones and 27% of industrial zones are located in areas deemed very highly or highly vulnerable (Molinaroli et al. 2019; Sperotto et al. 2016).

3 Major Natural Hazards in the Coastal Zone

Coastal changes are natural processes; however, they become hazardous when they pose risks to human lives, communities, or ecosystems. Some hazards occur abruptly, such as earthquakes, landslides, and tsunamis, while others develop over shorter time-frames, including hurricanes and extreme storms. Several regions experience coastal damage due to geographic conditions that are highly vulnerable to natural hazards. For instance, the Ring of Fire in the Asia–Pacific region significantly increases the risk of earthquakes and tsunamis along the coasts. In Asia, the Ring of Fire includes countries such as Indonesia, the Philippines, Japan, and Papua New Guinea (Jones et al. 2024). While Australia itself is not directly located within the Ring of Fire, its neighbouring regions, such as Papua New Guinea and the South Pacific islands, are part of this zone. As a result, the risk of earthquakes and tsunamis along the coasts of Asia and Australia, particularly in areas near tectonic plate boundaries, is notably high. This is evident from the destructive events in these regions, which often require prolonged recovery periods. The local geology, geography, and tectonic structure play a crucial role in assessing tsunami risks in a region. It also considers secondary factors, including funnelling effects, bay formations, submarine landslides, liquefaction, and landslides, which may have increased the tsunami's impact (Terry et al. 2022). Several natural disaster events from various coastal areas around the world are given in Table 1.

4 Pollution and Health Effects

Coastal erosion, driven by the combined effects of wave action and sediment movement, progressively alters shorelines, resulting in loss of land and natural habitats (Hecht and Oguchi 2017; Martyr-Koller et al. 2021; Wu 2021). Human activities, including coastal development and alterations to natural landscapes, can exacerbate these effects by undermining shoreline stability (Nunn et al. 2021). Sea-level rise, driven by global climate change and the melting of polar ice, intensifies these risks by increasing the frequency and severity of coastal flooding (Griggs and Reguero 2021; Ramasamy and Surendran 2011). Sea-level rise significantly impacts health by increasing risks such as injuries, drownings, exposure to pathogens, contaminants, and mental health issues (Niazi et al. 2022). It exacerbates existing environmental and socio-economic disparities, particularly affecting vulnerable coastal communities, including rural, remote, and indigenous populations in British Columbia, the Atlantic provinces, and northern Canada (Tracey 2022).

Pollution from human activities greatly affects coastal ecosystems. Numerous substances and materials released into marine environments can become pollutants when their concentrations surpass certain limits, harming biological components (Cabral et al. 2019). These pollutants released from various human activities, including agriculture, industry, urban and port development, transportation, fisheries,

Table 1 Destructive natural disasters that damage and/or originate from coastal areas

Events	Locations/area	Negative impact to the coastal area	Source
Strom surge and tropical cyclones	Sundarbans, India	Coastal slope impacts stormwater speed and exposure. The analysis shows 22.62% of Sundarbans villages which are highly exposed, 19.70% highly sensitive, and only 7.07% have strong adaptive capacity. Southern coastal villages are highly vulnerable due to proximity to the Bay of Bengal, frequent cyclones, and storm surge-induced floods, causing embankment breaches and land salinisation, which reduces agricultural productivity	(Hasnine and Nagdeve 2024)
Flooding	Languedoc, France	River flooding from extreme rainfall is a significant hazard in the coastal zone of southeastern France. A 2000-year flood record from the Hérault River basin, reconstructed using lagoon deposits and historical flood data, highlights the impact of human activities, such as forest clearings for agriculture, on increasing flood risks	(Degeai et al. 2022)
Algal bloom	Kamchatka coast, Russia	In the fall of 2020, a harmful algal bloom (HAB) covering 30,000 km^2 occurred along the Kamchatka Peninsula, causing significant damage to marine life, including sea urchins, octopuses, and bivalves. This is the first assessment of HAB impacts in Russian coastal waters. The bloom was caused by Karenia selliformis, with rDNA sequences identical to those found in Hokkaido in 2021	(Orlova et al. 2022)
Tsunami	Indonesia	The 2004 Indian Ocean earthquake and tsunami claimed 172,761 lives in Indonesia. The deadliest year was 1966, with 9264 fatalities from various disasters, including a drought in Lombok (8000 deaths), and the 7.7 Mw earthquake that struck on 17 July 2006, triggered the Tasikmalaya tsunami, resulting in approximately 802 fatalities	(Parwanto and Oyama 2014)

(continued)

Table 1 (continued)

Events	Locations/area	Negative impact to the coastal area	Source
Tidal wave attacks, coastal erosion, and flooding	Volta region, Ghana	The key socio-economic factors affecting the vulnerability of communities are poverty and insufficient access to social services and infrastructure. Community awareness of disaster risk is shaped by the combined effects of environmental hazards, including livelihood destruction (44%), property damage (19%), displacement (31%), and impacts on human life (5%)	(Mattah et al. 2024)
Tropical cyclone	Beira, Mozambique	Tropical cyclone Idai, among the most destructive storms in Southern Africa, severely impacted Mozambique, Zimbabwe, and Malawi. In Mozambique, it resulted in 598 deaths, 1600 injuries, the destruction of nearly 198,000 homes, widespread crop devastation, a cholera outbreak, and affected an estimated 1.85 million people	(Goulart et al. 2024)
Flooding	Kuala Selangor, Malaysia	The coastal area of Kuala Selangor, Malaysia, faces significant Natech risks due to emerging climate hazards and inadequate stakeholder awareness. Manufacturing industries in this region are vulnerable to flooding, with insufficient flood protection and unclear measures for hazardous material containment and storage	(Yahaya et al. 2024)

(continued)

Table 1 (continued)

Events	Locations/area	Negative impact to the coastal area	Source
Saltwater intrusion and flooding	Jakarta, Indonesia	Coastal areas of Jakarta are increasingly exposed to climate change impacts, including regular flooding from rivers and the sea, which heighten disaster risks. The region faces more frequent storms, intensified saltwater intrusion, and worsening inundation, leaving communities highly vulnerable. The documented destruction of homes, disruption of fishing activities, and the rise of disease outbreaks highlight the urgent to mitigate these challenges	(Purnomo et al. 2024)
Typhoon	Orchid Island, Taiwan	The entire island experienced power and water outages, along with extensive destruction to homes, ports, stores, gas stations, and fishing boats, resulting in an estimated loss of $13 million. Orchid island has also faced severe impacts from other typhoons, including Typhoon Pabuk in 2007, Typhoon Nepartak in 2016, and Typhoon Chanthu in 2012, all of which caused similar widespread damage to infrastructure and essential services	(Kao et al. 2023)

aquaculture, and recreation (Astuti et al. 2022; Beiras 2018). Coastlines are facing severe impacts from rising sea levels and extreme weather, while coastal waters are increasingly contaminated with nutrients, pollutants, and litter from various sources. These combined effects of pollution and climate change are causing significant transformations in global coastal zones, particularly in the Indian subcontinent (Lincoln et al. 2023). Elevated salinity in drinking water has been associated with a range of adverse health outcomes, including an increased risk of preeclampsia and gestational hypertension in pregnant women, higher infant mortality rates, and the prevalence of diseases such as cholera, renal disease, cirrhosis, and skin and diarrheal infections (Hossain et al. 2024; Khan et al. 2011, 2014). Communities that lack access to alternative water sources or effective treatment methods may face severe consequences, including displacement, due to deteriorating water security.

India is currently experiencing serious climate change impacts, which are expected to intensify unless greenhouse gas emissions are reduced. This country is grappling with increasing marine litter pollution, which is overwhelming both authorities and communities, and obstructing the achievement of sustainable development goals (Agarwal and Deb 2023; Lincoln et al. 2023; Lu et al. 2018). Indonesia, as an archipelago nation, is also threatened by pollution, particularly in coastal regions. The majority of coastal residents are from low socio-economic backgrounds, which makes accessing clean water and healthcare facilities particularly difficult. Predictions from

2020 to 2040 indicate that sea-level rise in northern Jakarta waters could be influenced by La Niña, Dipole Mode, MJO, and storm surges, resulting in increases of 14.5 cm (August–February), 14.9 cm (July–November), 5.8 cm (December–February), and 38 cm (July–November). Coastal areas are also affected by inland floods, especially in January and February, with average flood heights of 60 cm. The estimated land subsidence rate in northern Jakarta is 5.7–5.4 cm/year (Latief et al. 2018).

Sea-level rise will intensify coastal flooding and expand mosquito habitats, heightening the risk of vector-borne diseases like West Nile virus, which the Government of Canada monitors and surveys. Coastal inundation and flooding due to rising sea levels create more water-based environments for disease vectors such as mosquitoes (Tracey 2022). In Australia, the incidence of Ross River (RR) virus is influenced by socio-biophysical interactions, climate variability, and the potential increased threat posed by climate change. The existing and emerging socio-biophysical interactions in these regions contribute to the future risks and management challenges of RR virus (Lyth et al. 2005). Interventions by local authorities and the Ministry of Public Health (MPH) in Ecuador have effectively raised awareness about the importance of keeping yards and patios free of mosquito-breeding hazards (Harris and Carter 2019). In sub-Saharan Africa (sSA), fever frequently prompts medical consultation, but non-malarial causes are often not considered in differential diagnoses. The findings demonstrated a notable prevalence of DENV (dengue virus) among the studied population. Consequently, the results suggest that DENV should be included in the differential diagnosis for febrile adults in coastal Kenya, alongside AHI (acute HIV infection) and malaria (Ngoi et al. 2016). Mosquito control has become integrated into the washing, cleaning, and cooking routines predominantly managed by women in the area. However, the unreliable supply of safe water, combined with social customs related to gendered water practices, results in the collection and storage of water in open containers in peridomestic spaces, inadvertently creating mosquito-breeding sites.

While metals are naturally present in marine environments, their concentrations have significantly increased due to human activities such as coastal urbanisation and industrialisation (Hien et al. 2022; Marchellina et al. 2024). This rise in metal pollution has prompted extensive research over the past century, focusing on heavy metal concentrations in coastal zones, particularly estuaries (Mallongi et al. 2023). Studies have examined metal levels in sediments, water, suspended particulates, and marine organisms to evaluate the impact of pollution on coastal ecosystems and assess potential risks to environmental and human health (Chandía et al. 2024; Tracey 2022). Additionally, the weathering and leaching of certain rocks can release metals and other minerals, exacerbating the issue (Astuti et al. 2022; Mallongi et al. 2022). Elevated levels of runoff and flooding can further transport and disseminate these pollutants across broader areas.

Urban planning must incorporate comprehensive spatial strategies to enhance resilience and reduce vulnerabilities, particularly in disadvantaged regions. Although spatial distribution can sometimes lead to passive outcomes like urban sprawl, it is integral to resilience policies (Lincoln et al. 2023; Verschuur et al. 2020). Addressing environmental risks and hazards requires considering the uneven distribution of settlements in city planning and policy development to effectively improve resilience in vulnerable areas (Busayo and Kalumba 2020; Jabareen 2013). The physical layout of a city in a coastal area affects its habitats, ecosystems, and the daily lives of its residents, as well as its contribution to climate change. This component explores how spatial planning, architecture, and design influence the city's ecological health and functionality, including the effects of buildings and neighbourhoods (Busayo and Kalumba 2020; Jabareen 2013).

5 Strategies for Enhancing Coastal Resilience

Building seawalls, levees, and replenishing eroded beaches are traditional defences against flooding and damage, but they may not be sustainable long-term solutions (Kisacik et al. 2022; Nunn et al. 2021). Preserving or restoring natural buffers such as barrier islands, tidal wetlands, and mangroves can provide coastal protection and additional ecosystem services (Damastuti et al. 2023; Kroeger et al. 2017). The tidal wetland carbon accounting rate (CAR) is expected to increase this century and maintain its carbon sequestration capacity across all climate change scenarios, indicating a strong resilience to sea-level rise (Wang et al. 2019). Strategies like elevating buildings, flood-proofing, using land temporarily, and constructing channels can address both temporary and gradual inundation. Some vulnerable coastal areas may need to contemplate retreating from the encroaching seas (Griggs and Reguero 2021; Ramasamy and Surendran 2011). In Colombia, the detection of coastal subsidence through GPS data is important for accurately evaluating the seismic and tsunami hazards associated with mechanical interactions at subduction zones. To enhance accuracy, a viscoelastic earthquake cycle model is being used to investigate time-dependent, three-dimensional deformation in the Caribbean subduction zone in northwestern Colombia (Lizarazo et al. 2024).

Increasing concerns about climate change have led to more research into vulnerability assessments related to sea-level rise (SLR), which has been linked to severe flooding damage. Assessing flood risks in coastal areas is complex and requires a comprehensive approach that considers natural resources, physical conditions, socio-economic factors, and coastal hazards (Chen et al. 2021; Mattah et al. 2023). An integrated coastal vulnerability index was developed by analysing various parameters, including coastal features, environmental forces, and socio-economic conditions (Wu 2021). A study based on actual clinical data has shown an increasing risk of dengue fever in Europe, particularly with the anticipated effects of climate change by the end of the century. To mitigate this, it is crucial to develop precise risk projection models that incorporate various factors, including environmental changes. Public health agencies in high-risk areas should establish early warning systems, conduct active entomological monitoring, and set up sentinel clinical surveillance. Increasing awareness among healthcare professionals and the general public about the rising dengue fever risk is essential for effective prevention and response measures (Bouzid et al. 2014). The use of direct computation followed by normalisation provides an accurate assessment of vulnerable areas at different scales. Detailed data on municipal risk ratings, localities, populations, and households were also collected (Gallego Perez and Selvaraj 2019; Godwyn-Paulson et al. 2022).

Nature-based solutions offer a promising alternative to traditional engineering methods for disaster risk reduction, but they are often neglected in planning and implementation. Coral reefs, for example, function as highly effective natural breakwaters, reducing wave energy by up to 97% (Sims et al. 2024). Unfortunately, they are under severe threat from climate change, runoff, pollution, and overfishing. According to a 2019 United Nations report, approximately 50% of live coral cover has been lost since the 1870s, which could place 100–300 million people in coastal areas at increased risk due to the diminished protection from coastal habitats. An analysis of coral loss hot spots highlights where conservation efforts should be prioritised (Sims et al. 2024; Turay et al. 2024).

6 Community-Based Adaptation Strategies and Policies

The development of the New Manila International Airport (NMIA) is designed to ease congestion at Ninoy Aquino International Airport (NAIA) and enhance the efficiency of Manila's transportation infrastructure. However, this ambitious project has triggered significant concern among the Taliptip communities due to its potential environmental impacts on Manila Bay and the displacement of local residents. The resistance is compounded by factors such as established livelihoods, residential immobility, and socioeconomic challenges, including poverty and limited land ownership (Catane et al. 2024). Understanding these dynamics is crucial for developing effective community-based adaptation strategies that balance development with local needs and vulnerabilities. In a similar vein, in Banjarmasin, the old city area relies on wooden stilt houses that are highly vulnerable to flash floods. With 24% of the population living below the poverty line, the socioeconomic impact of climate-related events is profound. Poverty exacerbates vulnerability by limiting access to critical information, knowledge, and recovery resources. Additionally, droughts, recurrent floods, and storms can exacerbate poverty by damaging livelihoods, homes, businesses, and infrastructure (Gaborit 2022). To address these challenges, integrated coastal management should incorporate environmental protection and community engagement in Manila Bay, while improving flood management and enhancing community resilience in Banjarmasin. Socioeconomic support and environmental conservation are crucial for both areas to mitigate the impacts of relocation and climate-related risks.

Research in Singapore, Jakarta, and Manila highlights three main trends in coastal adaptation planning. These include an increased focus on disaster risk reduction and new policies, improvements to green spaces along the coast, and the development of coastal areas through reclamation and upscale projects. These trends are shaping urban development and governance, influencing the social and political futures of coastal cities (Herbeck and Flitner 2019). Practical measures can reduce community exposure and vulnerability to coastal hazards while enhancing adaptive capacities. Understanding the drivers and root causes of vulnerability is essential. For instance, degrading coastal ecosystems undermines livelihoods and increases hazard risks by reducing natural storm buffers. Rapid, unplanned coastal development exacerbates social, economic, and environmental impacts, leading to higher vulnerability and costly disasters (Glavovic 2021). Root causes often include weak governance, poverty, and inequity.

Effective risk reduction requires collaboration among government, civil society, the private sector, and the scientific community, with each playing distinct roles. In coastal Bangladesh, quantifying welfare losses from tropical cyclones under present and future climatic and socio-economic conditions is crucial (Verschuur et al. 2020). Understanding the relationship between future risks and the effectiveness of adaptation strategies requires modelling socio-economic growth and climate change while linking them to evolving resilience. Resilience is not static but develops alongside socio-economic progress. In Ghana, communities with many children and elderly residents and a lack of social amenities like roads, health facilities, markets, and schools have demonstrated high adaptive capacity. These communities have managed to withstand the impacts of ocean and river systems by developing various adaptation measures. Households have resorted to relocating to nearby communities or higher grounds, utilising open spaces, and placing belongings on elevated objects such as tables (Mattah et al. 2023). Adaptation options vary greatly depending on socio-economic contexts, emphasising the importance of high-resolution spatial analyses to avoid generic solutions and optimise disaster risk reduction spending (Almutairi et al. 2020; Jabareen 2013; Niazi et al. 2022).

Older individuals have reported feeling less prepared for emergency condition compared to younger people. Research attributes this to younger individuals often having children at home, while older residents, having lived in the same area longer, may feel more confident about their ability to survive an earthquake without injury or home damage. Regarding tornado preparedness, people over 60 participate less in tornado drill activities than those aged 20–59 (Chaney et al. 2013; Cisternas et al. 2024). Younger people are more frequently exposed to preparedness activities in schools or workplaces, whereas older adults have fewer opportunities to engage in such drills. When considering multiple threats, Bronfman et al. found that both young and older adults reported the lowest levels of household preparedness for earthquakes and floods (Bronfman et al. 2019).

To improve disaster preparedness and awareness in coastal regions, it is essential to identify the specific vulnerabilities associated with different age demographics (Bronfman et al. 2019; Niazi et al. 2022). Research indicates that older individuals may perceive themselves as less prepared for natural disasters, such as earthquakes and floods, despite their prolonged residence in the area. This lack of preparedness is often attributed to limited participation in preparedness drills. Conversely, younger families, who are more likely to have children, generally demonstrate greater engagement in acquiring preparedness information and participating in activities through educational institutions and workplaces (Niazi et al. 2022). Addressing these disparities necessitates the development of age-specific preparedness programs. For older adults, it is beneficial to organise training sessions at community centres or similar venues, employing methods that are accessible and relevant to their needs. This could include live demonstrations and light physical exercises designed to enhance their preparedness skills.

For younger families, targeted workshops conducted in schools and workplaces can effectively address family evacuation planning, emergency kit assembly, and disaster response drills involving children (Almutairi et al. 2020; Herbeck and Flitner 2019). Enhancing household emergency response protocols is crucial for effective self-evacuation. Protocols should include defining individual roles during earthquakes, assigning evacuation tasks, and planning for scenarios such as family members being outside the home. Integrating early warning systems, like tsunami sirens, highlights the urgency for timely evacuation, thereby improving community preparedness and mitigating potential risks (Turay et al. 2024).

Regular community-wide disaster simulations that involve all age groups can ensure everyone understands their roles during emergencies, focusing particularly on common coastal disasters like tsunamis, floods, and hurricanes (Almulhim 2022; Bronfman et al. 2019; Chaney et al. 2013; Wu 2021). Collaborating with local schools and businesses can help integrate regular preparedness drills into curriculums and workplace programs, while engaging community organisations and NGOs can extend the reach of these initiatives. Evaluating and following up on these programs through regular surveys and assessments can help adjust and improve them based on feedback, ensuring they remain relevant and effective. Engaging influential community leaders and role models can also encourage wider participation, leveraging their influence to promote disaster preparedness (Benazir et al. 2023; Damastuti et al. 2023). Many policymakers involved in coastal management often lack specialised knowledge in coastal processes, which can impact the effectiveness of their strategies. A workshop with six communities in Great Lakes revealed significant differences in how local decision-makers interpreted coastal hazards and ranked their impacts (Bunting and Theuerkauf 2024). To address these disparities, targeted training for policymakers on coastal hazards, the development of standardised terminology, and regular community engagement in hazard assessment are recommended.

A significant challenge lies in developing robust methodologies for assessing coastal vulnerability to climate-induced hazards at the regional level. This requires integrating the most accurate and detailed geographical data available for the specific study area, to identify the most vulnerable regions and to inform the development of operational adaptation strategies (Torresan et al. 2012). High population density along coasts accelerates erosion due to increased activity and tourism. It also emphasises the need to train coastal residents to manage hazards such as sea-level rise and storm surges. Therefore, quantifying population density is essential for reducing vulnerability and improving resilience (Gaborit 2022; Gallego Perez and Selvaraj 2019; Godwyn-Paulson et al. 2022). Reducing uncertainties can be achieved by correlating historical experiences with hazards, adaptation efforts, and mitigation strategies to current and future scenarios (Almulhim 2022).

The integration of grey and green infrastructure for water pollution mitigation is increasingly emphasised in urban water management policies worldwide. This approach is evident in various best management practices (BMPs) in USA (Meals et al. 2010), sustainable urban drainage systems (SUDSs) in the United Kingdom, the low-impact urban design and development (LIUDD) programme in New Zealand (Ignatieva et al. 2008), the active, beautiful, clean waters programme (ABC) in Singapore, and sponge city development (SCD) in China (Ren et al. 2017). Research conducted in a coastal megacity has shown that while grey infrastructure is more effective in reducing overall pollutant loadings in nearshore waters, green infrastructure offers a more cost-effective solution. According to Fig. 1, urban water system is an advanced conceptual model that depicts the flows within urban water management. In this model, blue arrows indicate water flow, grey arrows represent wastewater flow, brown arrows denote resource flows, and yellow arrows signify energy flows. This comprehensive framework aims to integrate various components of urban water management to enhance efficiency, sustainability, and resilience in urban environments (Ren et al. 2017).

Integrating technology and innovation, such as mobile apps and online platforms for real-time information and emergency alerts, can make preparedness resources more accessible. Using virtual reality (VR) technology to create immersive disaster simulation experiences can enhance engagement during training. Forming volunteer emergency response teams within the community can offer specialised training for a coordinated disaster response. Maintaining early warning systems and clearly marking evacuation routes and shelters can ensure timely alerts and safe evacuations. Through these comprehensive strategies, coastal communities can greatly enhance

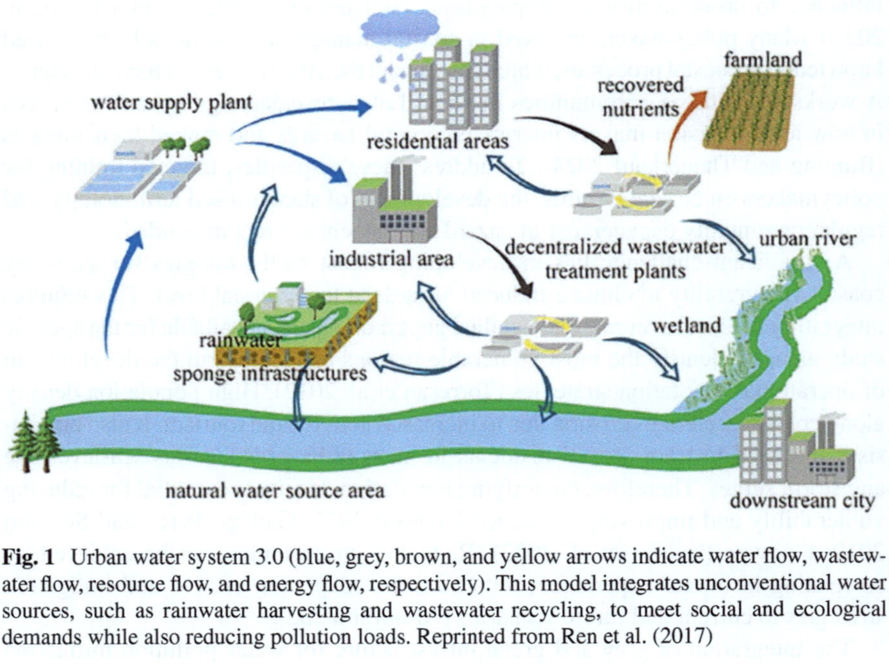

Fig. 1 Urban water system 3.0 (blue, grey, brown, and yellow arrows indicate water flow, wastewater flow, resource flow, and energy flow, respectively). This model integrates unconventional water sources, such as rainwater harvesting and wastewater recycling, to meet social and ecological demands while also reducing pollution loads. Reprinted from Ren et al. (2017)

disaster preparedness and awareness, ensuring all residents, regardless of age, are better prepared to handle natural disasters.

7 Conclusion

Coastal zones, characterised by their dynamic interfaces between land and sea, face heightened vulnerabilities to natural hazards including tropical cyclones, tsunamis, storm surges, and accelerated sea-level rise, exacerbated by climate change. These hazards significantly impact both human health and environmental integrity. Elevated sea levels and increased flooding create favourable conditions for vector-borne diseases, such as malaria and dengue fever, due to the proliferation of breeding sites for disease-carrying mosquitoes. Concurrently, pollution from industrial, agricultural, and urban sources contributes to severe water contamination, leading to gastrointestinal illnesses and chronic health conditions. To address these challenges, a multifaceted approach is required, encompassing the deployment of advanced early warning systems, the implementation of sustainable coastal development practices, and the restoration of natural protective barriers such as mangroves and coral reefs. Integrating nature-based solutions with conventional infrastructure measures, including elevating buildings and improving flood management systems, is essential for enhancing coastal resilience. Additionally, effective community engagement and intersectoral collaboration among policymakers, urban planners, and local stakeholders are crucial for the development and execution of targeted adaptation strategies. Public health initiatives must prioritise improvements in water quality management, pollution control, and vector management to mitigate health risks and bolster community resilience against the adverse effects of climate change.

References

Agarwal S, Deb T (2023) Climate change induced migration with special emphasis on the coastal zones in India. In: Singh P, Ao B, Yadav A (eds.) Global climate change and environmental refugees. Springer International Publishing, pp 105–119. https://doi.org/10.1007/978-3-031-24833-7_7

Alencar RM, Sepulveda CCP, Martinez-Villegas L, Bahia AC, Santana RA, De Souza IB, D'Elia GMA, Duarte APM, De Lacerda MVG, Monteiro WM, Secundino NFC, Pimenta PFP, Koerich LB (2023) Unravelling the genome of the brackish water malaria vector Anopheles aquasalis. Sci Rep 13(1):20472. https://doi.org/10.1038/s41598-023-47830-1

Al-Kahtany K, Nour HE, Giacobbe S, Alharbi T, El-Sorogy AS (2023) Heavy metal pollution in surface sediments and human health assessment in southern Al-Khobar coast Saudi Arabia. Mar Pollut Bull 187:114508. https://doi.org/10.1016/j.marpolbul.2022.114508

Almulhim AI (2022) Household's awareness and participation in sustainable electronic waste management practices in Saudi Arabia. Ain Shams Eng J 13(4):101729. https://doi.org/10.1016/j.asej.2022.101729

Almutairi A, Mourshed M, Ameen RFM (2020) Coastal community resilience frameworks for disaster risk management. Nat Hazards 101(2):595–630. https://doi.org/10.1007/s11069-020-03875-3

Astuti RDP, Mallongi A, Choi K, Amiruddin R, Hatta M, Tantrakarnapa K, Rauf AU (2022) Health risks from multiroute exposure of potentially toxic elements in a coastal community: a probabilistic risk approach in Pangkep Regency, Indonesia. Geomat Nat Hazards Risk 13(1):705–735. https://doi.org/10.1080/19475705.2022.2041110

Balasubramanian R, Nadh V, Sahina S (2021) Ecology of breeding habitats of mosquito population and screening for virus of Japanese encephalitis and West Nile in the coastal area of Kerala, India. J Vector Borne Dis 58(3):232. https://doi.org/10.4103/0972-9062.318307

Basheer Ahammed KK, Karang IWGA, Nuarsa IW, Indrawan IGS, Hendrawan IG, Dewi NMNBS, Pandey AC (2024) Mapping of the spatio-spectral dynamics of mangrove chlorophyll concentrations via sentinel-2 satellite imagery. Forum Geogr 38(2):244–256. https://journals2.ums.ac.id/fg/article/view/4518

Basheer Ahammed KK, Pandey AC (2019) Coastal social vulnerability and risk analysis for cyclone hazard along the Andhra Pradesh, east coast of India. KN-J Cartogr Geogr Inf 69(4):285–303

Beiras R (2018) Basic concepts. In: Marine pollution. Elsevier, pp 3–20. https://doi.org/10.1016/B978-0-12-813736-9.00001-5

Benazir S, Idris Y, Putra NP (2023) Connecting community's perspectives on tsunami risk to anticipated future tsunamis: a reflection from a progress of tsunami preparedness from a coastal community in Aceh-Indonesia after 19 years of the 2004 Indian Ocean Tsunami. Geoenviron Disasters 10(1):21. https://doi.org/10.1186/s40677-023-00252-7

Bouzid M, Colón-González FJ, Lung T, Lake IR, Hunter PR (2014) Climate change and the emergence of vector-borne diseases in Europe: case study of dengue fever. BMC Public Health 14(1):781. https://doi.org/10.1186/1471-2458-14-781

Bronfman NC, Cisternas PC, Repetto PB, Castañeda JV (2019) Natural disaster preparedness in a multi-hazard environment: characterizing the sociodemographic profile of those better (worse) prepared. PLoS ONE 14(4):e0214249. https://doi.org/10.1371/journal.pone.0214249

Bunting EL, Theuerkauf EJ (2024) What is a coastal hazard? Perceptions of critical coastal hazards amongst decision makers in communities across the Great Lakes. Ocean Coast Manag 251:107065. https://doi.org/10.1016/j.ocecoaman.2024.107065

Busayo ET, Kalumba AM (2020) Coastal climate change adaptation and disaster risk reduction: a review of policy, programme and practice for sustainable planning outcomes. Sustainability 12(16):6450. https://doi.org/10.3390/su12166450

Buzzi NS, Menéndez MC, Truchet DM, Delgado AL, Severini MDF (2022) An overview on metal pollution on touristic sandy beaches: is the COVID-19 pandemic an opportunity to improve coastal management? Mar Pollut Bull 174:113275. https://doi.org/10.1016/j.marpolbul.2021.113275

Cabral H, Fonseca V, Sousa T, Costa Leal M (2019) Synergistic effects of climate change and marine pollution: an overlooked interaction in coastal and estuarine areas. Int J Environ Res Public Health 16(15):2737. https://doi.org/10.3390/ijerph16152737

Camacho-Jiménez L, González-Ruiz R, Yepiz-Plascencia G (2023) Persistent organic pollutants (POPs) in marine crustaceans: bioaccumulation, physiological and cellular responses. Mar Environ Res 192:106184. https://doi.org/10.1016/j.marenvres.2023.106184

Catane SG, Flora JRR, Carag JWM, Capino JB, Go CM, Panganiban ALL (2024) Development in an environmentally critical coastal area: the risk perception on natural hazards and the New Manila International Airport by fishing communities in Taliptip Philippines. Ocean Coast Manag 253:107127. https://doi.org/10.1016/j.ocecoaman.2024.107127

Chandía C, Salamanca M, Hernández A, Urrutia R (2024) Mercury distribution in the coastal zone of Central Chile, Southeast Pacific: a comprehensive assessment of seawater, sediment, and biota. Mar Pollut Bull 199:116005. https://doi.org/10.1016/j.marpolbul.2023.116005

Chaney PL, Weaver GS, Youngblood SA, Pitts K (2013) Household preparedness for tornado hazards: the 2011 disaster in DeKalb county, Alabama. Weather Clim Soc 5(4):345–358. https://doi.org/10.1175/WCAS-D-12-00046.1

Chen Y-H, Kim J, Mueller N (2021) Estimating the economic impact of natural hazards on shared accommodation in coastal tourism destinations. J Destin Mark Manag 21:100634. https://doi.org/10.1016/j.jdmm.2021.100634

Cisternas PC, Cifuentes LA, Bronfman NC, Repetto PB, Castañeda JV (2024) Household preparedness for multi-natural hazards in coastal communities. Int J Disaster Risk Reduct 109:104584. https://doi.org/10.1016/j.ijdrr.2024.104584

Cuaton GP (2019) Post-disaster relocation of urban coastal communities in the Philippines. Andalas J Int Stud (AJIS) 8(2):143. https://doi.org/10.25077/ajis.8.2.141-151.2019

Damastuti E, Van Wesenbeeck BK, Leemans R, De Groot RS, Silvius MJ (2023) Effectiveness of community-based mangrove management for coastal protection: a case study from Central Java Indonesia. Ocean Coast Manag 238:106498. https://doi.org/10.1016/j.ocecoaman.2023.106498

Degeai J-P, Blanchemanche P, Tavenne L, Tillier M, Bohbot H, Devillers B, Dezileau L (2022) River flooding on the French Mediterranean coast and its relation to climate and land use change over the past two millennia. CATENA 219:106623. https://doi.org/10.1016/j.catena.2022.106623

Farrell S, Dates J, Ramirez N, Hausknecht-Buss H, Kolosov D (2024) Voltage-gated ion channels are expressed in the Malpighian tubules and anal papillae of the yellow fever mosquito (*Aedes aegypti*), and may regulate ion transport during salt and water imbalance. J Exp Biol 227(3):jeb246486. https://doi.org/10.1242/jeb.246486

Freitas FB, Novo MT, Esteves A, De Almeida APG (2012) Species composition and WNV screening of mosquitoes from lagoons in a wetland area of the Algarve, Portugal. Front Physiol 2. https://doi.org/10.3389/fphys.2011.00122

Gaborit P (2022) Climate adaptation to multi-hazard climate related risks in ten Indonesian cities: ambitions and challenges. Clim Risk Manag 37:100453. https://doi.org/10.1016/j.crm.2022.100453

Gallego Perez BE, Selvaraj JJ (2019) Evaluation of coastal vulnerability for the district of Buenaventura, Colombia: a geospatial approach. Remote Sens Appl Soc Environ 16:100263. https://doi.org/10.1016/j.rsase.2019.100263

Glavovic BC (2021) Community guide for community members interested in risk reduction efforts. How to reduce coastal hazard risk in your community: a step-by-step approach. IOC-UNESCO. https://unesdoc.unesco.org/ark:/48223/pf0000375768

Godwyn-Paulson P, Jonathan MP, Rodríguez-Espinosa PF, Abdul Rahaman S, Roy PD, Muthusankar G, Lakshumanan C (2022) Multi-hazard risk assessment of coastal municipalities of Oaxaca, Southwestern Mexico: an index based remote sensing and geospatial technique. Int J Disaster Risk Reduct 77:103041. https://doi.org/10.1016/j.ijdrr.2022.103041

Goulart HMD, Athanasiou P, Van Ginkel K, Van Der Wiel K, Winter G, Pinto I, Van Den Hurk B (2024) Exploring coastal climate adaptation through storylines: insights from cyclone Idai in Beira, Mozambique. Cell Rep Sustain 100270. https://doi.org/10.1016/j.crsus.2024.100270

Griggs G, Reguero BG (2021) Coastal adaptation to climate change and sea-level rise. Water 13(16):2151. https://doi.org/10.3390/w13162151

Harris ML, Carter ED (2019) Muddying the waters: a political ecology of mosquito-borne disease in coastal Ecuador. Health Place 57:330–338. https://doi.org/10.1016/j.healthplace.2019.05.010

Hasnine M, Nagdeve DA (2024) Natural hazard induced coastal vulnerability in Indian Sundarbans: a village-level study by using geospatial and statistical techniques. Nat Hazards Res S2666592124000763. https://doi.org/10.1016/j.nhres.2024.10.004

Hecht H, Oguchi T (2017) Global evaluation of erosion rates in relation to tectonics. Prog Earth Planet Sci 4(1):40. https://doi.org/10.1186/s40645-017-0156-3

Herbeck J, Flitner M (2019) Infrastructuring coastal futures: key trajectories in Southeast Asian megacities. Die ERDE. https://doi.org/10.12854/erde-2019-451

Hien TT, Chi NDT, Huy DH, Le HA, Oram DE, Forster GL, Mills GP, Baker AR (2022) Soluble trace metals associated with atmospheric fine particulate matter in the two most populous cities in Vietnam. Atmos Environ: X 15:100178. https://doi.org/10.1016/j.aeaoa.2022.100178

Hossain S, Hasan MA, Faruk MO, Salam MA (2024) Association between the salinity level with miscarriage and unintended pregnancy in Bangladesh: impact of salinity level on miscarriage and unintended pregnancy in Bangladesh. Heliyon 10(1):e23858. https://doi.org/10.1016/j.heliyon.2023.e23858

Ignatieva M, Meurk C, Stewart G (2008) Low impact urban design and development (LIUDD): matching urban design and urban ecology. Landsc Rev 12:61–73. https://doi.org/10.34900/LR.V12I2.254

Jabareen Y (2013) Planning the resilient city: concepts and strategies for coping with climate change and environmental risk. Cities 31:220–229. https://doi.org/10.1016/j.cities.2012.05.004

Jones BD, Collings B, Dickson ME, Ford M, Hikuroa D, Bickler SH, Ryan E (2024) Regional implementation of coastal erosion hazard zones for archaeological applications. J Cult Herit 67:430–442. https://doi.org/10.1016/j.culher.2024.04.007

Kao S-M, Wu W-N, Gu M-E (2023) Coastal disaster protection and hazard perceptions of the residents of Orchid Island Taiwan. Mar Policy 148:105459. https://doi.org/10.1016/j.marpol.2022.105459

Khan AE, Ireson A, Kovats S, Mojumder SK, Khusru A, Rahman A, Vineis P (2011) Drinking water salinity and maternal health in coastal Bangladesh: implications of climate change. Environ Health Perspect 119(9):1328–1332. https://doi.org/10.1289/ehp.1002804

Khan AE, Scheelbeek PFD, Shilpi AB, Chan Q, Mojumder SK, Rahman A, Haines A, Vineis P (2014) Salinity in drinking water and the risk of (Pre)eclampsia and gestational hypertension in coastal Bangladesh: a case-control study. PLoS ONE 9(9):e108715. https://doi.org/10.1371/journal.pone.0108715

Kisacik D, Tarakcioglu GO, Cappietti L (2022) Adaptation measures for seawalls to withstand sea-level rise. Ocean Eng 250:110958. https://doi.org/10.1016/j.oceaneng.2022.110958

Kroeger KD, Crooks S, Moseman-Valtierra S, Tang J (2017) Restoring tides to reduce methane emissions in impounded wetlands: a new and potent Blue Carbon climate change intervention. Sci Rep 7(1):11914. https://doi.org/10.1038/s41598-017-12138-4

Latief H, Putri MR, Hanifah F, Afifah IN, Fadli M, Ismoyo DO (2018) Coastal hazard assessment in northern part of Jakarta. Procedia Eng 212:1279–1286. https://doi.org/10.1016/j.proeng.2018.01.165

Levi T, Bausch D, Katz O, Rozelle J, Salamon A (2015) Insights from hazus loss estimations in Israel for dead sea transform earthquakes. Nat Hazards 75(1):365–388. https://doi.org/10.1007/s11069-014-1325-y

Lincoln S, Chowdhury P, Posen PE, Robin RS, Ramachandran P, Ajith N, Harrod O, Hoehn D, Harrod R, Townhill BL (2023) Interaction of climate change and marine pollution in Southern India: implications for coastal zone management practices and policies. Sci Total Environ 902:166061. https://doi.org/10.1016/j.scitotenv.2023.166061

Lizarazo SC, Sagiya T, Mora-Páez H (2024) Viscoelastic earthquake cycle model for the Caribbean subduction zone in northwestern Colombia: implications of coastal subsidence for seismic/tsunami hazards. J S Amn Earth Sci 141:104931. https://doi.org/10.1016/j.jsames.2024.104931

Lu Y, Yuan J, Lu X, Su C, Zhang Y, Wang C, Cao X, Li Q, Su J, Ittekkot V, Garbutt RA, Bush S, Fletcher S, Wagey T, Kachur A, Sweijd N (2018) Major threats of pollution and climate change to global coastal ecosystems and enhanced management for sustainability. Environ Pollut 239:670–680. https://doi.org/10.1016/j.envpol.2018.04.016

Lyth A, Holbrook NJ, Beggs PJ (2005) Climate, urbanisation and vulnerability to vector-borne disease in subtropical coastal Australia: sustainable policy for a changing environment. Environ Hazards 6(4):189–200. https://doi.org/10.1016/j.hazards.2006.10.001

Mallongi A, Rauf AU, Daud A, Hatta M, Al-Madhoun W, Amiruddin R, Stang S, Wahyu A, Astuti RDP (2022) Health risk assessment of potentially toxic elements in Maros karst groundwater:

a Monte Carlo simulation approach. Geomat Nat Hazards Risk 13(1):338–363. https://doi.org/10.1080/19475705.2022.2027528

Mallongi A, Rauf AU, Astuti RDP, Palutturi S, Ishak H (2023) Ecological and human health implications of mercury contamination in the coastal water. Glob J Environ Sci Manag 9(2). https://doi.org/10.22034/gjesm.2023.02.06

Marchellina A, Soegianto A, Putranto TWC, Mukholladun W, Payus CM, Irnidayanti Y (2024) An assessment of the potential health hazards associated with metal contamination in a variety of consumable species living along the industrialized coastline of East Java Indonesia. Mar Pollut Bull 202:116375. https://doi.org/10.1016/j.marpolbul.2024.116375

Martyr-Koller R, Thomas A, Schleussner C-F, Nauels A, Lissner T (2021) Loss and damage implications of sea-level rise on small Island developing states. Curr Opin Environ Sustain 50:245–259. https://doi.org/10.1016/j.cosust.2021.05.001

Mattah MM, Mattah PAD, Mensah A, Babanawo D, Brempong E, Mensah J, Appeaning Addo K (2023) Assessment of social factors that promote the vulnerability of communities to coastal hazards in the Volta estuary in Ghana. Int J Disaster Risk Reduct 95:103896. https://doi.org/10.1016/j.ijdrr.2023.103896

Mattah MM, Mattah PAD, Mensah AM, Adarkwah F, Mensah J, Addo KA (2024) Community perceptions, knowledge, and coping mechanisms concerning perennial climate change-related disasters along the Volta estuary of Ghana West Africa. Sci Afr 25:e02333. https://doi.org/10.1016/j.sciaf.2024.e02333

Meals DW, Dressing SA, Davenport TE (2010) Lag time in water quality response to best management practices: a review. J Environ Qual 39(1):85–96. https://doi.org/10.2134/jeq2009.0108

Molinaroli E, Guerzoni S, Suman D (2019) Do the adaptations of Venice and Miami to sea level rise offer lessons for other vulnerable coastal cities? Environ Manag 64(4):391–415. https://doi.org/10.1007/s00267-019-01198-z

Nghia BPQ, Pal I, Chollacoop N, Nguyen LH, Van Thinh L, Tuan TM, Van Tuong N (2024) Integrated approach for drought and saline intrusion severity assessment on the coastal Mekong Delta of Vietnam contextualizing physical change to risk management and policy development. Prog Disaster Sci 23:100338. https://doi.org/10.1016/j.pdisas.2024.100338

Ngoi CN, Price MA, Fields B, Bonventure J, Ochieng C, Mwashigadi G, Hassan AS, Thiong'o AN, Micheni M, Mugo P, Graham S, Sanders EJ (2016) Dengue and Chikungunya virus infections among young febrile adults evaluated for acute HIV-1 infection in Coastal Kenya. PLOS ONE 11(12):e0167508. https://doi.org/10.1371/journal.pone.0167508

Niazi IUHK, Rana IA, Arshad HSH, Lodhi RH, Najam FA, Jamshed A (2022) Psychological resilience of children in a multi-hazard environment: an index-based approach. Int J Disaster Risk Reduct 83:103397. https://doi.org/10.1016/j.ijdrr.2022.103397

Nicolodi JL, Pettermann RM (2012) Vulnerability of the Brazilian coastal zone in its environmental. Soc Technol Asp 64:1373

Nunn PD, Klöck C, Duvat V (2021) Seawalls as maladaptations along island coasts. Ocean Coast Manag 205:105554. https://doi.org/10.1016/j.ocecoaman.2021.105554

Orlova TY, Aleksanin AI, Lepskaya EV, Efimova KV, Selina MS, Morozova TV, Stonik IV, Kachur VA, Karpenko AA, Vinnikov KA, Adrianov AV, Iwataki M (2022) A massive bloom of Karenia species (Dinophyceae) off the Kamchatka coast, Russia, in the fall of 2020. Harmful Algae 120:102337. https://doi.org/10.1016/j.hal.2022.102337

Parwanto NB, Oyama T (2014) A statistical analysis and comparison of historical earthquake and tsunami disasters in Japan and Indonesia. Int J Disaster Risk Reduct 7:122–141. https://doi.org/10.1016/j.ijdrr.2013.10.003

Purnomo AH, Kurniawan T, Farandy AR, Apriliani T, Nurlaili Imron M, Sajise AJ (2024) Revisiting the climate change adaptation strategy of Jakarta's coastal communities. Ocean Coast Manag 253:107112. https://doi.org/10.1016/j.ocecoaman.2024.107112

Ramasamy R, Surendran SN (2011) Possible impact of rising sea levels on vector-borne infectious diseases. BMC Infect Dis 11(1):18. https://doi.org/10.1186/1471-2334-11-18

Ren N, Wang Q, Wang Q, Huang H, Wang X (2017).Upgrading to urban water system 3.0 through sponge city construction. Front Environ Sci Eng 11(4):9. https://doi.org/10.1007/s11783-017-0960-4

Renaud FG, Le TTH, Lindener C, Guong VT, Sebesvari Z (2015) Resilience and shifts in agro-ecosystems facing increasing sea-level rise and salinity intrusion in Ben Tre Province Mekong Delta. Clim Change 133(1):69–84. https://doi.org/10.1007/s10584-014-1113-4

Sanganyado E, Chingono KE, Gwenzi W, Chaukura N, Liu W (2021) Organic pollutants in deep sea: occurrence, fate, and ecological implications. Water Res 205:117658. https://doi.org/10.1016/j.watres.2021.117658

Sims J, Bausch D, Hoke A, Lindeman C, Kelly M, Zuzak C (2024) Mapping the risk reduction benefits of coral reef conservation—Hawai'i case study. Nat-Based Solut 5:100128. https://doi.org/10.1016/j.nbsj.2024.100128

Sperotto A, Torresan S, Gallina V, Coppola E, Critto A, Marcomini A (2016) A multi-disciplinary approach to evaluate pluvial floods risk under changing climate: the case study of the municipality of Venice (Italy). Sci Total Environ 562:1031–1043. https://doi.org/10.1016/j.scitotenv.2016.03.150

Terry JP, Goff J, Winspear N, Bongolan VP, Fisher S (2022) Tonga volcanic eruption and tsunami, January 2022: globally the most significant opportunity to observe an explosive and tsunamigenic submarine eruption since AD 1883 Krakatau. Geosci Lett 9(1):24. https://doi.org/10.1186/s40562-022-00232-z

Torresan S, Critto A, Rizzi J, Marcomini A (2012) Assessment of coastal vulnerability to climate change hazards at the regional scale: the case study of the North Adriatic Sea. Nat Hazards Earth Syst Sci 12(7):2347–2368. https://doi.org/10.5194/nhess-12-2347-2012

Tracey W (2022) Health risks associated with sea level rise. National collaborating centre for environmental health. https://ncceh.ca/sites/default/files/Final%20Draft%20-%20Health%20impacts%20of%20SLR_EN%20Dec%207_1.pdf

Turay B, Conteh I, Gbetuwa S (2024) Evaluating the values and limitations for coral and oyster reefs in coastal disaster risk reduction: a literature review. Environ Dev Sustain. https://doi.org/10.1007/s10668-024-04582-w

Verschuur J, Koks EE, Haque A, Hall JW (2020) Prioritising resilience policies to reduce welfare losses from natural disasters: a case study for coastal Bangladesh. Glob Environ Chang 65:102179. https://doi.org/10.1016/j.gloenvcha.2020.102179

Wang F, Lu X, Sanders CJ, Tang J (2019) Tidal wetland resilience to sea level rise increases their carbon sequestration capacity in United States. Nat Commun 10(1):5434. https://doi.org/10.1038/s41467-019-13294-z

Wu T (2021) Quantifying coastal flood vulnerability for climate adaptation policy using principal component analysis. Ecol Indic 129:108006. https://doi.org/10.1016/j.ecolind.2021.108006

Yahaya NS, Pereira JJ, Taha MR, Yaacob WZW (2024) Natech potential due to emerging climate hazards and awareness of the local stakeholders in the coastal industrial site of Kuala Selangor Malaysia. Chemosphere 367:143551. https://doi.org/10.1016/j.chemosphere.2024.143551

Quantitative Study of Beach Profiles on Field Observation and Laboratory Experiments

Nagamani Katukotta, Packialakshmi Shanmugam, and R. Nagalakshmi

Abstract Beach profiles provide useful information for coastal monitoring studies and management processes. Shoreline changes induced by erosion and accretion are natural processes that take place over a range of time scales. They may occur in response to smaller-scale (short-term) events, such as storms, regular wave action, tides, and winds, or in response to large-scale (long-term) events such as acrogenic cycles that may significantly alter sea levels (rise/fall) and tectonic activities that cause coastal land subsidence or emergence. The present research paper's main aim is to investigate the field measurements of beach profile datasets and lab experiments to understand the physical processes such as wind speed, wind velocity, wind direction, wave heights, and tidal heights that are responsible for the shoreline deterioration. Field methodology integrates beach profile, wave height, and tide height data. Seasonal changes of beach profiles were studied using monthly data collected at 4 stations (2 km of coast spilt into 4 stations equally 0.5 km each station) (0: Station 1, 500: Station 2, 1500: Station 3, and 2000: Station 4) along the Kalpakkam coast. The beach profile changes indicate that the erosion is prominent in Station 3 and Station 4 and deposition in Station 1 and Station 2. Wave height was observed visually through wooden stacks every month at all the 4 stations. In which it is observed that the highest wave height was 1.3 m and the minimum height 0.3 m. As far as Kalpakkam shore protection measures are concerned, structures such as sea wall are already constructed. Despite erosion occurs on Stations 3 and 4 to minimize the erosion, a suggestion on coastal protection measures for strengthening shore by planting natural vegetation like casuarinas and fixing soil bags of geotubes from Station 4 to Station 3 will be a cost-effective solution to control erosion. Further, two

N. Katukotta (✉)
Centre for Remote Sensing and Geoinformatics, Sathyabama Institute of Science and Technology, Sholinganallur, Chennai, Tamil Nadu, India
e-mail: nagamani@sathyabama.ac.in

P. Shanmugam
Department of Agricultural Engineering, SRM Valliammai Engineering College, Kattankulathur, Chennai, Tamil Nadu, India

R. Nagalakshmi
Department of Civil Engineering, SRM Institute of Science and Technology, Kattankulathur, Kanchipuram, Tamil Nadu, India

K. K. Basheer Ahammed et al. (eds.), *Monitoring Coastal Marine Ecosystems Under Climate Change*, https://doi.org/10.1007/978-3-032-09137-6_10

layers of geotubes of one above the other will be an optimal solution for the erosion sites in Kalpakkam coast, especially in Station 3 and Station 4.

Keywords Beach profile · Shoreline changes · Field measurements · GPS · RS · GIS

1 Introduction

Beach profiles are crucial parameters for assessing changes in coastal zones (Klemas 2011; Morton et al. 1993), as they illustrate the spatial and temporal variability of these areas. While numerous scientific articles and studies employ beach profiling as a monitoring method, numbering in the thousands (e.g., Carrasco et al. 2011; Ekwurzel 1990; Freire et al. 2020), significantly fewer focus on evaluating specific methodologies. Most of the existing research is conducted in high-income countries, which have easier access to expensive measurement equipment. In contrast, middle- and low-income nations, such as those in Latin America, face challenges in consistent monitoring due to the high costs of importing necessary equipment (Andrade and Ferreira 2006; Navarrete-Ramírez 2014).

The choice of methodology for beach profiling is directly influenced by the study's requirements concerning precision and accuracy, available economic resources, accessibility to the study area, and climatic conditions (Łabuz 2016; Moloney et al. 2018). Beach profiling methods serve various purposes, including assessing shoreline retreat (Caravaca-Colina et al. 2015; Cuevas-Jiménez and Euán-Ávila 2009; Franco Ochoa et al. 2019), inundation modeling (Freire et al. 2020), and analyzing the impacts of extreme weather events (Terefenko et al. 2020).

Despite the extensive study of beach profiles by various authors, there remains a gap in understanding the effectiveness of different methods across diverse environments. The methods for field measurement of beach profiles can be categorized into two main classifications. Manual methods are often named after the authors who pioneered and published them, including Emery (1961), Delgado and Lloyd (2004), Krause (2004), Andrade and Ferreira (2006), Puleo et al. (2008), and Chowdhury et al. (2014).

Shoreline changes due to erosion and accretion are natural processes that occur over various time scales. The dynamics of coastal processes and beach behavior have been extensively documented in the literature (Veerayya 1972; Murthy et al. 1975). These changes can result from short-term events like storms, wave action, tides, and winds, as well as long-term phenomena such as orogenic cycles that significantly affect sea levels and tectonic activities that lead to coastal subsidence or emergence. Consequently, many coastlines are inherently dynamic, with erosion cycles playing a vital role in shaping their ecological character. Natural forces like wind, waves, and currents readily mobilize unconsolidated sand and soil, causing rapid shifts in shoreline position.

With increasing development in coastal areas, there is heightened concern regarding erosion, prompting significant efforts to manage these issues and restore coastal resilience against both human-induced changes and natural events such as extreme weather and rising sea levels. The challenge of erosion is often exacerbated when countermeasures—whether hard or soft structures—are poorly designed, constructed, or maintained, and when their impacts on adjacent shores are not thoroughly assessed. Frequently, erosion is addressed in a localized manner, focusing on specific sites or jurisdictional boundaries rather than considering broader natural processes. To make the recovery simulations and the overall regression methodology computationally feasible, a morphological acceleration factor (morfac) (Ranasinghe et al. 2011; Roelvink 2006; Splinter et al. 2011; Vousdoukas et al. 2012) of ten was used for all recovery simulations. This disconnect often stems from a lack of understanding of coastal dynamics and the protective roles of coastal systems.

To address these challenges, the study proposes implementing shore protection measures—either hard or soft—along the Kalpakkam Coast to enhance its resilience and safeguard against erosion.

2 Scope of Work

The study aims to investigate the physical processes influencing shoreline deterioration along the Kalpakkam coast, including wind speed, wind direction, wave heights, and tidal levels. To analyze these shoreline dynamics, the research integrates satellite data, field surveys, and other physical reports. Utilizing this comprehensive data, a regression model has been developed specifically for the Kalpakkam coast to identify extreme shoreline events and their contributing factors. The primary objective of the study is to create a robust regression model to enhance understanding of the coastal deterioration mechanisms at this site.

3 Study Area

The coastline of northern Tamil Nadu, located on the east coast of India, stretches in a northeast–southwest direction from Pondicherry to Chennai. Kalpakkam, situated approximately 68 km south of Chennai, is illustrated in Fig. 1. The coastal area facing the Bay of Bengal is primarily a fluvial-alluvial plain, characterized by coarse sand near the shoreline that transitions to finer sand and silt further inland. The landforms along this coast include open beaches and sandy areas, with the coastline oriented in a NNE–SSW direction. Notably, manmade structures in the study area include a jetty connecting to the shore and an outfall structure located on the beach.

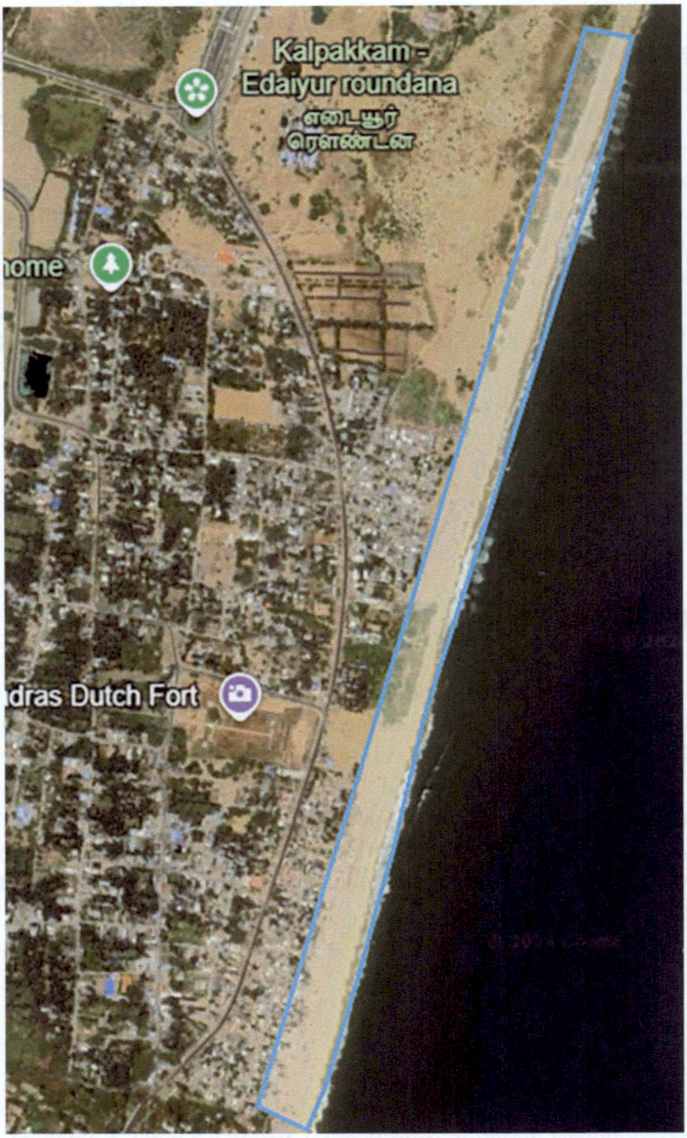

Fig. 1 Study area

4 Methodology

4.1 Field-Based Methodology

4.1.1 Beach Profile Survey

The "Emery board" method is a practical and effective technique for measuring beach profiles. It allows researchers to capture detailed topographic changes over time with relative ease and accuracy. By using two rods connected by a rope of a known length, the method establishes a consistent measurement interval, which is crucial for obtaining reproducible results. The inclusion of GPS for precise location tracking further enhances the data quality.

The establishment of benchmarks on stable structures is a smart way to ensure that measurements remain consistent over the study period. The use of vertical rods allows for accurate readings of elevation changes relative to the beach profile, and the method of assigning positive or negative values based on the slope is intuitive.

By collecting monthly data over several months, researchers can effectively analyze seasonal variations in beach morphology. This long-term approach helps in understanding the dynamic nature of coastal environments and can inform management strategies for erosion control and habitat preservation.

If you need more information or have specific questions about this method or its applications, feel free to ask! (Fig. 2).

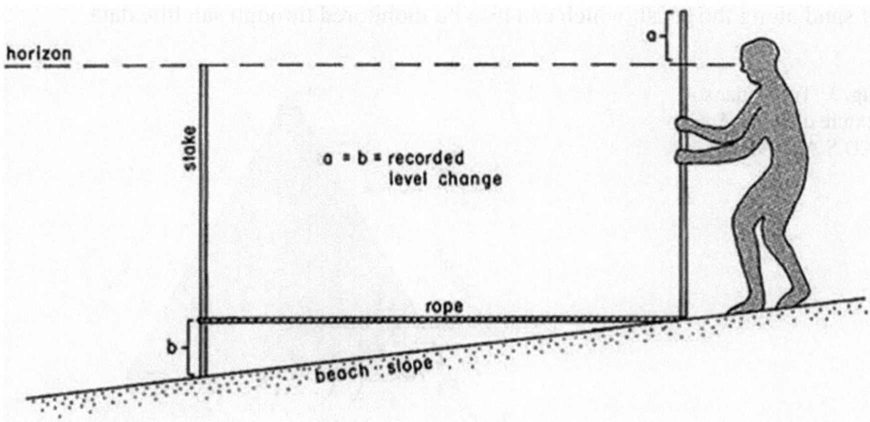

Fig. 2 Emery method for beach profiling (Emery 1961)

4.2 Sedimentation of Beach Sand Sample Collection

Sediment samples were collected from each pole along the beach profile, and grain size analysis was performed to examine changes in beach processes (see Fig. 3).

The soil texture triangle is used to classify different combinations of sand, silt, and clay. Coarse-textured soils, or sandy soils, mainly consist of medium to coarse sand particles, while fine-textured soils, or clayey soils, are characterized by their dominance of tiny clay particles. Interestingly, even soils with only 20% clay can display the sticky, gummy qualities typical of clay, owing to its strong physical properties. Loam, on the other hand, represents a balanced mix of sand, silt, and clay-sized particles, effectively combining the beneficial attributes of each component.

Soil with 5% clay, 94% sand, and 1% silt is classified as sandy, as shown in Fig. 3. According to the soil texture triangle, loamy soil consists of 40% sand, 20% clay, and 40% silt, whereas sandy loam is composed of 60% sand, 10% clay, and 30% silt (Source: U.D.S.A).

4.2.1 Wave Measurement

Wave heights in the intertidal zone were recorded using a fixed permanent benchmark to ensure reliable data collection. A leveling instrument was employed to measure both maximum and minimum wave heights for analysis, as illustrated in Fig. 4. Tide data were collected similarly during the full moon and half-moon phases. The effects of wave heights and tidal levels on the shoreline can result in erosion and deposition of sand along the coast, which can also be monitored through satellite data.

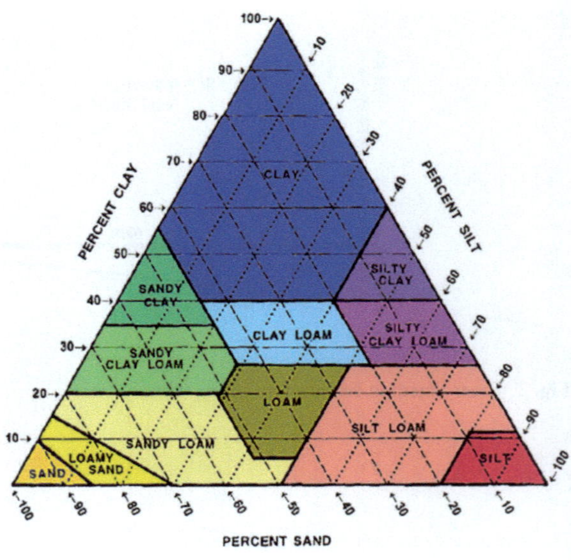

Fig. 3 Triangular soil texture diagram. *Source* U.D.S.A (1993)

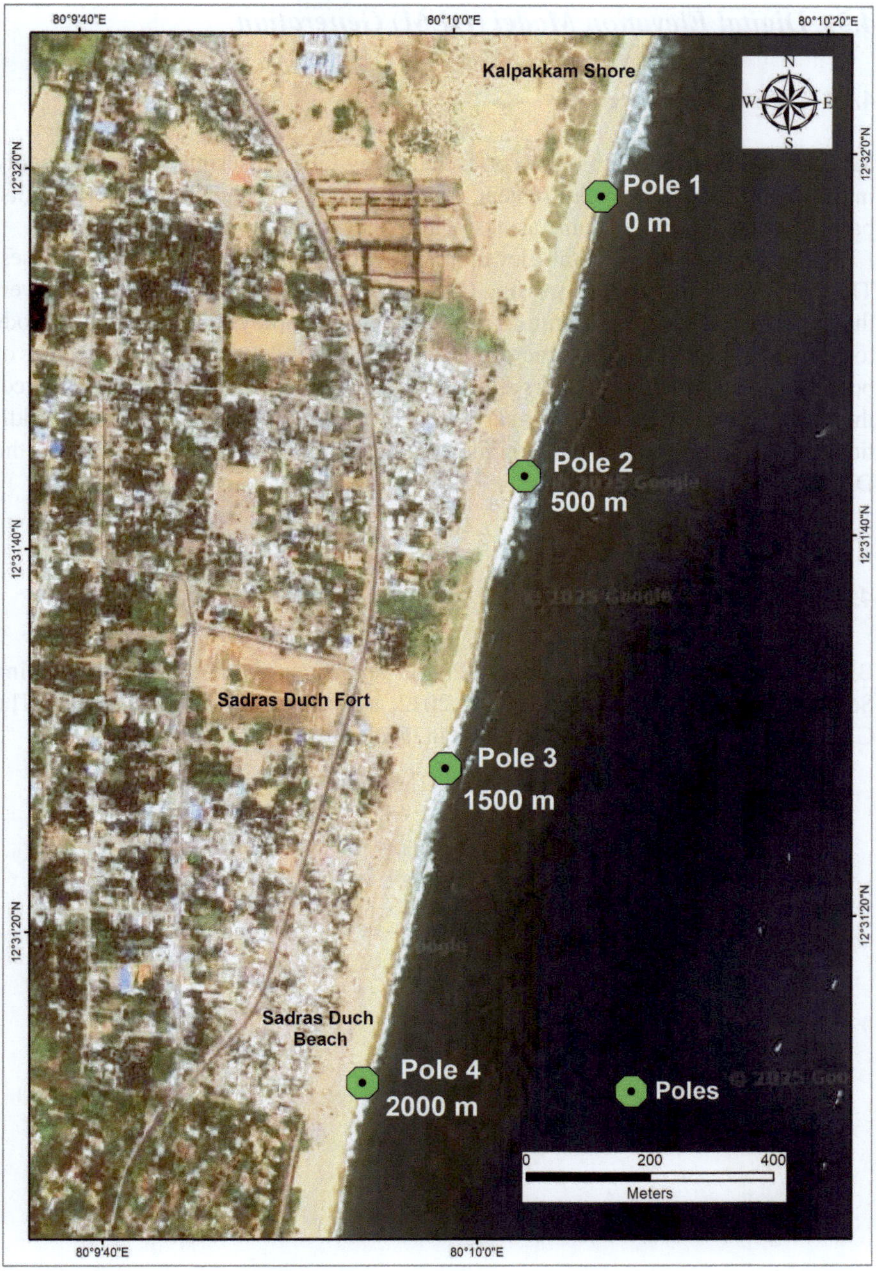

Fig. 4 Location points for beach profile, wave data

4.3 Digital Elevation Model (DEM) Generation

4.3.1 Surveying Procedures

Monitoring of sandy shores was conducted using GPS, ensuring high data quality in the field. Minimizing measurement errors is crucial for achieving reliable results, particularly in morphodynamic assessments.

The generation of a Digital Elevation Model (DEM) involves two main phases. The first phase consists of processing GPS data with specialized software to convert the collected points into shapefiles. The second phase employs interpolation methods to generate the DEM. The accuracy of the final DEM is influenced by the outcomes of both phases, with three key factors affecting data collection: the techniques employed, the characteristics of the ground surface, and the density of collected data. Additionally, the choice of interpolation method significantly impacts the quality of the DEM.

4.4 Premcrest Field Methods

BREM data were collected using GPS on specific new moon days, including September 15, 2012, and October 29, 2012, as well as during a full moon. The GPS data were processed and presented in tabular form.

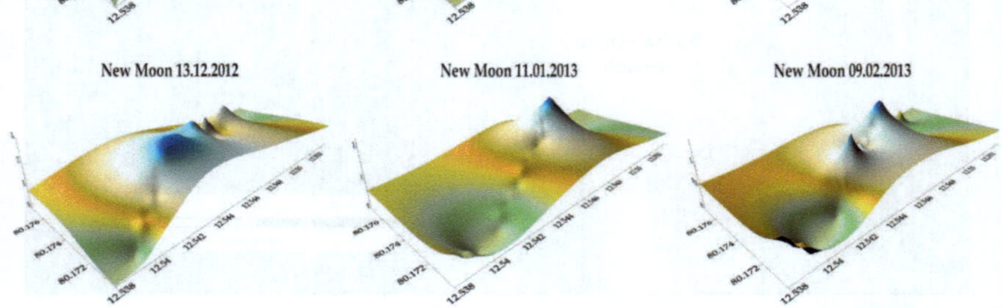

4.5 Satellite-Based Methodology

4.5.1 Satellite Data Processing

Satellite data for this study were obtained from the National Remote Sensing Center (NRSC), specifically from the Cartosat-2 satellite imagery dated April 25, 2012. This cloud-free dataset, with less than 10% cloud cover, represents the most recent acquisition by NRSC for 2012. Cartosat provides panchromatic 10-bit data in grayscale. The data were geo-referenced using ground control points collected with a Trimble Geo XM and subsequently projected to UTM in the WGS 84 datum. Various thematic analyses were conducted, including land use and land cover, geology, geomorphology, vegetation mapping, and shoreline changes.

4.5.2 Land Use and Land Cover Analysis

The satellite imagery from April 25, 2012, was utilized for visual interpretation of current land use and land cover. The digital data for mapping these categories were geo-referenced with Survey of India (SOI) topographical sheets. Variations in image characteristics—such as tone, texture, and pattern—facilitated the identification of different land use and land cover classes. The information derived from the imagery was transferred to a base map created from topographical data.

The boundaries of the study area were defined using both topographic and land use and land cover maps. Spatial data were digitized through onscreen visual digitization. Creating topology was essential for making the spatial data usable, with errors corrected through editing processes involving arc, label, move, and intersect commands. Topology was constructed using CLEAN and BUILD commands and reconstructed as needed. Database coverage from digitized units was converted to real-world coordinates using the UTM system.

At this stage, graphic analysis was conducted for the 2012 land use and land cover map, and the area for each classified category was calculated. A flow diagram illustrating the methodology for land use and land cover mapping is presented in Fig. 5. This research framework serves as the foundation for deriving land use statistics and informing the overall findings.

4.5.3 Shoreline Changes

Raster data from Landsat 2005 and Cartosat-2 2012 were acquired and geometrically corrected using an open-source dataset. Over 25 ground control points were utilized, yielding a Root Mean Square (RMS) error of 0.002 for geometric correction. Landsat TM Band 5 (2005) and Cartosat-2 Band 1 data were selected due to their enhanced contrast between land and ocean. These satellite images were vectorized through onscreen digitization using ERDAS Imagine 8.4 software. The vector layers for the

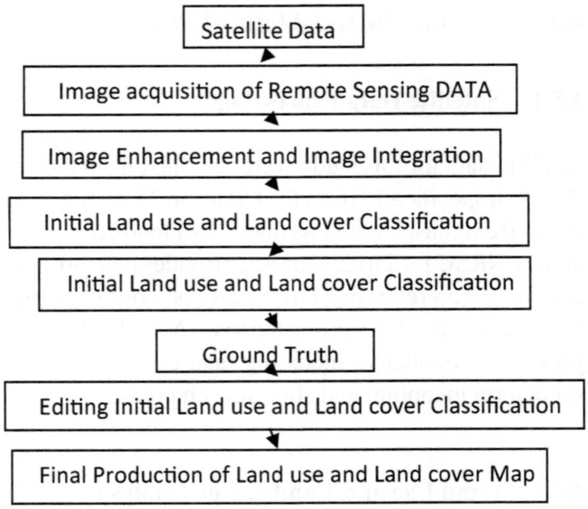

Fig. 5 Methodology for land use and land cover mapping

shorelines were created in ERDAS Imagine and then imported into ARCGIS as Arc coverage for both datasets. Each dataset was assigned a polygon ID—1 for land and 2 for ocean. The shorelines from the Landsat 2005 and Cartosat-2 2012 datasets were stored in separate coverages with consistent projections and map coordinates. These coverages were overlaid in ARCGIS to generate a shoreline change map. Despite differences in resolution, edge detection techniques allowed for clear demarcation of the land–water boundary, and shoreline features were further analyzed in ArcView GIS.

5 Results and Discussion

5.1 Field-Based Analysis

5.1.1 Beach Profile Analysis

Seasonal changes along the Kalpakkam coast were studied using monthly beach profiles collected at four stations (Station 1: 0 m, Station 2: 500 m, Station 3: 1500 m, and Station 4: 2000 m). These profiles represent two-dimensional vertical sections that illustrate variations in beach elevation relative to distance across the shore, measured perpendicular to the coastline. Below is a description of the observed changes in beach profiles over a 12-month period.

Located near the jetty (Fig. 4), beach profiles at this station were recorded from October 2011 to October 2012. The coastal area exhibited significant changes, particularly before and after the Thane cyclone, with noticeable erosion following the event. In January, the highest sand elevation reached approximately 1.54 m. From October

2011 to October 2012, erosion predominated, with accretion observed from February to April. These patterns suggest the beach is in a state of dynamic equilibrium, with the most significant erosion occurring after the Thane cyclone. A lack of dry sandy areas was noted in the upper foreshore, replaced by wetter sands. Notably, the erosion and deposition patterns at this station contrast sharply with those at Station 4, where erosion predominates. Beach profile minimum values ranged from − 0.01 m in December 2011 to a maximum of 1.54 m in January 2012, indicating fluctuating beach width over time.

Situated 500 m from Station 1, beach profiles were also collected from October 2011 to October 2012. December recorded the maximum beach elevation at 2.05 m, indicating a significant seasonal peak. The minimum elevation varied between 0.2 and 2 m, highlighting considerable variability in the profiles (Fig. 6). The upper and lower foreshore displayed the highest dynamism, with no long-term trends in erosion or accretion observed, suggesting a constant readjustment of sand between these zones. At Stations 2 and 3, the upper and lower foreshore regions were the most dynamic, but Station 2 demonstrated greater morphological fluctuations than Station 1 (Fig. 7).

The profiles from this location displayed significant spatial and temporal variability in beach morphology. Vertical fluctuations at the upper beach reached 0.1–0.97 m, with higher fluctuations noted at Station 2 (0.2–2.05 m). Large horizontal distances of beach change were evident, indicating net erosion of the shoreline.

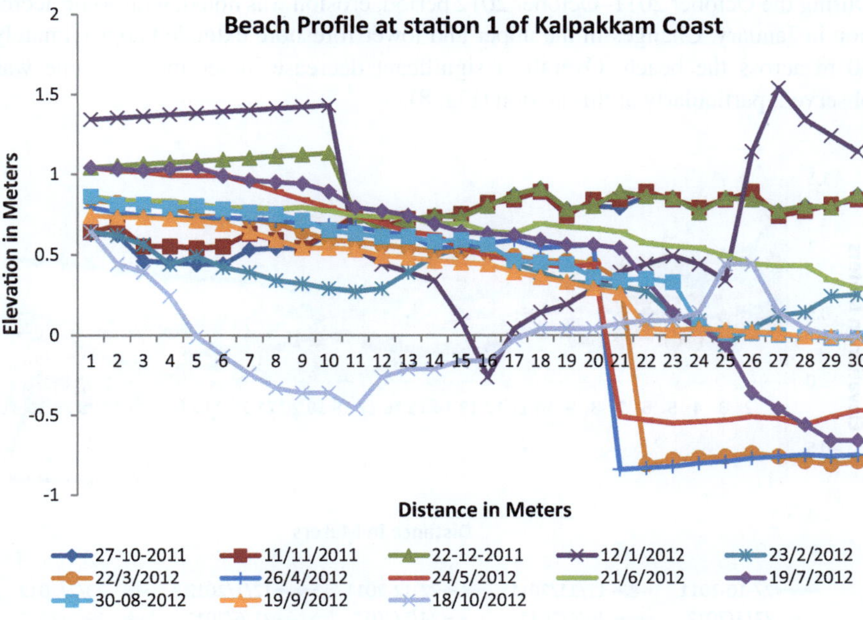

Fig. 6 Beach profile for Station 1

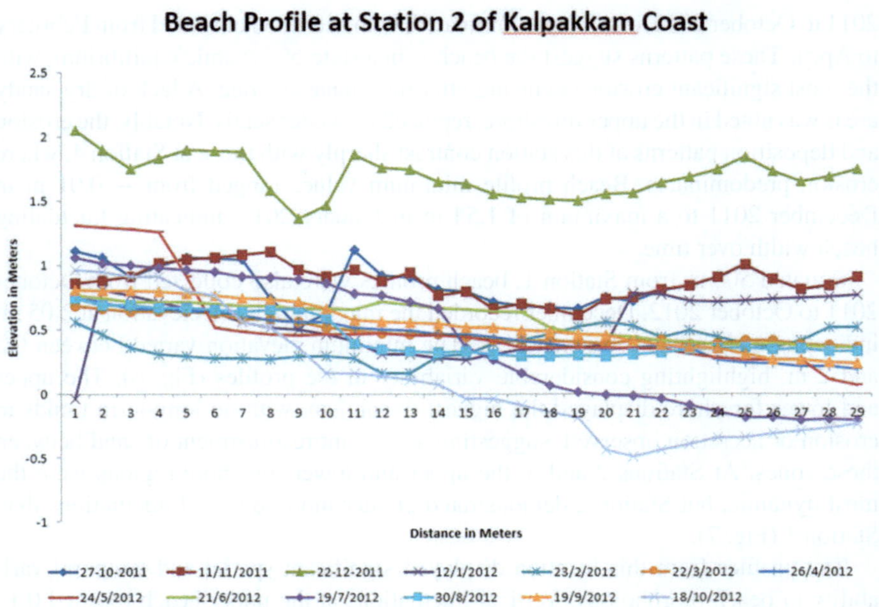

Fig. 7 Beach profile for Station 2

During the October 2011–October 2012 period, erosion was noted, with some accretion in January. Changes in the upper and lower foreshore extended approximately 30 m across the beach. Overall, a significant decrease in sediment volume was observed, particularly at this station (Fig. 8).

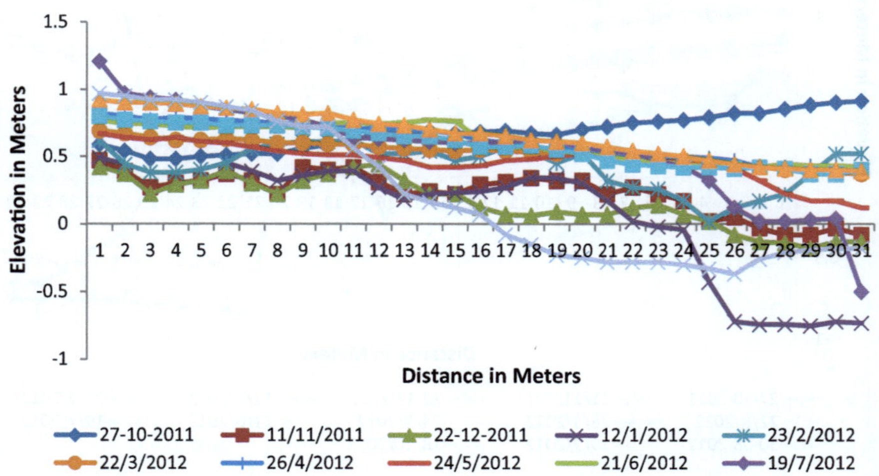

Fig. 8 Beach profiles for Station 3

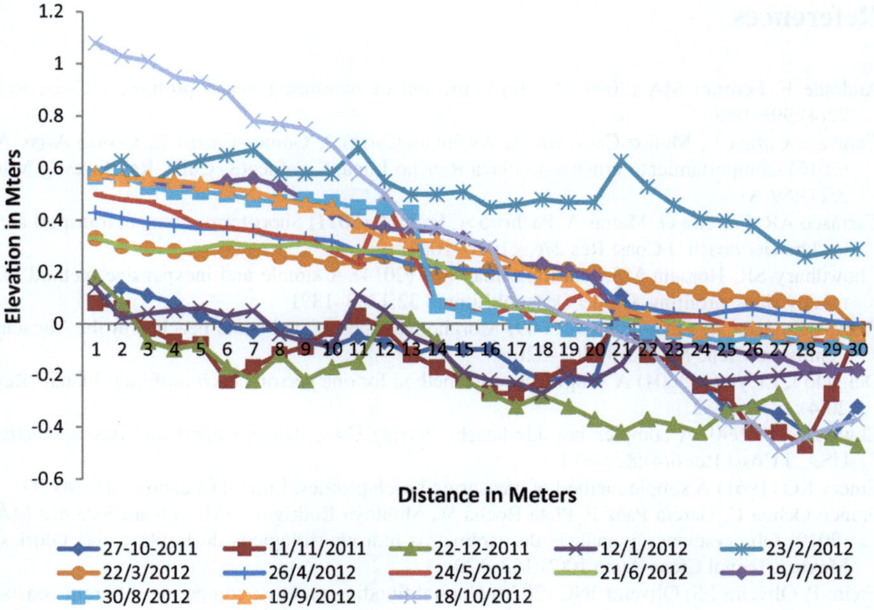

Fig. 9 Beach profile for Station 4

As the endpoint of the Station 4 (2000 m) coast, this station showed distinct patterns of erosion. Beach profiles were recorded from October 2011 to October 2012, with maximum erosion evident. The minimum elevation reached 0.02 m, while the maximum was 1.1 m (Fig. 9). This station exhibited an inverse pattern compared to Stations 1 and 2: erosion occurred here while deposition was observed at the northern side. This indicates a cyclical relationship where accretion on one side coincides with erosion on the other, likely influenced by major storm events.

6 Conclusions

The beach profiles from Stations 1 through 4 illustrate varying patterns of beach-level dynamics. A common feature across all four stations is the pronounced dynamism observed in the upper foreshore, particularly at Stations 1 and 2, with maximum vertical fluctuations reaching 0.2–2.5 m. The lower foreshore demonstrated relatively less variability, typically less than 1 m. This study highlights the complexity of beach processes along the coast, influenced by seasonal changes, storm events, and sediment dynamics.

References

Andrade F, Ferreira MA (2006) A simple method of measuring beach profiles. J Coast Res 22(4):995–999

Caravaca-Colina L, Muñoz-Caravaca A, Alcántara-Carrió J, Gómez-Castro E, Osorio-Arias A (2015) Comportamiento erosivo de Playa Rancho Luna, Cienfuegos, Cuba. Rev Investig Mar 35(1):69–81

Carrasco AR, Ferreira Ó, Matias A, Pacheco A, Freire P (2011) Short-term sediment transport at a backbarrier beach. J Coast Res 27(6):1076–1084

Chowdhury SR, Hossain MS, Sharifuzzaman SM (2014) A simple and inexpensive method for muddy shore profiling. Chin J Oceanol Limnol 32:1383–1391

Cuevas-Jiménez A, Euán-Ávila J (2009) Morphodynamics of carbonate beaches in the Yucatán Peninsula. Cienc Mar 35(3):307–320

Delgado I, Lloyd G (2004) A simple low cost method for one person beach profiling. J Coast Res 20(4):1246–1252

Ekwurzel B (1990) A complex bayside beach: Herring Cove Beach, Cape Cod, Massachusetts, USA. J Coast Res 6(4):879–891

Emery KO (1961) A simple method of measuring beach profiles. Limnol Oceanogr 6(1):90–93

Franco Ochoa C, García Paéz F, Plata Rocha W, Montoya Rodríguez JM, Vergara Sánchez MÁ (2019) Observaciones y análisis de parámetros hidromorfológicos de la playa Las Glorias, México. Tecnol Cienc Agua 10(2):153–170

Freire P, Oliveira FS, Oliveira JNC (2020) Coastal flooding process: comparing different coastal typologies response to extreme hydrodynamic conditions. In: Malvárez G, Navas F (eds) Global coastal issues of 2020. J Coast Res 95:797–802

Klemas V (2011) Beach profiling and LIDAR bathymetry: an overview with case studies. J Coast Res 27(6):1019–1028

Krause G (2004) The "Emery-method" revisited—Performance of an inexpensive method of measuring beach profiles and modifications. J Coast Res 20(1):340–346

Łabuz TA (2016) A review of field methods to survey coastal dunes—experience based on research from South Baltic coast. J Coast Conserv 20:175–190

Moloney JG, Hilton MJ, Sirguey P, Simons-Smith T (2018) Coastal dune surveying using a low-cost remotely piloted aerial system (RPAS). J Coast Res 34(5):1244–1255

Morton RA, Leach MP, Paine JG, Cardoza MA (1993) Monitoring beach changes using GPS surveying techniques. J Coast Res 702–720

Murthy CS, Veerayya M, Varadachari VVR (1975) Littoral and rip currents off Calangute beach, Goa (India). Indian J Mar Sci 4(1):1–4

Navarrete-Ramírez S (2014) Protocolo indicador variación línea de costa: perfiles de playa. Indicadores de monitoreo biológico del subsistema de áreas marinas protegidas (SAMP). Ser Publ Gen Invemar 73:36

Puleo JA, Pearre NS, He L, Schmied L, O'Neal M, Pietro LS, Fowler M (2008) A single-user subaerial beach profiler. J Coast Res 24(4):1080–1086

Ranasinghe R, Callaghan D, Stive MJF (2011) Estimating coastal recession due to sea level rise: beyond the Bruun rule. Clim Change 110(3–4):561–574. https://doi.org/10.1007/s10584-011-0107-8

Roelvink JA (2006) Coastal morphodynamic evolution techniques. Coastal Eng 53(2–3):277–287. https://doi.org/10.1016/j.coastaleng.2005.10.015

Splinter KD, Holman RA, Plant NG (2011) A behavior-oriented dynamic model for sandbar migration and 2DH evolution. J Geophys Res: Oceans 116(C7):C07008. https://doi.org/10.1029/2010JC006547

Terefenko P, Giza A, Paprotny D, Walczakiewicz S (2020) Characteristic of winter storm Xavier and its impacts on coastal morphology: results of a case study on the Polish coast. In: Malvárez G, Navas F (eds) Global coastal issues of 2020: proceedings from the international coastal symposium (ICS) (Seville, Spain). J Coast Res 95:684–688

United States Department of Agriculture (USDA) (1993) Soil survey manual. U.S. Department of Agriculture, Soil Conservation Service, Washington DC
Veerayya M (1972) Textural classification of Calangute beach sediments, Goa coast (India). Indian J Mar Sci 1(1):28–44
Vousdoukas MI, Ranasinghe R, Mentaschi L, Plomaritis TA, Athanasiou P, Luijendijk A, Feyen L (2012) A statistical framework for estimating the impact of climate change on coastal erosion. Coastal Eng 66:14–24. https://doi.org/10.1016/j.coastaleng.2012.03.009

Beyond the Surface: Effects of Natural Hazards on Coastal Marine Ecosystems

Namrata Kislay, Roli Mishra, Mandar D. Kirkire, Geeta Malbhage, and Rahul Kumar

Abstract Coastal marine ecosystems face significant susceptibility to natural hazards, resulting in consequences that go beyond just the immediate aftermath of events like hurricanes, tsunamis, and coastal erosion. This chapter delves into the various effects of natural hazards on these ecosystems, underscoring the intricacies and interconnectedness that shape their ability to bounce back and their susceptibility. Important factors to consider are the disturbance of habitats, modifications in water quality and chemistry, shifts in sedimentation patterns, and the arrival of invasive species. Moreover, natural hazards frequently cause chain reactions in food chains and ecosystem functions, impacting the ways of life of coastal populations who depend on these assets. It is crucial to implement effective strategies for mitigation and adaptation in order to enhance the resilience of ecosystems and safeguard biodiversity against increasing climate threats. Examples from different areas demonstrate both effective and difficult reactions to natural hazards, providing recommendations for successful and creative strategies in coastal management that promote sustainability. By thoroughly comprehending these effects, policymakers, researchers, and stakeholders can promote a more robust future for coastal marine ecosystems.

Keywords Coastal marine ecosystems · Natural hazards · Resilience · Biodiversity · Climate adaptation · Ecosystem services

---

N. Kislay · M. D. Kirkire · G. Malbhage · R. Kumar (✉)
Amity School of Applied Sciences, Amity University, Bhatan, Navi Mumbai, Maharashtra, India
e-mail: rahul.khadwalia@gmail.com

R. Mishra
Department of Environmental Science, Institute of Science, Mumbai, Maharashtra, India

K. K. Basheer Ahammed et al. (eds.), *Monitoring Coastal Marine Ecosystems Under Climate Change*, https://doi.org/10.1007/978-3-032-09137-6_11

1 Introduction

1.1 Coastal Marine Ecosystems and Their Types

Coastal marine ecosystems are crucial and varied habitats where the land and sea intersect, fostering a diverse array of life essential for global ecology and human health (Trégarot et al. 2024). These habitats vary due to factors such as water salinity, temperature, nutrient availability, and physical geography within the ecosystems (Potts et al. 2020). Coastal marine environments consist of various habitats, each having distinct features and roles in the ecosystem (Barbier 2017; Henseler et al. 2019). Table 1 lists the primary coastal marine ecosystems, which consist of estuaries, salt marshes, mangroves forests, coral reefs, rocky shores, sandy beaches, and seagrass meadows.

Table 1 Types and descriptions of coastal marine ecosystems

S. No.	Ecosystem type	Description	Citation
1.	Estuaries	Partially enclosed coastal bodies of water, where freshwater from rivers and streams mixes with saltwater from the ocean	Kennish (2001)
2.	Mangroves forests	Coastal ecosystems found in tropical and subtropical regions, characterized by the presence of mangrove trees	Alongi (2002)
3.	Coral reefs	Shallow, sunlit ecosystems found in tropical and subtropical oceans, characterized by coral polyps and a diverse array of marine life	Hughes et al. (2012)
4.	Seagrass meadows	Shallow, coastal ecosystems dominated by seagrass, a type of marine grass	Hemminga and Duarte (2000)
5.	Salt marshes	Coastal grasslands found in temperate and subtropical regions, characterized by a mix of grasses and shrubs	Adam (1990)
6.	Dunes	Coastal ecosystems characterized by sand dunes and a mix of grasses and shrubs	Hesp (2002)
7.	Sandy beaches	Loose deposits of sand along shorelines	Schlacher et al. (2008)
8.	Rocky marshes	Rugged wetlands with rocky outcrops and marsh vegetation, supporting diverse ecosystems	Piccolo (2021)

Estuaries, the meeting point of rivers and the sea, are some of the most productive ecosystems on Earth (Bianchi 2013). They go through changes in tides and different levels of salt in the water, which supports a diverse array of organisms (Brown et al. 2023; Jung et al. 2024). Estuaries are important as vital breeding grounds for marine creatures and crucial feeding spots for migrating birds and fish (Zhang et al. 2023). Salt marshes, which are located in the intertidal area, contain grasses and sedges that can tolerate high levels of salt (Adam 2023). They serve as barriers to coastal erosion, purify pollutants, and offer essential habitats and feeding areas for marine creatures (Hernández-Delgado 2024). Salt marshes are also important for capturing carbon and cycling nutrients (Jordan and Fröhle 2022). Mangrove forests, located in tropical and subtropical coastal regions offer homes and breeding sites for numerous species (Alongi 2008; Kumar et al. 2024). Shorelines are safeguarded by them, they store carbon, and they face risks from habitat destruction and climate change, requiring conservation actions (Chatting et al. 2022).

Coral reefs, constructed by coral polyps in transparent tropical waters, sustain a variety of marine organisms and offer ecosystem benefits such as safeguarding coastlines and supporting fisheries (Hoegh-Guldberg et al. 2017; Wasim et al. 2024). Their vulnerability to ocean warming and pollution underscores the importance of ensuring their protection (Hoegh-Guldberg et al. 2019). Rocky shores are coastal areas with solid surfaces that are exposed at low tide and are home to a diverse range of organisms that have evolved to survive in challenging environments. They are vital for coastal ecosystems and biodiversity, serving as dynamic environments (Underwood and Chapman 2022). Sandy shores are formed by waves and currents and host special communities of digging creatures while also serving as breeding grounds for sea turtles and homes for shoreline birds, even though they may appear desolate (McLachlan and Brown 2006). Seagrass meadows, located in shallow coastal waters, create dense underwater habitats that sustain a variety of marine creatures (Orth et al. 2006). They help support sediments, enhance water quality, and store carbon, leading to increased coastal resilience (Fourqurean et al. 2012).

These interconnected coastal marine ecosystems create intricate ecological networks that sustain biodiversity and offer crucial advantages to human communities (Halpern et al. 2019). Nevertheless, they are confronted with dangers like habitat loss, pollution, excessive fishing, climate change, and invasive species (Trégarot et al. 2024). Efforts to conserve these ecosystems are crucial in order to safeguard them and ensure their restoration for generations to come (TEEB 2023).

1.2 Importance of Understanding Natural Hazards

It is essential to comprehend the natural hazards that impact coastal marine ecosystems because these ecosystems are susceptible to a variety of environmental stressors due to their dynamic nature (Spalding et al. 2014a). Coastal marine ecosystems consist of various habitats such as coral reefs, mangroves, estuaries, and sandy beaches (Barbier et al. 2011). They are crucial for biodiversity, climate regulation, and ecosystem services for people (Moberg and Folke 1999).

Understanding natural hazards is crucial because it helps reduce the impact of these hazards on coastal communities and ecosystems (PreventionWeb 2024). Coastal regions are especially vulnerable to natural disasters like hurricanes, tsunamis, storm surges, and coastal erosion (Banholzer et al. 2014). Such occurrences can result in major harm to structures, loss of lives, and disturbance of means of living (Albino and Sasanelli 2019). Through comprehension of the frequency, magnitude, and impacts of these dangers, experts, decision-makers, and local residents can create plans to reduce risks by enhancing coastal planning, implementing early warning systems, and designing infrastructure (McGranahan et al. 2007).

Additionally, coastal marine ecosystems are frequently the first to experience the effects of climate change, such as rising sea levels and increasing ocean acidity (IPCC 2023). Elevated sea levels may result in higher occurrences of coastal flooding and erosion, putting at risk ecosystems like mangroves and salt marshes, which act as natural defenses against storm surges (Trégarot et al. 2024). The absorption of higher levels of carbon dioxide is causing ocean acidification and putting marine species like corals and shellfish at risk due to their sensitivity to pH level changes (Hoegh-Guldberg et al. 2007). Comprehension of these dangers enables specific conservation actions and flexible management plans to improve ecosystem resilience and promote biodiversity preservation (WMO 2023).

Another important factor in comprehending natural disasters in coastal marine ecosystems is how they impact socioeconomic activities. Coastal areas serve as key centers for economic activities like fishing, tourism, and transportation (Tuttle and Donahue 2022). Risks such as harmful algal blooms causing fish deaths and impacting human health, as well as oil spills damaging marine ecosystems and communities, highlight the need for monitoring and reducing hazards to promote sustainable progress (Lan et al. 2024).

Furthermore, these ecosystems offer essential services that enhance human welfare, such as coastal defense, nutrient recycling, and carbon storage (Hernández-Blanco et al. 2022). Healthy coastal ecosystems support food security, recreational activities, and cultural heritage. By recognizing the dangers facing these services, stakeholders can focus on preserving them and implementing sustainable resource management practices to protect ecosystem health and benefit local communities (TEEB 2023).

1.3 Harnessing AI and ML for Marine Ecosystem Monitoring

The utilization of AI and ML is transforming how we conduct natural resource mapping and impact assessment. As the need for sustainable management of Earth's resources increases, these cutting-edge technologies provide effective tools to improve our knowledge of ecosystems, especially in marine areas (Kamyab et al. 2023). By analyzing and interpreting large amounts of data, AI and ML can offer important insights into the health and dynamics of marine ecosystems, aiding in the protection of these crucial habitats for future generations (Ditria et al. 2022).

Marine ecosystems play a key role in preserving biodiversity, sustaining fisheries, and controlling climate patterns (Lotze 2021). Nevertheless, they face growing risks from human activities such as pollution, overfishing, and habitat destruction. Conventional monitoring methods frequently do not adequately capture the complexity and magnitude of these changes. On the other hand, AI and ML have the ability to analyze satellite images, underwater sensors, and acoustic data in order to uncover patterns that might be overlooked (Essamlali et al. 2024). Using these technologies allows researchers to have a better understanding of the condition of marine environments, which helps in making timely and well-informed decisions.

Moreover, AI and ML have the potential to conduct thorough impact assessments that analyze the effects of human activities on marine ecosystems overlooked (Essamlali et al. 2024). These tools are able to simulate potential future situations, forecasting the results of planned projects and aiding stakeholders in making well-informed decisions that reconcile economic expansion with environmental conservation. By combining ecological data with socioeconomic factors, AI and ML improve comprehension of marine ecosystems and encourage sustainable practices (Kaymaz Mühling 2023). This investigation into the use of AI and ML in mapping natural resources and assessing their impact emphasizes how important they are in promoting resilience in marine environments, thereby helping to create a more sustainable future.

2 Types of Natural Hazards Affecting Coastal Marine Ecosystems

Coastal marine ecosystems are vulnerable to various natural hazards, each posing significant threats to their health and stability (Table 2).

2.1 Storms and Hurricanes

Storms and hurricanes are severe weather occurrences known for their strong winds, significant rainfall, and sometimes accompanied by thunderstorms (National Oceanic and Atmospheric Administration [NOAA] 2023a). Hurricanes are specifically tropical cyclones that form over warm ocean waters, usually in the Atlantic or Eastern Pacific Oceans (National Hurricane Center [NHC] 2024). These occurrences have significant effects on various aspects of coastal marine ecosystems.

To begin with, coastal habitats such as coral reefs, mangroves, and seagrass beds suffer physical damage from storms and hurricanes. The powerful winds from

Table 2 Natural hazards and their impact on coastal marine ecosystems

S. No.	Natural hazard type	Description	Impact on marine ecosystem	Example case study
1.	Storms	High winds, heavy rainfall, storm surges	Habitat destruction, sedimentation, changes in water quality	Typhoon Haiyan, 2013
2.	Hurricanes	Intense tropical storms with high winds and heavy rainfall	Erosion, sedimentation, coral damage, altered salinity	Hurricane Katrina, 2005
3.	Tsunamis	Large ocean waves caused by underwater earthquakes or volcanic eruptions	Coral reef destruction, sediment redistribution, loss of habitat	Indian Ocean Tsunami, 2004
4.	Sea-level rise	Gradual increase in the average level of the world's oceans	Coastal erosion, loss of wetlands, saltwater intrusion	Miami Sea-Level Rise Observations
5.	Coastal erosion	Gradual wearing away of coastal land by the relentless action of currents, waves, and wind	Nutrient enrichment, potentially triggering eutrophication and harmful algal blooms (HABs)	Erosion of the Outer Banks in North Carolina, USA
6.	Earthquake	Seismic activity, tsunamis	Habitat destruction, changes in water quality, loss of biodiversity	Tōhoku Earthquake, 2011
7.	Ocean acidification	Excess carbon dioxide (CO_2) from the atmosphere dissolves into seawater, forming carbonic acid	Coral reef and shellfish destruction, disrupt predator–prey relationship	Oyster Mortality in the Pacific Northwest, 2005

these occurrences can remove vegetation and disturb ecosystems, with storm surges causing coastal flooding, erosion, and sediment buildup, changing the landscape and habitat layout (Gardner et al. 2005; Basheer Ahammed and Pandey 2019).

Furthermore, storms have a significant impact on the water quality of coastal marine environments. Intense rain leads to runoff from the land, transporting pollutants like sediment, nutrients, and chemicals into the sea. The increase in pollutants is harming the water quality, endangering marine life and ecosystems that rely on healthy environments for their existence (Levin et al. 2020).

Additionally, hurricanes have a direct effect on marine life through causing species mortality and displacement. For example, coral reefs are especially at risk during hurricanes because they face pressure from powerful currents and shifts in water temperature, resulting in coral bleaching and a decline in the ecosystem that follows (NOAA 2019). These disruptions impact not just corals but also the various groups of sea creatures that depend on these reefs for protection and food (Simmons et al. 2021).

Hurricanes not only affect the physical and biological aspects but also impact ocean dynamics through changes in currents and mixing patterns. The strong winds and waves from these storms can cause deep-water mixing and carry nutrients to the surface, possibly helping some marine species while disturbing the normal ocean conditions that promote ecosystem stability (Miles et al. 2023).

2.2 *Tsunamis*

Tsunamis are massive sea waves caused by underwater earthquakes, volcanic eruptions, landslides, or sometimes meteorite impacts (NOAA 2018). Although they can move quickly through ocean basins at speeds up to 500 mph, they are usually invisible in deep waters (Geist et al. 2007). When tsunamis reach near shallow coastlines, their powerful energy can cause devastating floods and significant destruction (Titov et al. 2023).

Physical destruction is a key effect of tsunamis on coastal marine ecosystems. Such occurrences have the ability to significantly alter coastal terrains by wearing down shorelines, flooding environments like coral reefs, seagrass beds, and mangroves, and changing the shape of the coast (Devi and Shenoi 2012). The power of a tsunami can displace plants, carry sediments inland, and destroy vital infrastructure necessary for ecosystem operations (NOAA 2023b).

Tsunamis also have a direct impact on marine organisms. The great power of tsunami waves can lead to movement, harm, or death of sea creatures (Devi and Shenoi 2012). Coral reefs are easily affected by physical disruption, leading to significant harm from waves breaking coral formations and displacing species living in the reef (Vasiliev 2012). Marine creatures like fish may be carried landward or further to sea, disturbing their population and habitat dynamics (Koester et al. 2023).

Furthermore, tsunamis have the potential to affect the quality of water in coastal regions. Similar to storms, these occurrences can result in an increase in sedimentation and nutrient runoff into marine ecosystems (Bunson and Hu 2023). Changes like these can cause a temporary decline in water quality, impacting marine organisms that rely on consistent nutrient levels and clear waters to survive and reproduce (ISRS 2004).

Coastal environments like mangroves and salt marshes, essential for marine life, are especially at risk of being harmed by tsunamis. The destruction of these habitats could result in significant decreases in species abundance and diversity as they are important for nurturing and feeding (Wallace et al. 2009). Restoring these habitats, crucial for the ability of coastal ecosystems to withstand challenges, may take a considerable amount of time and effort, particularly for long-living organisms such as corals that are fundamental to a variety of marine ecosystems (Scheffers et al. 2009; Majumdar et al. 2019).

Apart from immediate effects, tsunamis can also result in lasting ecological effects. Alterations in habitat characteristics and the types of species present may impact how ecosystems function and their ability to withstand future disruptions (Urabe and Nakashizuka 2016). Actions taken by humans in response to tsunamis, like building seawalls and breakwaters, may unintentionally interfere with natural sediment movement and affect marine habitats in the long run (Angnuureng et al. 2023).

Strategies like early warning systems, coastal zoning, land-use planning, habitat restoration, and ecosystem-based coastal management are being implemented to minimize the effects of tsunamis on coastal marine ecosystems. These actions are focused on increasing the strength of human communities and protecting natural ecosystems from the destructive impacts of powerful natural events (UNEP 2010).

2.3 Sea-Level Rise

The rise of sea levels and erosion along the coast are closely connected events that have significant effects on coastal marine ecosystems (Storlazzi 2020). The increase in global sea levels, known as sea-level rise, is mainly caused by the warming of seawater and the melting of glaciers and ice sheets resulting from climate change (IPCC 2023). These alterations have notable impacts on coastlines, especially in low-lying regions at risk of flooding and erosion (Sweet et al. 2022).

Coastal habitats like salt marshes, mangroves, and intertidal zones are at high risk of losing their habitats due to increasing sea levels (Mimura 2013). These environments could be flooded or worn away, putting at risk the essential roles they play for many sea creatures. The decrease in these environments not only lowers biodiversity but also lessens the ecosystem benefits they provide, like protecting shorelines and cycling nutrients (Guerry et al. 2022).

The movement of species due to increasing sea levels creates extra difficulties. Although some species may try to relocate inland to adjust, human-made structures such as coastal development often hinder the natural coastal processes and limit migration pathways (Garner et al. 2015). Moreover, increased sea levels accelerate coastal erosion by enhancing wave energy and storm surges (Dong et al. 2024). The heightened erosive power worsens habitats, weakening shorelines and increasing the risk of erosion-related dangers for coastal communities and infrastructure.

2.4 Coastal Erosion

Coastal erosion occurs when coastal land is slowly worn away by the continuous forces of currents, waves, and wind. This natural phenomenon greatly affects coastal marine ecosystems, influencing habitats and ecosystem dynamics in various ways

(Dong et al. 2024). The main outcome of coastal erosion is the immediate disappearance of important ecosystems like beaches, dunes, and coastal cliffs (Angnuureng et al. 2023). These environments act as essential locations for many sea creatures to reproduce, feed, and build nests, such as birds, turtles, and coastal plants.

Furthermore, the resilience of coastal ecosystems is decreased by coastal erosion, which reduces natural buffers against storms and high waves. As coastal defenses deteriorate, inland regions are increasingly vulnerable to floods and destruction during severe weather, endangering both ecosystems and human settlements near the coast (Sunkur et al. 2023). Moreover, eroded sediments may also impact nutrient cycling in coastal waters. Rising sediment levels may result in nutrient buildup, potentially causing eutrophication and harmful algal blooms (Akinnawo 2023). These occurrences can greatly affect marine environments by lowering levels of oxygen and causing harm to aquatic organisms relying on good water quality (Chislock et al. 2013).

Efforts to minimize the effects of rising sea levels and erosion on coastal marine ecosystems require comprehensive strategies for coastal management. These tactics involve actions such as beach nourishment for rejuvenating eroded shorelines, dune restoration endeavors to boost coastal resilience, and planting more mangroves to safeguard coastal habitats (Youdon 2024). Establishing protected zones and promoting sustainable coastal development practices are essential measures in protecting coastal environments, with the goal of reviving natural processes and minimizing human-caused pressures on vulnerable coastal ecosystems (Dhiman et al. 2014). By putting these measures into practice, we can improve the ability of coastal marine ecosystems to withstand ongoing environmental challenges and support their continued sustainability in the long run.

2.5 *Earthquakes*

Earthquakes have significant impacts on coastal marine ecosystems, influencing them in multiple interconnected ways due to their geological nature. These are abrupt, forceful motions of the Earth's crust resulting from the movement of tectonic plates beneath the surface. These seismic occurrences may happen on fault lines and are often linked to volcanic events, increasing the complexity of their potential effects (Legrand 2022).

Tsunamis can be initiated by earthquakes in coastal ecosystems, which is a major impact. The occurrence of earthquakes beneath the sea floor can produce huge oceanic waves that flood coastal regions, leading to extensive destruction of marine ecosystems like coral reefs, mangroves, and intertidal zones (Devi and Shenoi 2012). Tsunamis possess a powerful ability to alter coastlines and cause significant disturbances to ecosystems that depend on steady coastal environments for their existence (Devi and Shenoi 2012).

In addition, seismic activity can disrupt shoreline environments through shaking. Seismic events have the potential to cause landslides, harm reef formations, and disrupt sediment layers, changing habitat availability and affecting marine species' distribution (Zhou et al. 2022). Habitat fragmentation and loss caused by these disturbances can decrease biodiversity and affect ecosystem functions (Haddad et al. 2015).

Furthermore, seismic activity can impact the quality of water along coastlines by disturbing underground water systems and causing alterations in salinity levels or the discharge of contaminants into the sea (Akhtar et al. 2021). These alterations can harm marine organisms' health by impacting their physiology, reproduction, and ecosystem resilience (Akhtar et al. 2021).

Addressing the effects of earthquakes on coastal marine ecosystems necessitates the use of comprehensive strategies. Possible approaches could involve implementing early detection systems for tsunamis, launching projects to restore habitats and improve ecosystem endurance, and implementing actions to oversee and control the effects of water quality on the environment (Jr. Tolentino 2007). By recognizing and dealing with these geological processes, we can aim to protect coastal marine ecosystems and support their long-term health in the face of natural environmental obstacles.

2.6 Ocean Acidification

Ocean acidification, caused by rising levels of CO_2 in the atmosphere, presents serious risks to marine ecosystems near the coast. When carbon dioxide mixes with seawater, it creates carbonic acid, leading to a decrease in the ocean's pH level (IPCC 2023). This change in pH levels upsets the fragile equilibrium of ocean habitats, impacting numerous marine species and the ecosystems they depend on (Hall-Spencer and Harvey 2019).

Coral reefs, commonly known as the "rainforests of the ocean," are especially at risk from ocean acidification. Corals depend on calcium carbonate to construct their skeletons and create reef formations (Sun et al. 2020). Corrosive waters decrease the presence of carbonate ions, hindering corals in forming their skeletons. This leads to corals being weaker, more fragile, and growing at a slower pace (Hoegh-Guldberg et al. 2017). As a result, coral reefs' structural integrity is weakened, causing a decline in habitat complexity and biodiversity. The reduction in coral cover can also have a negative effect on the many species that rely on reefs for food and protection, disturbing the entire ecosystem of the reef (Eddy et al. 2021).

Shellfish are also greatly affected by ocean acidification. Oysters, clams, and mussels utilize calcium carbonate in the formation of their shells. Elevated levels of acidity disrupt the process of shell formation, resulting in shells that are thinner, weaker, and more vulnerable to damage and predators (Mackenzie et al. 2014). This could lead to significant impacts on shellfish populations, impacting their ability to survive and reproduce. Moreover, shellfish play a vital role in numerous coastal economies, serving as a staple food source and supporting aquaculture operations.

The decrease in shellfish populations can have significant economic consequences for communities that depend on these resources (Cobacho et al. 2020).

Marine algae, such as kelp and seagrass, are also encountering difficulties as a result of shifts in ocean chemistry. Acidification can change the growth rates and well-being of these primary producers, crucial for coastal ecosystems (Hall-Spencer and Harvey 2019). Kelp forests and seagrass beds play essential roles as habitats and help with nutrient cycling and carbon storage. Changes in these plant communities can impact the overall efficiency of coastal ecosystems and the species that depend on them, such as fish and invertebrates (Smale et al. 2013).

In addition, the behavior, sensory perception, and development of marine fish and invertebrates can be affected by ocean acidification. For example, certain fish young may struggle to find appropriate environments or evade predators in environments with increased acidity, resulting in increased death rates and changed population patterns. Alterations in the number and range of these species can cause disturbances in food chains and ecological relationships, leading to ripple effects across the marine environment.

3 Climate Change Impacts

Climate change is characterized by lasting changes in Earth's temperature, rainfall patterns, and other weather conditions over time. The primary causes of these changes are mainly due to human activities like burning fossil fuels, deforestation, and industrial processes. These actions emit greenhouse gases, such as carbon dioxide (CO_2), which lead to an increase in Earth's temperature through the greenhouse effect (IPCC 2023).

Coastal marine environments are especially at risk from the effects of global warming. The melting of polar ice caps and glaciers, driven by global warming, contributes to a significant phenomenon: the rise in sea levels. This occurs due to the addition of freshwater to the oceans and the thermal expansion of seawater (Bindoff et al. 2019). Increased sea levels result in more coastal erosion, loss of vital habitats such as mangroves and salt marshes, and greater vulnerability to storm surges and flooding (Morris et al. 2013). Coastal species that are adapted to specific tidal ranges and water depths may struggle to migrate or adjust to changing conditions (Doney et al. 2009).

Ocean acidification, another consequence of elevated atmospheric CO_2, leads to increased acidity in ocean waters. This acidification poses a threat to sea creatures like corals, shellfish, and plankton, which rely on calcium carbonate to build their shells and skeletons (USEPA 2024). Weakened shells and skeletons make these organisms more vulnerable to predation and disrupt the marine food web, potentially causing broader changes in marine ecosystems (Hoegh-Guldberg et al. 2007).

Climate change also results in warmer ocean temperatures, which have significant effects on marine life. Many species, including corals and fish, are sensitive to temperature fluctuations. Elevated water temperatures can lead to coral bleaching, where corals expel their symbiotic algae, resulting in widespread coral death and loss of essential habitats for numerous marine species (van Woesik et al. 2022).

Additionally, climate change has the potential to alter ocean circulation patterns and currents on a global scale. These changes affect nutrient availability, water temperature, and the distribution of marine species, leading to disruptions in marine ecosystems and impacts on fisheries that depend on specific oceanographic conditions (Hasselmann 1991). Climate change is also linked to the increased frequency and intensity of extreme weather events such as hurricanes and heavy rainfall, which can further degrade water quality and harm marine life by increasing pollutant runoff into coastal waters (Knutson et al. 2020).

4 Direct Impacts on Coastal Marine Ecosystems

The direct effects of natural disasters on coastal marine ecosystems can be significant and varied, impacting both the habitats and the species that rely on them (Table 3). This is how natural disasters can have a direct effect on these environments.

4.1 Physical Damage to Habitats (Coral Reefs, Mangroves, Etc.)

Coral Reefs: Coral reefs can suffer physical harm from natural disasters like hurricanes, cyclones, and typhoons due to powerful waves, storm surges, and sedimentation. These events have the potential to cause significant damage to coral branches, detach corals from their base, and cover reefs with sediment, impacting their overall

Table 3 Comparative impacts of different natural hazards on coral reefs

S. No.	Type of hazards	Immediate impact	Long-term impact	Recovery time (years)
1.	Hurricane	Physical damage to corals, sedimentation	Coral mortality, changes in community structure	5–10
2.	Tsunami	Severe physical destruction	Loss of coral colonies, disruption of reef dynamics	10–20
3.	Flooding	Increased sedimentation, nutrient overload	Algal overgrowth, decreased coral cover	3–7
4.	Sea-level rise	Altered light availability, increased thermal stress	Coral bleaching, habitat loss	30+ years

structure and well-being (Neal 2021; Great Barrier Reef Marine Park Authority 2011).

Mangroves: Strong winds and storm surges from natural disasters have the ability to remove mangrove trees, strip off leaves, and harm root systems. The mangroves' capacity to stabilize coastlines, offer marine species habitat, and serve as nurseries for young fish and invertebrates is diminished by this physical harm (Spalding et al. 2014b).

4.2 *Disruption of Species Distributions and Migrations*

Natural disasters have the potential to change ocean conditions like temperature, salinity, and currents, which are crucial for the movement and distribution of marine organisms. Species that depend on certain environmental conditions for reproduction, eating, or moving may encounter disturbance (Smyth and Elliott 2016). Changes in water temperature and salinity can influence the breeding patterns of fish and other marine creatures, which can ultimately impact population dynamics and reproductive outcomes (Shinwari et al. 2023).

4.3 *Loss of Biodiversity and Ecosystem Services*

Destruction of habitats and disturbance of species distributions from natural disasters can result in reduced biodiversity in coastal marine ecosystems. A decrease in biodiversity weakens the ability of ecosystems to bounce back from disturbances, affecting their capacity to offer crucial ecosystem services (Rahman 2023). Coastal marine ecosystems offer important services like safeguarding coastlines against storms, supporting fisheries, cycling nutrients, and providing tourism chances. Harm and destruction of these habitats reduce these benefits, impacting the economies of nearby areas and the welfare of coastal residents (Carlson et al. 2021).

Natural disasters can have a direct effect on coastal marine ecosystems by causing physical harm to habitats such as coral reefs and mangroves, disturbing species movement and migrations, and ultimately resulting in a decline in biodiversity and ecosystem functions (Lacambra et al. 2024). Habitat restoration, sustainable coastal development practices, and conservation efforts are essential to improve the resilience of these ecosystems against natural hazards (Keller et al. 2009).

5 Indirect Impacts on Coastal Marine Ecosystems

The lasting effects of natural disasters on coastal marine ecosystems can have a significant impact on ecological processes and dynamics, often in complex ways.

5.1 Changes in Water Quality (Salinity, Sedimentation)

Natural disasters like hurricanes, tsunamis, and heavy rainfall can significantly alter water quality conditions in coastal marine areas. Intense rainfall can lead to increased freshwater inflow into coastal zones, which decreases salinity levels and impacts marine organisms that are adapted to specific salinity ranges (Davis et al. 2022). Such changes can disrupt the ecological balance, affecting species distribution and health.

Sedimentation often rises notably as a result of erosion caused by storm surges and strong currents during natural disasters. Excessive sedimentation can suffocate coral reefs and other bottom-dwelling habitats, leading to decreased overall health and productivity of these ecosystems (Tuttle and Donahue 2022). High sediment loads can block sunlight essential for photosynthetic organisms, alter habitat structures, and diminish biodiversity (Erftemeijer et al. 2012).

5.2 Nutrient Runoff and Pollution

Natural calamities can worsen the flow of nutrients from coastal regions to marine ecosystems. One example is when intense rain can carry nutrients like nitrogen and phosphorus from farms, cities, and sewage systems into the ocean. This increase in nutrients can cause eutrophication, bloom of algae, and subsequent lack of oxygen (hypoxia) in impacted regions (Chislock et al. 2013; Devlin and Brodie 2023).

Pollution sources, such as litter and harmful substances, can be carried into marine ecosystems during natural disasters, worsening water quality and ecosystem health (Lincoln et al. 2022). These contaminants can exacerbate the impacts of nutrient overload, contributing to the degradation of marine habitats and posing additional risks to marine life.

5.3 Altered Food Webs and Trophic Interactions

Natural calamities have the potential to disturb food chains and relationships between different trophic levels in coastal marine environments. Changes in the arrangement of habitats, prey availability, and nutrient supply can impact the spread and number of marine organisms at different trophic levels (Bishop et al. 2017; Ullah et al. 2018). Disturbances to primary producers like phytoplankton and macroalgae can have a domino effect on the food chain, impacting herbivores, carnivores, and higher trophic levels including fish and marine mammals (Ullah et al. 2018; Mittelbach and McGill 2019).

These secondary effects emphasize the interdependence of natural processes and ecosystem dynamics in coastal marine ecosystems. Methods of management that encourage resilience, like restoring coastal habitats, practicing sustainable land use, and implementing pollution reduction measures, are crucial for lessening the effects of natural disasters on these fragile ecosystems (Sunkur et al. 2023; Powell et al. 2019).

6 Recent Case Studies of Natural Hazard Events

Below are four significant and recent instances of natural disasters impacting coastal ecosystems.

6.1 Hurricane Ian (USA, 2022)

In late September 2022, a strong Category 4 hurricane named Ian hit the southeastern United States. On September 28, it hit Florida and caused severe destruction, especially in Fort Myers. The storm resulted in a large amount of rain, powerful winds, and major storm surge, causing severe flooding and destruction. Ian subsequently relocated along the Eastern Coast, impacting various areas of the Carolinas. It caused more than 100 deaths and billions of dollars in destruction, ranking it as one of the most expensive hurricanes in U.S. history.

Effects on Coastal Ecosystems

Hurricane Ian had a major impact on coastal ecosystems, especially in Florida. The powerful winds and storm surge from the storm led to the erosion of beaches and wetlands, impacting habitats for marine and coastal species. The intrusion of saltwater impacted freshwater sources, changing the salinity levels of estuaries and causing harm to fish populations. Moreover, the flooding displaced vegetation, causing harm to mangroves and other important habitats that offers protection and breeding areas for wildlife. The ecosystems have taken a long time to recover, impacting biodiversity and coastal resilience in the long run.

Methodologies Used for Monitoring

The combination of remote sensing, modeling, and community surveys was extremely valuable in monitoring the impacts of disasters. Satellite images offered important information on the extent of damage and flooding, enabling quick evaluation of impacted regions. Hydrodynamic models were utilized to forecast flooding patterns, improving readiness and response tactics (Schumann et al. 2023). Furthermore, surveys conducted within affected communities after disasters provided valuable insights from residents, allowing for a thorough assessment of the impact on both people and infrastructure.

Findings

The findings of Hurricane Ian in 2022 highlighted the crucial need for prompt evacuations and the importance of strong infrastructure. Efficient evacuation procedures greatly decreased the threat to lives, while strong infrastructure aided in minimizing harm and expediting recovery. These observations underscore the importance of readiness and fortitude in reducing the effects of upcoming hurricanes (Alam et al. 2022).

6.2 Earthquake and Tsunami (Turkey, 2023)

A powerful earthquake with a magnitude of 7.8 on the Richter scale hit southern Turkey in February 2023. It resulted in extensive damage in various cities, causing a significant number of fatalities and injuries. The earthquake caused a tsunami that affected coastal regions, worsening the damage. The damaged infrastructure and severe weather conditions posed challenges for rescue and recovery efforts. The catastrophe led to tens of thousands of injuries and caused millions to be displaced, revealing weaknesses in the region's readiness for such disasters.

Effects on Coastal Environments

The coastal ecosystems in Turkey were greatly impacted by the earthquake and tsunami in February 2023. The tsunami resulted in the erosion of coastlines and the destruction of habitats, affecting coastal plants and animals (Chunga-Llauce and Pacheco 2021). Sediment movement changed nearby marine environments, impacting areas where fish and other marine species breed. Moreover, water quality and marine life were further threatened by pollution from damaged infrastructure and debris runoff. In general, the disaster disturbed the equilibrium of coastal ecosystems, resulting in lasting ecological effects (Chunga-Llauce and Pacheco 2021).

Methodologies Used for Monitoring

A varied approach was used to monitor seismic events. Monitoring seismic activity in real-time provided crucial information for obtaining a rapid understanding of the earthquake's qualities and assessing its effects promptly. Mathematical models were utilized in tsunami simulation to forecast wave movement and assess hazards to coastal regions. Furthermore, drills were carried out within the community to improve local readiness and response abilities, ensuring that residents were better prepared to react in case of an emergency (Rafliana et al. 2022).

Findings

The findings of the 2023 Turkey earthquake and tsunami underscored the importance of early warning systems in minimizing casualties. These systems issued timely notifications enabling residents to evacuate and find safety before the disaster hit, greatly reducing the number of casualties. The importance of investing in and enhancing early warning technologies for future preparedness was highlighted by the effectiveness of these warnings.

6.3 Typhoon Noru (Philippines, 2022)

In September 2022, the Philippines was hit by Typhoon Noru, a strong tropical cyclone that landed in Luzon. It caused heavy precipitation, powerful winds, and extensive flooding, mainly impacting farming regions. The storm caused extensive destruction to buildings and houses, forcing thousands to evacuate and causing multiple deaths. Recovery efforts were difficult because of the widespread damage and continued effects on communities that were already susceptible to natural disasters.

Effects on Coastal Environments

The effects of Typhoon Noru on coastal ecosystems in the Philippines were substantial. The storm resulted in coastal erosion and habitat loss, especially affecting mangroves and coral reefs. Water quality was negatively impacted by sediment runoff and pollution from flooding, causing harm to marine life and disturbance to breeding grounds. The decrease in protective coastal vegetation reduced the ability to bounce back from future storms, putting biodiversity and the livelihoods of communities relying on healthy marine ecosystems at even greater risk (Nioko et al. 2023).

Methodologies Used for Monitoring

Tracking Typhoon Noru in the Philippines involved various techniques such as using satellite imagery to monitor the storm's movements and strength, weather radar to evaluate rain and wind patterns, and automated weather stations to gather information on temperature and air pressure. Remote sensing methods were used to examine alterations in land and water before and after the typhoon, while on-site surveys conducted after the storm evaluated the impact on infrastructure and ecosystems. Collectively, these techniques offered a thorough insight into the typhoon's effects and guided response initiatives (Nioko et al. 2023).

Findings

Findings from Typhoon Noru showed considerable effects on both societies and environments in the Philippines. The typhoon resulted in extensive flooding and landslides, causing significant harm to infrastructure and agriculture, forcing thousands of residents to evacuate. It led to many deaths and exposed weaknesses in disaster readiness and response systems. From an environmental perspective, the storm worsened coastal erosion, damaged habitats of mangroves and coral reefs, and negatively impacted water quality through heightened sedimentation and runoff. In general, the event highlighted the importance of enhancing resilience strategies and improving disaster management practices to reduce future risks.

6.4 Cyclone Batsirai (Madagascar, 2022)

In February 2022, Cyclone Batsirai hit Madagascar as a strong tropical cyclone, reaching the Eastern Coast. Heavy rainfall, strong winds, and widespread flooding resulted in significant damage to infrastructure and agriculture. The cyclone caused thousands of people to be displaced and resulted in numerous deaths. Ongoing obstacles, such as food insecurity and the region's susceptibility to extreme weather events, impeded recovery efforts.

Effects on Coastal Environments

Cyclone Batsirai negatively impacted the coastal ecosystems of Madagascar. The storm led to significant shore erosion and harmed mangroves, which are vital for protecting coasts and biodiversity. The pollution of marine habitats due to increased sedimentation and runoff has had a detrimental effect on coral reefs and fish populations. The flooding disturbed the fragile equilibrium of coastal ecosystems, jeopardizing the sustenance of communities reliant on these resources for fishing and agriculture. In general, the cyclone worsened weaknesses in coastal environments that were already under pressure (Herrera-Silveira et al. 2022).

Methodologies Used for Monitoring

Tracking Cyclone Batsirai required using different methods, such as satellite images, to monitor its progression, intensity, and potential effects. Weather radar systems were utilized to evaluate precipitation and wind patterns, whereas automated weather stations gathered surface-level information on temperature, humidity, and air pressure. Remote sensing methods were used to evaluate land alterations and assess damage after the cyclone. Moreover, post-storm field surveys were carried out to assess the impact on impacted communities and ecosystems, aiding in gaining a thorough understanding of the cyclone's effects.

Findings

Findings from Cyclone Batsirai exposed widespread devastation throughout Madagascar, especially in the eastern areas. The cyclone resulted in substantial flooding, landslides, and infrastructure damage, forcing tens of thousands of residents to evacuate and causing dozens of deaths. Significant agricultural losses worsened food insecurity in a country already dealing with difficulties. The cyclone also brought attention to weaknesses in disaster preparedness and response systems, emphasizing the importance of enhancing resilience strategies to lessen the effects of upcoming extreme weather events. Environmental evaluations showed significant harm to coastal ecosystems, notably mangroves and coral reefs, impacting biodiversity and the livelihoods of nearby communities.

7 AI/ML in Monitoring and Managing Coastal Marine Ecosystems

The analysis of data from coastal marine ecosystems is being transformed by machine learning, deep learning, and remote sensing technologies, leading to improved monitoring and management. These methods make it easier to uncover useful information from complicated data sets, increasing our comprehension of how ecosystems function and boosting our ability to withstand environmental threats. AI and machine learning are changing the way we observe and control coastal marine environments (Bakker 2022).

7.1 Data Collection and Analysis

Remote Sensing: AI algorithms can examine satellite images to identify alterations in coastal habitats like mangroves or coral reefs, enabling continuous monitoring of ecosystem well-being.

Automated Sensors: IoT devices with AI can collect environmental data (such as temperature and salinity) non-stop.

7.2 Predictive Modeling

Ecosystem Reactions: By examining past data, machine learning models can forecast the way ecosystems will react to environmental stressors like pollution and climate change.

AI can be utilized for biodiversity monitoring by analyzing audio and visual data to identify where conservation efforts are most essential.

7.3 Risk Assessment

Flood Risk Modeling: AI can assist in predicting potential flood situations caused by sea level increases and storms, contributing to planning for potential disasters.

Pollution Tracking: Machine learning can pinpoint pollution sources and patterns in coastal regions, enabling precise interventions.

7.4 Decision Support Systems

Adaptive Management: Policymakers can receive data-driven recommendations for sustainable management practices from AI-powered decision support tools.

Community Engagement: AI-powered apps can deliver immediate notifications and updates to residents regarding changes in the environment or potential hazards.

7.5 Restoration Projects

Optimal Resource Allocation: AI can assist in the planning and overseeing of restoration projects to ensure efficient allocation of resources and measurable results.

8 Resilience and Adaptation of Coastal Marine Ecosystems

8.1 Natural Resilience Mechanisms in Ecosystems

Coastal marine environments possess natural resilience mechanisms that allow them to adjust to and recover from disruptions. Biodiversity is a crucial factor, as having a variety of species with different tolerances to disturbances can improve the resilience of ecosystems (Cleland 2011; Hooper et al. 2005). Furthermore, these ecosystems have the ability to adjust by evolving, which may involve alterations in the types of species present or modifications in the layout of the habitat, showcasing their ability to adapt (Western 2001). Additionally, coastal marine ecosystems are typically well-suited to common natural disturbances such as storms or tidal fluctuations, which play a significant role in influencing their structure and function (Wang et al. 2016). Ultimately, the interconnection of various environments like mangroves, coral reefs, and seagrass beds enables strength and flexibility through activities such as seed distribution, nutrient circulation, and species movement (Kandasamy 2014).

8.2 Human Interventions and Restoration Efforts

Human actions have a two-fold influence on the ability of coastal marine ecosystems to withstand challenges, resulting in both adverse and beneficial consequences. Human actions like pollution, overfishing, habitat destruction, and climate change can greatly decrease ecosystem resilience by harming habitats and biodiversity (Malhi et al. 2020). Alternatively, human actions can boost ecosystem resilience by implementing restoration projects. One example is the restoration of habitats, like planting mangroves, renovating seagrass beds, and reconstructing coral reefs, to enhance habitat quality and ecosystem function. Furthermore, adopting sustainable fishing techniques and creating marine protected zones can help fish numbers to rebound and uphold ecosystem equilibrium (Ojea et al. 2020). Additionally, improving water quality and ecosystem health can be achieved by decreasing pollution levels and adopting climate adaptation measures like coastal zone planning for rising sea levels and storm surges (Almeida and Mostafavi 2016). By being proactive, people can have a critical impact on aiding the resilience of coastal marine ecosystems.

8.3 *Role of Marine Protected Areas and Coastal Management*

Marine protected areas (MPAs) and efficient coastal management techniques are crucial for boosting the resistance and supporting the sustainable utilization of coastal marine ecosystems. MPAs are essential for preserving biodiversity by safeguarding crucial habitats and species, thereby enabling populations to bounce back and flourish (Kriegl et al. 2021). Additionally, MPAs play a crucial role in providing important ecosystem services such as fishing, safeguarding coastlines, and promoting tourism, all of which help sustain local communities and economies (Erskine et al. 2021). During climate change, properly managed MPAs can act as safe havens for at-risk species, aiding in their resilience and longevity (Gonçalves 2023). Integrated coastal management frameworks are required to balance socioeconomic needs and ecosystem health, by including conservation, sustainable development, and stakeholder participation (Dhiman et al. 2014). In addition, good governance and involvement of stakeholders are crucial for overseeing MPAs and putting in place adaptive management tactics that can adapt to evolving environmental circumstances, guaranteeing the enduring resilience and sustainability of coastal marine ecosystems (Gonçalves 2023).

9 Future Challenges and Considerations

Looking into the future, there are significant challenges and considerations regarding the effects of natural hazards on coastal marine ecosystems. Here's an exploration of predicted impacts of climate change, emerging hazards, and policy implications for safeguarding these ecosystems.

9.1 Predicted Impacts of Climate Change on Coastal Ecosystems

Climate change is on the brink of causing severe harm to coastal marine ecosystems, endangering the survival of these fragile environments. A major worry is the rising sea levels due to the rapid melting of polar ice caps and glaciers, posing a threat to coastal habitats like mangroves, salt marshes, and low-lying islands (Mimura 2013). Furthermore, the elevated carbon dioxide levels in the air are resulting in ocean acidification, which can severely impact marine creatures with calcium carbonate shells and skeletons, like corals and shellfish (USEPA 2024). Moreover, the increase in sea surface temperatures is causing disturbances in coral reefs, resulting in coral bleaching incidents and a decrease in biodiversity (van Woesik et al. 2022). Furthermore, increased occurrences of severe weather events like stronger hurricanes, storms, and heat waves are causing more damage to coastal ecosystems and disrupting the fragile equilibrium of ecosystem functions (Mawren et al. 2022).

9.2 Emerging Hazards and Their Potential Impacts

Coastal marine environments are experiencing various new risks that put their health and ability to recover in danger. One of the biggest worries is the excessive buildup of plastic waste in the oceans, which poses a serious threat to marine animals by being swallowed or caught in it, leading to negative impacts on the health of the entire ecosystem (Pothiraj et al. 2023). Shipping and aquaculture pose another significant threat by introducing invasive species that can disturb native ecosystems and change species composition, resulting in decreased biodiversity (Gallardo et al. 2024). Moreover, the quick urban growth in coastal regions is leading to the loss of habitats, pollution, and changes in hydrology, which can greatly affect ecosystems' structure and function (McGrane 2016; Xu and Zhang 2022). With the increasing severity of these new threats, it is crucial to implement preventative actions to reduce their impact and safeguard the fragile equilibrium of coastal marine ecosystems (Ward et al. 2022).

9.3 Policy and Management Implications for Safeguarding Coastal Marine Ecosystems

Developing and implementing effective policies and management strategies is essential to protect coastal marine ecosystems from the harmful effects of natural disasters. An all-encompassing strategy is needed, which involves implementing integrated coastal zone management (ICZM) techniques that achieve a harmony between preservation and sustainable growth, in turn reducing habitat destruction and decline.

Furthermore, it is crucial to both extend and properly oversee marine protected areas (MPAs) in order to preserve biodiversity, safeguard important habitats, and strengthen resistance to the effects of climate change (Hoppit et al. 2022). Additionally, adaptive governance structures need to be improved to support stakeholder involvement, adaptive decision-making, and resilience-building strategies (Munene et al. 2018). Incorporating climate change adaptation strategies into coastal planning and management is crucial to improve ecosystem resilience and facilitate species migration and habitat shifts (Adger et al. 2005). Ultimately, enhancing global collaboration and treaties is crucial in dealing with transboundary concerns like pollution and fisheries management that impact coastal marine ecosystems, guaranteeing a unified and thorough strategy for their preservation (Elliott et al. 2023).

9.4 Advancements in Technology for Climate Change Monitoring and Management

The monitoring and management of climate change have been completely transformed by technological advancements, improving the ability to collect and analyze data (Ma et al. 2024). Satellite images and remote sensing offer real-time information on environmental changes, like deforestation, glacier melting, and rising sea levels. Machine learning algorithms are utilized to examine extensive datasets for forecasting climate trends and evaluating potential hazards, in order to improve decision-making through informed choices (Sarker 2021). Furthermore, IoT devices allow for real-time monitoring of local climate conditions and emissions (Ramadan et al. 2024). These technologies help implement flexible management approaches, enhance ability to withstand climate effects, and encourage sustainable methods, ultimately playing a critical role in combatting the worldwide climate emergency.

10 Conclusion

This chapter has emphasized the extensive and harmful impacts of natural disasters on coastal marine ecosystems, such as reduced biodiversity, habitat deterioration, and degraded water quality. Coastal marine ecosystems are important and diverse environments that sustain a wide range of organisms vital for worldwide ecology and human well-being. These environments face different natural threats like storms, tsunamis, rising sea levels, coastal erosion, earthquakes, and ocean acidification, leading to harmful effects on their well-being and balance. Recognizing the significance of these ecosystems and the natural risks that impact them is essential in order to create successful measures to reduce the impacts of these risks and enhance ecosystem resilience. This necessitates a thorough strategy that includes habitat restoration, sustainable coastal development practices, conservation efforts, and adaptive management strategies. By understanding how natural processes and ecosystem dynamics are linked in coastal marine ecosystems, we can strive to protect these crucial ecosystems for future generations.

Utilizing AI and machine learning methods to track effects on marine ecosystems is a hopeful way to improve environmental stewardship. These technologies can identify alterations in habitat health, species distribution, and pollution levels in real-time by examining various data sources like satellite imagery, underwater sensors, and acoustic monitoring. Additionally, AI algorithms have the capability to recognize patterns and forecast upcoming situations, allowing for proactive management approaches to address arising dangers. The ability to conduct thorough analysis enhances our knowledge of marine ecosystems and aids in creating successful policies and conservation initiatives to protect these important resources.

References

Adam P (1990) Saltmarsh ecology. Cambridge University Press

Adam P (2023) Salt marshes: ecology and restoration. Cambridge University Press

Adger WN, Arnell NW, Tompkins EL (2005) Successful adaptation to climate change across scales. Glob Environ Change 15(2):77–86. https://doi.org/10.1016/j.gloenvcha.2004.12.005

Akhtar N, Syakir Ishak MI, Bhawani SA, Umar K (2021) Various natural and anthropogenic factors responsible for water quality degradation: a review. Water 13(19):2660. https://doi.org/10.3390/w13192660

Akinnawo SO (2023) Eutrophication: causes, consequences, physical, chemical and biological techniques for mitigation strategies. Environ Chall 12:100733. https://doi.org/10.1016/j.envc.2023.100733

Alam MS, Horner MW, Ozguven EE, Ventimiglia B, Smith D (2022) Understanding people's safety perceptions during a recent evacuation: the case of Hurricane Ian (2022). Findings. https://doi.org/10.32866/001c.91268

Albino V, Sasanelli R (2019) The impact of natural disasters on infrastructure, lives, and livelihoods: a global assessment. Int J Disaster Risk Reduct 34:108–120. https://doi.org/10.1016/j.ijdrr.2018.11.011

Alongi DM (2002) Present state and future of the world's mangrove forests. Environ Conserv 29(3):331–349. https://doi.org/10.1017/S0376892902000231

Alongi DM (2008) Mangrove forests: resilience, protection from tsunamis, and responses to global climate change. Estuar Coast Shelf Sci 76(1):1–13. https://doi.org/10.1016/j.ecss.2007.08.024

Angnuureng BD, Adade R, Chuku EO, Dzantor S, Brempong EK, Mattah PAD (2023) Effects of coastal protection structures in controlling erosion and livelihoods. Heliyon 9(10):e20633. https://doi.org/10.1016/j.heliyon.2023.e20633

Azevedo de Almeida B, Mostafavi A (2016) Resilience of infrastructure systems to sea-level rise in coastal areas: impacts, adaptation measures, and implementation challenges. Sustainability 8(11):1115. https://doi.org/10.3390/su8111115

Bakker K (2022) Smart oceans: artificial intelligence and marine protected area governance. Earth Syst Gov 13:100141. https://doi.org/10.1016/j.esg.2022.100141

Banholzer S, Kossin J, Donner S (2014) The impact of climate change on natural disasters. In: Reducing disaster: early warning systems for climate change, pp 21–49. https://doi.org/10.1007/978-94-017-8598-3_2

Barbier EB (2017) Marine ecosystem services. Curr Biol 27(11):R507–R510. https://doi.org/10.1016/j.cub.2017.03.020

Barbier EB, Hacker SD, Kennedy C, Koch EW, Stier AC, Silliman BR (2011) The value of estuarine and coastal ecosystem services. Ecol Monogr 81(2):169–193

Basheer Ahammed KK, Pandey AC (2019) Shoreline morphology changes along the Eastern Coast of India, Andhra Pradesh by using geospatial technology. J Coast Conserv 23:331–353

Bianchi TS (2013) Estuaries: where the river meets the sea. Nat Educ Knowl 4(4):12

Bindoff NL, Willebrand JC, Artale V, Arndt KT (2019) Changing ocean, marine ecosystems, and dependent communities. In: Pörtner H-O, Roberts DC, Masson-Delmotte V, Houghton JE (eds) IPCC special report on the ocean and cryosphere in a changing climate. Cambridge University Press, pp 447–587

Bishop MJ, Mayer-Pinto M, Airoldi L, Firth LB, Morris RL, Loke LHL, Hawkins SJ, Naylor LA, Coleman RA, Chee SY, Dafforn KA (2017) Effects of ocean sprawl on ecological connectivity: impacts and solutions. J Exp Mar Biol Ecol 492:7–30. https://doi.org/10.1016/j.jembe.2017.01.021

Brown CJ, Ange CS, Connolly RM, Hasan S, Jackson S, McMahon JM, Smart JCR (2023) Ecosystem services in connected catchment to coast ecosystems: monitoring to detect emerging trends. Sci Total Environ 869:161670. https://doi.org/10.1016/j.scitotenv.2023.161670

Bunson S, Hu C (2023) Did tsunamis lead to changes in ocean properties? A revisit. Front Mar Sci 10:1275445. https://doi.org/10.3389/fmars.2023.1275445

Carlson RR, Evans LJ, Foo SA, Grady BW, Li J, Seeley M, Xu Y, Asner GP (2021) Synergistic benefits of conserving land-sea ecosystems. Glob Ecol Conserv 28:e01684. https://doi.org/10.1016/j.gecco.2021.e01684

Chatting M, Al-Maslamani I, Walton M, Skov M, Kennedy H, Husrevoglu S, Le Vay L (2022). Future mangrove carbon storage under climate change and deforestation. Front Mar Sci 9:781876. https://doi.org/10.3389/fmars.2022.781876

Chislock MF, Doster E, Zitomer RA, Wilson AE (2013) Eutrophication: causes, consequences, and controls in aquatic ecosystems. Nat Educ Knowl 4(4):1–10

Chunga-Llauce JA, Pacheco AS (2021) Impacts of earthquakes and tsunamis on marine benthic communities: a review. Mar Environ Res 171:105481. https://doi.org/10.1016/j.marenvres.2021.105481

Cleland EE (2011) Biodiversity and ecosystem stability. Nat Educ Knowl 3(10):1–14

Cobacho SO, Wanke S, Konstantinou Z, Serafy GE (2020) Impacts of shellfish reef management on the provision of ecosystem services resulting from climate change in the Dutch Wadden Sea. Mar Policy 119:104058. https://doi.org/10.1016/j.marpol.2020.104058

Davis TR, Larkin MF, Forbes A, Veenhof RJ, Scott A, Coleman MA (2022) Extreme flooding and reduced salinity causes mass mortality of nearshore kelp forests. Estuar Coast Shelf Sci 275:107960. https://doi.org/10.1016/j.ecss.2022.107960

Devi E, Shenoi SSC (2012) Tsunami and the effects on coastal morphology and ecosystems: a report. Proc Indian Natl Sci Acad 78(3):513–521

Devlin M, Brodie J (2023) Nutrients and eutrophication. In: Reichelt-Brushett A (ed) Marine pollution—monitoring, management and mitigation. Springer textbooks in earth sciences, geography and environment. Springer, Cham. https://doi.org/10.1007/978-3-031-10127-4_4

Dhiman R, Tirodkar J, Inamdar AB (2014) Integrating sustainable coastal development initiatives along Maharashtra coast, using ICM guidelines. https://wgbis.ces.iisc.ac.in/biodiversity/sahyadri_enews/newsletter/Issue63/article/ewrg/sec7.html

Ditria EM, Buelow CA, Gonzalez-Rivero M, Connolly RM (2022) Artificial intelligence and automated monitoring for assisting conservation of marine ecosystems: a perspective. Front Mar Sci 9:918104. https://doi.org/10.3389/fmars.2022.918104

Doney SC, Fabry VJ, Feely RA, Kleypas JA (2009) Ocean acidification: the other CO_2 problem. Ann Rev Mar Sci 2:169–192. https://doi.org/10.1146/annurev.marine.010908.163834

Dong WS, Ismailluddin A, Yun LS, Ariffin EH, Saengsupavanich C, Maulud KNA, Ramli MZ, Miskon MF, Jeofry MH, Mohamed J, Mohd FA, Hamzah SB, Yunus K (2024) The impact of climate change on coastal erosion in Southeast Asia and the compelling need to establish robust adaptation strategies. Heliyon 10(4):e25609. https://doi.org/10.1016/j.heliyon.2024.e25609

Eddy TD, Lam VWY, Rygondeau G, Cisneros-Montemayor AM, Greer K, Palomares MLD, Bruno JF, Ota Y, Cheung WWL (2021) Global decline in capacity of coral reefs to provide ecosystem services. One Earth 4(9):1278–1285. https://doi.org/10.1016/j.oneear.2021.08.016

Elliott M, Borja Á, Cormier R (2023) Managing marine resources sustainably—ecological, societal and governance connectivity, coherence and equivalence in complex marine transboundary regions. Ocean Coast Manag 245:106875. https://doi.org/10.1016/j.ocecoaman.2023.106875

Erftemeijer PLA, Riegl B, Hoeksema BW, Todd PA (2012) Environmental impacts of dredging and other sediment disturbances on corals: a review. Mar Pollut Bull 64(9):1737–1765. https://doi.org/10.1016/j.marpolbul.2012.05.008

Erskine E, Baillie R, Lusseau D (2021) Marine protected areas provide more cultural ecosystem services than other adjacent coastal areas. One Earth 4(8):1175–1185. https://doi.org/10.1016/j.oneear.2021.07.014

Essamlali I, Nhaila H, Khaili ME (2024) Advances in machine learning and IoT for water quality monitoring: a comprehensive review. Heliyon 10(6):e27920. https://doi.org/10.1016/j.heliyon.2024.e27920

Fourqurean JW, Duarte CM, Kennedy H, Marbà N, Holmer M, Mateo MA, Apostolaki ET, Kendrick GA, Krause-Jensen D, McGlathery KJ, Serrano O (2012) Seagrass ecosystems as a globally significant carbon stock. Nat Geosci 5(7):505–509. https://doi.org/10.1038/NGEO1477

Gallardo B, Bacher S, Barbosa AM, Gallien L, González-Moreno P, Martínez-Bolea V, Sorte C, Vimercati G, Vilà M (2024) Risks posed by invasive species to the provision of ecosystem services in Europe. Nat Commun 15:2631. https://doi.org/10.1038/s41467-024-46818-3

Gardner TA, Côté IM, Gill JA, Grant A, Watkinson AR (2005) Impacts of hurricanes on coral reefs: a review of recent literature. Coral Reefs 24(3):400–417

Garner KL, Chang MY, Fulda MT, Berlin JA, Freed RE, Soo-Hoo MM, Revell DL, Ikegami M, Flint LE, Flint AL, Kendall BE (2015) Impacts of sea level rise and climate change on coastal plant species in the central California coast. PeerJ 3:e958. https://doi.org/10.7717/peerj.958

Geist EL, Titov VV, Arcas D, Pollitz FF, Bilek SL (2007) Implications of the 26 December 2004 Sumatra–Andaman earthquake on tsunami forecast and assessment models for great subduction-zone earthquakes. Bull Seismol Soc Am 97(1A):249–270. https://doi.org/10.1785/0120050619

Gonçalves EJ (2023) Marine protected areas as tools for ocean sustainability. In: Garcia MDG, Cortês A (eds) Blue planet law. Sustainable development goals series. Springer, Cham. https://doi.org/10.1007/978-3-031-24888-7_11

Great Barrier Reef Marine Park Authority (2011) Impacts of tropical cyclone Yasi on the Great Barrier Reef: a report on the findings of a rapid ecological impact assessment. GBRMPA, Townsville

Guerry AD, Silver J, Beagle J, Wyatt K, Arkema K, Lowe J, Hamel P, Griffin R, Wolny S, Plane E, Griswold M, Papendick H, Sharma J (2022) Protection and restoration of coastal habitats yield multiple benefits for urban residents as sea levels rise. npj Urban Sustain 2:13. https://doi.org/10.1038/s42949-022-00056-y

Haddad NM, Brudvig LA, Clobert J, Davies KF, Gonzalez A, Holt RD, Lovejoy TE, Sexton JO, Austin MP, Collins CD, Cook WM, Damschen EI, Ewers RM, Foster BL, Jenkins CN, King AJ, Laurance WF, Levey DJ, Margules CR, Melbourne BA, Nicholls AO, Orrock JL, Song DX, Townshend JR (2015) Habitat fragmentation and its lasting impact on earth's ecosystems. Sci Adv 1(2):e1500052. https://doi.org/10.1126/sciadv.1500052

Hall-Spencer J, Harvey BP (2019) Ocean acidification impacts on coastal ecosystem services due to habitat degradation. Emerg Top Life Sci 3(2):197–206. https://doi.org/10.1042/ETLS20180117

Halpern BS, Longo C, Lester SE (2019) Human impacts on coastal ecosystems: the role of ecological and socioeconomic factors. Annu Rev Environ Resour 44:95–122. https://doi.org/10.1146/annurev-environ-101718-033119

Hasselmann K (1991) Ocean circulation and climate change. Tellus A Dyn Meteorol Oceanogr 43(4):82–103. https://doi.org/10.3402/tellusa.v43i4.11939

Hemminga MA, Duarte CM (2000) Seagrass ecology. Cambridge University Press

Henseler C, Nordström MC, Törnroos A, Snickars M, Pecuchet L, Lindegren M, Bonsdorff E (2019) Coastal habitats and their importance for the diversity of benthic communities: a species- and trait-based approach. Estuar Coast Shelf Sci 226:106272. https://doi.org/10.1016/j.ecss.2019.106272

Hernández-Blanco J, Costanza R, Chen H, deGroot D, Jarvis D, Kubiszewski I, Montoya J, Sangha K, Stoeckl N, Turner K, van't Hoff V (2022). Ecosystem health, ecosystem services, and the well-being of humans and the rest of nature. Glob Change Biol 1–14. https://doi.org/10.1111/gcb.16281

Hernández-Delgado EA (2024) Coastal restoration challenges and strategies for small island developing states in the face of sea level rise and climate change. Coasts 4(2):235–286. https://doi.org/10.3390/coasts4020014

Herrera-Silveira JA, Teutli-Hernandez C, Secaira-Fajardo F, Braun R, Bowman J, Geselbracht L, Musgrove M, Rogers M, Schmidt J, Robles-Toral PJ, Canul-Cabrera JA, Guerra-Cano L (2022) Hurricane damages to mangrove forests and post-storm restoration techniques and costs. The Nature Conservancy, Arlington, VA. https://www.nature.org/content/dam/tnc/nature/en/documents/TNC_MangroveHurricaneDamages_Report.pdf

Hesp PA (2002) Foredunes and blowouts: initiation, geomorphology and dynamics. Geomorphology 48(1–3):245–268. https://doi.org/10.1016/S0169-555X(02)00184-8

Hoegh-Guldberg O, Mumby PJ, Hooten AJ, Steneck RS, Greenfield P, Gomez E, Harvell CD, Sale PF, Edwards AJ, Caldeira K, Knowlton N, Eakin CM, Iglesias-Prieto R, Muthiga N, Bradbury RH, Dubi A, Hatziolos ME (2007) Coral reefs under rapid climate change and ocean acidification. Science 31:1737–1742. https://doi.org/10.1126/science.1152509

Hoegh-Guldberg O, Poloczanska ES, Skirving W, Dove S (2017) Coral reef ecosystems under climate change and ocean acidification. Front Mar Sci 4. https://doi.org/10.3389/fmars.2017.00158

Hoegh-Guldberg O, Pendleton L, Kaup A (2019) People and the changing nature of coral reefs. Reg Stud Mar Sci 30:100699. https://doi.org/10.1016/j.rsma.2019.100699

Hooper DU, Chapin FS, Ewel JJ, Hector A, Inchausti P, Lavorel S, Lawton JH, Lodge DM, Loreau M, Naeem S, Schmid B, Setälä H, Symstad AJ, Vandermeer J, Wardle DA (2005) Effects of biodiversity on ecosystem functioning: a consensus of current knowledge. Ecol Monogr 75(1):3–35. https://doi.org/10.1890/04-0922

Hoppit G, Schmidt DN, Brazier P, Mieszkowska N, Pieraccini M (2022) Are marine protected areas an adaptation measure against climate change impacts on coastal ecosystems? A UK case study. Nat-Based Solut 2:100030. https://doi.org/10.1016/j.nbsj.2022.100030

Hughes TP, Baird AH, Dinsdale EA, Moltschaniwskyj NA, Pratchett MS, Tanner JE, Willis BL (2012) Assembly rules of reef corals are flexible along a steep climatic gradient. Curr Biol 22(8):1–6. https://doi.org/10.1016/j.cub.2012.02.068

IPCC (2023) Climate change 2023: the physical science basis. Intergovernmental Panel on Climate Change. https://www.ipcc.ch/report/ar6/wg1

ISRS (2004) The effects of terrestrial runoff of sediments, nutrients and other pollutants on coral reefs. Briefing paper 3. International Society for Reef Studies, pp 1–18

Jordan P, Fröhle P (2022) Bridging the gap between coastal engineering and nature conservation? J Coast Conserv 26:4. https://doi.org/10.1007/s11852-021-00848-x

Jr. Tolentino AS (2007) The challenges of tsunami disaster response planning and management. Int Rev Environ Strateg 7(1):147–154

Jung NW, Lee G, Dellapenna TM, Jung Y, Jo TC, Chang J, Figueroa SM (2024) Economic development drives massive global estuarine loss in the Anthropocene. Earth's Future 12(4):e2023EF003691. https://doi.org/10.1029/2023EF003691

Kamyab H, Khademi T, Chelliapan S, SberiKamarposhti M, Rezania S, Yusuf M, Farajnezhad M, Abbas M, Jeon BH, Ahn T (2023) The latest innovative avenues for the utilization of artificial intelligence and big data analytics in water resource management. Results Eng 20:101566. https://doi.org/10.1016/j.rineng.2023.101566

Kandasamy K (2014) Interconnectivity of coastal ecosystems: an overview. Indian J Geo-Mar Sci 43(6):985–994

Kaymaz Mühling Ş (2023) Utilizing artificial intelligence (AI) for the identification and management of marine protected areas (MPAs): a review. J Geosci Environ Prot 11:118–132. https://doi.org/10.4236/gep.2023.119008

Keller BD, Gleason DF, McLeod E, Woodley CM, Airamé S, Causey BD, Friedlander AM, Grober-Dunsmore R, Johnson JE, Miller SL, Steneck RS (2009) Climate change, coral reef ecosystems, and management options for marine protected areas. Environ Manage 44(6):1069–1088. https://doi.org/10.1007/s00267-009-9346-0

Kennish MJ (2001) Coastal salt marsh systems in the United States: a review of anthropogenic impacts. J Coast Res 17(3):731–748

Knutson T, Camargo SJ, Chan JCL, Emanuel K, Ho C, Kossin J, Mohapatra M, Satoh M, Sugi M, Walsh K, Wu L (2020) Tropical cyclones and climate change assessment. BAMS 303–322. https://doi.org/10.1175/BAMS-D-18-0194.1

Koester A, Gordó-Vilaseca C, Bunbury N, Ferse SCA, Ford A, Haupt P, A'Bear L, Bielsa M, Burt AJ, Letori J, Mederic E, Nancy E, Sanchez C, Waller M, Wild C (2023) Impacts of coral bleaching on reef fish abundance, biomass and assemblage structure at remote Aldabra Atoll, Seychelles: insights from two survey methods. Front Mar Sci 10:1230717. https://doi.org/10.3389/fmars.2023.1230717

Kriegl M, Elías Ilosvay XE, von Dorrien C, Oesterwind D (2021) Marine protected areas: at the crossroads of nature conservation and fisheries management. Front Mar Sci 8:676264. https://doi.org/10.3389/fmars.2021.676264

Kumar KCVN, Ahammed KB, Lele N, Nimmi N, Kubo S, Rao KN et al (2024) Assessing the significance of leaf chlorophyll content for measuring the vegetation health in mangrove species along the Kerala Coast, India—a multi-proxy approach. Reg Stud Mar Sci 77:103702

Lacambra CL, Spencer T, Munera C, Pizarro V, Lozano-Rivera P, Esquivel C, Cardona OD (2024) Coastal ecosystems contribution to climate adaptation and disasters risk management in the tropical Americas. Nat-Based Solut 5:100112. https://doi.org/10.1016/j.nbsj.2024.100112

Lan J, Liu P, Hu X, Zhu S (2024) Harmful algal blooms in eutrophic marine environments: causes, monitoring, and treatment. Water 16(17):2525. https://doi.org/10.3390/w16172525

Legrand D (2022) Which earthquake can trigger a volcanic eruption? J Volcanol Geotherm Res 432:107698. https://doi.org/10.1016/j.jvolgeores.2022.107698

Levin PS, Howe ER, Robertson JC (2020) Impacts of stormwater on coastal ecosystems: the need to match the scales of management objectives and solutions. Philos Trans B 375(1814):20190460. https://doi.org/10.1098/rstb.2019.0460

Lincoln S, Andrews B, Birchenough SNR, Chowdhury P, Engelhard GH, Harrod O, Pinnegar JK, Townhill BL (2022) Marine litter and climate change: inextricably connected threats to the world's oceans. Sci Total Environ 837:155709. https://doi.org/10.1016/j.scitotenv.2022.155709

Lotze HK (2021) Marine biodiversity conservation. Curr Biol 31(19):R1190–R1195. https://doi.org/10.1016/j.cub.2021.06.084

Ma X, Li J, Guo Z, Wan Z (2024) Role of big data and technological advancements in monitoring and development of smart cities. Heliyon 10(15):e34821. https://doi.org/10.1016/j.heliyon.2024.e34821

Mackenzie CL, Ormondroyd GA, Curling SF, Ball RJ, Whiteley NM, Malham SK (2014) Ocean warming, more than acidification, reduces shell strength in a commercial shellfish species during food limitation. PLoS ONE 9(1):e86764. https://doi.org/10.1073/pnas.2012025117

Majumdar D, Chakraborty S, Saha S, Datta D (2019) Geospatial analysis of the effects of tsunami on coral and mangrove ecosystems of Mayabunder in Andaman Islands, India. Braz J Oceanogr 67:e19258. https://doi.org/10.1590/S1679-87592019025806715

Malhi Y, Franklin J, Seddon N, Solan M, Turner MG, Field CB, Knowlton N (2020) Climate change and ecosystems: threats, opportunities and solutions. Philos Trans R Soc B 375(1794):20190104. https://doi.org/10.1098/rstb.2019.0104

Mawren D, Hermes J, Reason CJC (2022) Marine heat waves and tropical cyclones—two devastating types of coastal hazard in south-eastern Africa. Estuar Coast Shelf Sci 277:108056. https://doi.org/10.1016/j.ecss.2022.108056

McGranahan G, Balk D, Anderson B (2007) Enhancing coastal resilience through integrated planning and risk reduction strategies. Environ Sci Policy 10(4):271–280. https://doi.org/10.1016/j.envsci.2006.12.002

McGrane SJ (2016) Impacts of urbanisation on hydrological and water quality dynamics, and urban water management: a review. Hydrol Sci J 61(13):2295–2311. https://doi.org/10.1080/02626667.2015.1128084

McLachlan A, Brown AC (2006) The ecology of sandy shores. Elsevier. https://doi.org/10.1016/B978-0-12-372569-1.X5000-9

Miles TN, Coakley SJ, Engdah JM, Rudzin JE, Tsei S, Glenn SM (2023) Ocean mixing during Hurricane Ida (2021): the impact of a freshwater barrier layer. Front Mar Sci 10:1224609. https://doi.org/10.3389/fmars.2023.1224609

Mimura N (2013) Sea-level rise caused by climate change and its implications for society. Proc Jpn Acad Ser B Phys Biol Sci 89(7):281–301. https://doi.org/10.2183/pjab.89.281

Mittelbach GG, McGill BJ (2019) Food chains and food webs: controlling factors and cascading effects. In: Community ecology, 2nd edn. Oxford Academic, Oxford. https://doi.org/10.1093/oso/9780198835851.003.0011

Moberg F, Folke C (1999) Ecological goods and services of coral reef ecosystems. Ecol Econ 29(2):215–233

Morris JT, Sundberg K, Hopkinson CS (2013) Salt marsh primary production and its responses to relative sea level and nutrients in estuaries at Plum Island, Massachusetts, and North Inlet, South Carolina, USA. Oceanography 26:78–84. https://doi.org/10.5670/oceanog.2013.48

Munene MB, Swartling ÅG, Thomalla F (2018) Adaptive governance as a catalyst for transforming the relationship between development and disaster risk through the Sendai Framework? Int J Disaster Risk Reduct 28:653–663. https://doi.org/10.1016/j.ijdrr.2018.01.021

National Hurricane Center (2024) Tropical cyclone report: Hurricane Fiona. National Oceanic and Atmospheric Administration. https://www.nhc.noaa.gov/data/tcr/AL072022_Fiona.pdf

National Oceanic and Atmospheric Administration (2018) Tsunamis basics. https://www.noaa.gov/education/resource-collections/ocean-coasts/tsunamis

National Oceanic and Atmospheric Administration (2023a) Hurricane basics. https://www.noaa.gov/hurricane-basics

National Oceanic and Atmospheric Administration (2023b) Tsunamis dangers. https://www.noaa.gov/jetstream/tsunamis/tsunami-dangers

Neal X (2021) Effect of tropical cyclones on coastal marine ecosystems. Theses, dissertations and culminating projects. Montclair State University, 681 pp. https://digitalcommons.montclair.edu/etd/681

Nioko LIJ, Tablang KMC, Tamondong AM, Ang MRCO, Bautista DM (2023) Typhoon Karding (Noru) storm surge analysis using the COAWST modeling system. In: The international archives of the photogrammetry, remote sensing and spatial information sciences, volume XLVIII-4/W8-2023 Philippine geomatics symposium (PhilGEOS) 2023, Diliman, Quezon City, Philippines, 6–7 Dec 2023. https://doi.org/10.5194/isprs-archives-XLVIII-4-W8-2023-403-2024

NOAA (2019) Coral reef ecosystems. National Oceanic and Atmospheric Administration, U.S. Department of Commerce. https://www.noaa.gov/education/resource-collections/marine-life/coral-reef-ecosystems

Ojea E, Lester SE, Salgueiro-Otero D (2020) Adaptation of fishing communities to climate-driven shifts in target species. One Earth 2(6):544–556. https://doi.org/10.1016/j.oneear.2020.05.012

Orth RJ, Carruthers TJB, Dennison WC, Duarte CM, Fourquerean JW, Heck KL, Hughes AR Jr, Kendrick GA, Kenworthy WJ, Olyarnik S, Short FT, Waycott M, Williams SL (2006) A global crisis for seagrass ecosystems. Bioscience 56:987–996

Piccolo MC (2021) Effects of rainfall extreme events on coastal marine ecosystems. In: Precipitation. https://doi.org/10.1016/B978-0-12-822699-5.00024-0

Pothiraj C, Gokul TA, Kumar KR, Ramasubramanian A, Palanichamy A, Venkatachalam K, Pastorino P, Barcelò D, Balaji P, Faggio C (2023) Vulnerability of microplastics on marine environment: a review. Ecol Indic 155:111058. https://doi.org/10.1016/j.ecolind.2023.111058

Potts LJ, Gantz JD, Kawarasaki Y, Philip BN, Gonthier DJ, Law AD, Moe L, Unrine JM, McCulley RL, Lee RE Jr, Denlinger DL, Teets NM (2020) Environmental factors influencing fine-scale distribution of Antarctica's only endemic insect. Oecologia 194:529–539. https://doi.org/10.1007/s00442-020-04714-9

Powell EJ, Tyrrell MC, Milliken A, Tirpak JM, Staudinger MD (2019) A review of coastal management approaches to support the integration of ecological and human community planning for climate change. J Coast Conserv 23:1–18. https://doi.org/10.1007/s11852-018-0632-y

PreventionWeb (2024) Coastal communities need resilient ecosystems and infrastructure, 6 Sept 2024. PreventionWeb. https://www.preventionweb.net/news/coastal-communities-need-resilient-ecosystems-and-infrastructure

Rafliana I, Jalayer F, Cerase A, Cugliari L, Baiguera M, Salmanidou D, Necmioğlu Ö, Ayerbe IA, Lorito S, Fraser S, Løvholt F, Babeyko A, Salgado-Gálvez MA, Selva J, Risi RD, Sørensen MB, Behrens J, Aniel-Quiroga I, Zoppo MD, Belliazzi S, Pranantyo IR, Amato A, Hancilar U (2022) Tsunami risk communication and management: contemporary gaps and challenges. Int J Disaster Risk Reduct 70:102771. https://doi.org/10.1016/j.ijdrr.2021.102771

Rahman MM (2023) Habitat loss and biodiversity loss go hand in hand in Bangladesh: causes and consequences. SSRN Electron J 1–13. https://doi.org/10.2139/ssrn.4443842

Ramadan MNA, Ali MAH, Khoo SY, Alkhedher M, Alherbawi M (2024) Real-time IoT-powered AI system for monitoring and forecasting of air pollution in industrial environment. Ecotoxicol Environ Saf 283:116856. https://doi.org/10.1016/j.ecoenv.2024.116856

Sarker IH (2021) Machine learning: algorithms, real-world applications and research directions. SN Comput Sci 2:160. https://doi.org/10.1007/s42979-021-00592-x

Scheffers SR, Haviser J, Browne T, Scheffers A (2009) Tsunamis, hurricanes, the demise of coral reefs and shifts in prehistoric human populations in the Caribbean. Quat Int 195(1–2):69–87. https://doi.org/10.1016/j.quaint.2008.07.016

Schlacher TA, Schoeman DS, Dugan J, Lastra M, Jones A, Scapini F, McLachlan A (2008) Sandy beach ecosystems: key features, sampling issues, management challenges and climate change impacts. Mar Ecol 29(suppl. 1):70–90

Schumann G, Giustarini L, Tarpanelli A, Jarihani B, Martinis S (2023) Flood modeling and prediction using earth observation data. Surv Geophys 44:1553–1578. https://doi.org/10.1007/s10712-022-09751-y

Shinwari Z, Jan SA, Habib N, Abdullah A, Sher A, Afridi M, Khan M (2023) Impact of climate change on marine biodiversity: current challenges and future perspectives. Proc Pak Acad Sci B Life Environ Sci 60(1):29–47. https://doi.org/10.53560/PPASB(60-1)768

Simmons KR, Eggleston DB, Bohnenstiehl DR (2021) Hurricane impacts on a coral reef soundscape. PLoS ONE 16(2):e0244599. https://doi.org/10.1371/journal.pone.0244599

Smale DA, Burrows MT, Moore P, O'Connor N, Hawkins SJ (2013) Threats and knowledge gaps for ecosystem services provided by kelp forests: a northeast Atlantic perspective. Ecol Evol 3(11):4016–4038. https://doi.org/10.1002/ece3.774

Smyth K, Elliott M (2016) Effects of changing salinity on the ecology of the marine environment, chap 9. In: Stressors in the marine environment. Open University Press. https://doi.org/10.1093/acprof:oso/9780198718826.003.0009

Spalding M, McIvor A, Tonneijck FH, Tol S, van Eijk P (2014a) Mangroves for coastal defence. Guidelines for coastal managers & policy makers. Wetlands International and The Nature Conservancy, pp 1–42. https://www.nature.org/media/oceansandcoasts/mangroves-for-coastal-defence.pdf

Spalding MD, Ruffo S, Lacambra C, Meliane I, Hale L, Shepard C, Beck M (2014b) The role of ecosystems in coastal protection: adapting to climate change and coastal hazards. Ocean Coast Manag 90:50–57. https://doi.org/10.1016/j.ocecoaman.2013.09.007

Storlazzi C (2020) Sea-level rise and climate change impacts to reefs. Pacific Coastal and Marine Science Center. https://www.usgs.gov/centers/pcmsc/science/sea-level-rise-and-climate-change-impacts-reefs

Sun CY, Stifler CA, Chopdekar RV, Schmidt CA, Parida G, Schoeppler V, Fordyce BI, Brau JH, Mass T, Tambutté S, Gilbert PUPA (2020) From particle attachment to space-filling coral skeletons. Proc Natl Acad Sci USA 117(48):30159–30170. https://doi.org/10.1073/pnas.2012025117

Sunkur R, Kantamaneni K, Bokhoree C, Ravan S (2023) Mangroves' role in supporting ecosystem-based techniques to reduce disaster risk and adapt to climate change: a review. J Sea Res 196:102449. https://doi.org/10.1016/j.seares.2023.102449

Sweet WV, Hamlington BD, Kopp RE, Weaver CP, Barnard PL, Bekaert D, Brooks W, Craghan M, Dusek G, Frederikse T, Garner G, Genz AS, Krasting JP, Larour E, Marcy D, Marra JJ, Obeysekera J, Osler M, Pendleton M, Roman D, Schmied L, Veatch W, White KD, Zuzak C (2022) Global and regional sea level rise scenarios for the United States: updated mean projections and extreme water level probabilities along U.S. coastlines. NOAA technical report NOS 01. National Oceanic and Atmospheric Administration, National Ocean Service, Silver Spring, MD, 111 pp. https://oceanservice.noaa.gov/hazards/sealevelrise/noaa-nostechrpt01-global-regional-SLR-scenarios-US.pdf

TEEB (2023) The economics of ecosystems and biodiversity: coastal and marine ecosystems. Earthscan

Titov VV, Meinig C, Stalin S, Wei Y, Moore C, Bernard E (2023) Technology transfer of PMEL tsunami research protects populations and expands the new blue economy. Oceanography 36(2–3):186–195. https://doi.org/10.5670/oceanog.2023.205

Trégarot E, D'Olivo JP, Botelho AZ, Cabrito A, Cardoso GO, Casal G, Cornet CC, Cragg SM, Degia AK, Fredriksen SK, Furlan E, Heiss G, Kersting DK, Maréchal JP, Meesters E, O'Leary BC, Pérez G, Seijo-Núñez C, Simide R, Geest MVD, Juan S (2024) Effects of climate change on marine coastal ecosystems—a review to guide research and management. Biol Conserv 289:110394. https://doi.org/10.1016/j.biocon.2023.110394

Tuttle LJ, Donahue MJ (2022) Effects of sediment exposure on corals: a systematic review of experimental studies. Environ Evid 11:4. https://doi.org/10.1186/s13750-022-00256-0

Ullah H, Nagelkerken I, Goldenberg SU, Fordham DA (2018) Climate change could drive marine food web collapse through altered trophic flows and cyanobacterial proliferation. PLoS Biol 16(1):e2003446. https://doi.org/10.1371/journal.pbio.2003446

Underwood AJ, Chapman MG (2022) Rocky intertidal ecosystems: ecology and management. In: Reference module in life sciences. Routledge. https://doi.org/10.1016/B978-0-12-822562-2.00061-X

United Nations Environment Programme (UNEP) (2010) Ecosystem-based approaches to disaster risk reduction: an overview. UNEP. https://www.unep.org/topics/disasters-and-conflicts/disaster-risk-reduction/ecosystem-based-disaster-risk-reduction

Urabe J, Nakashizuka T (2016) Ecological impacts of tsunamis on coastal ecosystems: lessons from the Great East Japan earthquake. In: Ecological research monographs. https://doi.org/10.1007/978-4-431-56448-5

USEPA (2024) Effects of ocean and coastal acidification on ecosystems. https://www.epa.gov/ocean-acidification/effects-ocean-and-coastal-acidification-ecosystems

van Woesik R, Shlesinger T, Grottoli AG, Toonen RJ, Vega Thurber R, Warner ME, Marie Hulver A, Chapron L, McLachlan RH, Albright R, Crandall E, DeCarlo TM, Donovan MK, Eirin-Lopez J, Harrison HB, Heron SF, Huang D, Humanes A, Krueger T, Madin JS, Manzello D, McManus LC, Matz M, Muller EM, Rodriguez-Lanetty M, Vega-Rodriguez M, Voolstra CR, Zaneveld J (2022) Coral-bleaching responses to climate change across biological scales. Glob Change Biol 28(14):4229–4250. https://doi.org/10.1111/gcb.16192

Vasiliev D (2012) The role of biodiversity in ecosystem resilience. IOP Conf Ser Earth Environ Sci 1072(1):012012. https://doi.org/10.1088/1755-1315/1072/1/012012

Wallace J, Stewart L, Hawdon A, Keen R, Karim F, Kemei J (2009) Flood water quality and marine sediment and nutrient loads from the Tully and Murray catchments in north Queensland, Australia. Mar Freshw Res. https://doi.org/10.1071/MF08356

Wang X, Wang W, Tong C (2016) A review on impact of typhoons and hurricanes on coastal wetland ecosystems. Acta Ecol Sin 36(1):23–29. https://doi.org/10.1016/j.chnaes.2015.12.006

Ward D, Melbourne-Thomas J, Pecl GT, Evans K, Green M, McCormack PC, Novaglio C, Trebilco R, Bax N, Brasier MJ, Cavan EL, Edgar G, Hunt HL, Jansen J, Jones R, Lea MA, Makomere R, Mull C, Semmens JM, Shaw J, Tinch D, van Steveninck TJ, Layton C (2022) Safeguarding marine life: conservation of biodiversity and ecosystems. Rev Fish Biol Fisher 32(1):65–100. https://doi.org/10.1007/s11160-022-09700-3

Wasim M, Pandey AC, Ahammed KB, Dwivedi CS (2024) Two decades of coral bleaching in selected islands of Pacific Ocean: a holistic impact assessment. Reg Stud Mar Sci 77:103648

Western D (2001) Human-modified ecosystems and future evolution. PNAS 98(10):5458–5465. https://doi.org/10.1073/pnas.101093598

WMO (2023) State of the global climate 2023. World Meteorological Organization. https://wmo.int/publication-series/state-of-global-climate-2023

Xu W, Zhang Z (2022) Impact of coastal urbanization on marine pollution: evidence from China. Int J Environ Res Public Health 19(17):10718. https://doi.org/10.3390/ijerph191710718

Youdon C (2024) Coastal erosion and adaptation measures: an assessment. https://maritimeindia.org/coastal-erosion-and-adaptation-measures-an-assessment/

Zhang T, Du N, Geng Z, Wang S, Gao Y, Yang G, Huang X, Zhang T, Zhuang P, Zhao F (2023) Estimation of estuarine habitat degradation and its influence on the reproduction process of the crab *Eriocheir sinensis* in the Yangtze River Estuary. Ecol Process 12:59. https://doi.org/10.1186/s13717-023-00473-6

Zhou Q, Wang S, Liu J, Hu X, Liu Y, He Y, He X, Wu X (2022) Geological evolution of offshore pollution and its long-term potential impacts on marine ecosystems. Geosci Front 13(5):101427. https://doi.org/10.1016/j.gsf.2022.101427

Natural Hazards in the Coastal Zones of India, Associated Vulnerability and Risk Assessment

Jagabandhu Panda, Geo Tom, Asmita Mukherjee, Ipshita Bhasi, and Debjyoti Roy

Abstract The coastal zones are highly susceptible to the impacts of natural hazards. The topographic characteristics, societal features, and economic status define the extent of risk and exposure in a location. India has a longer coastline, and a higher percentage of the population resides in the coastal zones, which exacerbates the vulnerability to natural hazards. This chapter concisely reports various natural hazards occurring over the coastal zones, mainly emphasizing the coastal states of India. Besides, the impacts of such natural hazards are deliberated under the future climate scenarios, and the relevant contributors defining the regional risk. The eastern coastal states are revealed to be highly hazardous to the effects of tropical cyclone strikes and associated storm surge events that can lead to coastal flooding. On the other hand, the western coastal states, bounded by the Western Ghats, are severely prone to landslides and frequent severe flooding due to extreme rainfall events. The increased rate of sea-level rise and the rapid warming of the northern Indian Ocean might accelerate the risk of coastal zones to frequent and intensified tropical cyclone hazards in the future. Also, rapid urbanization, weak economic status, and higher population density further increases the possible exposure of coastal zones to risks. The comprehensive deliberations encompassed in this chapter could aid in framing disaster management strategies to mitigate the risks and enhance resilience.

Keywords Coastal hazards · Climate change · Risk · Vulnerability

1 Introduction

Coastal hazards endanger communities, infrastructure, and coastal habitats. These dangers include floods, tsunamis, storm surges, coastal erosion, sea-level rise, and flooding, which may have disastrous effects on coastal regions. Coastal development, population expansion, and climate change are some causes of coasts' vulnerability to

---

J. Panda (✉) · G. Tom · A. Mukherjee · I. Bhasi · D. Roy
Department of Earth and Atmospheric Sciences, National Institute of Technology Rourkela, Rourkela, Odisha 769008, India
e-mail: jagabandhu@gmail.com

K. K. Basheer Ahammed et al. (eds.), *Monitoring Coastal Marine Ecosystems Under Climate Change*, https://doi.org/10.1007/978-3-032-09137-6_12

these threats. Protecting coastal ecosystems, local economies, and residents' safety depends on understanding and effort to mitigate these threats. These hazards are extremely dangerous for India's eastern and western coasts, especially when it comes to storm surges from cyclones and the resulting floods in coastal cities that are major economic centers. Low-lying coastal zones are particularly susceptible to flooding and active processes like shoreline erosion and sediment deposition, which can have profound implications for local populations. In many parts of the world, these challenges are aggravated due to rapidly expanding populations, which frequently lead to unplanned or poorly considered development in areas highly vulnerable to natural hazards (Szlafsztein and Sterr 2007). The study by Bird (1985) documented that ~ 70% of the countries are suffering from coastal erosion, and it is mostly profound in Brazil (around 30%). Leatherman et al. (2000) determined that the long-term recession of sandy shorelines occurs at a rate approximately 150 times greater than that of sea-level rise. This phenomenon is especially pronounced along soft coastlines composed of chalk and clay, and coral reef systems are highly susceptible to erosion driven by rising sea levels. In addition to directly contributing to coastal erosion, sea-level rise indirectly exacerbates erosion by altering the dynamics of lagoons and tidal basins, further intensifying the degradation of coastal zones.

The warming of the earth can significantly increase the risks of meteorological hazards in the coastal zones as increase of sea surface temperature can supply more energy for cyclogenesis and can supply more moisture for the formation of intense thunderstorm activities in the coastal cities (Mukhopadhyay et al. 2012). Different climatic phenomena like El-Niño southern oscillation (ENSO), Quasi-biennial Oscillation (QBO), Madden Jullien Oscillation (MJO), and Indian Ocean Dipole (IOD) can further accelerate the storms' frequency (Felton et al. 2013; Zhao and Li 2019; Wahiduzzaman et al. 2019, 2022; Panda et al. 2023, 2025) over the coastal zones and can cause coastal flooding, which can affect the life in the coastal cities and neighborhood significantly. An important review by McInnes et al. (2016) highlighted the substantial rise of sea levels in the coming years, posing a significant risk of storm surges to the coastal habitats.

Considerable research has already been done along the Bay of Bengal (BoB), and Arabian Sea (AS) coast for tropical cyclones, storm surges, coastal flooding, alteration of sea-level rise, coastal erosion, and tsunamis (Dube et al. 2006; Jain et al. 2006; Kumar et al. 2008). INCCA (2010) utilized the 23 years of data from 1987 to 2010 at Chennai tidal station and found that the sea level is rising at a rate 1.3 mm/year along the eastern coast of India. The Great Sumatra Earthquake (Mw 9.3) caused severe storm surge along the Tamil Nadu coast and the flooding reached several hundred kilometers inland, severely affecting the Nagapattinam and Cuddalore area. In the Cuddalore district, the maximum run-up height was 2.5–3.3 m, and the flooding distance was 330–1360 m (Subramanian 2006).

Indian coastal planning is governed by the coastal regulation zone (CRZ) notification under the Environmental Act of 1986, which was further modified in 2011. The CRZ was first declared in 1991 and was defined as the region between the low tide line (LTL) and the high tide line (HTL) that extends up to 500 m in the interior and the intertidal zone. Specific activities were limited within this specified zone, while others were permitted subject to restrictions based on predetermined distance limits and their closeness to the beach (Saxena et al. 2013).

Many Indian studies suggest that remote sensing and GIS techniques can be used considerably to assess coastal vulnerability over the Indian coastal cities to avoid further casualties (Mujabar and Chandrasekar 2011; Kumar and Kunte 2012; Jana and Bhattacharya 2013; Kumar et al. 2023). Also several approaches, viz., statistical techniques, multi-criteria decision making (MCDM), machine learning (ML) frameworks are generally utilized for the identification and assessment of risk over the coastal zones. The following sections provides a discussion on the vulnerability and risks over the coastal zones of India by considering the physical, societal, and economic perspectives.

2 Natural Hazards in Coastal Zones of India

The Indian subcontinent is characterized by a triangular landmass surrounded by the Indian Ocean, which encompasses two significant sub-basins, including the Arabian Sea (AS) and the Bay of Bengal (BOB). These regions are recognized as breeding grounds for cyclonic activities and play a fundamental role in moisture transport, impacting multiple states (Basheer Ahammed et al. 2023). Consequently, several coastal states are exposed to various natural hazards, including extreme rainfall events, floods, tropical cyclone landfalls, and storm surges. The subsequent sections offer a detailed consideration of the impacts of these hazards on terrestrial and marine ecosystems, socio-economic conditions, infrastructure, and urban areas.

2.1 Tropical Cyclones

Tropical cyclones (TCs) develop over warm ocean surfaces, supported by several favorable atmospheric conditions, including low vertical wind shear, reduced sea-level pressure, high humidity in the lower and middle troposphere, and increased atmospheric instability over both the AS and BOB. Significantly higher cyclonic disturbances (CDs) are hoisted by BOB compared to AS (TCs are intensified CDs) leading to more landfalls in the eastern coast compared to the western coasts of India (Fig. 1). These cyclones are characterized by strong winds and intense rainfall, causing significant damage at landfall. Studies by Mohanty et al. (2012), Roxy et al. (2014), Singh et al. (2019a, 2019b), Balaji et al. (2018), and Deshpande et al. (2021) have shown a shift in TC trends due to changing climate conditions over the NIO, resulting in increased cyclone activity, including larger and more severe storms (Kumar et al. 2021, 2023; Basheer Ahammed and Pandey 2021).

Singh et al. (2019a, 2019b) conducted an explicit analysis of climatological track variation and land falling trends of NIO CDs during a changing climate scenario. The results highlighted that regions which are most susceptible to landfalls due to CDs

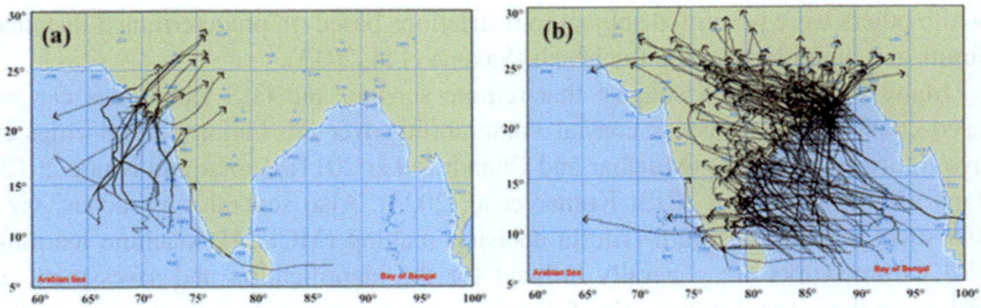

Fig. 1 Track of the cyclonic disturbances making landfall over the western coast (**a**) and eastern coast (**b**) of India during 1980–2023. The figures are produced from the India Meteorological Department (IMD) cyclone e-atlas (http://www.rmcchennaieatlas.tn.nic.in)

differ between both pre- and current-warming scenarios. During the current-warming period (CWP), Andhra Pradesh (AP), Tamil Nadu (TN), and West Bengal (WB) are most frequently impacted by severe cyclonic storms (SCS) in the BOB, while Gujarat and the Arabian Peninsula are the primary targets in the AS side. Seasonal analysis indicates AP, TN, and WB are at greater risk during the post-monsoon months. A significant proportion of CS in the AS dissipates over the ocean (approximately 61%) or recurves. The zonal increase in sea surface temperature (SST) from west to east promotes eastward movement of CDs more effectively than meridional SST gradients, resulting in a higher frequency of SCS category CDs making landfall in Gujarat. Besides the impact of landfalls, heavy rainfall accompanied with TCs impacts the coastal and neighboring areas, the seasonal variation and distribution. Singh et al. (2020) conducted an analysis of rainfall contributions and distribution from CDs across India's coastal states highlighting higher average annual rainfall received by the eastern coastal regions during the pre-monsoon months. However, the accumulated rainfall contributions and distribution are significantly higher in the interior coastal areas during the post-monsoon season. AP consistently receives over 100 mm of CD contributed rainfall annually during both seasons, followed by north-eastern TN, Gujarat (~ 70%), western Rajasthan, and coastal Maharashtra. The rainfall contributions by CDs may help positively in the agriculture over the inland areas; however, the coastal zone would often experience floods. Besides, the significant winds would bring in property losses and damages along the coastal zones too.

2.2 Storm Surges

The advancement of a TC toward the coastline induces a rapid sea-level rise, also known as storm surge, leading to sudden momentous flooding, which significantly impacts coastal communities, both environmentally and infrastructural manner. Thus, accurate prediction and implementation of an early warning system is essential for mitigating the potential damage. Several studies (Sahoo and Bhaskaran 2019; Rao et al. 2020a, 2020b) conducted a comprehensive analysis regarding storm surges and associated inundation characteristics that are impacted by factors such as wind speed, beach slope, and angle of incidence. The analysis concerning the power dissipation index (PDI) and storm surge height revealed a positive correlation between increasing cyclone intensity and surge height in recent decades. The analysis by Sahoo and Bhaskaran (2019) also indicated that across different coastal states, a positive correlation trend is observed between the continental shelf width and slope with the storm surge height. The low altitude central zones of the coastal state are particularly vulnerable to inland water surges. The existence of prominent riverine systems exacerbates the probability of flooding in the central area of several eastern coastal states. Besides, a decrease in continental shelf further amplifies surge height, leading to a greater risk of extensive inundation.

During the storm surge, the cumulative effects of high tides and sea-level rise cause flooding in river deltas, which penetrate upstream into the rivers. Simultaneously, excess water from heavy rainfall associated with the cyclone attempts to flow seaward through the rivers, aggravating the flooding. The study by Rao et al. (2020a, 2020b) assessed the combined impact of storm surges along the east coast of NIO adopting the ADvanced CIRCulation modeling system (ADCIRC) (Luettich et al. 1992). Focusing on TN, AP, WB, and Odisha, the results identified the river deltaic regions of Godavari, Hooghly, and Mahanadi as highly susceptible to elevated water levels and extensive inundation. In extreme climate change scenarios, districts such as Jagatsinghpur, Kendrapara, and Bhadrak could experience water levels exceeding 9–11 m. Also, the coastal WB has been identified as the most vulnerable area, with inundation extending up to 100–120 km inland. However, during the changing climate scenario, northern Odisha would experience significant inundation, with a projected 5–8% increase in the inundation extent.

2.3 *Tsunami*

Tsunami waves are generated by a sudden vertical displacement of the ocean floor, which causes the water column to rise or fall abruptly. This phenomenon resembles the disturbance created when an object is rapidly submerged in water. Tsunamis are most commonly triggered by seismic activity, such as earthquakes, as a consequence of tectonic plates movements. If tectonic plates display horizontally across one another without vertical displacement, the overlying water column does not experience the strong vertical impulse required to generate a tsunami. However, if the plates move vertically, the water surface is either elevated or depressed, leading to the formation of a tsunami. These types of movements occur most frequently near subduction zones, where one tectonic plate is forced beneath another. Unlike surface waves generated by wind, tsunami waves involve the movement of the entire water column, propagating from the point of origin. In deep ocean waters, nearly at a depth of 5000 m, a tsunami can travel unimpeded at speeds reaching up to 800 kmph.

Fig. 2 The giant wave surges of tsunami caused inundation and damages at Marina Beach, Tamil Nadu, on December 26, 2004. The image is retrieved from *The Atlantic news article* (https://www.theatlantic.com/photo/2014/12/ten-years-since-the-2004-indian-ocean-tsunami/100878/)

In the year 2004, a tsunami was triggered by the Sumatra–Andaman Earthquake (with magnitude 9.1). It caused calamitous damage across countries surrounding the NIO basin, resulting in nearly 200,000 fatalities and economic losses estimated at $575 million (USGS Earthquake Hazards Program; Murty et al. 2006). The seafloor near the earthquake epicenter was elevated by several meters, displacing the overlying water column and triggering the tsunami. The tsunami caused widespread devastation to infrastructure, houses and buildings, crops, fishery, and several assets across countries, including India, Sri Lanka, Indonesia, and Thailand. Its effects also reached the eastern coastal areas of Africa. The 2004 tsunami significantly impacted inhabitants and natural habitat of coastal states such as Pondicherry, TN, AP, Kerala, and the islands of Andaman and Nicobar (e.g., Fig. 2).

2.4 *Periodic Flooding*

Flooding is primarily caused by storm surges, tsunamis, and extreme rainfall, which cause river delta regions to overflow, leading to inundation of surrounding areas and significant economic and livelihood losses. Coastal regions with flat topography, shallow shelves, bays, and estuaries are particularly susceptible to flooding. The southern coastal states of India consist of topography and bathymetric features, prone to periodic flooding during the monsoon season due to increased convective activity and multiple heavy rain spells. During monsoon, the strong westerlies prevail over the south Indian peninsula due to the low pressure developed over the north-eastern BOB. This enables the massive moisture transport from AS to Indian landmass in the form of rain. Roxy et al. (2017) documented that there has been a positive trend in heavy to extremely heavy precipitation events across the Indian mainland in recent decades. The precipitation pattern exhibits significant regional variability, which can be attributed to differences in cloud formation mechanisms across India (Vijaykumar et al. 2017).

In the central Indian plains, cloud formations typically reach significant depths, whereas clouds over the Western Ghats and much of southern Peninsular India tend to be shallower. India's west coast is highly susceptible to flooding, triggered by medium to heavy precipitation following a prolonged wet period, accompanied by extreme occurrences such as cloudbursts that deliver substantial rainfall in a short period of time. The resulting heavy runoff often exceeds river capacities, and the sloping topography from the Western Ghats to the low-lying coastal areas accelerates the floodwaters, making several coastal states prone to periodic flooding. This was evident in Kerala, where multiple heavy rain spells, with a cumulative rainfall of more than 1250 mm, led to floods across various districts (Fig. 3; Vijaykumar et al. 2017).

Rapid industrialization and urbanization have led to the increased concentration of the population in megacities like Mumbai, Kolkata, Chennai, Vishakhapatnam, etc. (Rath et al. 2022; Mukherjee and Panda 2024; Mukherjee et al. 2024). This demographic explosion further enhances the vulnerability of these urban conglomerations mainly during the monsoon extreme rainfall events triggering floods. Besides the monsoon cloud burst event, the coastal cities are further prone to flooding during cyclonic landfall. TC Michaung, which formed as a low-pressure system in the Gulf of Thailand within the NIO, on December 2, 2023, gradually developed and progressed

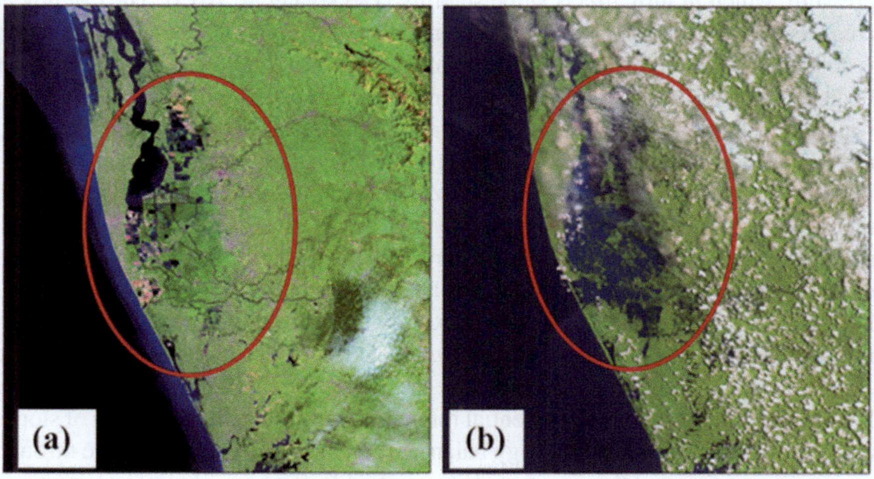

Fig. 3 Satellite images illustrating the extent of inundation across the coast of Kerala, before (**a**) and after (**b**) the 2018 flood. The images are adopted from the National Aeronautics and Space Administration (NASA) archives (https://earthobservatory.nasa.gov/images/92669/before-and-after-the-kerala-floods)

toward the north west of BOB reaching the east coast of India and poured heavy rainfall in Chennai and AP (Radhakrishnan et al. 2024). The coastal areas experienced both heavy rainfall and strong winds, which induced extensive flooding in Chennai. About 15 casualties were reported and numerous individuals were relocated in both TN and AP (https://www.bbc.com/news/world-asia-67617727).

2.5 *Landslides*

Coastal states located at the foothills of the Western Ghats are highly susceptible to landslides, particularly during the monsoon season when extreme precipitation events occur. The factors contributing to landslides in these regions were carefully selected based on the specific features. These factors encompass geological characteristics such as lithology, proximity to lineaments, and soil texture. Slope stability is further influenced by the degree of weathering and soil strata thickness, which are primarily determined by the underlying rock. Additionally, factors such as slope angle, profile curvature, plan curvature, and slope aspect, play a crucial role. Curvature, in particular, affects parameters like flow velocity, erosion, deposition, and runoff concentration, all of which significantly impact slope stability (Ayalew et al. 2004). Hydrological factors influencing slope stability include the topographic wetness index (TWI), stream power index (SPI), and proximity to streams. TWI, which depends on both the upslope contributing area and local slope, determines the concentration of surface runoff and the level of soil moisture (Beven and Kirkby 1979). Consequently, the concentration of runoff and soil moisture levels have a direct impact on pore-water pressure, which in turn affects slope stability (Devkota et al. 2013). The catchment area, also called a drainage basin or watershed, is the region where all precipitation is collected and it drains into a common outlet, such as a river, stream, or other water bodies. When both the catchment area and slope gradient increase, the volume of water contributed by the upslope region and the velocity of the flowing water also rises. This increased water flow leads to erosion and saturation of the slope materials in the lower reaches, which can destabilize the slope. Given that most landslides in the region are "debris flows," triggered by ephemeral streams during heavy rainfall, streams play a significant role in the occurrence of landslides in this area (Kuriakose et al. 2008). Besides, climatic and environmental factors, including rainfall, NDVI, land use and land cover (LULC), and proximity to roads, play a crucial role in determining slope stability. The foothill regions, situated in a monsoon-dominated climatic zone, when experience high-intensity, but short-duration rainfall events, may trigger landslides. Thus, vegetation is essential for maintaining the natural stability of slopes and altering slope natural hydrology balance, as plants extract soil moisture through evapotranspiration and their roots and stems provide mechanical reinforcement and support.

3 Vulnerability and Risks Under Climate Change Scenarios

Globally, the warming climate poses a significant threat to the population residing in the coastal and near-coastal zones. These zones are highly exposed and susceptible to the impacts of extreme events than the inland regions. The increase in air and ocean temperatures, ocean acidification, and sea-level rise owing to anthropogenic activities accelerates and fuels extreme weather events. Subsequently, it increases the magnitude, frequency, and spatial extent of extreme events, in effect, making the population more vulnerable to the hazards (Gallina et al. 2016). However, the coastal zones that share boundaries with the ocean are expected to experience maximum impacts from climate change due to the compound effects of multiple hazards. For instance, rather than extreme weather events, several factors, including coastal erosion, coastal inundation, saltwater intrusion, etc., make the zones more prone to the upcoming impacts of hazardous events. The major factors that increase the impacts and risks of coastal zones under the changing climate scenarios are discussed as follows.

3.1 Sea-Level Rise

Sea-level rise (SLR) is considered to be the most significant consequence of climate change. The low-lying coastal zones are expected to suffer from the impacts of SLR. The Indian sub-continent has a higher population density, and a large coastline of ~ 7516 km, which increases the risk exposure. Besides, it is also reported that the SLR over the Indian coast is rising at a higher rate than the global average (WMO 2022). In the NIO basin, there is a linear rise in the sea level at a rate of 0.31 mm/year between 1958 and 2000 (Thompson et al. 2007; Rabbani et al. 2010); however, for the near-past epoch (1977 and 2002), the north-eastern coast underwent an SLR at a rate of ~ 5.3 mm/year (Rahman et al. 2007). The projected SLR majorly leads to the inundation of millions of settlements and populations residing in the low-lying coastal zones of India. Also, Wassmann et al. (2004) unveiled that a rise in one meter of sea level leads to the inundation of ~ 5763 km^2 in India.

Globally, most coastal zones exhibit higher population growth rates and urbanization than the inland regions, which makes the situation worse (Neumann et al. 2015). The highly populated mega-urban agglomerations located in the coastal areas viz., Vishakhapatnam, and Chennai in the east and Trivandrum, Kochi, Mangalore, Mumbai, and Surat in the west are highly prone to the direct impacts of SLR. The SLR primarily leads to the inundation of land areas, saltwater intrusion into inland water bodies, and disruption of the coastal ecosystem. However, it indirectly enhances the effects and extent of storm surges and flooding. A recent study by Sahu and Mehta (2024) over the urban district of Surat revealed that more than 80 lakh population and assets worth ~ 2628 billion USD are exposed to coastal flooding in the far future due to SLR.

The projected changes in SLR under different climates via., shared socio-economic pathway (SSP) under four scenarios, i.e., SSP1-1.9, 2.6, SSP2-4.5, and SSP5-8.5, over the NIO basin are examined and depicted in Fig. 4. In the least emission scenario with a higher mitigation scheme (SSP1-1.9), the majority of the Indian coast experiences the SLR at a rate of ~ 4 mm/year in the current epoch (Fig. 4a). However, it is projected to decrease significantly in the near (Fig. 4b) and far future (Fig. 4c). Conversely, in the lower emission scenario, with a lag in mitigation scheme (SSP1-2.6), the SLR is observed to have a similar trend in the current era (Fig. 4d) but tends to rise in the near (Fig. 4e) and far future (Fig. 4f). Following that, the two emission pathways without the mitigation schemes (SSP2-4.5 and SSP5-8.5) show a significantly higher rate of SLR along the Indian coast. The prior scenario (SSP2-4.5) displays a greater rate of SLR, which exceeds 5 mm/year and 7 mm/year in the near (Fig. 4d) and far future (Fig. 4i), respectively. In the highest emission pathway (SSP5-8.5), the SLR exceeds the prior threshold rate in the near future (Fig. 4k). However, it attains a maximum rate of > 13 mm/year in the far future (Fig. 4l) compared to the current period (Fig. 4j). It is also evident that the rate of SLR changes over the Indian coast is not constant. The north-eastern and north-western coastal zones are highly susceptible to the impacts of SLR. Hence, the impacts and vulnerability associated with the neighboring coastal zones also vary.

The temporal variability of total sea level changes along the Indian coastal zones compared to the global scenario are depicted in Fig. 5. The north-eastern coast (Fig. 5b) and the north-western coast (Fig. 5d) are expected to experience a higher SLR relative to the global scenario (Fig. 5a) in the near future. Nevertheless, the southern coast (Fig. 5c) is projected to be less prone to SLR than other coastal zones and the global scenario. A similar trend can also be observed for the far-future time scales with a higher rate of SLR. Moreover, under strict mitigation strategies (SSP1-1.9, 2.6), the rate of SLR changes is observed to be constrained. However, the exclusion of the mitigation and adaptation schemes (SSP2-4.5 and SSP5-8.5) leads to an unprecedented level of SLR, which would cause extensive impacts on the coastal and near-coastal environments.

3.2 Sea Surface Temperature

The sea surface temperature (SST) has a crucial influence on the weather system and has a dominant impact on the marine and terrestrial environment. The rise in SST affects ocean circulations, increases the intensity and frequency of TCs, causes thermal expansion that leads to SLR, and disrupts the marine ecosystem. Primarily, the Indian coastal states are highly vulnerable to the impacts of TCs, and several studies have addressed the effects of changing SST on their intensity and frequency (Singh et al. 2019a, 2019b). For instance, a significant association between SST and various parameters including PDI (Emanuel 2005) and maximum surface wind

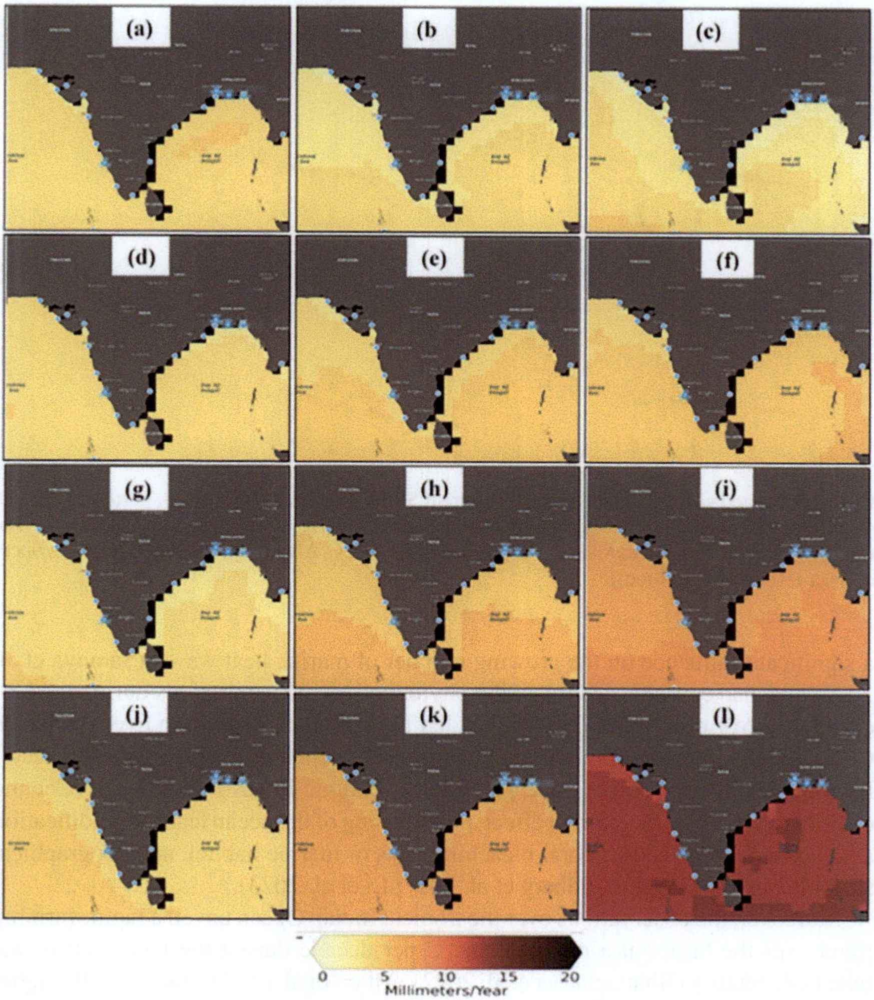

Fig. 4 Projected sea-level rise rate (mm/year) over NIO under different SSP scenarios, i.e., SSP1-1.9 (**a–c**), SSP1-2.6 (**d–f**), SSP2-4.5 (**g–i**), and SSP5-8.5 (**j–l**) for the current era-2020 (first column), near-future-2050 (second column) and far future-2100 (third column). The map is adopted from NASA archive hoisting sea level projection for IPCC-AR6 (https://sealevel.nasa.gov/ipcc-ar6-sea-level-projection-tool)

speeds of TCs (Arora and Dash 2016), and associated processes like the intensification of TCs (Foltz et al. 2018), are few of them disclosing the effects of SST in governing TC characteristics.

However, a consistent rise in SST over the tropical Indian Ocean makes the basin a breeding zone for TCs (Arora et al. 2016; Singh et al. 2019a, 2019b). Subsequently, the impacts of highly intensified TCs in the near and far future might put the coastal zones under severe threat. The higher warming trends over the Indian Ocean have

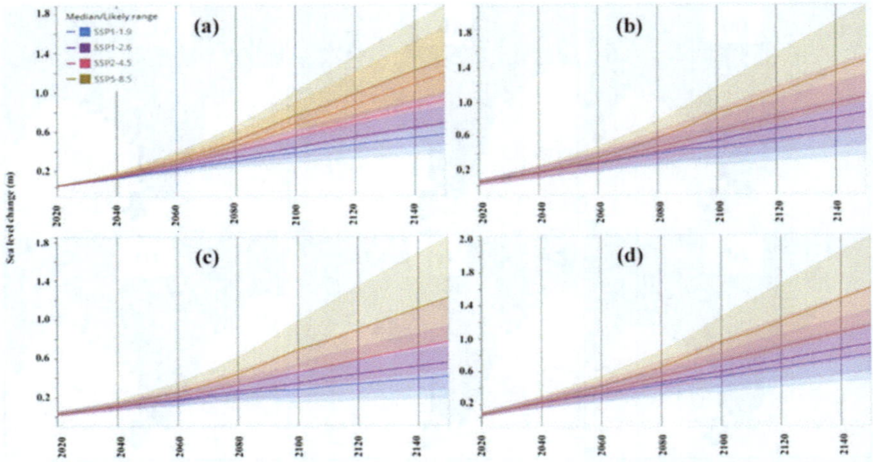

Fig. 5 Projected sea-level change under different SSP scenarios for global (**a**), north-eastern (**b**), south-eastern (**c**), and north-western (**d**) coast of India relative to a baseline of 1995–2014. The figure is adopted from NASA sea level projection for IPCC-AR6 (https://sealevel.nasa.gov/ipcc-ar6-sea-level-projection-tool)

a significant influence on the growing number of marine heat waves (Saranya et al. 2022) and also caused a decline in the phytoplankton concentrations during the past decades (Roxy et al. 2016), which has led to ecological disruption over the basin. Consequently, the coastal communities primarily depending on the fisheries sector for their livelihoods, might undergo higher economic losses in the future (Cheung et al. 2021). Also, as an indirect effect, the warming of the ocean leads to acidification and, consequently, causes coral bleaching, loss of marine habitat, and geographical shift of species (Hoegh-Guldberg et al. 2017; Li et al. 2023).

Moreover, the observations over the tropical Indian Ocean unveil a rapid warming trend over the basin, at a rate of 0.15 °C per decade during the latter half of the twentieth century (Bhattacharya et al. 2022). Albert et al. (2023) discovered a higher sub-surface warming over the AS sector in recent decades. Also, the projected climate scenarios show a higher warming trend over the north-western parts of the Indian Ocean (Naskar et al. 2024). The projected changes in the SST over the NIO basin under four scenarios are depicted in Fig. 6. The spatial changes in the SST patterns for two future time slices are examined, where the prior represents the near future (2020–2049) and the latter for the far future (2070–2099) relative to a historical period of 1970–1999.

Similar to the previous suggestions (Bhattacharya et al. 2022; Naskar et al. 2024), maximum warming has been observed over the north-western sector of the Indian Ocean. In the near-future scenarios, under the lower range of emission (Fig. 6a, c), SST shows a significantly greater warming trend over the north-western sector; however, in the higher emission scenarios (Fig. 6e, g), southern parts of NIO also show

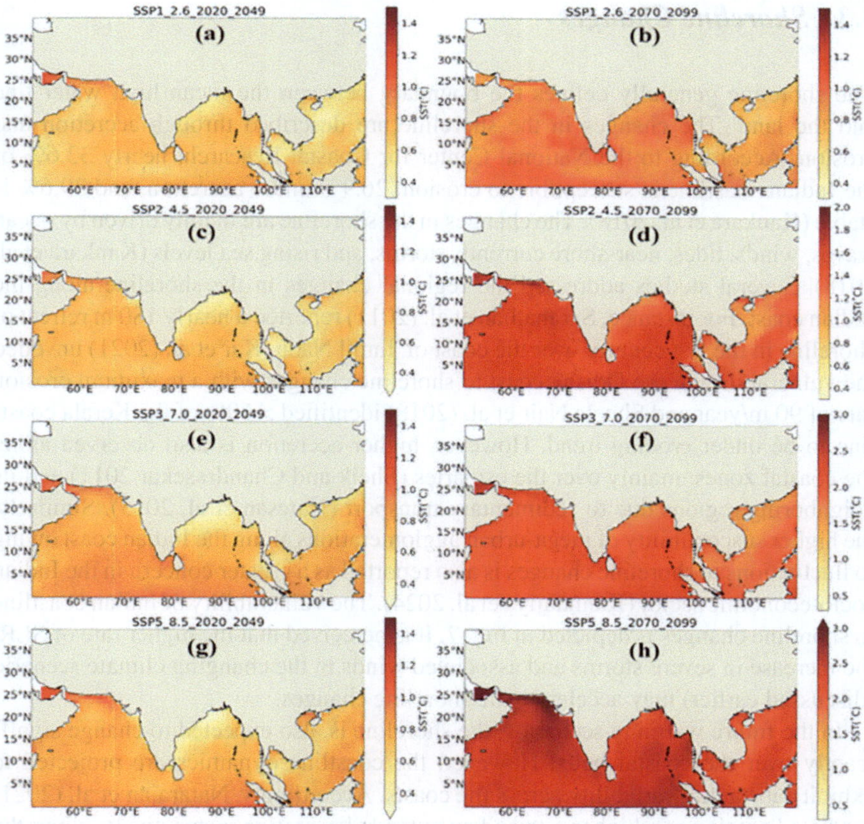

Fig. 6 Spatial distribution of future changes in SST (°C) for different scenarios: SSP1-2.6 (**a**, **b**), SSP2-4.5 (**c**, **d**), SSP3-7.0 (**e**, **f**), and SSP5-8.5 (**g**, **h**) and for the two future time slices (2020–2049 and 2070–2099) as compared to the base period (1970–1999) using IITM-ESM model

considerably higher values. Distinctively, in the far-future scenarios, with an exception to the SSP1-2.6 (Fig. 6b), a significantly higher range of warming is observed across the NIO basin, with more substantial warming over the AS (Fig. 6d, f, h). Also, in the higher concentration pathways (Fig. 6f, h), the SST increased by 3 °C compared to the historical era. It is also evident that, in the far-future scenarios, the rise in SST over the AS is relatively higher than BOB, similar to the observations of Naskar et al. (2024). Hence, the coastal zones bounded by the AS might experience more intense and frequent TCs in the near to far-future periods, making the region more hazardous.

3.3 Shoreline Changes

The shoreline generally defines the boundary between the mean high water line and the land. The changes in the shoreline are described through accretion and erosion. According to the National Center for Coastal Research, nearly 33.6% of the Indian coastline is susceptible to erosion, 26.9% under accretion, and 39.6% is stable (Kankara et al. 2018). The changes in the shoreline are mainly driven by ocean waves, winds, tides, near-shore currents, storms, and rising sea levels (Kankara et al. 2018). Several studies addressed the regional changes in the shoreline along the Indian coast. For instance, Saranathan et al. (2011) reported a nearly 180 m retreat of shoreline in the last century over the coast of Tamil Nadu, Kar et al. (2021) unveiled the vulnerability of the Odisha coast to shoreline changes with a maximum erosion rate of 90 m/year and Sheela Nair et al. (2018) identified > 60% of the Kerala coastline to be under eroding trend. However, higher accretion is also observed along the coastal zones, mainly over the estuaries (Sheik and Chandrasekar 2011) and its neighboring regions due to sedimentary transport (Natesan et al. 2015). Similarly, the higher susceptibility of mega-urban agglomerations along the Indian coast owing to fluctuations in shoreline changes is also reported as a greater concern in the Indian socio-economic sector (Kannaujiya et al. 2024). The vulnerability of Indian coastline to shoreline changes is depicted in Fig. 7. It is perceived that the higher rate of SLR, the increase in severe storms and associated winds in the changing climate scenario (discussed earlier) may accelerate the shoreline changes.

In the future warming scenarios, the shoreline is also expected to change significantly over the coastal belts. However, the coastline dynamics are projected to exhibit substantial variability across the coasts. Accordingly, Natarajan et al. (2021) predicted erosion at a higher rate and an extended accretion, respectively, along the two parts of the southern coast for the succeeding two decades by following the historical trend. Rajasree et al. (2016) indicated an accelerated erosion along the west coast in the future using the ANN framework integrated with wave simulations. However, (Gopikrishna and Deo 2018) reported an accretion rather than erosion, projected on the northern-part of the eastern coast due to the changes in sediment drift driven by wave activity in the near future. While there are changes in SLR, the shoreline is expected to retreat nearly 29 m and 45 m under RCP 4.5 and RCP 8.5 scenarios in the far future, respectively (Patil and Deo 2020) along the Mangalore coast. Thus, shoreline changes are determined to follow a heterogeneous trend significantly driven by regional features, wave characteristics, surface currents, and SLR. And it is evident that most parts of the Indian coastline is projected to be unstable either by accretion or erosion, which might lead to ecosystem disruption and infrastructural damages in the future.

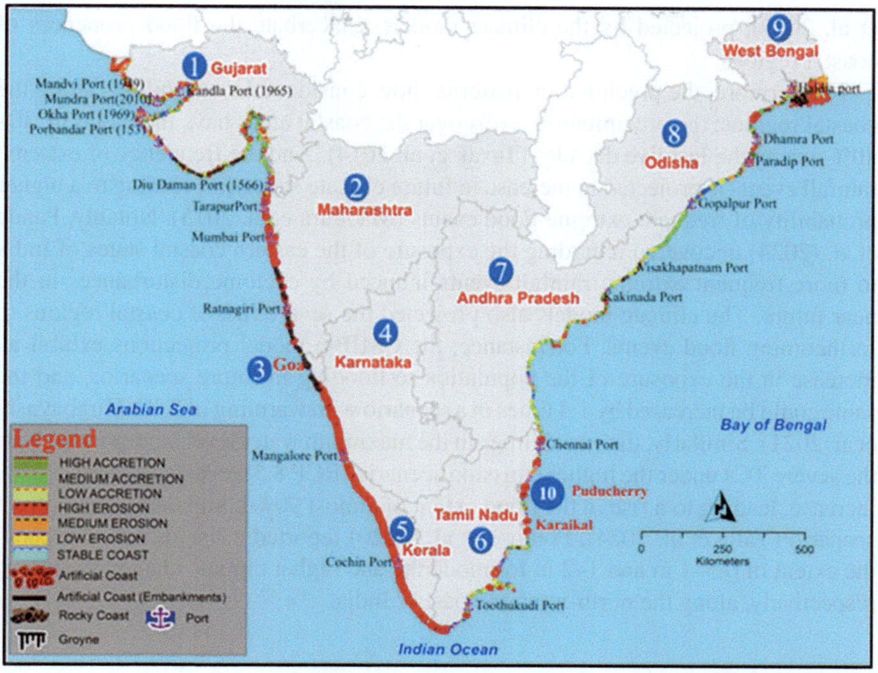

Fig. 7 Shoreline change map along the coasts of India for the period of 1990–2016. The map is adopted from the National Center for Coastal Management (https://ncscm.res.in/shoreline-change-mapping)

3.4 Coastal Flooding

The coastal zones are more prone to flooding than the hinterland due to lower elevation, proximity to the ocean, susceptibility to torrential rainfall, and storm surges. However, in the changing climate scenario, the SLR, increase in the frequency and intensity of storms, higher storm surges, variabilities in the wave directions, and surface currents (Karambas 2015) will accelerate the exposure of coastal regions to frequent flooding events. Notably, the mega-urban agglomerations located in the coastal parts of India are highly vulnerable to the impacts of floods in the future. For instance, an expected increase in the spatial extent of inundation areas in Surat city (Parikh et al. 2017), the higher exposure of a dominant percentage of the Mumbai population to floods (Ranger et al. 2011), and an increase in the susceptibility of Kolkata city to floods driven by extreme storm surges and high tides (Ghosh 2015) in the future climate change scenario, are some of the studies that highlight the vulnerability of urban areas to coastal floods. Also, an increase in the number of precipitation days, a decrease in the drainage gradient due to SLR, and an increase in extreme storm surge events and TC risks under the era of climate change (Dhiman et al. 2019), projected by the climate models, exacerbate the flood proneness of coastal regions.

Furthermore, the precipitation patterns show considerable variabilities along the coastal regions; the precipitation levels over the coastal areas have increased by 30–40% during the last five decades (Taxak et al. 2014). And the frequency of extreme rainfall events is projected to increase in future climate scenarios, leading to a higher probability of frequent extreme flood events (Madhura et al. 2015). Notably, Panda et al. (2024) uncovered regarding the exposure of the eastern coastal states of India to more frequent extreme rainfall events induced by cyclonic disturbances in the near future. The climate models also projected the severe risk of coastal regions to forthcoming flood events. For instance, the CMIP-6 model projections exhibit an increase in the exposure of the population to flooding in future scenarios, and the same could be increased by 1.4 times in a scenario with warming of 3 °C (Hirabayashi et al. 2021). Similarly, the variabilities in the maximum water level incorporated with the severe TCs under the higher emission scenario (RCP 8.5) revealed a nearly 150% increase, leading to a rise in the flood extent of almost 90% compared to the current scenario (Yalla et al. 2024). Poulose et al. (2020) reported a rise in water levels to the extent of 0.5–1 m and 1–2 m for moderate and higher climate change scenarios, respectively, along the north-western coast of India.

3.5 Ecosystem Disruption

The coastal ecosystem has a significant role in regulating the global climate by governing the carbon cycle. It comprises mangroves, salt marshes, coral reefs, seagrass beds, and coastal shelf ecosystems. They are considered the primary carbon sink over the land, and the carbon storage rates per area are considerably higher than the terrestrial ecosystem. Also, compared to the terrestrial and tropical forests, the coastal vegetation has higher sequestration rates, i.e., nearly 10 and 50 times greater than the terrestrial and tropical forests, respectively (Pidgeon 2009). Annually, this zone captures and stores more than 1.3 lakh tons of carbon worldwide (Heckbert et al. 2011). Notably, the carbon sequestration potential of the coastal ecosystems is projected to be near stable owing to the uninterrupted sedimentation. In contrast, the sequestration capacity of terrestrial ecosystems is anticipated to reduce in this century (Piao et al. 2009). Therefore, the contribution of the coastal ecosystem in conserving the carbon balance is significant, and the disruption of these zones has a higher impact on the climate system.

In the changing climate scenario, the increase in SLR, ocean acidification, increased inundation, and changes in precipitation patterns lay the ecosystem under threat. The acidification of the ocean, owing to higher CO_2 concentrations, leads to coral bleaching and, thereby, damage the coral reefs (Lough 2008). The increase in SLR leads to saltwater intrusion into the estuaries, altering the sediment influx, and frequent coastal flooding accelerates the degradation (Kirwan and Temmerman 2009) of these biodiversity hotspots. Moreover, soil salinization changes the nitrogen

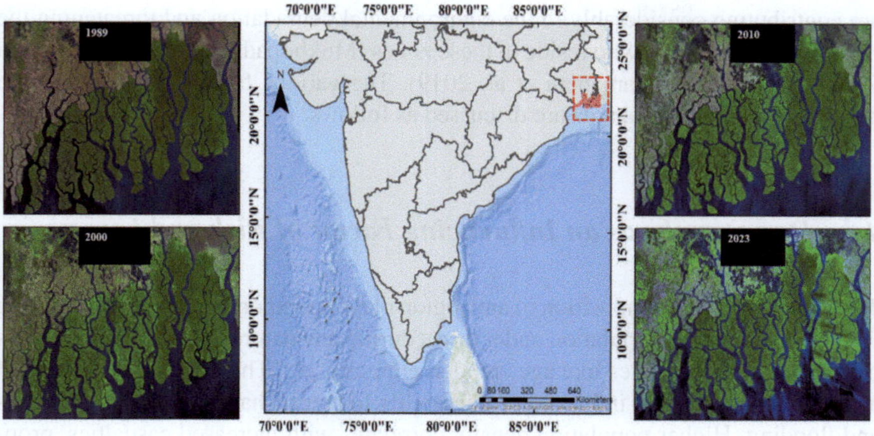

Fig. 8 The location of Sundarbans, the largest mangrove forest in the world and changes from 1989 to 2023. The satellite images are adopted from the UNDP environmental hotspots platform (https://hotspots.unepgrid.ch/)

cycle, causes the extinction of species, and enhances the dominance of salt-tolerant species, decreasing biodiversity (Noto and Shurin 2017). For instance, Raha et al. (2012) revealed the vulnerability of Sundarbans of West Bengal to the impacts of climate change; the changes in salinity significantly altered the mangrove growth and spatial variability (Fig. 8), and also the freshwater species are expected to be adversely affected in the near future.

Also, the increase in the frequency and intensity of TCs in the warming climate is a threat to the ecosystem; the direct TC strike over the coastal zones causes massive destruction and even leads to the conversion of its state (Lu et al. 2018). As mentioned earlier, the climate models projected a rapid rise in the rate of SLR, changes in the dynamics of the coastline owing to an increase in erosion and accretion, a higher number of storms, and variabilities in the tidal levels, catalyzing the vulnerability of the coastal ecosystem. Notably, the reduction in the spatial extent of these zones diminishes the carbon capture capacity and provides positive feedback to the climate system, which would further enhance the warming rates.

4 Assessing Risks and Losses

The Indian coastal regions are home to numerous beaches, ports, marshlands, dunes, mangroves, etc., supporting tourism and other economic activities. Nearly 76% of the Indian coastal regions are exposed to natural hazards and becoming increasingly vulnerable, leading to significant risks and economic losses (https://socialissuesindia.wordpress.com/2014/01/06/climatic-vulnerable-of-indias-coastal-regions/). Under the changing climate scenario, the enhanced frequency and intensity of such hazards are contributing considerably to the environmental degradation and threatening the livelihoods of millions who reside in these zones (Mukhopadhyay et al. 2012; Sudha Rani et al. 2015; Kantamaneni et al. 2019). The various factors that amplify the magnitude of coastal hazards are discussed as follows.

4.1 Demography as an Influencing Factor of Risk and Losses

The demographical composition of any region plays an important role in determining the proportion of the population under threat during any natural hazard. This includes population characteristics like age, gender, ethnicity, etc. The population density of any region significantly influences vulnerability to natural hazards, such as cyclones and flooding. Higher population density correlates with increased casualties, property damage, and loss of livelihood during such events. This trend is particularly concerning in densely populated areas, as the concentration of people and infrastructure amplifies the impact of natural disasters, resulting in more severe human and economic losses (Ahammed and Pandey 2019). Also, dense populations in coastal areas stress local health infrastructure that is crucial during and after natural disasters (Azevedo de Almeida and Mostafavi 2016). During a natural disaster, the demand for medical care spikes due to injuries, trauma, and the need for emergency services (Lane et al. 2013). Regions with well-developed health infrastructure can provide timely care, reducing mortality and morbidity rates.

Figure 9 illustrates the growth of population density (person/km^2) from 2000 to 2020, where it can be observed that coastal or near-coastal regions have attracted more migrated population from inland areas. The maximum displacement is observed in some of the major cities like Mumbai, Chennai, Kolkata, Kochi, Surat, etc., owing to better income opportunities, standard of living, and facilities, thereby increasing the vulnerability of such hazards. On the contrary, limited or negative displacement is observed within the inland regions. The population density data has been acquired from the Gridded Population of the World, Version 4 (https://doi.org/10.7927/H49C6VHW) developed by the Center for International Earth Science Information Network, Columbia University at 30 arc-second (~ 1 km at the equator) resolution (Doxsey-Whitfield et al. 2015).

The age distribution within coastal communities further affects their resilience to natural hazards. Regions with a higher proportion of elderly and children are more vulnerable to natural hazards, as these groups may have limited mobility and require special assistance during evacuations. This heightens the susceptibility for disaster mitigation or recovery (Felsenstein and Lichter 2014; Malak et al. 2020). On the contrary, a higher proportion of working-age individuals might contribute to quicker recovery post-disaster, but during the event, the economic activity disruption can lead to significant long-term impacts. The ratio of dependents (both young-0–14; and old-65+) upon the working age population data has been acquired from the WorldPop Whole Continent Age Structures data series (https://hub.worldpop.org/geodata/summary?id=1279). The relative levels of multidimensional deprivation and

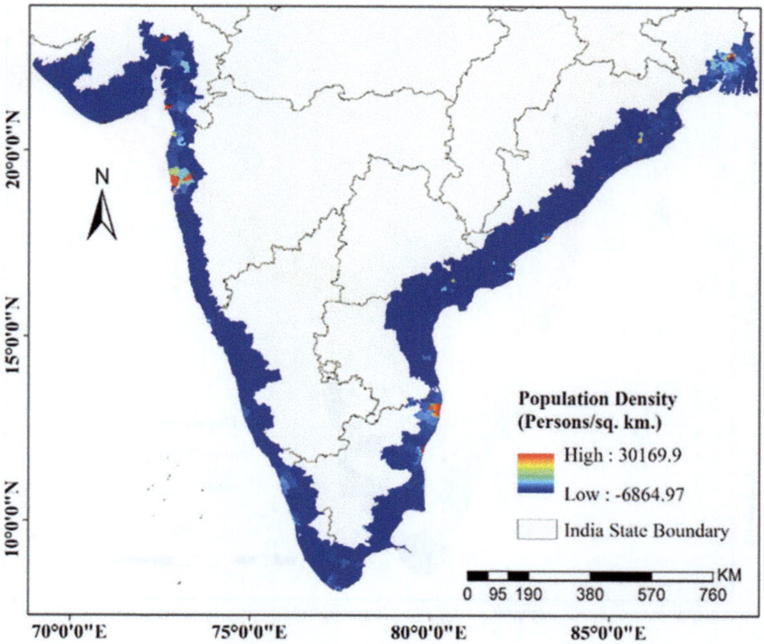

Fig. 9 Spatial variation of population density (person/km^2) growth from 2000 to 2020 over the coastal regions of India

poverty values demonstrate that relatively higher dependency ratios are observed in the states of Odisha and Maharashtra, moderate in West Bengal and Gujarat, followed by other coastal states (Fig. 10).

Socio-economic status plays a critical role in how communities prepare for and respond to natural hazards. Marginalized communities (or communities with lower income and inadequate housing) in coastal areas are particularly vulnerable to risks and losses from natural hazards due to pre-existing socio-economic disparities. These communities often lack the financial resources needed for disaster preparedness and recovery, making them more susceptible to the impacts of extreme weather events and subsequent economic hardships (Azadegan 2018). The limited access to resources, including effective disaster warning systems, further exacerbates the vulnerability of these communities. Many marginalized individuals may not receive timely information or sufficient support to evacuate or protect themselves during disasters. This lack of preparedness can lead to higher casualty rates and increased property damage within these populations (https://www.prb.org/resources/disaster-risk/). The economic consequences of natural disasters are significantly harsher for low-income families in coastal regions (Deria et al. 2020; Cappelli et al. 2021). Households that experience displacement often face mounting debts as they struggle to rebuild their lives after losing their homes and jobs. Moreover, these individuals may rely heavily on low-wage jobs, which can be lost during recovery periods, further entrenching

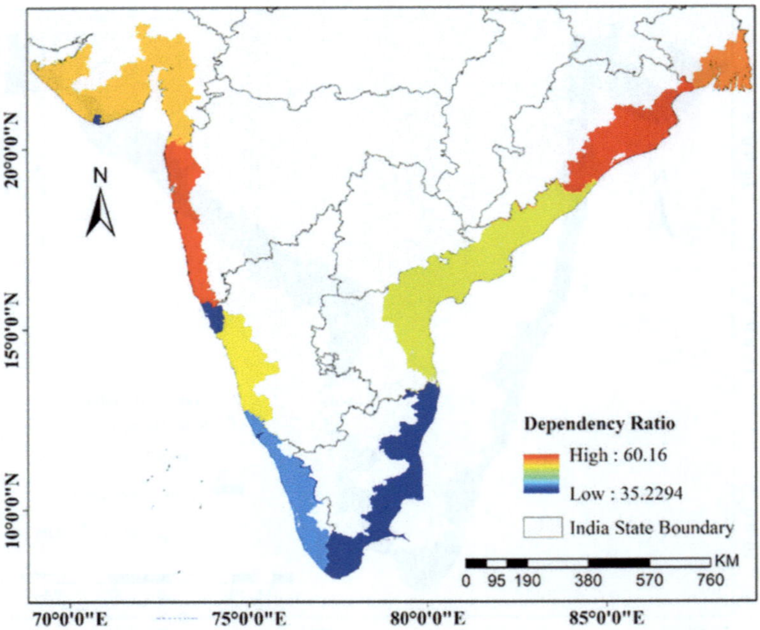

Fig. 10 Spatial variation of dependency ratio of the year 2010 over the coastal regions of India

their economic difficulties. The relative levels of multi-dimensional deprivation and poverty indicate that it is mostly low in the bigger cities like Chennai, Kolkata, Mumbai, Kochi, Thiruvananthapuram, Surat, Ahmedabad, etc. (Fig. 11). On the contrary, it is comparatively higher in other coastal regions, especially remote areas like the Sundarbans, and some parts of the Odisha and Gujarat coast. The data has been obtained from the Gridded Relative Deprivation Index, Version 1 (https://doi.org/10.7927/3xxe-ap97).

Educated individuals tend to have a better understanding of natural hazards, which can significantly reduce the number of casualties and moderate potential damages (Frankenberg et al. 2013; Cerulli et al. 2020). This awareness is vital for identifying the early signs of an approaching disaster and grasping its potential consequences, which encourages proactive measures. The ability to understand and react to information from awareness campaigns, such as the importance of evacuation orders, weather warnings, and the need for preparation, is enhanced by formal education. It also fosters a greater sense of willingness to deduce evacuation routes, track weather alerts, and accumulate emergency supplies. In fact, higher-educated communities are more likely to engage in drills and simulations concerning real-world emergencies. Besides, technological proficiency allows them to utilize online resources and weather applications and respond to emergency alerts more efficiently.

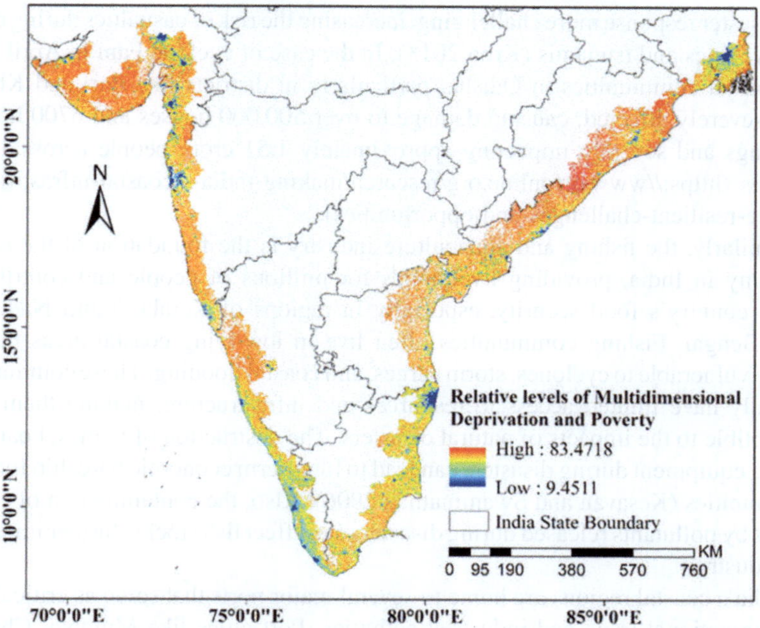

Fig. 11 Spatial variation of relative levels of multidimensional deprivation and poverty

4.2 *Economic Activities as Factors Influencing Risk and Losses*

Coastal regions of India are hubs of economic activities such as tourism, fishing, and port operations, which are vital to the local and national economy. However, these activities have intensified the pressure on the sustainability of the region and increased the vulnerability to natural hazards. India's extensive coastline, spanning over 7500 km, is characterized by vibrant economic activities that significantly contribute to the country's GDP and provide livelihoods to millions of people. Tourism is a major economic driver, attracting millions of domestic and international visitors each year to popular destinations like Goa, Kerala, Andaman, and Nicobar Islands, etc., which rely heavily on tourism for economic growth. Consequently, to accommodate the influx of tourists, extensive development has taken place, including the construction of hotels, resorts, restaurants, and recreational facilities. These developments often occur in ecologically sensitive areas, leading to the destruction of natural buffers such as sand dunes, mangroves, and coral reefs (Zahedi 2008; Saha and Paul 2021; Dhivya et al. 2023). The loss of these natural defenses increases the vulnerability of coastal areas to storm surges, coastal erosion, and flooding. Besides, tourist hotspots experience seasonal population surges, significantly increasing the number of people exposed to natural hazards. During peak tourist seasons, the dense concentration of visitors in coastal areas makes evacuation and disaster response more challenging, increasing the risk of casualties during events like cyclones and tsunamis (Kron 2013). In the case of cyclone Fani in April 2019, the coastal communities in Odisha, particularly in districts like Puri and Khurda, were severely affected, causing damage to over 500,000 houses and 6700 hospital buildings and severely impacting approximately 1.51 crore people across 16,659 villages (https://www.orfonline.org/research/making-india-s-coastal-infrastructure-climate-resilient-challenges-and-opportunities).

Similarly, the fishing and aquaculture industry is the foundation of the coastal economy in India, providing livelihoods for millions of people and contributing to the country's food security, especially in regions of Kerala, Tamil Nadu, and West Bengal. Fishing communities often live in low-lying coastal areas that are highly vulnerable to cyclones, storm surges, and coastal flooding. These communities typically have limited access to resources and infrastructure, making them more susceptible to the impacts of natural disasters. The destruction of homes, boats, and fishing equipment during disasters can lead to long-term economic hardship for these communities (Kesavan and Swaminathan 2006). Also, the contamination of coastal waters by pollutants released during disasters can affect fish stocks, further impacting the industry.

India's coastal regions are home to several major ports that serve as critical hubs for international trade and industrial activities. Port cities like Mumbai, Chennai, Kolkata, and Visakhapatnam handle a significant proportion of India's trade, and the industries clustered around these ports contribute substantially to the economy. However, in most cases, they are located in low-lying areas, thereby making them susceptible to flooding, storm surges, and cyclones (Sudha Rani et al. 2015; Sajith et al. 2024) too. The availability of valuable infrastructure and goods inside these areas increases the probability of economic losses, disrupting supply chains.

4.3 LULC Change as an Influencing Factor of Risk and Loss

The LULC pattern, particularly relating to urbanization and agriculture, plays a crucial role in modulating the impact and severity of coastal hazards. The alteration of urban areas and agricultural land in these regions has profoundly transformed the natural landscape and exacerbated the associated risks. For example, Fig. 12 exhibits LULC changes in the coastal and near-coastal districts of India during the last three decades, i.e., from 1992 to 2023. It can be observed that maximum expansion of urban areas is observed in the coastal districts of Kerala, followed by West Bengal and Gujarat. However, a significant decrease in agricultural area can be seen, much of which has transformed into built areas and bare grounds. The same can be observed in Fig. 13, too. Notably, the LULC dataset for 1992 was acquired from the European Space Agency (ESA) Climate Change Initiative (CCI) open data portal (https://www.esa-landcover-cci.org/?q=node/164). The datasets contain 29 years of annual land cover maps (Harper et al. 2023). Similarly, the LULC map of 2023 was obtained from the Environmental Systems Research Institute (ESRI) land cover portal (https://livingatlas.arcgis.com/landcover/). The ESRI datasets are available from 2017 to 2023, and are developed using Sentinel-2 satellite products. Although the two different LULC datasets used in the current study have different spatial resolutions (300 m and 10 m, respectively) and land cover classes, the ESA-CCI data was resampled to 10 m and both have been reclassified into six land use classes (i.e., water, trees and flooded vegetation, agriculture, built area, bare ground, and rangeland) to maintain uniformity.

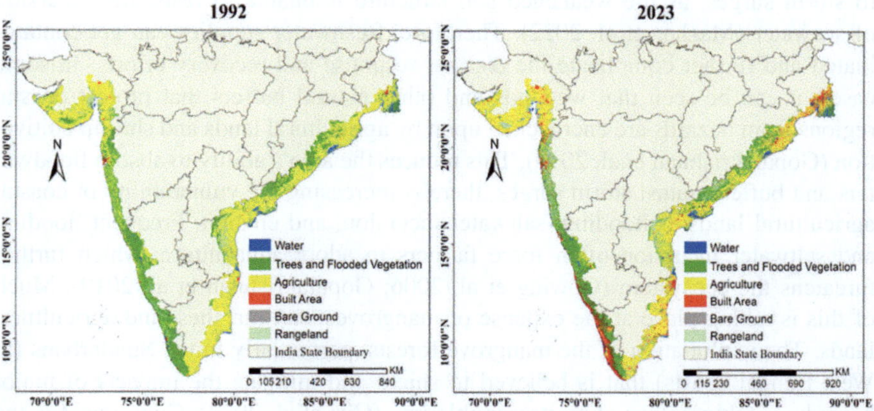

Fig. 12 LULC change from 1992 to 2023 observed in the coastal and near-coastal districts of India

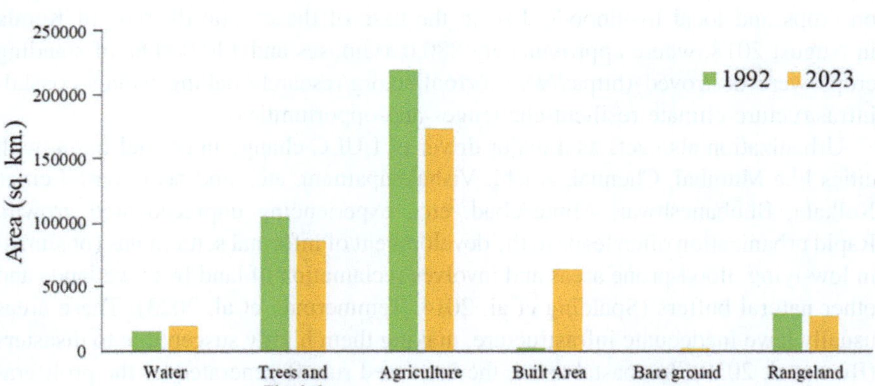

Fig. 13 Variation in land use and land cover classes between 1992 and 2023 in the coastal and near-coastal districts of India

Agricultural modification often involves the clearing of forests and vegetation, leading to deforestation and soil erosion. In coastal areas, this practice can destabilize the land, making it more prone to erosion and landslides, particularly during heavy rains and cyclonic events. Besides, the construction of irrigation channels and drainage systems can affect the natural flow of rivers and streams, leading to waterlogging, increased flood risks, and coastal erosion (https://www.fao.org/4/W4347E/w4347e10.htm). This can further aggravate due to the accretion of agricultural waste in rivers and estuaries, enhancing the severity of coastal flooding. Often intensive agriculture can cause salinization of soil and water resources. This not only degrades the fertility of the land but also increases the vulnerability of these areas to storm surges, as the weakened soil structure is unable to resist the incursion of seawater (Mazhar et al. 2022). Therefore, freshwater supplies can get contaminated and further complicate the disaster response and recovery efforts. In some cases, it can be seen that wetlands and other natural buffers that protect coastal regions from hazards are encroached upon by agricultural lands and shrimp cultivation (Gopalakrishnan et al. 2019). This reduces the area's ability to absorb floodwaters and buffer against storm surges, thereby increasing the vulnerability of coastal agricultural lands to flooding, saltwater incursion, and erosion. Frequent flooding and saltwater incursion often force farmers to adopt aquaculture, which further threatens the ecosystem (Gowing et al. 2006; Gopalakrishnan et al. 2019). Much of this is taking place at the expense of mangroves, salt marshes, and agricultural lands. The exploitation of the mangrove forests, particularly in the Sundarbans (in West Bengal coasts) that is believed to shield and mitigate the impacts of major hazards, is taking place at an irreversible rate (Giri et al. 2022). Consequently, the coastal communities in West Bengal, face the dual challenges of rising sea levels and increased cyclonic activity. Also, floods have a significant devastating impact on crops and local livelihoods, like in the case of the coastal districts of Kerala in August 2018, where approximately 280,000 houses and 140,000 ha of standing crops were destroyed (https://www.orfonline.org/research/making-india-s-coastal-infrastructure-climate-resilient-challenges-and-opportunities).

Urbanization also acts as a major driver of LULC change in coastal India, with cities like Mumbai, Chennai, Kochi, Vishakhapatnam, etc., and near-coastal cities Kolkata, Bhubaneshwar, Ahmedabad, etc., experiencing unprecedented growth. Rapid urbanization often leads to the development of informal settlements (or slums) in low-lying, flood-prone areas and involves reclamation of land from wetlands and other natural buffers (Spalding et al. 2014; Temmerman et al. 2023). These areas usually have inadequate infrastructure, making them highly susceptible to disasters (Burt et al. 2019). In coastal areas, the increased runoff generated by the proliferation of impervious surfaces exacerbates the risk of flash floods, particularly during cyclones and storm surges, as water is unable to infiltrate or drain efficiently into the sea. For instance, in Mumbai, coastal slum dwellers are particularly susceptible to the impacts of sea-level rise and intense flooding, which have become more frequent with changing climate.

4.4 Mortality as an Influencing Factor of Risk and Losses

As mentioned earlier, the coastal natural hazards including cyclones, tsunamis, storm surges, and flooding, pose significant threats to life along the vast coastal regions of India. Various factors, including population density, preparedness levels, socio-economic status, and limited availability of early warning systems influence the mortality associated with these hazards. The spatial variation of mortality risk associated with TCs, floods, landslides triggered by precipitation, and exposure to tsunamis have been downloaded from the Global Risk Data Platform hosted by the UN Environment Program GRID-Geneva. The mortality risk datasets were designed for the Global Assessment Report on Risk Reduction (GAR). The estimation of exposure to tsunami data was designed by the International Center for Geo-hazards for the GAR. The TC mortality risk suggests that the coastal and near-coastal regions of the eastern coast of India are more vulnerable (medium to high mortality risk) as compared to the western coast, where only Gujarat has medium risk (Fig. 14a). A similar scenario is observed in the case of floods (Fig. 14b). The eastern coast has a high to extreme mortality risk in most of the regions, while it is realized in only some portions of Kerala and Gujarat along the west coast. Surprisingly, a completely opposite scenario is observed in the case of landslides triggered by precipitation. The entire western coast has a higher risk than the eastern coast (Fig. 14c). This could be possibly due to the rising instability along the Western Ghats in recent years with increasing frequency and intensity of heavy rainfall events (Sreelash et al. 2018; Nathan and Ganapathy 2024). In the case of tsunamis, the physical exposure is mostly present along the eastern coastline with maximum exposure observed in the southern parts of West Bengal (Fig. 14d). This might be due to the presence of tectonic plate boundaries underneath this region, lower elevation, higher population density, etc. (Biswas and Sil 2023). Therefore, vulnerable populations, such as coastal fishing communities, urban poor, women, aged, and children, face particularly high risks. Nonetheless, to reduce mortality, it is essential to strengthen disaster management strategies that prioritize early warning systems, resilient infrastructure, community preparedness, and targeted support for vulnerable groups.

5 Concluding Remarks

This article reviews different catastrophic hazards that occurred in the coastal region of India, like tropical cyclones, tsunamis, coastal erosion, coastal flooding due to storm surges, etc., in the changing climate scenario and highlights the vulnerability of these coastal hazards in the coastal society and economy. The study highlights that changing climate has positive feedback on the cyclogenesis process in the NIO basin. Increased SST supplies more energy to tropical storms, which makes them more intensified. West Bengal, Tamil Nadu, and Andhra Pradesh on the eastern coast and Gujarat on the western coast, is particularly vulnerable to TC landfalls. Due to storm

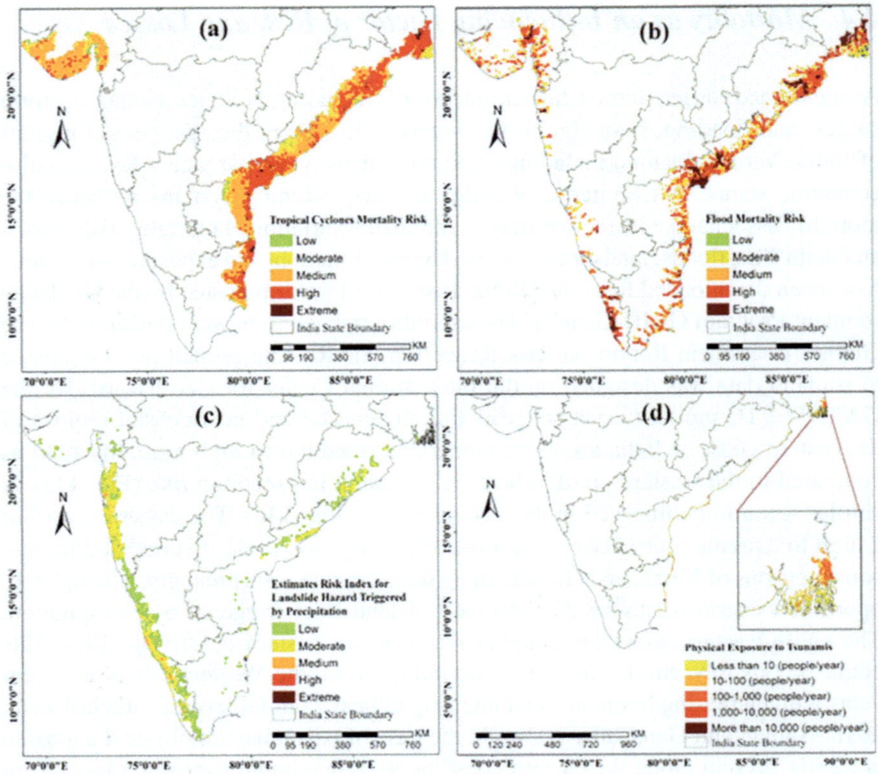

Fig. 14 Spatial variation of tropical cyclone mortality risk (**a**), flood mortality risk (**b**), landslide hazard triggered by precipitation risk (**c**), and physical exposure to tsunamis (**d**) along the coastal zones of India

surges, the river delta regions in the eastern coasts are highly in danger, and proper coastal management plans need to be undertaken at the earliest possible. The study underscores the urgent need for a high-precision tsunami early warning system to protect the coastal zones of NIO from future casualties. The CMIP6 future projections indicate a linear rise in the sea level along the coastlines of India due to the changing climate and global warming, which makes the coastal region more susceptible to flooding and highlights the urgent need for effective planning in coastal cities. The study suggests, the NIO temperature will be increased by 3 °C in the far future, which will further increase the risk of flooding exposure by ~ 1.4 times compared to the current scenario, leading to a loss of nearly 2000 billion USD in different parts of the country especially in coastal urban areas. Demographic analysis implicates that in major coastal cities like Mumbai and Chennai, where population density exceeds 20,000 per km^2, an effective coastal management plan is needed since they contribute significantly to India's economy. Moreover, rapid urbanization in coastal cities increases the risk of exposure to coastal disasters to common people, thus needs an appropriate strategy to maintain a balance between the economic advancement and losses due to natural hazards under changing climate.

References

Ahammed BKK, Pandey AC (2019) Coastal social vulnerability and risk analysis for cyclone hazard along the Andhra Pradesh, East Coast of India. KN-J Cartogr Geogr Inf 69(4):285–303

Albert J, Gulakaram VS, Vissa NK, Bhaskaran PK, Dash MK (2023) Recent warming trends in the Arabian Sea: causative factors and physical mechanisms. Climate 11(2):35

Arora K, Dash P (2016) Towards dependence of tropical cyclone intensity on sea surface temperature and its response in a warming world. Climate 4(2):30

Arora A, Rao SA, Chattopadhyay R, Goswami T, George G, Sabeerali CT (2016) Role of Indian Ocean SST variability on the recent global warming hiatus. Glob Planet Change 143:21–30

Ayalew L, Yamagishi H, Ugawa N (2004) Landslide susceptibility mapping using GIS-based weighted linear combination, the case in Tsugawa area of Agano River, Niigata Prefecture, Japan. Landslides 1:73–81

Azadegan S (2018) Dimensions of vulnerability, resilience, and social justice in a low income Hispanic neighborhood during disaster recovery. In: Natural hazards center quick response grant report series, p 284

Azevedo de Almeida B, Mostafavi A (2016) Resilience of infrastructure systems to sea-level rise in coastal areas: impacts, adaptation measures, and implementation challenges. Sustainability 8(11):1115. https://doi.org/10.3390/su8111115

Balaji M, Chakraborty A, Mandal M (2018) Changes in tropical cyclone activity in north Indian Ocean during satellite era (1981–2014). Int J Climatol 38(6):2819–2837

Basheer Ahammed KK, Pandey AC (2021) Characterization and impact assessment of super cyclonic storm AMPHAN in the Indian subcontinent through space borne observations. Ocean Coast Manag 205:105532

Basheer Ahammed KK, Pandey AC, Parida BR, Wasim, Dwivedi CS (2023) Impact assessment of tropical cyclones Amphan and Nisarga in 2020 in the northern Indian Ocean. Sustainability 15(5):3992

Beven KJ, Kirkby MJ (1979) A physically based, variable contributing area model of basin hydrology/Un modèle à base physique de zone d'appel variable de l'hydrologie du bassin versant. Hydrol Sci J 24(1):43–69

Bhattacharya B, Mohanty S, Singh C (2022) Assessment of the potential of CMIP6 models in simulating the sea surface temperature variability over the tropical Indian Ocean. Theoret Appl Climatol 148(1):585–602

Bird EC (1985) Coastline changes: a global review. Wiley, 219 pp. ISBN: 9780471906469, 0471906468

Biswas S, Sil A (2023) Tsunami vulnerability assessment and multi-criteria decision making analysis of eastern coast of India using GIS-based tools. KSCE J Civ Eng 27(3):1270–1287

Burt JA, Killilea ME, Ciprut S (2019) Coastal urbanization and environmental change: opportunities for collaborative education across a global network university. Reg Stud Mar Sci 26:100501

Cappelli F, Costantini V, Consoli D (2021) The trap of climate change-induced "natural" disasters and inequality. Glob Environ Change 70:102329

Cerulli D, Scott M, Aunap R, Kull A, Pärn J, Holbrook J, Mander Ü (2020) The role of education in increasing awareness and reducing impact of natural hazards. Sustainability 12(18):7623

Cheung WW, Frölicher TL, Lam VW, Oyinlola MA, Reygondeau G, Sumaila UR et al (2021) Marine high temperature extremes amplify the impacts of climate change on fish and fisheries. Sci Adv 7(40):eabh0895

Deria A, Ghannad P, Lee YC (2020) Evaluating implications of flood vulnerability factors with respect to income levels for building long-term disaster resilience of low-income communities. Int J Disaster Risk Reduct 48:101608

Deshpande M, Singh VK, Ganadhi MK, Roxy MK, Emmanuel R, Kumar U (2021) Changing status of tropical cyclones over the north Indian Ocean. Clim Dyn 57:3545–3567

Devkota KC, Regmi AD, Pourghasemi HR, Yoshida K, Pradhan B, Ryu IC et al (2013) Landslide susceptibility mapping using certainty factor, index of entropy and logistic regression models in GIS and their comparison at Mugling–Narayanghat road section in Nepal Himalaya. Nat Hazards 65:135–165

Dhiman R, VishnuRadhan R, Eldho TI, Inamdar A (2019) Flood risk and adaptation in Indian coastal cities: recent scenarios. Appl Water Sci 9(1):5

Dhivya S, Karthi N, Balamurugan S, Ramesh DA (2023) Valuing ecologically sensitive area's ecosystem services in Bhitharkanika: implications for sustainable management. In: Water, land, and forest susceptibility and sustainability. Academic Press, pp 97–122

Doxsey-Whitfield E, MacManus K, Adamo SB, Pistolesi L, Squires J, Borkovska O, Baptista SR (2015) Taking advantage of the improved availability of census data: a first look at the gridded population of the world, version 4. Pap Appl Geogr 1(3):226–234

Dube SK, Jain I, Rao AD (2006) Numerical storm surge prediction model for the north Indian Ocean and the south China Sea. Disaster Dev 1:47–63

Emanuel K (2005) Increasing destructiveness of tropical cyclones over the past 30 years. Nature 436(7051):686–688

Felsenstein D, Lichter M (2014) Social and economic vulnerability of coastal communities to sea-level rise and extreme flooding. Nat Hazards 71:463–491

Felton CS, Subrahmanyam B, Murty VSN (2013) ENSO-modulated cyclogenesis over the Bay of Bengal. J Clim 26(24):9806–9818

Foltz GR, Balaguru K, Hagos S (2018) Interbasin differences in the relationship between SST and tropical cyclone intensification. Mon Weather Rev 146(3):853–870

Frankenberg E, Sikoki B, Sumantri C, Suriastini W, Thomas D (2013) Education, vulnerability, and resilience after a natural disaster. Ecol Soc J Integr Sci Resil Sustain 18(2):16

Gallina V, Torresan S, Critto A, Sperotto A, Glade T, Marcomini A (2016) A review of multi-risk methodologies for natural hazards: consequences and challenges for a climate change impact assessment. J Environ Manage 168:123–132

Ghosh A (2015) Kolkata and climate change. In: Climate change policy paper IV. WWF. Available at https://awsassets.wwfindia.org/downloads/kolkata_and_climate_change.pdf

Giri S, Samanta S, Mondal PP, Basu O, Khorat S, Chanda A, Hazra S (2022) A geospatial assessment of growth pattern of aquaculture in the Indian Sundarbans Biosphere Reserve. Environ Dev Sustain 24(3):4203–4225

Gopalakrishnan T, Hasan MK, Haque AS, Jayasinghe SL, Kumar L (2019) Sustainability of coastal agriculture under climate change. Sustainability 11(24):7200

Gopikrishna B, Deo MC (2018) Changes in the shoreline at Paradip Port, India in response to climate change. Geomorphology 303:243–255

Gowing JW, Tuong TP, Hoanh CT (2006) Land and water management in coastal zones: dealing with agriculture-aquaculture-fishery conflicts. Environment and livelihoods in tropical coastal zones: managing agriculture-fishery-aquaculture conflicts. CABI, Wallingford, UK, pp 1–16

Harper KL, Lamarche C, Hartley A, Peylin P, Ottlé C, Bastrikov V et al (2023) A 29-year time series of annual 300 m resolution plant-functional-type maps for climate models. Earth Syst Sci Data 15(3):1465–1499

Heckbert S, Costanza R, Poloczanska ES, Richardson AJ (2011) 12.10–Climate regulation as a service from estuarine and coastal ecosystems. In: Treatise on estuarine and coastal science, pp 199–216

Hirabayashi Y, Tanoue M, Sasaki O, Zhou X, Yamazaki D (2021) Global exposure to flooding from the new CMIP6 climate model projections. Sci Rep 11(1):3740

Hoegh-Guldberg O, Poloczanska ES, Skirving W, Dove S (2017) Coral reef ecosystems under climate change and ocean acidification. Front Mar Sci 4:158

INCCA Indian Network for Climate Change Assessment (2010) Climate change and India: a 4 × 4 assessment—a sectoral and regional analysis for 2030s. http://moef.nic.in/downloads/public-information/fin-rpt-incca.pdf

Jain I, Chittibabu P, Agnihotri N, Dube SK, Sinha PC, Rao AD (2006) Numerical storm surge prediction model for Gujarat coast of India and adjoining Pakistan coast. Nat Hazards 42:67–73

Jana A, Bhattacharya AK (2013) Assessment of coastal erosion vulnerability around Midnapur-Balasore Coast, Eastern India using integrated remote sensing and GIS techniques. J Indian Soc Remote Sens 41(3):675–686. https://doi.org/10.1007/s12524-012-0251-2

Kankara RS, Murthy MR, Rajeevan M (2018) National assessment of shoreline changes along Indian coast: status report for 26 years (1990–2016). NCCR Publication

Kannaujiya VK, Rai AK, Malakar S (2024) Coastal shoreline change in eastern Indian metropolises. PFG J Photogramm Remote Sens Geoinf Sci 1–14

Kantamaneni K, Sudha Rani NNV, Rice L, Sur K, Thayaparan M, Kulatunga U, Rege R, Yenneti K, Campos LC (2019) A systematic review of coastal vulnerability assessment studies along Andhra Pradesh, India: a critical evaluation of data gathering, risk levels and mitigation strategies. Water 11(2):393

Kar PK, Mohanty PK, Pradhan S, Behera B, Padhi SK, Mishra P (2021) Shoreline change along Odisha coast using statistical and geo-spatial techniques. J Earth Syst Sci 130:1–20

Karambas TV (2015) Modelling of climate change impacts on coastal flooding/erosion, ports and coastal defence structures. Desalin Water Treat 54(8):2130–2137

Kesavan PC, Swaminathan MS (2006) Managing extreme natural disasters in coastal areas. Philos Trans R Soc A Math Phys Eng Sci 364(1845):2191–2216

Kirwan M, Temmerman S (2009) Coastal marsh response to historical and future sea-level acceleration. Quat Sci Rev 28(17–18):1801–1808

Kron W (2013) Coasts: the high-risk areas of the world. Nat Hazards 66(3):1363–1382

Kumar AA, Kunte PD (2012) Coastal vulnerability assessment for Chennai, east coast of India using geospatial techniques. Nat Hazards. https://doi.org/10.1007/s11069-012-0276-4

Kumar CS, Murugan PA, Krishnamurthy RR, Batvari BPD, Ramanamurthy MV, Usha T, Pari Y (2008) Inundation mapping—a study based on December (2004) tsunami hazard along Chennai coast, Southeast India. Nat Hazards Earth Syst Sci 8:617–626

Kumar S, Panda J, Singh K, Guha BK, Kant S (2021) Structural characteristics of north Indian Ocean tropical cyclones during 1999–2017: a scatterometer observation-based analysis. Theoret Appl Climatol 143:227–240

Kumar S, Panda J, Paul D, Guha BK (2023) Impact of environmental variables on the north Indian Ocean tropical cyclones radial parameters. Clim Dyn 60(3–4):813–830

Kuriakose SL, Jetten VG, Van Westen CJ, Sankar G, Van Beek LPH (2008) Pore water pressure as a trigger of shallow landslides in the Western Ghats of Kerala, India: some preliminary observations from an experimental catchment. Phys Geogr 29(4):374–386

Lane K, Charles-Guzman K, Wheeler K, Abid Z, Graber N, Matte T (2013) Health effects of coastal storms and flooding in urban areas: a review and vulnerability assessment. J Environ Public Health 2013(1):913064

Leatherman SP, Zhang K, Douglas BC (2000) Sea level rise shown to drive coastal erosion. EOS Trans Am Geophys Union 81(6):55–57

Li J, Chai G, Xiao Y, Li Z (2023) The impacts of ocean acidification, warming and their interactive effects on coral prokaryotic symbionts. Environ Microb 18(1):49

Lough JM (2008) 10th anniversary review: a changing climate for coral reefs. J Environ Monit 10(1):21–29

Lu Y, Yuan J, Lu X, Su C, Zhang Y, Wang C, Sweijd N (2018) Major threats of pollution and climate change to global coastal ecosystems and enhanced management for sustainability. Environ Pollut 239:670–680

Luettich RA Jr, Westerink JJ, Scheffner NW (1992) ADCIRC: an advanced three-dimensional circulation model for shelves, coasts, and estuaries. Report 1. Theory and methodology of ADCIRC-2DDI and ADCIRC-3DL (No. CERC-TR-DRP-92-6). Coastal Engineering Research Center, Vicksburg

Madhura RK, Krishnan R, Revadekar JV, Mujumdar M, Goswami BN (2015) Changes in western disturbances over the Western Himalayas in a warming environment. Clim Dyn 44:1157–1168

Malak MA, Sajib AM, Quader MA, Anjum H (2020) "We are feeling older than our age": vulnerability and adaptive strategies of aging people to cyclones in coastal Bangladesh. Int J Disaster Risk Reduct 48:101595

Mazhar S, Pellegrini E, Contin M, Bravo C, De Nobili M (2022) Impacts of salinization caused by sea level rise on the biological processes of coastal soils—a review. Front Environ Sci 10:909415

McInnes KL, White CJ, Haigh ID et al (2016) Natural hazards in Australia: sea level and coastal extremes. Clim Change 139:69–83. https://doi.org/10.1007/s10584-016-1647-8

Mohanty UC, Osuri KK, Pattanayak S, Sinha P (2012) An observational perspective on tropical cyclone activity over Indian seas in a warming environment. Nat Hazards 63:1319–1335

Mujabar PS, Chandrasekar N (2011) Coastal erosion hazard and vulnerability assessment for southern coastal Tamil Nadu of India by using remote sensing and GIS. Nat Hazards. https://doi.org/10.1007/s11069-011-9962-x

Mukherjee A, Panda J (2024) A study on the urban growth and dynamics over 16 major cities of India. J Earth Syst Sci 133(2):66. https://doi.org/10.1007/s12040-024-02280-9

Mukherjee A, Panda J, Choudhury A, Singh S, Bhattacharyya S (2024) Analyzing urban footprints over four coastal cities of India and the association with rainfall and temperature using deep learning models. Urban Clim 102123. Accepted for publication

Mukhopadhyay A, Dasgupta R, Hazra S, Mitra D (2012) Coastal hazards and vulnerability: a review. Int J Geol Earth Environ Sci 2(1):57–69

Murty CVR, Jain SK, Sheth AR, Jaiswal A, Dash SR (2006) Response and recovery in India after the December 2004 great Sumatra earthquake and Indian Ocean tsunami. Earthq Spectr 22(3_suppl):731–758

Naskar PR, Mohapatra M, Singh GP (2024) CMIP6 projections of surface latent heat flux over the north Indian Ocean. Theoret Appl Climatol 1–10

Natarajan L, Sivagnanam N, Usha T, Chokkalingam L, Sundar S, Gowrappan M, Roy PD (2021) Shoreline changes over last five decades and predictions for 2030 and 2040: a case study from Cuddalore, southeast coast of India. Earth Sci Inf 14:1315–1325

Natesan U, Parthasarathy A, Vishnunath R, Kumar GEJ, Ferrer VA (2015) Monitoring longterm shoreline changes along Tamil Nadu, India using geospatial techniques. Aquat Procedia 4:325–332

Nathan CS, Ganapathy GP (2024) Quantifying the impact of changing rainfall patterns on landslide frequency and intensity in the Nilgiris district of Western Ghats, India. Prog Disaster Sci 100351

Neumann B, Vafeidis AT, Zimmermann J, Nicholls RJ (2015) Future coastal population growth and exposure to sea-level rise and coastal flooding—a global assessment. PLoS ONE 10(3):e0118571

Noto AE, Shurin JB (2017) Early stages of sea-level rise lead to decreased salt marsh plant diversity through stronger competition in Mediterranean-climate marshes. PLoS ONE 12(1):e0169056

Panda J, Paul D, Routray A, Giri RK (2023) Atmospheric and Ocean characteristics associated with NIO tropical cyclones: a comprehensive review vis-à-vis the intensity and movement. Vayu Mandal 49(1):112–137

Panda J, Singh K, Tom G, Mahapatra SR, Paul D (2024) Rainfall contribution by cyclonic disturbances over India in two epochs during 1901–2021 and risk assessment, chap 3. In: Collins J, Done J, Zhu YJ, Wilson P (eds) Advances in hurricane risk in a changing climate. Springer Nature Switzerland, Cham, pp 43–75. https://doi.org/10.1007/978-3-031-63186-3_3

Panda J, Paul D, Sarkar A, Mukherjee A, Bhasi I, Tom G, Das S, Lohan N, Roy D, Kumar S (2025) Severe storms in changing climate scenario with a distinctive emphasis on South Asia and India. In: Das S, Tao W-K (eds) Severe storms: anatomy, early warning systems and aftermath

in changing climate scenarios. Springer, Singapore, pp 563–606. https://doi.org/10.1007/978-981-97-7075-5_22
Parikh K, Parikh J, Kumar M (2017) Vulnerability of Surat, Gujarat to flooding from Tapi River: a climate change impact assessment. Vayu Mandal 43(2):120–129
Patil RG, Deo MC (2020) Sea level rise and shoreline change under changing climate along the Indian Coastline. J Waterw Port Coast Ocean Eng 146(5):05020002
Piao S, Ciais P, Friedlingstein P, de Noblet-Ducoudré N, Cadule P, Viovy N, Wang T (2009) Spatiotemporal patterns of terrestrial carbon cycle during the 20th century. Glob Biogeochem Cycl 23(4)
Pidgeon E (2009) Carbon sequestration by coastal marine habitats: important missing sinks. In: The management of natural coastal carbon sinks, p 47
Poulose J, Rao AD, Dube SK (2020) Mapping of cyclone induced extreme water levels along Gujarat and Maharashtra coasts: a climate change perspective. Clim Dyn 55(11):3565–3581
Rabbani MG, Rahman AA, Islam N, Michel D, Pandya A (2010) Climate change and sea level rise: issues and challenges for coastal communities in the Indian Ocean region. In: Coastal zones and climate change, pp 17–29
Radhakrishnan S, Duraisamy Rajasekaran SK, Sujatha ER, Neelakantan TR (2024) A comparative study on 2015 and 2023 Chennai flooding: a multifactorial perspective. Water 16(24):1–20, Article 2477. https://doi.org/10.3390/w16172477
Raha A, Das S, Banerjee K, Mitra A (2012) Climate change impacts on Indian Sunderbans: a time series analysis (1924–2008). Biodivers Conserv 21:1289–1307
Rahman AA, Alam M, Alam SS, Uzzaman MR, Rashid M, Rabbani G (2007) Risks, vulnerability and adaptation in Bangladesh. Hum Dev Rep 8
Rajasree BR, Deo MC, Nair LS (2016) Effect of climate change on shoreline shifts at a straight and continuous coast. Estuar Coast Shelf Sci 183:221–234
Ranger N, Hallegatte S, Bhattacharya S, Bachu M, Priya S, Dhore K et al (2011) An assessment of the potential impact of climate change on flood risk in Mumbai. Clim Change 104:139–167
Rao AD, Upadhaya P, Pandey S, Poulose J (2020a) Simulation of extreme water levels in response to tropical cyclones along the Indian coast: a climate change perspective. Nat Hazards 100:151–172
Rao AD, Upadhaya P, Ali H, Pandey S, Warrier V (2020b) Coastal inundation due to tropical cyclones along the east coast of India: an influence of climate change impact. Nat Hazards 101:39–57
Rath SS, Mohanty S, Panda J (2022) Analyzing the fragmentation of urban footprints in eastern and southern Indian cities and driving factors. J Indian Soc Remote Sens 50(8):1499–1517
Roxy MK, Ritika K, Terray P, Masson S (2014) The curious case of Indian Ocean warming. J Clim 27(22):8501–8509
Roxy MK, Modi A, Murtugudde R, Valsala V, Panickal S, Prasanna Kumar S et al (2016) A reduction in marine primary productivity driven by rapid warming over the tropical Indian Ocean. Geophys Res Lett 43(2):826–833
Roxy MK, Ghosh S, Pathak A, Athulya R, Mujumdar M, Murtugudde R et al (2017) A threefold rise in widespread extreme rain events over central India. Nat Commun 8(1):1–11
Saha J, Paul S (2021) An insight on land use and land cover change due to tourism growth in coastal area and its environmental consequences from West Bengal, India. Spat Inf Res 29(4):577–592
Sahoo B, Bhaskaran PK (2019) Prediction of storm surge and inundation using climatological datasets for the Indian coast using soft computing techniques. Soft Comput 23(23):12363–12383
Sahu RT, Mehta DJ (2024) Impact of coastal inundation due to rise in sea level: a case study of Surat City, India. Water Pract Technol 19(5):1753–1768
Sajith S, Raju TB, Aswani RS (2024) Are Indian ports safe? Identifying, analysing and prioritizing the risks affecting India's major ports. Maritime Transp Res 6:100108
Saranathan E, Chandrasekaran R, Soosai Manickaraj D, Kannan M (2011) Shoreline changes in Tharangampadi village, Nagapattinam district, Tamil Nadu, India—a case study. J Indian Soc Remote Sens 39:107–115

Saranya JS, Roxy MK, Dasgupta P, Anand A (2022) Genesis and trends in marine heatwaves over the tropical Indian Ocean and their interaction with the Indian summer monsoon. J Geophys Res Oceans 127(2):e2021JC017427

Saxena S, Purvaja R, Mary Divya Suganya G et al (2013) Coastal hazard mapping in the Cuddalore region, South India. Nat Hazards 66:1519–1536. https://doi.org/10.1007/s11069-012-0362-7

Sheela Nair L, Prasad R, Rafeeque MK, Prakash TN (2018) Coastal morphology and long-term shoreline changes along the southwest coast of India. J Geol Soc India 92:588–595

Sheik M, Chandrasekar (2011) A shoreline change analysis along the coast between Kanyakumari and Tuticorin, India, using digital shoreline analysis system. Geo-Spat Inf Sci 14(4):282–293

Singh K, Panda J, Sahoo M, Mohapatra M (2019a) Variability in tropical cyclone climatology over north Indian Ocean during the period 1891 to 2015. Asia-Pac J Atmos Sci 55(2):269–287

Singh K, Panda J, Rath SS (2019b) Variability in landfalling trends of cyclonic disturbances over north Indian Ocean region during current and pre-warming climate. Theoret Appl Climatol 137(1–2):417–439

Singh K, Panda J, Kant S (2020) A study on variability in rainfall over India contributed by cyclonic disturbances in warming climate scenario. Int J Climatol 40(6):3208–3221

Spalding M, Mcivor A, Tonneijck F, Tol S, Eijk PV (2014) Mangroves for coastal defence. Available at https://www.nature.org/media/oceansandcoasts/mangroves-for-coastal-defence.pdf

Sreelash K, Sharma RK, Gayathri JA, Upendra B, Maya K, Padmalal D (2018) Impact of rainfall variability on river hydrology: a case study of Southern Western Ghats, India. J Geol Soc India 92:548–554

Subramanian BR (2006) Inundation of seawater due to Indian Ocean tsunami along the Indian coast. In: Rajamanickam GV, Subramaniyan BR, Baba M, Ramesh R, Elango L, Prithviraj M (eds) 26th December 2004 tsunami: causes, effects, remedial measures pre and post tsunami disaster management: a geoscientific perspective, pp 10–44

Sudha Rani NNV, Satyanarayana ANV, Bhaskaran PK (2015) Coastal vulnerability assessment studies over India: a review. Nat Hazards 77:405–428

Szlafsztein C, Sterr H (2007) A GIS-based vulnerability assessment of coastal natural hazards, state of Pará, Brazil. J Coast Conserv 11:53–66. https://doi.org/10.1007/s11852-007-0003-6

Taxak AK, Murumkar AR, Arya DS (2014) Long term spatial and temporal rainfall trends and homogeneity analysis in Wainganga basin, central India. Weather Clim Extrem 4:50–61

Temmerman S, Horstman EM, Krauss KW, Mullarney JC, Pelckmans I, Schoutens K (2023) Marshes and mangroves as nature-based coastal storm buffers. Ann Rev Mar Sci 15(1):95–118

Thompson KR, Ohashi K, Sheng J, Bobanovic J, Ou J (2007) Suppressing bias and drift of coastal circulation models through the assimilation of seasonal climatologies of temperature and salinity. Cont Shelf Res 27(9):1303–1316

Vijaykumar P, Abhilash S, Santhosh KR, Mapes BE, Kumar CS, Hu IK (2017) Distribution of cloudiness and categorization of rainfall types based on INSAT IR brightness temperatures over Indian subcontinent and adjoining oceanic region during south west monsoon season. J Atmos Solar Terr Phys 161:76–82

Wahiduzzaman M, Oliver EC, Klotzbach PJ, Wotherspoon SJ, Holbrook NJ (2019) A statistical seasonal forecast model of north Indian Ocean tropical cyclones using the quasi-biennial oscillation. Int J Climatol 39(2):934–952

Wahiduzzaman M, Cheung K, Luo JJ, Bhaskaran PK, Tang S, Yuan C (2022) Impact assessment of Indian Ocean dipole on the north Indian Ocean tropical cyclone prediction using a statistical model. Clim Dyn 58:1275–1292

Wassmann R, Hien NX, Hoanh CT, Tuong TP (2004) Sea level rise affecting the Vietnamese Mekong Delta: water elevation in the flood season and implications for rice production. Clim Change 66:89–107

WMO (2022) State of the global climate in 2021. Report No. WMO-No. 1290. World Meteorological Organization. Available at https://library.wmo.int/records/item/56300-state-of-the-global-climate-2021

Yalla V, Sriram V, Murali K (2024) Influence of maximum water level and coastal inundation on the east coast of India based on future tropical cyclones. Nat Hazards 120(2):1263–1294
Zahedi S (2008) Tourism impact on coastal environment. WIT Trans Built Environ 99:45–57
Zhao C, Li T (2019) Basin dependence of the MJO modulating tropical cyclone genesis. Clim Dyn 52(9):6081–6096

Blue Economy: Scope and Global Challenges

Climate Change Induced Natural Hazards of Agriculture in Coastal Zones

Suman Dutta, Priyanka Saha, Sucismita Satapathy, Shashank Patel, and Bishal Debnath

Abstract Climate change significantly impacts agriculture and coastal areas, increasing vulnerability to natural disasters. The combination of increasing sea levels, severe weather, and changing climate patterns poses serious problems for human settlements and ecosystems. Coastal areas are more vulnerable to extreme weather events such storm surges, floods, tsunamis, saltwater intrusion, and coastal erosion. These risks lead to increasing salinity in aquifers and estuaries, habitat loss, infrastructure degradation, and water pollution. Agricultural output and food security are particularly susceptible in coastal locations due to factors including floods, temperature variations, loss of arable land, pest and disease transmission, and declining biodiversity. Strategies for adaptation and mitigation are crucial to addressing these issues. Through the integration of social, economic, and environmental considerations, integrated coastal zone management (ICZM) and the use of remote sensing can sustainably manage coastal resources. Artificial intelligence and machine learning further support this by developing predictive models for flooding and drought events based on historical data, satellite imagery, and climate patterns. Developing climate-resilient crops that can endure harsh weather, drought, and salt helps safeguard food production. Resilient farming systems and reinforced coastal defences are essential for reducing natural disasters. Water shortage and salinity problems may be addressed with the use of water management techniques including desalination, effective irrigation, and rainwater collection. Early warning systems lessen the impact on agricultural and human settlements by enabling prompt evacuations and preparation. Agricultural resilience and biodiversity are improved by

S. Dutta (✉)
Ramakrishna Mission Vivekananda Educational and Research Institute, Narendrapur, West Bengal, India
e-mail: duttasumansiliguri@gmail.com

P. Saha · S. Patel
ICAR-Indian Agricultural Research Institute, New Delhi, India

S. Satapathy
Banaras Hindu University, Uttar Pradesh, Varanasi, India

B. Debnath
Punjab Agricultural University, Ludhiana, India

K. K. Basheer Ahammed et al. (eds.), *Monitoring Coastal Marine Ecosystems Under Climate Change*, https://doi.org/10.1007/978-3-032-09137-6_13

ecosystem-based adaptation, which makes use of wetlands and mangroves as natural infrastructure. Ultimately, the development of new technologies requires investment in research and innovation, robust governance, and financial incentives for resilient behaviours. The lives of future generations and the ecosystems in coastal agricultural zones may be safeguarded from the continued consequences of climate change by implementing comprehensive and coordinated solutions.

Keywords Climate change · Coastal zones · Natural hazards · Sea-level rise · Agriculture resilience · Soil salinity

1 Introduction

Climate change significantly impacts coastal zones and agriculture, leading to increased vulnerability and risk of natural hazards. Significant obstacles face both natural ecosystems and human livelihoods in these areas as a result of the confluence of increasing sea levels, intense weather, and changing climate patterns (Maity and Gonekar 2023). Because of the melting of polar ice caps brought on by rising global temperatures, coastal areas are especially vulnerable to a range of natural hazards that are made worse by climate change (Vousdoukas et al., 2018). Sea level rise results in low-lying regions flooding, increased salinity in aquifers and estuaries, and coastal erosion, all of which endanger natural ecosystems and human communities. Increases in hurricane, typhoon, and storm frequency and severity have resulted in catastrophic wind, flooding, and storm surge devastation that has destroyed towns, infrastructure, and natural environments (Needham et al., 2015). Coastal ecosystems and human settlement are put at risk by coastal erosion, which is exacerbated by increasing sea levels and increased wave activity (Ahammed et al., 2023). Sea level rise and decreased river flow also cause saltwater intrusion into freshwater systems, endangering natural ecosystems, agriculture, and drinking water sources (Tully et al., 2019). Increased frequency and severity of coastal and riverine floods are caused by intensified rainfall and storm events, which affect residential areas, infrastructure, and agricultural regions. Rising sea levels and weakening coastal defenses can intensify the impact of tsunamis on coastal zones, even though they are not directly caused by climate change (Talling et al., 2015). These natural dangers brought on by climate change provide particular problems for agriculture in coastal areas. Rising sea levels and saltwater intrusion cause soil salinity to rise, which lowers agricultural output and damages crops. Saline soils also inhibit plant development and lower crop yields. While protracted waterlogging can result in root infections and crop loss, frequent and severe floods harm crops, erode topsoil, and throw off planting and harvesting schedules (Sorensen et al., 2009). Temperature and precipitation variations affect agricultural viability, growing seasons, and water availability. Extreme weather events can cause droughts or excessive rainfall, which can have an adverse effect on crop yields and health (Prasad et al., 2015). Arable land is lost as a result of coastal erosion and sea level rise, which decreases the amount of land that can be used for agriculture and intensifies competition for resources. Climate change has the potential to increase pest and disease populations, endangering food security and agricultural productivity (Mukherjee et al., 2023). Coastal farming towns face challenges to their economic stability due to declining agricultural yield and rising expenses for disaster response and infrastructure maintenance (Roy et al., 2024). A reduction in biodiversity can also be caused by altered habitat conditions and elevated salinity, which can have an impact on pollinators, soil quality, and other ecosystem services that agriculture depends on. Several adaptation and mitigation techniques are necessary to meet the problems that climate change is posing to agriculture and coastal areas. When paired with artificial intelligence (AI) and machine learning (ML), remote sensing provides revolutionary approaches to mitigating the increasing effects of climate change on coastal agriculture (McCarthy et al., 2017). Remote sensing gives vital information on soil moisture, vegetation health, temperature fluctuations, and sea level rise by using real-time satellite photography (Mahrad et al., 2020). The utilisation of AI and ML facilitates the examination of extensive datasets, detecting trends and forecasting the occurrence and intensity of natural disasters. By providing accurate and localised forecasts of climate-induced occurrences, such as storm surges and saltwater intrusion, this integration improves decision-making and enables farmers to adopt adaptive agricultural techniques. There are still gaps in the availability of targeted, real-time solutions for saltwater intrusion, erosion, and soil salinity in coastal agriculture, despite advancements in the use of remote sensing, artificial intelligence, and machine learning. Additionally, there is not enough evidence to support the efficacy of AI-driven early warning systems in reducing the negative effects of extreme weather on agriculture. To assure cost-effective adoption in vulnerable coastal locations, evaluate long-term sustainability, and customise these technologies for smallholder farmers, more study is required.

Hence, this chapter aims to comprehensively explore the multifaceted impacts of climate change on coastal zones and agriculture, highlighting the heightened vulnerability and risk of natural hazards in these regions. The chapter looks at the particular problems caused by climate change, extreme weather, and rising sea levels to clarify the intricate interactions between these phenomena and how they affect both human livelihoods and natural ecosystems. The chapter also seeks to identify and address the particular difficulties that coastal agriculture faces, such as floods, temperature and precipitation variations, soil salinity, and biodiversity loss. The chapter aims to highlight the need to implement effective adaptation and mitigation strategies, like remote sensing and Integrated Coastal Zone Management (ICZM), to sustainably manage and protect coastal resources, agricultural productivity, and the economic stability of coastal communities. It does this by giving a thorough understanding of these issues.

2 Natural Hazards in Coastal Zones

The coastal zone areas are very susceptible to various types of natural calamities as they are situated in very close proximity to sea or ocean land. The description of occurrence of different natural hazard in coastal zones can be ranges widely based on various types of stresses due to changing climate (Table 1). However, there are some types of natural hazards which required to look into the most for environmental security.

2.1 *Sea Level Rise*

Addressing climate change and its repercussions, including sea-level rise, is a significant problem for the twenty-first century. Rising water levels and floods along US and global coasts impart substantial challenges (Moser et al., 2012). The global climate is transforming and temperature continue to increase, which eventually effects the rise of sea level. Global warming from greenhouse gas concentrations over the past century has resulted in rising sea levels due to the melting of ice shelves and ocean thermal expansion. Particularly the soaring sea levels poses a major threat to low-lying coastal zones (Nicholls and Cazenave, 2010). Climate change affects coasts in more ways than just rising sea level rather impacts coastal processes and dynamics through shifting wind flow, storm surges, and wave action throughout the period (Barnard et al., 2019). Among them, wind patterns, wave force, high waves, and sea levels significantly impact coastlines ecosystem (Reguero and Griggs, 2022; Morim et al., 2021). Regions and coastlines experience seasonal, interannual, and long-term changes, leading to local consequences such as flooding and erosion. Rising water temperatures and ocean acidification will have a significant influence on coastal ecosystems and their services, including fisheries, coastal protection, and carbon sequestration (Barbier and Burgess, 2021). There are also various factors which directly and indirectly aggravate the long term rises in mean sea level (Table 2). According to Mimura (2013), it is projected that 1–3 °C increase in sea water temperature is expected to increase the frequency of coral bleaching and eventually trigger coral reef death due to an increase in mean sea level. In expansive deltas such as those found in Bangladesh, the boundary between fresh and saline water has progressively moved inland. This shift is primarily due to rising sea levels and increased saltwater intrusion, which result from the combined effects of climate change and human activities (Minar et al. 2013).

Table 1 Major threats to coastal environment due to climate change

Sector	Climate change threat	Other human threat
Coastal ecosystem and coral reef	Loss of coral reefs due to coral bleaching, migration or loss of coastal ecosystems, coastal erosion and sedimentation, changes in the distribution of marine species, and the increased spread of exotic and invasive species	Intense coastal development leading to habitat loss, pollution causing marine dead zones, conversion of mangroves and wetlands for mariculture, damage to seagrass beds, coral mining, oil spills, and the spread of invasive species
Fisheries	Overall decline in ocean Productivity; Eutrophication and coral mortality; Loss of shift in critical fish habitat; shift in temperature causing migration of fishes	The overexploitation of resources; destructive fishing methods; pollution originating from land; and the sedimentation of the coastal system due to land-based causes
Mariculture	An increase in water temperature may cause unpredictably high or low levels of cultural output, increased stress and disease susceptibility, altered weather patterns, and extreme weather occurrences	Overexploitation of juveniles and larvae seedstock for fish farm; loss of protected habitats as a result of mariculture facilities being positioned incorrectly
Freshwater resource	Erosion, storms, and precipitation are damaging infrastructure and causing beach losses; water quality is compromised, leading to more beach closures; tourism insurance costs are rising; saltwater is intruding into freshwater sources; saltwater is encroaching into estuaries and coastal rivers; waves and storm surges are reaching further inland, increasing coastal flooding and inundation; and decreased precipitation is worsening saltwater intrusion and exacerbating water supply issues	Untreated sewage discharge and chemical pollution of coastal water; Unregulated freshwater extraction and withdrawal of groundwater; Upstream dams; Enlargement and dredging of waterways

2.2 *Storm Surges and Extreme Weather*

Storm surge, triggered by low air pressure and strong winds, is a significant contributor to coastal flooding (Resio and Westerink, 2008). Storm surges are generated by air pressure and wind conditions. Positive surges occur when water is lifted towards the beach, whereas negative surges occur when water is pushed away from the shore.

Table 2 Factors and causes of sea level rise

Factors	Causes of sea level rise
Changes in the volume of sea water contained in ocean basins	1. Geothermal increase of sea water 2. Growth and decay of ice caps on land 3. Terrestrial water storage, including dam lakes, reservoirs, and groundwater depletion
Changes in the shape and size of the ocean basins	1. Isostatic adjustment of land mass, particularly Glacier Isostatic Adjustment (GIA) 2. Tectonic movement, includes ground subsidence and elevation caused by earthquakes, compaction, ground water pumping, and sediment input from land
Other factors	1. Ocean currents changes and atmospheric pressure 2. Tsunami, tide, storm surge, and waves 3. Natural inter-annual changes, include the Pacific Decadal Oscillation (PDO)

Ref Cazenave and Llovel, 2010.

Surges are influenced by extreme weather events such as storms and the magnitude, path, dimension, and moving speed of the storms (Xuan et al., 2021). Storm surge characteristics are influenced by ocean dynamics and ailments including water depth, tides, ice, shelf width, and stratification (Idier et al., 2019; Wang and Bernier, 2023). In addition, overstepping the local tidal maxima can cause flooding and erosion. When water levels fall below local tidal minimums, vessels' drafts may exceed the depth of the water, posing a navigation risk (Mortensen et al., 2022). Research indicates that extreme rainfall and riverine floods can exacerbate flooding hazards and impacts (Couasnon et al., 2020; Wijetunge and Neluwala, 2023). Several studies have found that the combined effects of rivers, rain, surges, and waves can considerably impact flooding risk. In addition, the intensity of extreme precipitation also affects riverine and coastal floods (Heinrich et al., 2023). Therefore, in the coming years, elevated sea levels and coastal flooding will continue to be a major source of natural disasters as sea-level rise causes critical water levels to be exceeded more commonly (Griggs and Reguero, 2021).

2.3 Coastal Erosion

Coastline hazards are currently in increasing prominence globally due to rapid changes in geological and physical variables caused by dynamic coastal processes. Generally, the coastal processes involve natural influences that can able to alter shorelines and landforms (Kaliraj et al., 2015). Pararas-Carayannis (2018) claimed that landforms are greatly impacted by coastal hazards, which are intimately related to sedimentary occurrences. According to (2010), saltwater intrusion into aquifers and rivers, pollution, dune deterioration, and erosion are some of the serious environmental problems that coastal areas confront. Natural processes that erode and spread materials along the shoreline, such as waves, tidal currents, and deflation, are the main causes of coastal erosion. According to Prasad and Kumar (2014), coastal sediments are rearranged along the coast to produce dunes, beaches, marshes, and reefs. These sediments are also a result of interior erosion and river movement. It usually refers to the sea gradually encroaching on land over a long period of time, taking into account local sediment dynamics, weather, and storm occurrences (Vinayaraj et al., 2011). Because of their varied geographical and temporal dimensions, evaluating the mechanisms that influence shoreline existence and evolution is difficult (Alexandrakis and Poulos, 2014). Coastal erosion happens when the shoreline loses more sediment than it gains, which results in a negative sediment budget. A positive sediment budget shows that the shoreline gets more material than it loses (Bird, 2008). Sea level rise, storm intensity, soil subsidence, unplanned development, sand extraction for building and dam construction are the main causes of coastal erosion (Nicholls and Cazenave, 2010).

2.4 *Salt Water Intrusion*

Since that 80% of the world's population depends on local aquifers for drinking water, saltwater intrusion (SWI) poses a danger to the ecosystem. Bennett et al. (2002) state that SWI is becoming a global problem in coastal aquifers, particularly in cases where a permeable aquifer extends offshore and interacts with the ocean. When saltwater levels in aquifers approach 2%, this interaction can upset the equilibrium of the ecosystem and seriously degrade freshwater supplies. Rising sea levels are caused by several factors, including shifting air pressure, ocean thermal expansion, and the melting of ice caps and glaciers. By 2100, it is predicted that the global sea level would have risen by 0.6 m (Nicholls et al., 2007). Coastal aquifers that are restricted by topography may see a further reduction in their hydraulic gradient as a result of a combination of global and local vertical land movements (Michael et al., 2013). Reduced precipitation in warmer regions causes a hydraulic gradient to drop and groundwater recharge to decline (Holding and Allen, 2015). This eventually reduces fresh groundwater flow towards the ocean, aggravating SWI.

2.5 *Flooding*

One of the most common and damaging natural catastrophes in the globe is flooding (Jonkman, 2005). The frequency and intensity of floods are expected to increase due to climate change (Bruwier et al., 2015). Floods inflict enormous harm, with hundreds of millions of dollars in damages and thousands of fatalities every time (Guha-Sapir et al., 2015). For example, the Queensland floods of 2010–2011 caused significant indirect economic losses in addition to nearly AUD 2 billion in infrastructure damage. With between 7.3 and 9.7 billion people on the planet by 2050, flooding is predicted to become a more frequent and more severe natural disaster. Furthermore, it is predicted that 66% of people on Earth will live in cities, with 40% of urban land located in places vulnerable to flooding (Güneralp et al., 2015). Moreover, 90% of projected urban growth will take place in Asia and Africa, where a large population already resides in areas vulnerable to flooding (Muis et al., 2018). Extreme floods are predicted to occur more frequently as a result of climate change, which will increase the frequency and severity of heavy rainfall events. Significant precipitation is usually one of the main causes of floods, although other important variables include location, rainfall patterns, duration, catchment moisture, and soil hydraulic qualities (Johnson et al., 2018).

2.6 Tsunamis

A tsunami is an outbreak of ocean waves with exceptionally broad wavelengths and periods, typically triggered by an undersea earthquake, volcanic eruption, or landslide which is not directly related to the effect of climate change (Goff and Nunn, 2016). Coastal settlements are vulnerable to marine flooding caused by tsunamis and storm surges, which can harm the ecological and built environment. Neumann et al., (2015) forecasted that as coastal areas become more densely inhabited, the threat of tsunamis would increase. According to him, the population of low-lying coastal areas is projected to increase by 68–122% from 625 million in 2015 to 1052 to 1388 million by 2060. The Global Historical Tsunami Database (NGDC/WDS, 2015) reports around 940,000 deaths from tsunami strikes worldwide, including prehistorical, historical, and modern ages. In 2004, the Indian Ocean Tsunami (IOT) killed around 230,000 people, making it the most catastrophic tsunami in recent history (Telford et al., 2006). Tsunamis can have both immediate and long-term effects, including damage to power plants and salt-water intrusion in heavily cultivated delta plains (Nakamura et al., 2017), which can be minimized with hard or soft tsunami barriers.

3 Impact on Agriculture

Coastal agriculture is particularly vulnerable to the effects of climate change (Table 3). The difficulties already faced by agricultural producers are made worse by rising sea levels, harsh weather, and shifting precipitation patterns (Bhattacharyya et al., 2020). Key difficulties include excessive salt of the soil and water, recurrent flooding, inadequate drainage, and insufficient quality irrigation water (Fig. 1). Additional factors that lower yields include cyclonic disruptions, protracted monsoons,

Table 3 Effect of climate change on coastal agriculture and crop production

Climate change consequences	Direct/indirect impact on crop production and coastal agriculture
Sea level rise	Loss of land, flooding, coastal erosion, soil salinization increasing
Rainfall and river flow	Coastal agriculture by impeded aquaculture damage, drainage
Storm surges	The estuarine region is affected, coconut crop and aquaculture damage occur
Soil erosion	Soil degradation, nutrient loss, low productivity
Wetland loss and change	Low C sequestration, high GHGs emission
Frequent flooding	Estuarine soil salinity, sedimentation, and aquaculture damage
High-frequency drought	Moisture stress, low productivity, low yield
Cold wave	Desiccation, low yield
High temperature	Estuarine region affected, coconut crop and aquaculture damage
Heat wave	Desiccation, low yield

and issues with the soil, including as acidity and nutrient deficits (Eswar et al., 2021). In these areas, agricultural and livelihoods are threatened by high salinity, waterlogging, and frequent natural catastrophes (Rajaeian et al., 2022).

3.1 Soil Salinity

High salinity levels greatly reduce crop yield by hindering plant water uptake and nutrient absorption, leading to stunted growth and lower yields. The osmotic or water-deficit impact of salinity is the reduction in plant water uptake caused by a high concentration of salt in soil solution. When the concentration reaches a point where crop development is starting to be inhibited, damage happens. The osmotic effect of salinity induces metabolic changes in the plant identical to those caused by water stress-induced "wilting" (Munn et al., 2002). Soil salinity negatively impacts plant growth and productivity. This osmotic stress leads to poor seed germination, stunted growth, and lower yields. Sensitive crops like rice, beans, and many fruits and vegetables are particularly affected. High salt concentrations can also lead to the breakdown of soil aggregates, reducing soil porosity and permeability. Salinity can affect plant growth indirectly by sodium's effect on the degradation of the soil's physical condition and by increasing the soil's pH (Machado and Serralheiro, 2020). This degradation affects root growth and water infiltration, further impairing plant development. Over time, salinization can render soils unproductive and lead to land abandonment.

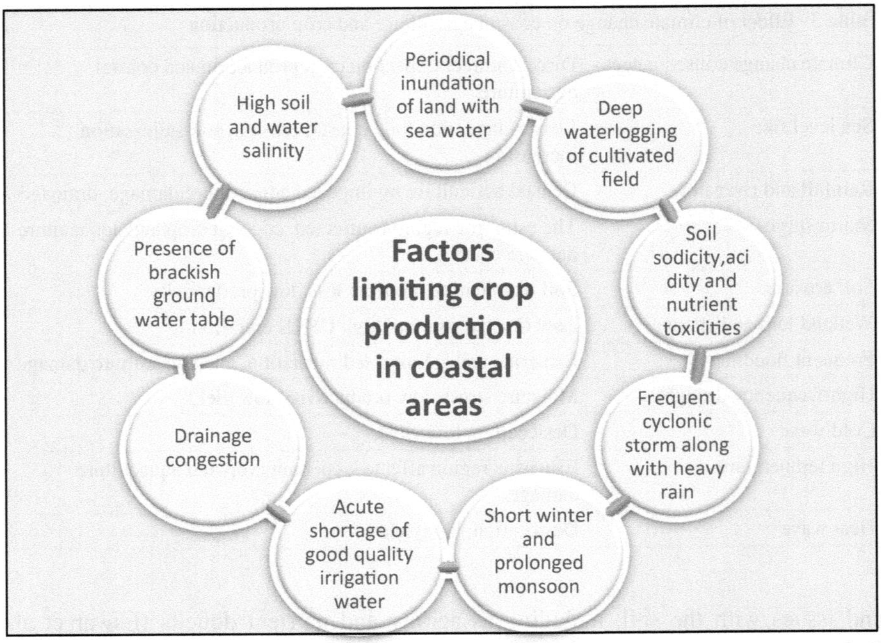

Fig. 1 Various aspects of climate change affecting the production of coastal crops

3.2 *Flooding and Waterlogging*

Flooding can submerge crops, leading to immediate physical damage and reduced oxygen availability to plant roots. Due to overcrowding and an increase in habitation, farming, and other developmental activities, India is seeing a surge in the real and intangible damages caused by floods. Massive losses have been suffered by Indian agriculture as a result of floods. These losses have included crop losses that have decreased yields, erosion of the most fertile soil, damage to machinery, stored inputs, and roads (Panwar and Sen, 2020). Prolonged waterlogging can cause root rot and other diseases, significantly reducing crop yields. Soil erosion is one of the disastrous phenomena which reduce the overall fertility and productivity of soil. Floodwaters can erode topsoil, which is rich in nutrients and organic matter. Prolonged waterlogging reduces soil aeration, leading to root suffocation and increased susceptibility to diseases. Excess water can leach essential nutrients from the soil, particularly nitrogen, potassium, and phosphorus, leaving the soil nutrient-poor and less productive (Aslam et al., 2023).

3.3 *Temperature and Precipitation Changes*

Higher temperatures can reduce crop yields by increasing heat stress and water demand. High temperatures can impair plant physiological processes, such as photosynthesis and respiration. Crops may experience wilting, leaf scorching, and fruit drop. Heat stress can also affect pollination and seed set, reducing overall productivity. Altered precipitation patterns can disrupt the water supply for irrigation. Inconsistent water availability can lead to reduced crop yields and increased reliance on irrigation, further straining water resources. According to Guiteras (2009), climate change will cause yields to drop between 4.5 and 9%; but, in the long term, if adaptation is not made, it will significantly cut yields by at least 25%. With the continued warming, there will probably be a greater need for agricultural water, necessitating additional irrigation (Venkateswarlu, 2015).

3.4 *Loss of Arable Land*

India, with its vast coastline of over 7500 km, is particularly vulnerable to the impacts of climate change. The loss of arable land in these regions poses a severe risk to agricultural productivity, food security, and the livelihoods of millions of people (Mohanty et al., 2017). One of the most direct impacts of climate change on coastal agriculture is sea level rise. According to the Intergovernmental Panel on Climate Change (IPCC), global sea levels could rise by up to one meter by the end of the century if greenhouse gas emissions continue unabated. In India, this would result in the inundation of low-lying coastal regions, rendering vast areas of agricultural land unusable (Oppenheimer et al., 2019). The loss of arable land and reduced agricultural productivity in coastal areas have far-reaching implications. Agriculture is a primary source of livelihood for millions of people in India. The degradation of agricultural land due to climate change threatens food security and the economic stability of farming communities. Due to the thermal expansion of the oceans and partial melting of glaciers and ice caps brought on by CO_2-induced warming, it is anticipated that sea levels will rise. This will have an impact on agriculture, primarily through the inundation of low-lying farmland and an increase in the salinity of coastal groundwater (Soora et al., 2013). Sea levels are expected to increase by 9 to 29 cm by 2030, with a best estimate of 18 cm, and by 28–96 cm by 2090, with a best estimate of 58 cm, according to IPCC predictions (Falchetta et al., 2019).

3.5 *Pest and Disease Proliferation*

Climate change alters the distribution and behavior of pests and diseases, increasing their impact on coastal agriculture. According to Kaur et al. (2018), global warming will lengthen the wet season and increase air humidity in places where rainfall would rise. These circumstances, when coupled with rising temperatures, may encourage the growth of fungal infections. In a similar vein, an increase in warmth and humidity may lead to a rise in the prevalence of insects and disease vectors. Elad and Pertot (2014) had talked about how plant diseases and pathogens are affected by climate change. They clarified that host specificity, the ideal circumstances for infection, and plant infection mechanisms will all be impacted by climate change. They also mentioned how the shifting environment will have an impact on host plants as well as diseases. Sharma et al. (2018) has brought attention to the ramifications of climate change with regard to insects, crop protection, and food security. Warmer climates can expand the range of pests such as insects, mites, and nematodes. This expansion leads to higher infestation levels and increased crop damage. The temporal and geographical distribution of plant diseases may be directly impacted by climate change. Due to the significant environmental effects of plant infections, the climate change scenario may have an impact on the pathogens' survival, rate of multiplication, vigour, sporulation, direction, and distance of inocula dissemination (Park et al., 2014). Changes in humidity and temperature can facilitate the spread of plant diseases, affecting crop health and yields. Fungal diseases, bacterial infections, and viral diseases can become more prevalent under favorable conditions. Plant diseases have a direct influence on crop productivity. Farmers may face increased costs for pest and disease management.

3.6 Economic Impact

The economic impacts of climate change on coastal agriculture are driven by reduced productivity, increased costs, and market disruptions. Reduced crop yields and loss of arable land led to lower agricultural income for farmers. This income loss can affect the livelihoods of farming communities and contribute to increased poverty. Farmers may face higher costs for inputs such as salt-tolerant seeds, irrigation systems, pest control measures, and soil amendments. These increased expenses can strain farmers' budgets and reduce profitability (Malhi et al., 2021). Lower yields and higher costs reduce the profitability of coastal agriculture. Reduced agricultural productivity can lead to higher food prices and increased food insecurity. Coastal communities that rely on local agriculture for sustenance may face challenges in accessing affordable and nutritious food. The economic well-being of coastal communities is closely tied to agriculture. Climate impacts can lead to broader socio-economic challenges, including loss of employment opportunities, migration, and social instability. Climate-induced crop failures can lead to supply shortages and price volatility.

3.7 Biodiversity Loss

Climate change impacts coastal ecosystems, leading to biodiversity loss that affects agricultural productivity and resilience. Sea-level rise and coastal development reduce natural habitats, affecting biodiversity. Changes in temperature and precipitation can disrupt ecosystem balance, affecting pollinators and natural pest control. These changes can also alter ecosystems and disrupt the balance of species interactions. Some species may struggle to adapt to these changes, leading to population declines and extinctions (Valkengoed and Steg, 2019). Climate change has resulted in variations in the ecosystem, reducing net productivity, encroaching on agricultural and construction lands, depriving natural habitats, altering migration times and routes, mixing species, extinction, loss of biodiversity, and impacts on livelihood, according to reports from the Indian Network for Climate Change Assessment (INCCA) and the Centre for Climate Change Research (CCCR) (Singh et al., 2021). Loss of biodiversity can reduce ecosystem services such as pollination, pest control, and soil fertility. Biodiversity provides essential ecosystem services such as pollination, pest control, nutrient cycling, and soil fertility. Reduced biodiversity can increase vulnerability to pest and disease outbreaks. Biodiverse ecosystems are more resilient to environmental stresses and disturbances. Biodiversity loss can reduce the resilience of agricultural systems, making them more vulnerable to climate impacts. The main issues facing the Indian agricultural industry are the lack of hardy crop types, the lack of incentives to implement soil and water conservation techniques, the lack of a strong support structure to diversify agricultural revenue, and the difficulty of obtaining loans and access to markets (Chen and Jin, 2017).

4 Adaptation and Mitigation Strategies

Several approaches are essential to addressing the issues posed by climate change in coastal zones and agriculture. Crops that are robust to climate change guarantee food production even under extreme circumstances, and integrated coastal zone management (ICZM) safeguards the sustainable management of coastal resources. Well-built infrastructure, such as seawalls, guards against natural disasters, while effective water management fights shortages. Preparedness is improved by early warning systems, education, and community involvement. Resilience is supported by ecosystem-based adaptation, robust regulations, and agricultural innovation and research.

4.1 Integrated Coastal Zone Management (ICZM)

An integrated approach to policy-making involves different elements under a single, unifying concept. A pertinent definition of integration in the context of marine policy, which at the time primarily focused on fisheries and mineral extraction. In contrast, the management of coastal environments was heavily influenced by terrestrial land use planning for recreation, tourism, and urban development (Portman et al., 2012). ICZM is a process for managing coastal areas to balance environmental, economic, and social objectives. It involves integrating policies and activities across different sectors and levels of governance to achieve sustainable development of coastal zones (Cicin-Sain and Knecht, 2013). Key components of ICZM are as follows:

Environmental Considerations: ICZM emphasizes the protection and conservation of coastal ecosystems, including mangroves, coral reefs, and wetlands. It addresses issues such as coastal erosion, habitat loss, and pollution, implementing measures like shoreline stabilization and habitat restoration (Harvey and Caton, 2010).

Economic Factors: Economic activities in coastal zones, such as tourism, fisheries, and shipping, are managed to ensure they do not compromise the health of the environment. ICZM promotes sustainable economic development by encouraging practices that support both economic growth and environmental preservation (Vinge and Sorensen, 2020).

Social Aspects: The involvement of local communities and stakeholders is crucial in ICZM, as their livelihoods and well-being are directly linked to coastal resources. Public participation and awareness campaigns are essential to garner support for ICZM initiatives and ensure compliance with regulations (Christie et al., 2007).

4.2 Climate-Resilient Crops

Climate change is undeniably real, and its effects are increasingly evident worldwide, marked by a rise in average maximum temperatures. In India, predictions suggest temperature changes ranging from 0.87 °C to 6.31 °C and rainfall variations from − 24.83% to + 15.18% by the 2080s (Jat et al., 2018). While crops like pulses and oilseeds might initially benefit from elevated CO_2 levels, these advantages are likely to be negated by warmer climates and altered rainfall patterns. This reduction is attributed to a shortened growing period and crop duration, faster crop development leading to less efficient resource use, decreased radiation interception, lower harvest index, reduced biomass accumulation, and heightened water stress due to increased evapotranspiration demands (Dimes et al., 2015). Therefore, there is an urgent need to identify climate-resilient crops and cultivars tailored to different regions (Table 4). Pigeon pea and sorghum, as opposed to maize and peanuts, were more robust to climate change shocks in Zimbabwe's semi-arid areas, according to simulation experiments using APSIM by Dimes et al. (2015). This was mainly because of their enhanced harvest index and water-use efficiency. Therefore, it is essential to recommend technologies that enhance farmers' resilience to climate change. To optimize the benefits of elevated CO_2 levels, some strategies to consider are: cultivating early-maturing, photo-insensitive, high-tillering cultivars with optimal root traits and tolerance to abiotic and biotic stresses; mulching with crop residues; planting more seedlings per hill for heat stress; improving soil nutrient and water management; conserving moisture for late monsoon onset; and using stored rainwater for life-saving irrigation during mid-season droughts. Crop intensification and diversification with high-value crops have helped households produce basic staples and generate surplus income in model watersheds adopted by ICRISAT (Wani and Garg, 2018).

Table 4 Examples of abiotic stress tolerant varieties suitable for saline coastal zone

Crop	Characteristics	Varieties	References
Tamarind	Salt-tolerant, requires minimal water once established	PKM 1, Prathisthan, Urigam, Yogeshwari	https://www.sciencedirect.com/science/article/pii/S0378112707000083
Coconut	Thrives in coastal areas, high salt tolerance, high humidity	West Coast Tall, East Coast Tall, Lakshadweep Ordinary, Dwarf Orange	https://www.fao.org/3/AD089E/AD089E03.htm
Mango	Tolerates coastal conditions, moderately salt-tolerant	Alphonso, Banganapalli, Himsagar, Kesar	https://www.researchgate.net/publication/279095845_Salt_tolerance_in_Mango
Cashew	Prefers sandy soils, salt-tolerant	Vengurla 1, Vengurla 4, VRI 3, Ullal 3	https://www.researchgate.net/publication/285908114_Cashew
Paddy	Varieties like CSR-10 and MTU-1010 for saline-alkaline soils	CSR-10, MTU-1010, IR 36, IR 64	https://www.nabard.org/auth/writereaddata/tender/1811200876Saline-Alkaline%20Rice.pdf
Sorghum	Drought-resistant, handles saline soils	CSV 15, CSV 17, SSV 84, M 35–1	https://www.ncbi.nlm.nih.gov/pmc/articles/PMC3273450/
Pearl millet	Drought-resistant, grows in poor soil conditions	Pusa 605, Raj 171, HHB 67, ICTP 8203	https://www.cabi.org/isc/abstract/20103272288
Okra	Adaptable to various soils, including saline	Arka Anamika, Varsha Uphar, Parbhani Kranti, Pusa Sawani	https://www.sciencedirect.com/science/article/pii/S0378429017303390
Seaweed	Grown in coastal waters, provides food and income	Kappaphycus alvarezii, Gracilaria, Ulva, Sargassum	https://www.sciencedirect.com/science/article/pii/S0168160512002619
Groundnut	Grows in sandy and saline soils, well-suited for coastal areas	JL 24, K 6, TAG 24, TMV 2	https://www.ncbi.nlm.nih.gov/pmc/articles/PMC5885517/

4.3 Improved Infrastructure

Protecting coastal regions from natural dangers requires improving coastal defences, such as seawalls and levees. Furthermore, putting money into sturdy agricultural infrastructure protects food production from severe weather occurrences. The following actions support long-term sustainability and climate adaption in addition to safeguarding agricultural areas:

Seawalls and levees: Seawalls and levees are essential for protecting coastal areas from storm surges and flooding. They can make a big difference in managing the impact of rising sea levels and extreme weather events. Studies show that these structures are very effective in reducing flood risks, though we need to handle challenges like high costs and possible environmental impacts (Linham and Nicholls, 2012).

Nature-Based Solutions: Different nature-based solutions, such as restoration of mangroves, wetlands, and other coastal ecosystems, can provide sustainable and cost-effective defense mechanisms. This nature based solutions not only protect against floods but also enhance biodiversity and support local livelihoods (Morris et al., 2018; Kumar et al., 2020).

4.4 Sustainable Water Management

Water requirements vary significantly depending on the crop, soil type, growth stage, and weather conditions. With depleting water resources, it is crucial to adopt methods that optimize water use. To promote sustainable water management, we had to consider the following options:

- Use micro-irrigation systems like sprinklers, drip and sub-surface irrigation for higher water-use efficiency (Enciso et al., 2007).
- Adopt water-saving techniques and practices such as furrow-irrigated raised beds (FIRBs), skip-furrow/alternate furrow irrigation, aerobic rice cultivation, direct drilling into stubble, mulching, anti-transpirants and hydrogels (Rana et al., 2020).
- Rainwater harvesting techniques, such as surface runoff harvesting, rooftop rainwater harvesting, and contour trenches, are also vital for sustainable water use (Agarwal et al., 2001).
- Desalination technologies, including reverse osmosis (RO), electrodialysis, and solar desalination, are essential for converting saline water into freshwater. RO uses a membrane-based process to remove salts and impurities, producing high-quality water suitable for desalinating seawater and brackish water, though it requires significant energy and maintenance. Electrodialysis employs electric current to remove ions, making it efficient for brackish water and less energy-intensive than RO at certain salinity levels. Solar desalination uses solar energy to

evaporate and condense water, leaving salts behind, and is best suited for small-scale applications in areas with high solar intensity due to its renewable energy source and low operational costs (Greenlee et al., 2009; Strathmann, 2010).

4.5 Community Engagement and Education

Engaging local communities in climate adaptation planning and educating farmers on sustainable practices and hazard preparedness are essential components of building resilient agricultural systems. By involving local stakeholders in the planning process, communities can develop adaptation strategies that reflect their unique environmental and socio-economic realities. Imagine a small farming village where local farmers gather to share their experiences and insights. Through educational programs, these farmers learn about sustainable practices like soil conservation, water management, and crop diversification. They start implementing these techniques, seeing firsthand how these changes improve their yields and protect their land. Additionally, picture farmers participating in hazard preparedness training sessions. They learn to anticipate extreme weather events, like floods or droughts, and develop plans to protect their crops and livelihoods. This training helps them feel more secure and less vulnerable to the unpredictable impacts of climate change. These efforts create a strong sense of community as farmers and local stakeholders work together to develop and implement effective adaptation plans. This collaboration not only enhances sustainable agricultural productivity but also strengthens food security for the entire community. Through shared knowledge and collective action, these farming communities become more resilient and better prepared for the future.

4.6 Early Warning Systems

Developing and implementing robust early warning systems for extreme weather events to allow timely evacuation and preparation, reducing the impact on agriculture and human settlements. Early warning systems (EWS) are essential for protecting people and their livelihoods from the devastating impacts of extreme weather events. Understanding the risks associated with extreme weather events is the first step in developing effective EWS. This involves creating detailed maps and assessments to identify high-risk areas and vulnerable populations, as emphasized by the United Nations Office for Disaster Risk Reduction (UNDRR) (Fakhruddin, 2020). Continuous monitoring of weather patterns and environmental conditions using advanced technologies like satellites and weather stations is crucial for spotting early signs of extreme weather. The World Meteorological Organization (WMO) highlights the critical role of accurate forecasting in EWS. Ensuring that warnings and forecasts reach everyone who needs them is vital. Effective communication strategies involve using various channels such as television, radio, social media, text messages, and local community networks. This includes training communities and emergency responders to ensure everyone knows what to do in an emergency, as noted by UNDRR. EWS provides numerous benefits, including protecting agriculture. In India, initiatives like the Agromet Advisory Services offer weather forecasts and farming advice to help farmers mitigate weather-related risks. Early warnings allow people to evacuate and find safety before disasters strike, reducing casualties. For instance, India's cyclone warning system, managed by the India Meteorological Department (IMD), has significantly reduced fatalities during cyclones. Preventing or minimizing damage to homes, infrastructure, and crops can save substantial amounts of money that would otherwise be spent on recovery and rebuilding (Canton, 2021). Moreover, continuous improvement of EWS is crucial for building resilient communities, leading to a more secure and stable environment. Real-world examples highlight the effectiveness of EWS. The IMD's cyclone warning system uses satellite data and advanced forecasting models to provide early warnings and coordinate evacuation plans, significantly reducing deaths and damage from cyclones in India. The Central Water Commission (CWC) in India uses river basin models and real-time data to provide flood forecasts and warnings, helping in timely evacuations and preparedness activities, thus reducing the impact of floods on communities and agriculture. Additionally, the Indian government collaborates with international organizations like the Famine Early Warning Systems Network (FEWS NET) to monitor drought conditions and food security, helping predict droughts and coordinate relief efforts to support vulnerable populations (Brooks et al., 2023).

4.7 Ecosystem-Based Adaptation (EbA)

Ecosystem-Based Adaptation (EbA) involves leveraging natural infrastructures like mangroves, wetlands, and dunes to safeguard coastal regions, enhance biodiversity, and bolster agricultural resilience. These ecosystems provide essential services such as flood protection, erosion control, and habitats for diverse species. For instance, mangroves act as natural barriers against storm surges and help trap sediments, thereby reducing coastal erosion. Wetlands mitigate flood impacts by absorbing excess rainfall and offer habitats for various aquatic species. Dunes buffer inland areas from high tides and storm waves, preventing seawater intrusion. EbA strategies are especially advantageous for agricultural resilience. Coastal wetlands improve water quality by filtering pollutants, crucial for irrigation and soil health. Furthermore, healthy ecosystems sequester carbon, aiding in climate change mitigation. Integrating EbA into coastal management plans delivers multiple socio-economic and environmental benefits. EbA supports sustainable development by preserving the natural balance and health of ecosystems, ensuring they continue to provide valuable services long-term. Local knowledge and participation can enhance strategy effectiveness, ensuring they are tailored to specific environmental and social contexts. Educational programs and capacity-building initiatives empower communities to manage and protect their natural resources, fostering ownership and stewardship.

The importance of EbA is increasingly acknowledged in international climate policy, aligning with the goals of the Paris Agreement and the Sustainable Development Goals (SDGs), particularly those related to climate action, life below water, and life on land. As climate change impacts intensify, EbA's role in enhancing resilience and adaptive capacity becomes even more critical (Spiegel et al., 2023).

4.8 *Remote Sensing, AI, and ML for Managing Natural Hazards in Coastal Agriculture*

In coastal zones affected by climate change, remote sensing, artificial intelligence (AI), and machine learning (ML) are essential tools for tracking and reducing the effects of natural hazards on agriculture (Himeur et al., 2022). By providing cutting-edge instruments to evaluate environmental factors, forecast risks, and create flexible plans, these technologies guarantee the sustainability of agriculture in these susceptible areas. Remote sensing provides real-time data on important environmental factors such soil moisture, temperature, vegetation health, and sea level rise, enabling continuous, large-scale monitoring of coastal agricultural zones (McCarthy et al., 2017). Storms, floods, and droughts can be detected early with the use of Synthetic Aperture Radar (SAR) sensors in conjunction with satellite photography. Farmers may utilise this data to make educated decisions in real time by using it to assess damage, monitor recovery, and apply precise mitigating measures. Through the analysis of massive datasets, AI and ML improve remote sensing's prediction power by spotting patterns that conventional techniques would overlook (Mahrad et al., 2020). Natural disasters like cyclones, storm surges, and salt intrusion may be predicted in terms of frequency and intensity using machine learning algorithms that analyse historical climate data in conjunction with real-time satellite imagery. These forecasts contribute to the development of early warning systems related to climate change, allowing farmers enough time to save resources and crops (Tahsin et al. 2018). AI may help with decision-making related to adaptive agricultural practices, such as choosing crop types that are adaptable to climate change and allocating resources as efficiently as possible in light of anticipated dangers. These technologies also support the post-disaster evaluation process. The combination of AI, ML, and remote sensing becomes critical in coastal agriculture zones, where climate change exacerbates increasing sea levels, saltwater intrusion, and unpredictable weather patterns. By giving farmers and policymakers the capacity to anticipate, plan for, and react to dangers, they improve the sustainability and resilience of agriculture in these delicate environments.

4.9 Policy and Governance

Strengthening policies and governance frameworks to support climate adaptation measures, provide financial incentives for resilient practices, and ensure equitable resource distribution. India's coastal areas are increasingly at risk due to climate change, facing threats like rising sea levels, storm surges, and coastal erosion. To mitigate these risks, following robust policies and effective governance frameworks are essential to support climate adaptation efforts in coastal regions.

A. **Policy Framework**

National Action Plan on Climate Change (NAPCC): Initiated in 2008, the NAPCC outlines India's strategy for addressing climate change, focusing on eight national missions, including the National Mission for Sustainable Agriculture and the National Mission on Strategic Knowledge for Climate Change (Rani, 2023).

Coastal Regulation Zone (CRZ) Notification: The CRZ Notification regulates human activities along the coastline to protect coastal ecosystems and prevent erosion, providing guidelines for sustainable development in these areas (Agarwal, 2019).

National Cyclone Risk Mitigation Project (NCRMP): Aiming to reduce the vulnerability of coastal communities to cyclones and other hydro-meteorological hazards, the NCRMP enhances early warning systems and disaster management practices (Sarkar, 2023).

B. **Governance Strategies**

State Action Plans on Climate Change (SAPCC): Coastal states in India have developed SAPCCs to adapt national climate strategies to local conditions, addressing specific vulnerabilities within each state.

Community-Based Adaptation (CBA): CBA involves local communities in planning and implementing adaptation measures, ensuring that strategies are culturally appropriate and locally relevant (Rahman and Huq, 2021).

Public–Private Partnerships (PPPs): PPPs have been vital in mobilizing resources for climate adaptation projects, especially in infrastructure development, by leveraging private sector investment (Phuyal, 2020).

C. **Financial Support Mechanisms**

National Adaptation Fund for Climate Change (NAFCC): Established in August 2015, the NAFCC supports climate adaptation projects across various sectors. With a budget of ₹350 crores for 2015–16 and 2016–17 and an estimated ₹181.5 crores for 2017–18, it prioritizes projects that build climate resilience as identified in the SAPCC and NAPCC (Prasad and Sud, 2019).

Green Climate Fund (GCF) Projects in India: India utilizes the GCF to fund large-scale adaptation projects, particularly in vulnerable coastal areas, to enhance resilience against climate impacts (Bhandary, 2022).

Insurance and Risk Financing: Innovative financial products, such as parametric insurance and catastrophe bonds, are being developed in India to provide protection against climate-related disasters (Herrmann, 2017).

5 Conclusion

In conclusion, a comprehensive strategy that incorporates effective adaptation and mitigation techniques is needed to address the effects of climate change in coastal areas and agriculture. Because it protects coastal ecosystems and encourages sustainable development, Integrated Coastal Zone Management (ICZM) is essential to striking a balance between social, economic, and environmental goals. Ensuring food security in the face of severe weather patterns requires the development of climate-resilient crops that are suited to local circumstances. Effective defence mechanisms can be provided by improved infrastructure, such as seawalls and naturally occurring solutions, while sustainable water management maximises the use of finite resources. Resilience and preparation are increased by local involvement fostered by community engagement and education. Combined with remote sensing, artificial intelligence, and machine learning, early warning systems give farmers and governments even more tools to predict and efficiently address environmental disasters. Finally, improved financial support systems and legislative frameworks guarantee that adaptation strategies are carried out fairly and that resources are allocated effectively. When combined, these tactics offer a thorough response to the escalating problems brought about by climate change, strengthening the resilience and sustainability of agricultural systems as well as coastal communities. Future initiatives should concentrate on encouraging sustainable water management, enhancing community involvement, and integrating cutting-edge technology like AI, ML, and remote sensing for climate adaptation in agriculture. Improving financial support systems and legislative frameworks will also be essential for creating coastal areas that are resilient to climate change and maintaining food security in the face of environmental difficulties.

References

Agarwal A (2019) Climate change and coastal zone regulation: dilution of coastal protection, an analysis of CRZ notification, 2018. IOSR J Environ Sci, Toxicol Food Technol 13(8):49–56

Ahammed KKB, Pandey AC, Parida BR, Wasim, Dwivedi CS (2023) Impact assessment of tropical cyclones amphan and nisarga in 2020 in the northern Indian ocean. Sustainability 15(5):3992

Alexandrakis G, Poulos SE (2014) A holistic approach to beach erosion vulnerability assessment. Sci Rep 4(1):6078

Aslam A, Mahmood A, Ur-Rehman H, Li C, Liang X, Shao J, Negm S, Moustafa M, Aamer M, Hassan MU (2023) Plant adaptation to flooding stress under changing climate conditions: ongoing breakthroughs and future challenges. Plants 12(22):3824

Barbier EB, Burgess JC (2021) Economics of peatlands conservation, restoration and sustainable management. Restor Sustain Manage
Barnard PL, Erikson LH, Foxgrover AC, Hart JAF, Limber P, O'Neill AC, Van Ormondt M, Vitousek S, Wood N, Hayden MK, Jones JM (2019) Dynamic flood modeling essential to assess the coastal impacts of climate change. Sci Rep 9(1):4309
Bennett G, Bredehoeft J, Motz LH (2002) Saltwater intrusion and the minimum aquifer level in the southern water use caution area: hydrologic evaluation section, Southwest Florida water management district
Bhandary RR (2022) National climate funds: a new dataset on national financing vehicles for climate change. Clim Policy 22(3):401–410
Bhattacharyya P, Pathak H, Pal S, Bhattacharyya P, Pathak H, Pal S (2020) Impact of climate change on agriculture: evidence and predictions. In: Climate smart agriculture: concepts, challenges, and opportunities, pp 17–32
Bird EC (2008) Coastal geomorphology: an introduction. John Wiley & Sons, New Jersey
Brooks SK, Patel SS, Greenberg N (2023) Struggling, forgotten, and under pressure: a scoping review of experiences of sex workers during the COVID-19 pandemic. Arch Sex Behav 52(5):1969–2010
Bruwier M, Erpicum S, Pirotton M, Archambeau P, Dewals B (2015) Assessing the operation rules of a reservoir system based on a detailed modelling chain. Nat Hazards Earth Syst Sci 15:365–379
Canton H (2021) Pacific community. In: The Europa directory of international organizations 2021. Routledge, pp 735–738
Cazenave A, Llovel W (2010) Contemporary sea level rise. Ann Rev Mar Sci 2(1):145–173
Chen Z, Jin M (2017) Financial inclusion in China: use of credit. J Fam Econ Issues 38:528–540
Christie RM (2007) Environmental aspects of textile dyeing. Elsevier
Cicin-Sain B, Knecht R (2013) Integrated coastal and ocean management: concepts and practices. Island Press
Couasnon A, Eilander D, Muis S, Veldkamp TI, Haigh ID, Wahl T, Winsemius HC, Ward PJ (2020) Measuring compound flood potential from river discharge and storm surge extremes at the global scale. Nat Hazard 20(2):489–504
Dimes J, Rodriguez D, Potgieter A (2015) Raising productivity of maize-based cropping systems in eastern and southern Africa: step-wise intensification options. In: Crop physiology. Academic Press, pp 93–110
Elad Y, Pertot I (2014) Climate change impacts on plant pathogens and plant diseases. J Crop Improv 28(1):99–139
Enciso J, Jifon J, Wiedenfeld B (2007) Subsurface drip irrigation of onions: effects of drip tape emitter spacing on yield and quality. Agric Water Manage 92(3):126–130
Eswar D, Karuppusamy R, Chellamuthu S (2021) Drivers of soil salinity and their correlation with climate change. Curr Opin Environ Sustain 50:310–318
Agarwal A, Narain S, Khurana I (2001) Making water everybody's business: practice and policy of water harvesting. Centre for Science and Environment
Fakhruddin B, Clark H, Robinson L, Hieber-Girardet L (2020) Should I stay or should I go now? Why risk communication is the critical component in disaster risk reduction. Prog Disaster Sci 8:100139
Falchetta G, Pachauri S, Parkinson S, Byers E (2019) A high-resolution gridded dataset to assess electrification in sub-Saharan Africa. Sci Data 6(1):110
Goff J, Nunn PD (2016) Rapid societal change as a proxy for regional environmental forcing: evidence and explanations for Pacific Island societies in the 14–15th centuries. Island Arc 25(5):305–315
Greenlee LF, Lawler DF, Freeman BD, Marrot B, Moulin P (2009) Reverse osmosis desalination: water sources, technology, and today's challenges. Water Res 43(9):2317–2348
Griggs G, Reguero BG (2021) Coastal adaptation to climate change and sea-level rise. Water 13(16):2151

Guha-Sapir D, Below R, Hoyois P (2015) EM-DAT: international disaster database. Université Catholique de Louvain, Brussels, Belgium

Guiteras R (2009) The impact of climate change on Indian agriculture. In: Manuscript. Department of Economics, University of Maryland, College Park, Maryland, pp 1–54

Güneralp B, Güneralp İ, Liu Y (2015) Changing global patterns of urban exposure to flood and drought hazards. Glob Environ Chang 31:217–225

Harvey N, Caton B (2010) Coastal management in Australia. University of Adelaide Press, p 342

Heinrich P, Hagemann S, Weisse R, Gaslikova L (2023) Changes in compound flood event frequency in northern and central Europe under climate change. Front Clim 5:1227613

Herrmann VS (2017) Culture on the move: towards an inclusive framework for cultural heritage considerations in climate-related migration, displacement and relocation policies

Himeur Y, Rimal B, Tiwary A, Amira A (2022) Using artificial intelligence and data fusion for environmental monitoring: a review and future perspectives. Inf Fusion 86:44–75

Holding S, Allen DM (2015) Risk to water security for small islands: an assessment framework and application. Regional Environ Chang 16:827–839. https://doi.org/10.1007/s10113-015-0794-1

Idier D, Bertin X, Thompson P, Pickering MD (2019) Interactions between mean sea level, tide, surge, waves and flooding: mechanisms and contributions to sea level variations at the coast. Surv Geophys 40(6):1603–1630

Jat RK, Singh P, Jat ML, Dia M, Sidhu HS, Jat SL, Parihar CM, Kumar U, Ridaura SL (2018) Heat stress and yield stability of wheat genotypes under different sowing dates across agro-ecosystems in India. Field Crop Res 218:33–50

Johnson DM, Berry ZC, Baker KV, Smith DD, McCulloh KA, Domec JC (2018) Leaf hydraulic parameters are more plastic in species that experience a wider range of leaf water potentials. Funct Ecol 32(4):894–903

Jonkman SN (2005) Global perspectives on loss of human life caused by floods. Nat Hazards 34:151–175

Kaliraj S, Chandrasekar N, Magesh NS (2015) Evaluation of coastal erosion and accretion processes along the southwest coast of Kanyakumari, Tamil Nadu using geospatial techniques. Arab J Geosci 8:239–253

Kaur R, Shivay YS, Singh G, Virk HK, Sen S, Rajni R (2018) Increasing area under pulses and soil quality enhancement in pulse-based cropping systems-Retrospect and prospects. Indian J Agric Sci 88(1):10–21

Kumar P, Debele SE, Sahani J (2020) Towards an operationalisation of nature-based solutions for natural hazards. Sci Total Environ 731:138855

Linham MM, Nicholls RJ (2012) Adaptation technologies for coastal erosion and flooding: a review. In: Proceedings of the institution of civil engineers-maritime engineering, vol 165, no 3. Thomas Telford Ltd., pp 95–112

Machado R, Serralheiro R (2020) Salt stress alleviation through fertilization in fruit crops. Elsevier, pp 465–480

Mahrad BE, Newton A, Icely JD, Kacimi I, Abalansa S, Snoussi M (2020) Contribution of remote sensing technologies to a holistic coastal and marine environmental management framework: a review. Remote Sens 12(14):2313

Maity M, Gonekar VK (2023) Critical study and review of cyclones: with special emphasis on Gangetic West Bengal from 2001 to 2020. J Namibian Stud: Hist Polit Cult 39:857–871

Malhi GS, Rana MC, Kumar S, Rehmani MIA, Hashem A, Abd_Allah EF (2021) Efficacy, energy budgeting, and carbon footprints of weed management in blackgram (*Vigna mungo* L.) Sustainability 2021 13:13239

McCarthy MJ, Colna KE, El-Mezayen MM, Laureano-Rosario AE, Méndez-Lázaro P, Otis DB, Toro-Farmer G, Vega-Rodriguez M, Muller-Karger FE (2017) Satellite remote sensing for coastal management: a review of successful applications. Environ Manage 60:323–339

Michael HA, Russoniello CJ, Byron LA (2013) Global assessment of vulnerability to sea-level rise in topography-limited and recharge-limited coastal groundwater systems. Water Resour Res 49(4):2228–2240. https://doi.org/10.1002/wrcr.20213

Mimura N (2013) Overview of climate change impacts. In: Climate change and global sustainability. United Nations, New York, pp 46–57

Minar MH, Hossain MB, Shamsuddin MD (2013) Climate change and coastal zone of Bangladesh: vulnerability, resilience and adaptability. Middle-East J Sci Res 13(1):114–120

Mohanty B, Vivekanandan E, Mohanty S, Mahanty A, Trivedi R, Tripathy M, Sahu J (2017) The impact of climate change on marine and inland fisheries and aquaculture in India. Clim Chang Impacts Fish Aquac: Glob Anal 2:569–601

Morim J, Vitousek S, Hemer M, Reguero B, Erikson L, Casas-Prat M, Wang XL, Semedo A, Mori N, Shimura T, Mentaschi L, Timmermans B (2021) Global-scale changes to extreme ocean wave events due to anthropogenic warming. Environ Res Lett 16(7):074056

Morris RL, Konlechner TM, Ghisalberti M, Swearer SE (2018) From grey to green: efficacy of eco-engineering solutions for nature-based coastal defence. Glob Chang Biol 24(5):1827–1842

Mortensen S, Kazerooni RF, Harkin A, Womersley T, Jensen B (2022) Dynamic assessment of safety in manoeuvring through constricted navigational channels using an online operational system. Australasian Coasts & Ports 2021: Te Oranga Takutai, Adapt and Thrive: Te Oranga Takutai, Adapt and Thrive. New Zealand Coastal Society, Christchurch, NZ, pp 609–615

Moser SC, Jeffress Williams S, Boesch DF (2012) Wicked challenges at Land's end: managing coastal vulnerability under climate change. Annu Rev Environ Resour 37(1):51–78

Muis S, Haigh ID, Guimarães Nobre G, Aerts JC, Ward, PJ (2018) Influence of El Niño-Southern oscillation on global coastal flooding. Earth's Future 6(9):1311–1322

Mukherjee S, Dash PK, Das D, Das S (2023) Growth, yield and water productivity of tomato as influenced by deficit irrigation water management. Environ Proc 10(1):10

Nakamura M, Tanaka K, Tanaka F, Matsuura Y, Komi R, Niiyama M, Kawakami M, Koeda Y, Sakai T, Onoda T, Itoh T (2017) Long-term effects of the 2011 Japan earthquake and tsunami on incidence of fatal and nonfatal myocardial infarction. Am J Cardiol 120:352–358

Needham HF, Keim BD, Sathiaraj D (2015) A review of tropical cyclone-generated storm surges: global data sources, observations, and impacts. Rev Geophys 53(2):545–591

Neumann B, Vafeidis AT, Zimmermann J, Nicholls RJ (2015) Future coastal population growth and exposure to sea-level rise and coastal flooding-a global assessment. PLoS ONE 10:e0118571. https://doi.org/10.1371/journal.pone.0118571

NGDC/WDS (2015) National Geophysical Data Center/World Data Service, Global his torical tsunami database. Tsunami event data

Nicholls RJ, Cazenave A (2010) Sea-level rise and its impact on coastal zones. Science 328(5985):1517–1520

Nicholls RJ, Wong PP, Burkett V, Codignotto J, Hay J, McLean R, Ragoonaden S, Woodroffe CD, Saito Y (2007) Coastal systems and low-lying areas

Oppenheimer M, Glavovic BC, Hinkel J, van de Wal R, Magnan AK, Abd-Elgawad A, Cai R, Cifuentes-Jara M, DeConto RM, Ghosh T, Hay J, Isla F, Marzeion B, Meyssignac B, Sebesvari Z (2019) Sea level rise and implications for low-lying Islands, coasts and communities. In: IPCC special report on the ocean and cryosphere in a changing climate. Cambridge University Press, Cambridge, UK and New York, NY, USA, pp 321–445

Panwar V, Sen S (2020) Examining the economic impact of floods in selected Indian states. Climate Dev 12(3):281–296

Pararas-Carayannis G (2018) Brief history of early pioneering tsunami research-part A. Sci Tsunami Hazards 37(1)

Park YJ, Lee HJ, Kwak KJ, Lee K, Hong SW, Kang H (2014) MicroRNA400-guided cleavage of pentatricopeptide repeat protein mRNAs renders arabidopsis thaliana more susceptible to pathogenic bacteria and fungi. Plant Cell Physiol 55(9):1660–1668

Phuyal M (2020) Study on India's public-private partnerships (PPPS) in infrastructure: opportunities, trends, and challenges. Int J Adv Res Manage Soc Sci 9(7):1–24

Portman ME, Esteves LS, Le XQ, Khan AZ (2012) Improving integration for integrated coastal zone management: an eight-country study. Sci Total Environ 439:194–201

Prasad DH, Kumar ND (2014) Coastal erosion studies—a review. Int J Geosci

Prasad R, Kumar M, Varma A (2015) Role of PGPR in soil fertility and plant health. In: Plant-growth-promoting rhizobacteria (PGPR) and medicinal plants, pp 247–260
Prasad RS, Sud R (2019) Implementing climate change adaptation: lessons from India's national adaptation fund on climate change (NAFCC). Clim Policy 19(3):354–366
Rahman MF, Huq S (2021) Community based adaptation to climate change. Our Warm Planet: Clim Chang Impacts Adapt 2:144
Rajaeian S, Ketabchi H, Ebadi T (2022) Water resources assessment in Hashtgerd study area based on the system of environmental-economic accounting for water. Amirkabir J Civ Eng 54(9):3313–3332
Rana T, Fatima M, Khan AQ, Naeem Z, Javaid S, Sajid N, Habib A (2020) Hydrogels: a novel drug delivery system. J ISSN 2766:2276
Rani S (2023) Initiatives on climate change mitigation. Climate, land-use change and hydrology of the Beas River Basin, Western Himalayas. Springer International Publishing, Cham, pp 177–202
Reguero BG, Griggs G (2022) Adaptation to coastal climate change and sea-level rise. Water 14(7):996
Resio DT, Westerink JJ (2008) Modeling the physics of storm surges. Phys Today 61(9):33–38
Roy D, Soneji J, Francis A (2024) Institutional inertia in the maintenance of infrastructure assets: the case of Majerhat Bridge. SAGE Publications, SAGE Business Cases Originals
Sarkar S (2023) National disaster management authority: close encounters with COVID-19. In: International handbook of disaster research. Springer Nature Singapore, Singapore, pp 273–288
Sharma A, Jha P, Reddy GV (2018) Multidimensional relationships of herbicides with insect-crop food webs. Sci Total Environ 643:1522–1532
Singh RK, Singh A, Ksherchokpa L, Rallen O, Taniang B, Lego YJ, Padung E, Kumar A, Tag H, Mauerhofer V (2021) Grassroots approaches for sustaining biocultural diversity and livelihood Security: insights from Indian eastern Himalaya. Environ Manage 68:17–37
Soora NK, Aggarwal PK, Saxena R, Rani S, Jain S, Chauhan N (2013) An assessment of regional vulnerability of rice to climate change in India. Clim Change 118:683–699
Sorensen RB, Nuti RC, Butts CL (2009) Yield and plant growth response of peanut to midseason forage harvest. Agron J 101(5):1198–1203
Spiegel SJ, Kachena L, Gudhlanga J (2023) Climate disasters, altered migration and pandemic shocks:(im) mobilities and interrelated struggles in a border region. Mobilities 18(2):328–347
Strathmann H (2010) Electrodialysis, a mature technology with a multitude of new applications. Desalination 264(3):268–288
Tahsin S, Medeiros SC, Singh A (2018) Assessing the resilience of coastal wetlands to extreme hydrologic events using vegetation indices: a review. Remote Sens 10(9):1390
Talling PJ, Allin J, Armitage DA, Arnott RW, Cartigny MJ, Clare MA, Felletti F, Covault JA, Girardclos S, Hansen E, Hill PR (2015) Key future directions for research on turbidity currents and their deposits. J Sediment Res 85(2):153–169
Telford J, Cosgrave J, Houghton R (2006) Joint evaluation of the international response to the Indian Ocean tsunami. Synth Rep 110
Tully J, DeSantis MK, Schock MR (2019) Water quality–pipe deposit relationships in Midwestern lead pipes. AWWA Water Sci 1(2):e1127
Van Valkengoed AM, Steg L (2019) Meta-analyses of factors motivating climate change adaptation behaviour. Nat Clim Chang 9(2):158–163
Venkateswarlu B (2015) Climate change and dryland agriculture. MS Swaminathan Award 2015–16, 28
Vinayaraj P, Johnson G, Dora GU, Philip CS, Sanil Kumar V, Gowthaman R (2011) Quantitative estimation of coastal changes along selected locations of Karnataka, India: a GIS and remote sensing approach. Int J Geosci 2:385–393
Vinge H, Sørensen SØ (2020) From agri-culture to agri-nature: new allianced or farmland preservation in Norway. In: Finance or Food? The Role of Cultures, Values, and Ethics in Land Use Negotiations

Vousdoukas MI, Mentaschi L, Voukouvalas E, Bianchi A, Dottori F, Feyen L (2018) Climatic and socioeconomic controls of future coastal flood risk in Europe. Nat Clim Chang 8(9):776–780

Wang P, Bernier NB (2023) Adding sea ice effects to a global operational model (NEMO v3. 6) for forecasting total water level: approach and impact. Geosci Model Dev Discuss 2023:1–29

Wani SP, Garg KK (2018) Improving water use in tropical rain-fed systems: the situation in India crops research institute for the semi-arid tropics (ICRISAT), India. In: Water management for sustainable agriculture. Burleigh Dodds Science Publishing, pp 443–464

Wijetunge JJ, Neluwala NGPB (2023) Compound flood hazard assessment and analysis due to tropical cyclone-induced storm surges, waves and precipitation: a case study for coastal lowlands of Kelani River basin in Sri Lanka. Nat Hazards 116(3):3979–4007

Xuan J, Ding R, Zhou F (2021) Storm surge risk under various strengths and translation speeds of landfalling tropical cyclones. Environ Res Lett 16(12):124055

Integrating Remote Sensing and Geographic Information Science for Sustainable Coastal and Marine Ecosystem Management

Karankumar Kishorkumar Ramteke, Sahina Akter, P. Vineeth, Sagar Ronad, Manab Jyoti Barman, and S. Thanga Anusya

Abstract Coastal and marine ecosystems, including coral reefs, mangroves, and seagrass beds, are vital for maintaining biodiversity, supporting livelihoods, and mitigating climate change. However, increasing pressures from human activities and climate change necessitate robust strategies for their conservation and management. This chapter explores the integration of Remote Sensing (RS) and Geographic Information System (GIS) technologies for monitoring and managing these ecosystems. RS, through satellite and aerial imagery, provides crucial, real-time data for assessing ecosystem health, tracking harmful algal blooms (HABs), and mapping sensitive habitats. GIS enables spatial analysis and visualization, essential for marine spatial planning, resource allocation, and effective conservation. The chapter highlights the synergy between RS and GIS, demonstrating their combined power in producing spatially explicit insights for ecosystem-based management. Despite challenges like data accessibility and technical limitations, advancements in sensor technology, machine learning, and open-access data sharing create new opportunities for the sustainable management of coastal and marine environments. This integrated approach supports adaptive, evidence-based decision-making and helps mitigate the effects of socio-economic and environmental pressures.

Keywords Coastal and marine ecosystems · RS · GIS · Ecosystem monitoring · Sustainable management

K. K. Ramteke (✉) · S. Akter · P. Vineeth · S. Ronad · M. J. Barman
ICAR-Central Institute of Fisheries Education, Mumbai, India
e-mail: karankumar@cife.edu.in

S. Thanga Anusya
ICAR-Central Soil Salinity Research Institute, Karnal, Haryana, India

K. K. Basheer Ahammed et al. (eds.), *Monitoring Coastal Marine Ecosystems Under Climate Change*, https://doi.org/10.1007/978-3-032-09137-6_14

1 Introduction

Coastal and marine ecosystems are among the most productive and biologically diverse regions on Earth, playing an essential role in maintaining ecological balance and supporting human communities (McLean et al. 2001). These ecosystems include coral reefs, mangroves, estuaries, tidal flats and seagrass beds. They provide numerous services, such as climate regulation, support for fisheries and biodiversity, transportation, tourism, aquaculture and shoreline protection from natural hazards (Barbier et al. 2011; Biju Kumar et al. 2017). Furthermore, these sensitive coastal ecosystems serves as critical breeding, spawning and nursery grounds for various marine species (Roberts et al. 2003; Seitz et al. 2014). However, with more than half of the world's population residing within 100 km of coastal areas put immense pressure from numerous anthropogenic sources, including urbanization, mangroves deforestation, overexploitation, pollution, and climate change (Turner et al. 1996; Hinrichsen 2013). These impacts have led to habitat degradation, biodiversity loss and reduced productivity reduction, ultimately affecting food security, economic stability and human livelihoods (Macusi et al. 2011).

Given these ongoing challenges, the effective monitoring and management of coastal and marine ecosystems has become our immense responsibility to ensure their sustainable use and preservation. Monitoring helps in understanding the ecosystem state, identifying areas that are at risk and evaluating the effectiveness of different conservation efforts (Crain et al. 2009). Sustainable management involves implementation strategies and policies that balance human needs with the preservation of ecological integrity (Kennish 2022). To achieve these goals, robust and up-to-date information that is spatially comprehensive and temporally consistent is needed, enabling decision makes to develop informed strategies to protect these vital ecosystems.

Remote sensing (RS) and Geographic Information System (GIS) technologies have emerged as powerful tools to monitor and manage these dynamic environments, offering critical insights that enhance conservation and resource management practices (Lakra et al. 2024). Remote sensing technology allows researchers to acquire large-scale and real-time information about the Earth's surface from a distance using satellites, UAVs (unmanned aerial vehicles) and aircraft, all of which contribute monitoring coastal and marine ecosystems (Yang et al. 2022; Karang et al. 2024). Satellite imagery were captured through satellites equipped with specialized sensors, which help in collecting the information about physico-chemical parameters, such as Sea Surface Temperature (SST), Sea Surface Salinity (SSS) and Chlorophyll concentration (CHL) (Loza et al. 2013). This type of information is very much crucial for assessing ecosystem health, mapping the distribution of sensitive habitats and species (Akter et al. 2024), tracking coastal erosion and detecting the occurrence of Harmful Algal Bloom (HABs). GIS, on the other hand enables the visualization, integration and spatial analysis of these data, providing a comprehensive understanding of spatial patterns and processes (De Freitas and Tangliani 2009). GIS also plays a crucial role in marine spatial planning, allowing decision makers to allocate resources and activities like fishing, tourism and conservation zones to minimize conflicts and promote ecosystem health (Noble et al. 2019).

Integration of RS and GIS technologies offer a powerful framework for monitoring ecosystems, guiding conservation efforts, and supporting evidence-based policy-making to ensure the sustainable utilization and preservation of coastal and marine resources (Dwarakish et al. 2008; Nobre and Ferreira 2009; Cenci et al. 2018). Despite their effectiveness, there are challenges associated with using RS and GIS technologies. These include technical limitations, such as the cost of high-resolution data, and operational challenges, like the need for skilled personnel to operate UAVs or analyze satellite images (Shahbazi et al. 2014; Karang et al. 2024). Additionally, environmental factors like cloud cover can affect data quality, especially for optical remote sensing (Wei et al. 2016). However, technological advancements are creating opportunities for innovation. Improvements in sensor resolution, machine learning for data processing, and interdisciplinary collaboration can enhance the utility of RS and GIS in coastal management. There is also an increasing push towards open-access data, which can foster greater collaboration and data-sharing opportunities among researchers worldwide. This chapter presents a comprehensive discussion of how RS and GIS can be used to enhance the monitoring, management, and sustainable conservation of coastal and marine ecosystems, providing an advanced approach that combines cutting-edge technologies for effective decision-making.

2 Remote Sensing Technologies in Coastal and Marine Ecosystem Monitoring

2.1 Overview of Remote Sensing Technologies

The mapping of coastal and marine resources is necessary for ensuring their sustainable utilization and conservation (Bondur 2005). Remote sensing is an efficient method for acquiring frequent, large-scale data on the conditions of oceans and coastal regions (Chassot 2011). It has been providing excellent opportunities for various activities including habitat mapping, ecosystem monitoring, precision agriculture etc. With ever increasing developments and improvements over the decades, currently the Unmanned Armed Vehicles has taken over and is being used widely across the globe. The UAVs offer high versatility and flexibility over the conventional technologies and can be operated rapidly without planned scheduling (Pajares 2015).

2.1.1 Satellite Imagery

Satellite Remote Sensing (SRS) involves capturing image of the Earth's surface using satellites aided with specific sensors that orbit at high altitudes, ranging from a few

Table 1 Presents key parameters, associated sensors and their platforms for coastal and marine ecosystem monitoring

Parameter	Sensor	Platform
SST	MODIS	EOS AQUA
SST	Pathfinder V5	NOAA AVHRR
SST	SEVIRI	MSG, GOES-east
SST METOP	OSI-SAF EUMETSAT	AVHRR
SSS	MIRAS (Level 1/2)	SMOS
Chl *a*	MODIS	EOS AQUA
Chl *a*	SeaWiFS	SeaStar
Chl *a*	MERIS	ENVISAT
Windspeed	ERS	AMI
Wind direction	QuickScat	Seawind

hundred kilometers to thousands of kilometers. These sensors are sophisticated electronic device that detect electromagnetic radiation, either emitted or reflected and convert this information into physical values that can be recorded, processed and analysed (Chassot 2011). Depending on the spectral range of solar radiation Sensors are typically categorized into four main types: visible and near-infrared (which includes ocean-color radiometry), mid-infrared, thermal infrared, and microwave sensors (Martin 2014).

SRS systems are typically distinguished by three fundamental types of resolution: spatial, temporal and spectral (Campbell 2007). Spatial resolution describes the approximate size of each pixel in a satellite image, while temporal resolution refers to the frequency at which a satellite revisits the same location. Spectral resolution indicates the number, width and the position of spectral bands in the electromagnetic spectrum where it can collect reflected radiance. The application of SRS to coastal and marine ecosystems relies heavily on selecting appropriate parameters, sensors and platforms that suit the requirements for data precision and relevance. The most relevant parameters for monitoring coastal and marine ecosystems, along with the respective sensors and platforms are presented in the Table 1.

2.1.2 Aerial Photography

Aerial photography involves capturing images of the Earth's surface or specific target areas using manned aircraft flying at relatively lower altitudes compared to satellites. This approach provides high-resolution imagery that is invaluable for a wide range of applications, including agriculture, forestry, land use planning, and fire detection, mapping wetlands and beach erosion, oceanography and more (Klemas 2009). The advancement of digital aerial cameras has revolutionized the field of aerial remote sensing as by enabling enhanced photogrammetric accuracy and expansive area coverage. These cameras facilitate the acquisition of multispectral data with

Table 2 Remote sensing requirements for open ocean and coastal ecosystems

Requirement	Open ocean	Estuaries
Spatial resolution	1-10 km	20-200 m
Coverage area	2000 km × 2000 km	200 km × 200 km
Coverage frequency	Every 1-6 days	Every 0.5-6 h
Dynamic range	Limited	Extensive
Radiometric resolution	10 to 12 bits	10 to 12 bits
Spectral resolution	Multispectral	Hyperspectral

user-defined resolution, capable of reaching ground sampling distance as fine as 0.1 m (Klemas 2009).

Remote sensing via aircrafts is typically conducted at varying altitudes: high, medium, or low, depending on the specific requirements for resolution and area coverage. High altitude flights are generally conducted by government agencies and cover extensive regions, providing large-scale insights into environmental conditions. In contrast, medium and low altitude flights are employed for more localized studies, often involving small aircrafts to complement field data collection and provide detailed spatial information where high-resolution imagery is crucial (Table 2).

2.1.3 Unmanned Aerial Vehicle System (UAVs)

The term UAVs encompasses a wide variety of platform, including fixed-wing and rotary-wing UAVs, lighter-than-air vehicle such as balloon and airships, as well as more specialized forms like lethal aerial vehicles, decoys and targets, alternatively piloted aircrafts, and uninhabited combat aerial vehicles (Eisenbeiss 2009). UAVs have the capability to cover a large area in a relatively short period, are not subjected to cloud obstruction and ensures the freedom of positioning and timing limitations (Muchiri and Kimathi 2016). Compared to manned aircraft systems, UAVs provides the advantage of being able to use in inaccessible areas, without endangering human life and at low flight profiles (Ma et al. 2019). Although numerous types of UAVs exists, the rotocopter and the fixed wing types are the most common types of UAVs used for image capturing. The captured images travels through a processing step, regraded as 'Image stitching' (Ma et al. 2013) (Table 3).

2.2 Applications of Remote Sensing in Monitoring Coastal Erosion, Water Quality, and Habitat Mapping

With ongoing advancements in sensor technology and data analysis methods, remote sensing systems are increasingly applied to study and manage coastal ecosystems like

Table 3 General features of UAV and Satellites for remote sensing applications (Vanhard 2021)

Features	UAV	Satellite
Operational Flexibility	Very high	Limited
Dependence on Cloud Cover	Not dependent	Dependent
Weather Limitations	Impacted by wind and rain	No direct limitations
Data Preparation	Requires extensive pre-processing	Ready-to-use data
Human Involvement	Requires an operator	Operates autonomously
Data Handling Requirements	Complex and resource-intensive	Minimal
Equipment Configuration	Swappable payload options	Fixed payloads
Regulatory Concerns	Subject to strict regulations	Few to no regulations

wetlands, estuaries, and coral reefs. Imagery from both multispectral and hyperspectral sensors is instrumental in mapping coastal land cover and monitoring suspended particles – whether organic or inorganic – and dissolved substances in coastal waters. Thermal infrared scanners are employed to accurately map SSTs and to trace coastal currents, whereas microwave radiometers provide crucial data on parameters such as ocean salinity, soil moisture, and additional hydrological characteristics. Moreover, airborne LiDAR (Light Detection and Ranging) systems are highly effective for generating bathymetric maps in regions with low to moderate water turbidity. Given the inherent spatial complexity and temporal variability of coastal ecosystems, a blend of satellite and aerial data is often necessary to ensure optimal spatial, spectral, and temporal resolution (Klemas 2011).

2.2.1 Coastal Wetlands and Watersheds Mapping

To assess the effects of land runoff on estuarine and coastal ecosystems, a variety of modelling techniques are employed. These include watershed modelling, hydrodynamic simulations, water quality evaluations (Shivakrishna et al. 2020), and assessments of living resources (Li et al. 2007). The evaluation of land runoff's impact is supported by diverse approaches that encompass watershed models, hydrodynamic analysis, and water quality assessments. These models required detailed data on land use or cover as fundamental inputs. These models enable the estimation of runoff volume and composition entering rivers, bays, and estuaries, and help forecast their effects on ecosystems using data on land cover changes along with other variables such as slope and precipitation (Jensen 2009). With its 30 m resolution and wide range of spectral bands, Landsat TM has shown to be a trustworthy source of land cover data, allowing for the detection of changes in sizable coastal watersheds.

2.2.2 Submerged Aquatic Vegetation and Coral Reefs Mapping

High-resolution multispectral and hyperspectral imagery is crucial for mapping submerged aquatic vegetation (SAV), coral reefs, and various seabed features, usually within a resolution range of 1-4 m (Mishra et al. 2006). Coral reefs are generally found in shallow, clear coastal areas, making them easier to map compared to SAV that grows in more turbid waters. To map SAV, aerial hyperspectral scanners, high-resolution multispectral satellites, as well as systems like IKONOS and QuickBird, have been used effectively, achieving approximately 75% accuracy. These mapping initiatives have categorized regions into several classes, including high-density seagrass, low-density seagrass, and areas without vegetation (Mishra et al. 2006). Advanced hyperspectral sensors are also capable of detecting subtle spectral differences, thereby enhancing the precision of coral reef mapping and their health assessment (Porter et al. 2006). Airborne hyperspectral sensors are particularly beneficial in identifying spectral variations that reflect coral health, providing valuable insights into the condition of these important ecosystems.

2.2.3 Remote Sensing of Ocean Chlorophyll and Productivity

Remote sensing is widely used for monitoring ocean productivity due to its cost-effectiveness and extensive coverage capabilities. Multispectral and hyperspectral sensors, which primarily operate within the visible spectrum, are employed to measure ocean color, chlorophyll levels, and overall productivity. Thermal infrared (TIR) sensors and passive microwave radiometers are utilized for monitoring SSTs and ocean salinity, respectively (Klemas 2011). Instruments like the Sea-Viewing Wide Field-of-View Sensor (SeaWiFS) and MODIS are specifically engineered to track chlorophyll concentrations and SSTs globally, offering valuable insights into marine productivity (Oliver et al. 2004). Calibration using data from in-situ sources such as ships and buoys enhances the accuracy of chlorophyll concentration mapping, ensuring that satellite-derived data is robust and reliable for scientific analysis (Table 4).

2.2.4 Observing Eutrophication and Hazardous Algal Blooms

The distinct colour, location and seasonal occurrence of algal blooms make them detectable by satellites. Remote sensing is effectively utilized to forecast the timing and location of HABs by examining environmental conditions that promote bloom formation and monitoring their progression from offshore areas to coastal zones (Jernakoff et al. 1996). Indicators like chlorophyll-a levels and total suspended solids are helpful for assessing eutrophication and water turbidity. Changes in ocean color are often linked with elevated chlorophyll-a levels, which can signal the occurrence of algal blooms. Regions with increased SSTs are frequently associated with HABs,

Table 4 Oceanographic parameters and corresponding remote sensors (Klemas 2011)

Sensor type	Oceanographic application
Visible and near-infrared (NIR) Sensor	Evaluates turbidity levels in surface water; determines phytoplankton density
Thermal infrared (TIR) sensor	Monitors SSTs; calculates surface heat flux
Microwave sensor	Measures temperature of the ocean's surface, surface heat flux, salinity, sea state, and soil moisture levels
Nadir-pointing radar altimeter	Analyzes ocean current patterns and tidal movements; records significant wave heights
Side-looking radar scatterometer	Estimates wind speed and direction over ocean surfaces
High-resolution synthetic aperture radar (SAR)	Detects swell and internal wave formations; locates oil spills

and these can be identified using advanced very high resolution radiometer (AVHRR) data to monitor SST variations (Chang et al. 2002).

2.2.5 Sea Surface Temperature (SST) Monitoring

Thermal infrared sensors have been extensively used to generate images of both cloud tops and SSTs. Accurate SST determination (±0.5 °C) requires atmospheric corrections, as TIR radiance over the ocean varies based on surface temperatures (Martin 2014). Long term, large-scale observations of SST are crucial for various oceanographic studies, such as evaluating heat sources at air-sea interface and understanding the steric contribution to sea level rise. SST monitoring is also essential for assessing the health the coral reefs, particularly in relation to heat stress events that may lead to coral bleaching. SST data helps in analysing coastal upwelling zones and major oceanic currents, such as the Gulf Stream and Kuroshio Current (Klemas 2011).

2.2.6 Oil Spill Detection and Tracking

RS plays an important role in tracking and forecasting the spread of oil spills, offering critical information for emergency response and mitigation efforts, including the use of skimmer vessels and containment booms. An effective response to oil spills demands rapid data acquisition and frequent imaging of the impacted region. Satellites and aircraft equipped with suitable sensors are capable of collecting data at varying resolutions and temporal intervals, enabling thorough tracking of oil spills. Moreover, sound-based sensors placed on vessels, along with imaging devices on underwater robotic systems, can be utilized to examine the movement and dispersion of spilled oil below the water surface.

3 Geographic Information Systems in Ecosystem Management

ESRI describes GIS as a coordinated system comprising computer hardware, software, geographic data, and skilled personnel. This system is intended to efficiently capture, store, update, manipulate, analyze, and visualize geographically referenced information. Their capacity to map and model large areas with significant volumes of spatial data has led to a remarkable proliferation of applications (Escobar et al. 2008). GIS enables a better understanding of scientific data, patterns, and relationships from geographical contexts, which ultimately leads to improved communication, management, and decision-making. GIS can be used to perform various tasks such as identifying issues, delineating degraded areas, urban planning, market and trend analysis, prediction and forecasting, ecosystem monitoring, climate change modelling, impact analysis, locating disease outbreaks, landscape assessment, and many other applications. Spatial data analysis and GIS-based studies provide opportunities for illustrations, figures, and maps that enhance transparency regarding management measures and improve compliance among stakeholders, such as fishermen (Dineshbabu et al. 2016).

3.1 Components of GIS

GIS integrates five major components: Hardware, Software, Data, Personnel, and Methods.

Hardware: Hardware is the fundamental component of GIS, enabling data storage and analysis. The choice of hardware depends on the requirements of the organization and the tasks to be performed. Confidential units employ intranet connections between centralized servers and workstations, while other institutions use GIS via high-speed internet connections to collaborate with remote locations. GIS hardware can be divided into two categories: input devices (e.g., keyboards, scanners, and digitizers) and output devices (e.g., printers and plotters).

Software: Software provides the tools to facilitate data storage, analysis, and visualization. The core components of GIS software include a Database Management System (DBMS), Geographical User Interface (GUI), and tools that support data input, analysis, and visualization. GIS software packages are application-specific, and various packages exist today with diverse functional capabilities. Common GIS software includes ArcGIS, GRASS GIS, QGIS, SAGA GIS, MapInfo, and AutoCAD Map.

Data: Data is the most critical component of GIS, divided into spatial (geographic) data and non-spatial (attribute) data. Spatial data describes the absolute and relative location of a geographic entity, facilitating the representation of features such as topography, water bodies, land use patterns, forest cover, and infrastructure. Attribute data provides detailed information about the nature and characteristics of spatial data. Spatial and attribute data are entered into GIS through a process called digitizing. Integrating geographic data with its corresponding attribute data forms the basis for all GIS analysis and visualization applications.

People: The successful operation of GIS depends on the people who operate the system to find solutions to real-world problems. GIS technology plays a significant role in various sectors, providing employment opportunities to personnel across multiple industries globally.

Methods: Methods refer to the guidelines, specifications, procedures, operating practices, and standards necessary for the proper functioning of GIS.

3.2 Applications of GIS in Ecosystem Management

GIS has the potential to execute various functions associated with marine ecosystem management. Caddy and Garcia (1986) identified several relevant applications for GIS in this context, including mapping local environments, implementing management strategies, organizing scientific resource surveys and data collection, cataloguing resources, delineating fishing grounds in support of international agreements, and mapping fishing effort distribution.

3.3 Mapping of Resources

3.3.1 Demarcation of Areas of Decline

Coastal ecosystems and ecologically sensitive habitats, such as mangroves, are declining at an alarming rate due to deforestation for shrimp farming, timber extraction, and land use transformation. To assess the extent of this decline, mapping and monitoring resources is crucial. The mapping of mangroves is carried out using satellite images from Landsat-5 Thematic Mapper (TM), Landsat-7 Enhanced Thematic Mapper Plus (ETM +), and Landsat-8 Operational Land Imager (OLI). Selected spectral bands, including Green, Red, Near-Infrared (NIR), and Shortwave Infrared (SWIR), are converted from digital numbers to reflectance to enhance the accuracy of mangrove identification, as digital numbers can be similar for different objects while reflectance may differ (NASA 2011).

Reflective spectral characteristics, vegetation indices, or ratio indices, such as Normalized Difference Vegetation Index (NDVI), Soil-Adjusted Vegetation Index (SAVI), and Normalized Difference Water Index (NDWI), are derived from satellite images to locate mangrove forests. NDVI measures forest cover by calculating the difference between near-infrared light, which vegetation strongly reflects, and red light, which vegetation absorbs. NDWI is calculated as:

Table 5 Summary of remote sensing studies on land use, mangrove cover, and health assessment

Study	Indices used	Observation	Area of the study	Reference
Land use and land cover changes	NDVI, NDWI	Increase in aquaculture and a significant decrease in agriculture land and forest cover	Bangladesh	Faruque et al. (2022)
Detection of changes in mangrove cover	NDVI	Decline in the mangrove cover for a brief period from 2013–2016, followed by an increase from 2019–2021	Pulau Kukup	Nordin et al. (2022)
Detection of changes in mangrove cover	NDVI, SAVI, NDWI	Increase in the mangrove cover during the last three decades	Odisha coast	Roy et al. (2019)
Assessment of mangrove health	NDVI	Negative trend/poor health due to deforestation	Segara Anakan, Indonesia	Akbar et al. (2020)

$$\mathrm{NDWI} = \frac{\rho_{\mathrm{NIR}} - \rho_{\mathrm{SWIR}}}{\rho_{\mathrm{NIR}} + \rho_{\mathrm{SWIR}}}$$

where ρ_{NIR} represents the reflectance value of the Near Infrared band, and ρ_{SWIR} represents the reflectance value of the Shortwave Infrared band. NDWI is sensitive to the changes in the liquid water content of vegetation and can be used to monitor vegetation health (Gao 1996). SAVI, representing the non-vegetation land cover, is calculated by using the following formula (Huete 1988):

$$\mathrm{SAVI} = (1 + L) \times \frac{\mathrm{NIR} - \mathrm{RED}}{\mathrm{NIR} + \mathrm{RED} + L}$$

where, L is a soil brightness correction factor (Huete 1988). Subtracting the calculated NDVI value from SAVI helps to obtain a well-defined mangrove cover (Table 5).

3.3.2 Identification of Fishing Grounds

The essential data requirements for mapping the distribution of marine fishery resources, such as latitudinal and longitudinal coordinates of the area of fishing operation, can be obtained from the GPS data of the fishing vessels. Long-term data on the temporal distribution of a particular fish species can be employed in tracing the migration routes in different life history stages of the species and also identify the changes in the distribution and abundance of the species. Additionally, data on the distribution and abundance of a particular species also provides evident information on the type of habitat and substratum associated with the species. GIS enables the analysis of both anthropogenic and natural stressors on the marine environment within a spatial platform, facilitating the formulation of policies and management measures such as restriction on fishing effort, ban on the capture of juveniles, conservation of endangered species, identification of critically sensitive ecosystems, and demarcation of Marine protected areas (MPA) based on the complexity and severity of the problem (Dineshbabu et al. 2016).

3.3.3 Marine Spatial Planning

Marine spatial planning is often regarded as a practical tool that facilitates the implementation of Ecosystem-based management and helps achieve the goals of the Marine Strategy Framework Directive (Stelzenmuller et al. 2013; Santos et al. 2014). It is generally defined as a public process aimed at analyzing and allocating the spatial and temporal distribution of human activities within marine environments, aiming to balance ecological, economic, and social objectives, typically established through a political framework (Ehler and Douvere 2009). MSP involves conducting an integrated assessment of multiple scenarios, as well as analyzing and interpreting data with a spatial component so there are obvious advantages in employing MSP within a GIS framework (Stelzenmuller et al. 2013). GIS can be employed in all three key steps of MSP (Ehler and Douvere 2009), such as Data collection on present conditions, Analysis of existing conditions, and Prediction and projection of future conditions by identification and assessment of alternative management measures (Stamoulis and Delevaux 2015).

Data collection: Four different types of spatial data, such as ecological, administrative, environmental, and human use, are essential for MSP (Shucksmith and Kelly 2014). For the analysis and visualization of accumulated spatial data, a GIS application is necessary.

Data analysis: Different tools are available for the analysis, all of which fall under the domain of Geographic Information Science (GISc). The analyses include Spatial ecological modelling and Cumulative Impact Assessment. Spatial ecological modelling refers to the characterization of marine ecosystems and biological communities to highlight the ecologically sensitive areas based on species-habitat interactions (Mellin et al. 2009). GIS, coupled with modelling and machine learning, can be employed to extrapolate in-situ data on the diversity, distribution, absolute, and relative abundance of a species based on marine landscape characteristics (Bostrom et al. 2011). The resulting map provides spatially explicit information, which aids spatial managers and ecologists in gaining a better understanding of species-habitat interactions and assists in formulating the necessary management strategy (Stamoulis and Delevaux 2015). Cumulative impact assessment, on the other hand, is the analysis of human use data overlaid as spatial layers in a GIS to identify the potential areas of conflict between human activities and the natural environment (Maxwell et al. 2013). The analytical process and the subsequently created maps can be useful in assisting marine planning efforts.

Table 6 Overview of studies on marine spatial planning (MSP) and multi-criteria evaluation (MCE) in marine management

Study	Location of the study	Observation	Reference
Evaluation of MSP scenarios	German EEZ	Identification of suitable sites for co-locating offshore wind farms and aquaculture	Gimpel et al. (2015)
Development of multi-criteria evaluation (MCE) models	Port Stephens-great lakes marine park (PSGLMP)	Integrated assessment of ecological data with stakeholder feedback for reviewing the existing spatial plan of MPA	Noble et al. (2021)

Prediction and projection of future conditions: This step involves the identification and assessment of alternative management measures and in this context, Decision Support systems (DSS) play a significant role. DSS tools work by comparing the various alternative scenarios, locating the domains of synergy, recognizing potential cost-efficient solutions, and analyzing trade-offs (Collie et al. 2013; White et al. 2012). Commonly used GIS-based DSS tools are SITES, Marine Atlas, Habitat Suitability Modelling, OceanMap, and CORSET (Table 6).

3.3.4 Disaster Management

GIS plays a pivotal role in the entire disaster management cycle: Mitigation, Preparedness, Recovery, and Response (Krishnamoorthi 2016). Integrated analysis of data on the historical patterns of devastating phenomena, topography, hydrology, geology, demographics, and infrastructure of the potential area of disaster using GIS is used in mapping the disaster-prone and highly vulnerable areas, development of situational awareness and disaster management strategies, and coordinating rescue and response efforts.

4 Integrative Approaches Using RS and GIS for Sustainable Coastal and Marine Ecosystem Management

Coastal and marine environment management is a complex challenge that requires an integrative approach to address the myriad environmental, social and economic dimensions involved. RS and GIS are two complementary tools that have proven to be invaluable for sustainable coastal and marine ecosystems management (Dean and Populus 2013). The integration of these technologies enables comprehensive monitoring, analysis and management, offering insights that would be difficult to obtain using traditional methods.

4.1 Synergy Between RS and GIS for Comprehensive Monitoring and Management

The combination of RS and GIS creates a powerful synergistic effect, enabling comprehensive and efficient monitoring of coastal and marine ecosystems. RS provides the necessary data inputs, while GIS serves as a tool for processing, analyzing, and visualizing these inputs in a spatially explicit manner (De Freitas and Tagliani 2009). Together, they allow for the monitoring of coastal changes, identification of hotspots of ecological degradation, and analysis of long-term trends (Gabriele et al. 2023). The dynamic nature of coastal environments, which are subject to frequent and rapid changes, makes RS particularly useful. Satellite imagery can provide repeated coverage of large areas, allowing for the monitoring of variables such as sediment transport, shoreline dynamics, and algal blooms (Chassot et al. 2011; Loza et al. 2013). When integrated into a GIS environment, this information can be merge with other data layers, such as human infrastructure, protected areas, and socio-economic data, thereby providing a more holistic understanding of the ecosystem and the pressures it faces (Ouellette and Getinet 2016).

Remote Sensing (RS) provides detailed satellite and aerial imagery, as well as data on water quality, elevation, topography, bathymetry, land use, pollution, and sediment discharge (Petropoulos et al. 2015; Gholizadeh et al. 2016; Gutierres et al. 2016; Erena et al. 2020). These RS data can be integrated into Geographic Information Systems (GIS), enabling the accurate mapping of coastal habitats such as mangroves, coral reefs, and seagrass beds (Ouellette and Getinet 2016). This integration facilitates monitoring coastal erosion, assessing shoreline changes, evaluating ecosystem trophic health, mapping pollution, conducting disaster risk assessments, detecting and tracking HABs, bathymetric mapping and seabed characterization, as well as supporting marine spatial planning and fisheries management (Madkour 2015; Paul et al. 2016; Cham et al. 2020; Khan et al. 2021; Manno et al. 2022). This synergy between RS and GIS supports more accurate assessments, improved resource management, and the effective implementation of conservation strategies, ensuring the sustainability of coastal and marine ecosystems.

4.2 Integrative Methodologies Combining RS Data with GIS Tools for Enhanced Decision-Making

Integrative methodologies that combine RS data with GIS are diverse and cater to various aspects of coastal and marine ecosystem management. These methodologies exploit the strengths of RS in providing detailed spatial and temporal information, and the analytical power of GIS for processing, visualizing, and modelling these datasets. Here in table we have discussed some of the key integrative methodologies (Table 7).

Table 7 Overview of integrative methodologies and their applications in coastal and marine ecosystem management

Integrative methodology	Application
Multi-source data integration	This approach can identify SST changes and their correlation with coral bleaching events
Multi-temporal analysis	Monitoring changes in ecosystems over time, such as coastal erosion, land cover, and mangrove deforestation
Spatial modelling and predictive analysis	Used to develop spatial models for predicting the impacts of different environmental scenarios
Environmental sensitivity index (ESI) mapping	Identify vulnerable areas, especially for oil spill response and natural disaster preparedness
Marine spatial planning (MSP) and decision support systems (DSS)	MSP allocates marine space to optimize ecological and economic objectives, while DSS evaluate potential conflicts or synergies between activities like conservation, fisheries, and offshore energy development
Multi-criteria decision analysis (MCDA)	Identifying optimal areas for activities like wind farms or conservation by weighing environmental and socio-economic factors
Habitat suitability modelling	Identify critical habitats, aid conservation planning, and determine factors influencing species distribution
Mapping anthropogenic impacts	Overlay human activity layers, like urban development, agricultural runoff, or emissions, to evaluate impacts on ecological health and water quality for targeted interventions
Real-time monitoring and adaptive management	Continuous monitoring, enabling rapid analysis and adaptive management to address emerging environmental issues
Bathymetric mapping and seabed characterization	This approach is essential for understanding underwater topography and seabed composition

4.3 *Practical Examples and Case Studies Demonstrating Integrated RS and GIS Approaches*

The integrated RS and GIS approach has been effectively demonstrated in numerous practical case studies. Three case study were presented below to show successful integration of RS and GIS.

- **Mapping and predicting HABs**: Karki et al. (2018) analyzed time-series data from MODIS to determine spatial and temporal trends related to HABs. The researchers used remote sensing techniques, including spectral indices, to detect chlorophyll-a levels, which serve as indicators of algal presence, and to observe bloom patterns. By combining these datasets, they created a model to predict HAB occurrences, enabling the identification of bloom hotspots and issuing early warnings for local communities. Environmental factors were also assessed, such as SST and nutrient levels, to understand the primary drivers of HAB events. This predictive model offered critical insights for ecosystem managers and policymakers, aimed at reducing the adverse effects of HABs on public health, fisheries, and tourism.

- **Mangrove forest mapping and monitoring**: Giri et al. (2011) utilized RS data alongside GIS to assess and quantify global mangrove coverage, emphasizing temporal dynamics. By integrating these technologies, they conducted a comprehensive spatial analysis, allowing them to detect mangrove loss or gain over time. GIS was instrumental in merging and georeferencing satellite images from various years, facilitating the examination of changes across different regions. The study produced a valuable baseline dataset that supports policymakers, conservationists, and environmental managers in efforts to conserve and restore mangrove ecosystems.
- **Marine Spatial Planning (MSP)**: In their study, Stelzenmüller et al. (2013) utilized remote sensing (RS) to collect essential spatial information, including habitat classifications, oceanographic features, and human impacts. This data was processed and analyzed using GIS, enabling the integration of multiple spatial layers to evaluate potential conflicts, synergies, and the suitability of various marine activities. They developed spatial models that illustrated complex data relationships, thereby aiding stakeholders and planners in making informed decisions. This integrated approach proved effective in balancing ecological conservation goals with socio-economic activities, contributing to sustainable management strategies in marine spatial planning.

5 Challenges and Opportunities

5.1 Challenges

RS and GIS are an important tool in recent years for detailed local scale surveys and spatial data management by providing the highly accurate digital maps. Spatial data from various satellites used widely to produce panoptic landscape pattern, land use and land cover maps, sea scape maps, altimetry data, and many others. However in the initial days of remote sensing, satellite data like Landsat and SPOT faced challenges in capturing complex features like coastal habitats, ecosystems, and geology, which are highly dynamic both spatially and temporally. Considering the recent developments in the field of RS, we have observed major contributions in coastal dynamics, integrating land and sea for various environmental applications. However, certain shortcomings and limitations still exist, particularly regarding coastal management.

Implementing GIS and RS to coastal management brings numerous benefits, but it also faces several kinds of challenges, and one of the main challenge among them is cloud cover, which obstructs the spectral light and limits the effectiveness of obtained images that are gathered in the electromagnetic spectral range i.e., infrared and visible parts. The plausible cause behind it can be potentially mitigated, based on the purpose of its use, by utilizing the imagery captured in the microwave region of the electromagnetic spectrum, which is not impacted by any obstructions such as cloud cover. Alternatively, using high temporal resolution imagery can increase the chances of obtaining cloud-free data. Additionally, the coastal zone experiences significant variability in both the concentration and type of light-absorbing aerosols, which can complicate current atmospheric correction methods (McCarthy et al. 2017).

Satellite Remote Sensing (SRS) is one of the essential tool for assessing the current status and pressures on biodiversity across various spatial scales, including landscape, regional, ecosystem, continental, and global levels for about 3 decades (Horning et al. 2010). Satellite Remote Sensing (SRS) has the potential to aid in natural resource management is greater than ever, yet several challenges hinder its effectiveness. Even though many of the satellite products are available at free of cost, majority of them aren't free to access (Turner 2013). Additionally, conducting SRS-based data analyses can be costly due to the logistical needs for hardware, software, trained personnel, and training for processing large datasets. These expenses can be substantial and restrict the broader use especially in applied ecology and management by satellite monitoring (Strittholt et al. 2007; Leidner et al. 2012). The integration of data obtained from regional ecologists with insights from peer analysts with a remote sensing background is often inadequate, resulting in the underuse and undervaluation of SRS. Biodiversity monitoring experts must navigate the intersection of ecology and remote sensing, but they frequently find it challenging to comprehend the pros and cons of specific SRS products (Nagendra et al. 2013). A variation in the temporal data between the users need and the availability of SRS-derived products can mislead information among users. Environmental data is accessed by many users mainly scientists and managers desire for the current year data, but SRS-based products issue latent data which are often released years after the data is collected, although this issue is gradually overcoming in the recent years (Pettorelli et al. 2014a, b). SRS have the potential to enhance wildlife and environmental conservation management in most effective way and realized through strong collaborations between remote sensing specialists and biodiversity monitoring experts. However, such collaborative efforts are infrequent due to several factors such as no collaboration with interdisciplinary platforms, challenges related to varying spatial and temporal scales; (iv) hurdles in logistic information transfer and barriers in framing research objectives that are scientifically significant and beneficial to practitioners in various fields (Pettorelli et al. 2014a, b).

Similarly, we have Aerial Remote Sensing (ARS) commonly referred to as aerial photography or manned aircraft involves capturing photographs from aerial space with the help of crews in aircraft and saw significant advancements, particularly during the world wars (Skoog 2010). Manned aircraft have proven effective in various coastal research applications, including coastal managements such as assessments of wetlands, beach erosion management, coral reef studies, marine habitat evaluations, and land-use planning (Yang and Artigas 2009; Klemas 2013). But still, there is a scope of improvement as ARS uses aerial pathways with georeferenced lines, the resultant images produced must be compatable with data from other sensors so that it should be processed in standardized format (Pepe et al. 2018).

Unmanned Surface Vehicles (USV) are also known as autonomous surface crafts, autonomous surface vehicles, or unmanned surface vessels, it is operated on the water's surface without any crew member (Manley 2008; Durate et al. 2016). They have been employed in various marine management activities, including sea surface sampling, bathymetric surveys, pollutant tracking, HABs detection, meteorological monitoring, observation of marine fauna and even spotting the oil spills (Ryan et al. 2010; Meyer 2016; Verfuss et al. 2019). USV challenges possesses, ensuring its automatic traction and autonomous guidance and navigation for Unmanned Surface Vehicles (USVs) continues to be difficult. Additionally, USV sensors are hindered by issues such as water reflections and visual noise. A full understanding of hydrodynamic forces and moments has not yet been achieved, making stable USV missions challenging (Liu 2016). Conducting surveys in extreme weather conditions of marine ecosystem also remains a significant obstacle (Jorge et al. 2019).

Apart from the above-mentioned challenges, due to the rigid complexity in coastal dynamics and marine ecosystems there are many uncertainties from remote sensing derived images. Such as, due to the shallowness in the coastal regions the water has high reflectance rates, which creates biasness for data acquisition (Zheng and DiGiacomo 2017). Due to this, uncertainties that arise at different stages in processing satellite data can lead to faulty results and interpretations. Assigning uncertainty to specific variables when evaluating coastal dynamics and marine ecological parameters is particularly challenging. Therefore, it is critical to examine the anomalies associated with each method and parameter in greater depth (Mahrad et al. 2020). Even though Remote Sensing technologies have advanced significantly in assessing coastal and marine ecosystems, with improvements in data acquisition, processing, and analysis at various scales. However, challenges remain in terms of data accessibility and affordability, especially for countries with limited resources. The growing volume of high-resolution imagery poses difficulties for data access, as building the necessary infrastructure and tools is costly (Mahabir et al. 2018). The coastal and marine research community struggles with data dissemination and the adoption of results necessary for advancing integrated coastal and marine management. We can find difficulties in establishing data storage systems and standard protocols for archiving, particularly for marine ecological applications. Integrating data from various sources is complicated due to limited collaboration among various stakeholders i.e., data holders and the absence of uniform protocols. Apart from this communication between users end and data providers also poses challenges (Pelletier 2020). Countries with accessibility to remote sensing missions could plausibly share environmental data globally, but political and ethical barriers remain unresolved. Remote sensing technologies are vital for supporting integrated coastal and marine management. However, political obstacles continue to hinder international cooperation and for environmental purposes, there will be free exchange of data (Rao and Murthi 2006).

5.2 *Opportunities*

Among the various challenges, there exists a significant opportunity to utilize remote sensing (RS) for gaining deeper insights into sustainable coastal and marine ecosystem management. Emerging remote sensing technologies, including SAR, and LiDAR, have revolutionized sustainable coastal and ecosystem management by providing advanced tools for the monitoring and management of coastal regions. SAR satellites include many satellites among them one such satellite is Sentinel-1, facilitates localised weather patterns, diurnal and semi-diurnal dynamics of sea surface conditions, aiding in the spotting of oil spills, movements of oil rigs, and semantic wave dynamics (Kumar et al. 2025). The intricate details about coastal ecosystems helps to identify materials and that is critical in tackling complex and dynamic challenges while assuring the sustainable management and development of resilient coastal areas. LiDAR on the other hand provides a system generated data mainly, high-resolution topographic data, this aids in enabling assessments of coastal territories, shoreline alterations, underwater landscapes and seascapes.

Utilization of algorithms in RS data is valuable for enhancing our focus to better understand the environment around us. In recent times, artificial neural network algorithms have been employed for various statistical analysis such as regression, image segmentation, image classification, and various other facets of processing the remotely sensed data (Qdais and Shatnawi 2019). In the near future deep learning algorithms are gaining traction for analyzing remote sensing (RS) images, as they can execute tasks such detailed fragmentation of obtained images and even the assessment is accurate (Ma et al. 2019). Internet of Things (IoT) is one of the systems that is contributing to the enhancement of various management aspects and also utilized in coastal and marine environments. The integration of accessing the data, securing and controlling data infrastructure facilitates the monitoring of coastal and marine environmental conditions by consolidating data from different networks into a unified infrastructure. Additionally, they manage big data transmission, ensures data privacy among users, and integrate into large scale and process data through IoT offer significant opportunities for understanding the environment (Granell et al. 2020).

Model-based remote sensing data and collaborative platforms like the Copernicus Data and Information Access Services, have improved data accessibility at regional levels. This enhanced access has fostered stimulated user innovation among users and has led to new avenue for research opportunities (Groom et al. 2019). Remote sensing can be virtually employed to recreate or simulate coastal and marine environments and methods utilizing reality will offer a novel form of information. This will enable stakeholders and the general public to visualize, analyze, and engage with these environments (Bartolini et al. 2018). Such information can aid in analyzing these intricate ecosystems and offers a greater chance to attain sustainable development.

6 Future Directions and Recommendations

6.1 Emerging Trends and Prospects in RS and GIS Technologies

RS and GIS are ever-evolving technologies driven by advancements in satellite systems, big data-enabled next-generation GIS platforms, machine learning, and data analytics. These technologies are expected to play crucial roles in a plethora of fields, including urban planning in coastal regions, disaster risk reduction, and marine ecosystem management (Yue and Jiang 2014). Geospatial Artificial Intelligence (GeoAI) is an upcoming discipline that integrates artificial intelligence with spatial science (Hosena et al. 2023). AI has found its way into nearly every sector, revolutionizing how we extract knowledge from large spatial datasets. It can provide real information by processing large amounts of datasets, which will aid in evidence-based decision-making.

Machine learning algorithms can recognize complex patterns in geospatial data. This capability assists in identifying trends, assisting researchers and decision-makers to address complex problems effectively. Hyperspectral Images (HSI) and Multispectral Images (MSI) are being used for wetland species classification due to their complex composition and patterns. HSI, with its wide spectral range, helps distinguish coastal wetland species that have similar vegetation types, whereas MSI provides detailed spatial context. The fusion of HSI with MSI, therefore, offers complementary benefits, giving a more comprehensive representation (Liu et al. 2020).

Deep learning methods, such as Convolutional Neural Networks (CNNs) have also proven highly effective in vegetation remote sensing. CNNs can differentiate vegetation properties from remote sensing imagery, facilitating effective monitoring and proactive interventions if an imminent threat to vegetation is detected (Kattenborn et al. 2021).

6.2 Recommendations for Policy-Makers, Researchers, and Practitioners to Leverage These Technologies Effectively

Remotely sensed earth observation has brought about a major change in human consciousness. An image of the earth taken in 1968 by Apollo 8 mission crew member created a wave of human awareness towards environmental awareness and many acts concerning environment were put in place (Gore 2006). GIS and RS have been a great tool in providing spatial and temporal data enabling different stakeholders to take an informed decision and facilitate effective monitoring, potential threat detection resource management, and conservation strategies. Since the 1980's world has moved into what is called as Evidence Based Policy (EBP) approach to evaluate the effectiveness of policy and RS can be very much helpful in filling the existing knowledge gaps (Bell et al. 2023).

To effectively leverage RS and GIS for coastal and marine management, **policymakers** should invest in infrastructure that supports high-resolution data acquisition and processing capabilities. **Open-access data policies** should be encouraged to facilitate the sharing of RS and GIS data across regions and disciplines, which can enhance collaborative management efforts. **Researchers** are encouraged to explore new algorithms and analytical techniques that incorporate **AI and machine learning**, as these can automate and enhance the accuracy of data interpretation. Interdisciplinary research initiatives that combine ecological, social, and technological perspectives are also recommended, as these can generate more holistic insights into ecosystem dynamics. **Practitioners** should focus on building capacity in RS and GIS through targeted training programs, ensuring that staff can utilize advanced technologies effectively. Integrating RS and GIS tools into decision support systems (DSS) can help in assessing multiple scenarios and implementing informed management actions. Practitioners should also engage with local communities to communicate insights generated from RS and GIS analyses, enhancing stakeholder participation and compliance.

6.3 *Strategic Planning for Sustainable Coastal and Marine Ecosystem Management Using RS and GIS*

Strategic planning should completely utilize harness the full potential of RS and GIS to create data-driven frameworks that support adaptive and resilient coastal policies. Scenario analysis using GIS-based modelling tools can help stakeholders visualize potential outcomes under different management strategies, thereby aiding in informed decision-making (Stelzenmüller et al. 2013). RS technologies can provide real-time monitoring, which is critical for adaptive management, allowing responses to be adjusted as conditions change, such as during disaster events or seasonal environmental shifts (Hu et al. 2010).

It is also essential to incorporate Ecosystem-Based Management (EBM) principles into planning efforts, using RS and GIS to assess habitat connectivity, biodiversity hotspots, and areas of ecological significance (Giri et al. 2011). Such strategic use of technologies ensures that conservation efforts align with sustainable development goals, maintaining ecosystem services while supporting local livelihoods (Ehler and Douvere 2009). Ultimately, by embracing technological advancements and fostering cross-sector collaboration, RS and GIS can serve as pivotal tools in the proactive and sustainable management of coastal and marine ecosystems (Walker et al. 2005).

7 Conclusion

The combination of Remote Sensing (RS) and Geographic Information Systems (GIS) has significantly enhanced our ability to sustainably manage and conserve marine ecosystems. These ecosystems – including mangroves, coral reefs, estuaries, and seagrass beds – offer critical ecological services, contribute to biodiversity, help regulate the climate, and support the livelihoods of coastal communities. However, these ecosystems are increasingly under threat from human activities and climate change, necessitating comprehensive, targeted management strategies. RS and GIS technologies are essential in overcoming these challenges. Remote Sensing, using satellites, UAVs, and aerial imagery, provides extensive and real-time data about the physical, chemical, and biological characteristics of marine environments. This is essential for monitoring habitat conditions, detecting HABs, tracking changes SST, and mapping coastal erosion. GIS, in turn, serves as a powerful tool for integrating, analyzing, and visualizing this data, thereby supporting informed decisions, spatial planning, and conservation efforts.

The combined use of RS and GIS supports an evidence-based approach for adaptive and resilient management of coastal and marine environments. By employing tools like scenario analysis, habitat suitability modeling, and environmental sensitivity assessments, stakeholders are better equipped to foresee changes, allocate resources strategically, and implement preventive actions to address risks. Such a holistic approach is crucial for tackling present environmental challenges and planning for future issues related to climate change and socio-economic pressures on coastal areas. Though the implementation of RS and GIS is not without its obstacles – such as limited access to high-quality data, technical barriers, and the need for specialized expertise – advances in technology like artificial intelligence, machine learning, and cloud computing offer solutions to overcome these limitations. Furthermore, there are increasing opportunities for collaboration across disciplines and initiatives for open data sharing, both of which will enhance the effectiveness of RS and GIS in coastal and marine management.

Abbreviations

AMI	Active Microwave Instrument
AVHRR	Advanced Very High-Resolution Radiometer
CHL	Chlorophyll concentration
CNN	Convolutional Neural Network
DBMS	Database Management System
DSS	Decision Support Systems
EBM	Ecosystem-Based Management
EBP	Evidence-Based Policy
EEZ	Exclusive Economic Zone

ENVISAT	Environmental Satellite
EOS	Earth Observing System
ERS	European Remote-Sensing Satellite
ETM +	Enhanced Thematic Mapper Plus
GeoAI	Geospatial Artificial Intelligence
GIS	Geographic Information System
GOES	Geostationary Operational Environmental Satellite
GUI	Geographical User Interface
HAB	Harmful Algal Bloom
HSI	Hyperspectral Imaging
IoT	Internet of Things
LiDAR	Light Detection and Ranging
MCE	Multi-Criteria Evaluation
MERIS	Medium Resolution Imaging Spectrometer
MODIS	Moderate Resolution Imaging Spectroradiometer
MPA	Marine Protected Area
MSG	Meteosat Second Generation
MSI	Multispectral Imaging
MSP	Marine Spatial Planning
NDVI	Normalized Difference Vegetation Index
NDWI	Normalized Difference Water Index
NIR	Near-Infrared
NOAA	National Oceanic and Atmospheric Administration
OLI	Operational Land Imager
PSGLMP	Port Stephens-Great Lakes Marine Park
RS	Remote Sensing
SAR	Synthetic Aperture Radar
SAVI	Soil-Adjusted Vegetation Index
SeaWiFS	Sea-Viewing Wide Field-of-View Sensor
SEVIRI	Spinning Enhanced Visible and Infrared Imager
SMOS	Soil Moisture and Ocean Salinity
SRS	Satellite Remote Sensing
SSS	Sea Surface Salinity
SST	Sea Surface Temperature
SWIR	Shortwave Infrared
TIR	Thermal Infrared
TM	Thematic Mapper
UAV	Unmanned Aerial Vehicle
USV	Unmanned Surface Vehicle

References

Abu Qdais H, Shatnawi N (2019) Assessing and predicting landfill surface temperature using remote sensing and an artificial neural network. Int J Remote Sens 40(24):9556–9571

Akbar MR, Arisanto PAA, Sukirno, BA, Merdeka PH, Priadhi MM, Zallesa S (2020) Mangrove vegetation health index analysis by implementing NDVI (normalized difference vegetation index) classification method on sentinel-2 image data case study: Segara Anakan, Kabupaten Cilacap. Earth Environ Sci 584(1)

Akter S, Nakhawa A, Bhendekar S, Lal DM, Abidi ZJ, Nayak BB, Ramteke K (2024) Deciphering the distribution of Indian mackerel, Rastrelliger kanagurta (Cuvier, 1817) along the Northwest coasts of India. Thalassas: Int J Mar Sci 40:1481–1493

Alvarez-Vanhard E, Corpetti T, Houet T (2021) UAV & satellite synergies for optical remote sensing applications: a literature review. Sci Remote Sens 3:100019

Barbier EB, Hacker SD, Kennedy C, Koch EW, Stier AC, Silliman BR (2011) The value of estuarine and coastal ecosystem services. Ecol Monog 81(2):169–193

Bartolini S, Mecocci A, Pozzebon A, Zoppetti C, Bertoni D, Sarti G, Caiti A, Costanzi R, Catani F, Ciampalini A, Moretti S (2018) Augmented virtuality for coastal management: a holistic use of in situ and remote sensing for large scale definition of coastal dynamics. Int J Geogr Inf 7(3):92

Bell A, Klein D, Rieser J, Kraus T, Thiel M, Dech S (2023) Scientific evidence from space—a review of spaceborne remote sensing applications at the science–policy interface. Remote Sens 15(4):940

Biju Kumar A, Bhagyalekshmi V, Riyas A (2017) Climate change, fisheries and coastal ecosystems in India. J Aquat Bio Fish 5:7–17

Bondur VG (2005) Complex satellite monitoring of coastal water areas. In: Proceedings of the 31st international symposium on remote sensing of environment. St. Petersburg, Russia, 20–24 May 2005

Boström C, Pittman SJ, Simenstad C, Kneib RT (2011) Seascape ecology of coastal biogenic habitats: advances, gaps, and challenges. Mar Ecol Prog Ser 427:191–217

Caddy JF, Garcia SM (1986) Fisheries thematic mapping-a prerequisite for intelligent management and development of fisheries. Oceéanographie Tropicale 21(1):31–52

Campbell JB (2007) Introduction to remote sensing Guilford press, New York

Cenci L, Disperati L, Persichillo MG, Oliveira ER, Alves FL, Phillips M (2018) Integrating remote sensing and GIS techniques for monitoring and modeling shoreline evolution to support coastal risk management. Gisci Remote Sens 55(3):355–375

Cham DD, Son NT, Minh NQ, Thanh NT, Dung TT (2020) An analysis of shoreline changes using combined multitemporal remote sensing and digital evaluation model. Civ Eng J 6(1):1–10

Chang GC, Dickey TD, Schofield OM, Weidemann AD, Boss E, Pegau WS, Moline MA, Glenn SM (2002) Nearshore physical processes and bio-optical properties in the New York Bight. J Geophys Res Oceans 107(C9):16–21

Chassot E, Bonhommeau S, Reygondeau G, Nieto K, Polovina JJ, Huret M, Dulvy NK, Demarcq H (2011) Satellite remote sensing for an ecosystem approach to fisheries management. ICES J Mar Sci 68(4):651–666

Collie JS, Beck MW, Craig B, Essington TE, Fluharty D, Rice J, Sanchirico JN (2013) Marine spatial planning in practice. Estuar Coast Shelf Sci 117:1–11

Crain CM, Halpern BS, Beck MW, Kappel CV (2009) Understanding and managing human threats to the coastal marine environment. Ann NY Acad Sci 1162(1):39–62

De Freitas DM, Tagliani PRA (2009) The use of GIS for the integration of traditional and scientific knowledge in supporting artisanal fisheries management in southern Brazil. J Environ Manage 90(6):2071–2080

Dean AM, Populus J (2013) Remote sensing and GIS integration. FAO, Rome

Dineshbabu AP, Sujitha T, Dinesh AC (2016) Handbook on application of GIS as a decision support tool in marine fisheries. CMFRI special publication No.121. Central Marine Fisheries Research Institute, Kochi, p 104

Duarte M, Costa V, Gomes J, Rodrigues T, Silva F, Oliveira SM, Christensen AL (2016) Evolution of collective behaviors for a real swarm of aquatic surface robots. PloS One 11(3)

Dwarakish GS, Vinay SA, Dinakar SM, Pai JB, Mahaganesh K, Natesan U (2008) Integrated coastal zone management plan for Udupi coast using remote sensing, geographical information system and global position system. J Appl Remote Sens 2(1):023515

Ehler C, Douvere F (2009) Marine spatial planning: a step-by-step approach toward ecosystem-based management. In: Intergovernmental oceanographic commission and man and the biosphere programme

Eisenbeiss H (2009) UAV photogrammetry. Dissertation, Institut für Geodesie und Photogrammetrie, ETH-Zürich. Zürich, Switzerland

Erena M, Domínguez JA, Atenza JF, García-Galiano S, Soria J, Pérez-Ruzafa Á (2020) Bathymetry time series using high spatial resolution satellite images. Water 12(2):531

Escobar F, Hunter G, Bishop I, Zerger A (2008) Introduction to GIS. Department of Geomatics, The University of Melbourne, Parkville, Australia

Faruque MJ, Vekerdy Z, Hasan MY, Islam KZ, Young B, Ahmed MT, Monir MU, Shovon SM, Kakon JF, Kundu P (2022) Monitoring of land use and land cover changes by using remote sensing and GIS techniques at human-induced mangrove forests areas in Bangladesh. Remote Sens Appl: Soc Environ 25:100699

Gabriele M, Brumana R, Previtali M, Cazzani A (2023) A combined GIS and remote sensing approach for monitoring climate change-related land degradation to support landscape preservation and planning tools: the Basilicata case study. Appl Geomat 15(3):497–532

Gao BC (1996) NDWI—a normalized difference water index for remote sensing of vegetation liquid water from space. Remote Sens Environ 58(3):257–266

Gholizadeh MH, Melesse AM, Reddi L (2016) A comprehensive review on water quality parameters estimation using remote sensing techniques. Sensors 16(8):1298

Gimpel A, Stelzenmüller V, Grote B, Buck BH, Floeter J, Núñez-Riboni I, Pogoda B Temming A (2015) A GIS modelling framework to evaluate marine spatial planning scenarios: co-location of offshore wind farms and aquaculture in the German EEZ. Mar. Policy 55:02–115

Giri C, Ochieng E, Tieszen LL, Zhu Z, Singh A, Loveland T, Masek J, Duke N (2011) Status and distribution of mangrove forests of the world using earth observation satellite data. Glob Ecol Biogeogr 20(1):154–159

Gore A (2006) An inconvenient truth: The planetary emergency of global warming and what we can do about it. Rodale Press, New York

Granell C, Kamilaris A, Kotsev A, Ostermann FO, Trilles S (2020) Internet of things. In: Guo H, Goodchild MF, Annoni A (eds) Manual of digital earth, 1st edn. Springer, Singapore, pp 387–423

Groom S, Sathyendranath S, Ban Y, Bernard S, Brewin R, Brotas V, Brockmann C, Chauhan P, Choi JK, Chuprin A, Ciavatta S (2019) Satellite ocean colour: current status and future perspective. Front Mar Sci 6:485

Gutierres F, Teodoro AC, Reis E, Neto C, Costa JC (2016) Remote Sensing technologies for the assessment of marine and coastal ecosystems. In: Finkl C, Makowski C (eds) Seafloor mapping along continental shelves. Coastal Research Library, vol 13. Springer, Cham, p 69–104

Hinrichsen D (2013) Coastal waters of the world: trends, threats, and strategies. Island Press, California

Horning N, Robinson JA, Sterling EJ, Turner W, Spector S (2010) Remote sensing for ecology and conservation: a handbook of techniques. Oxford University Press, New York

Hosena B, Rahamanb M, Kumarc S, Sagard L, Akhtard MN (2023) Leveraging artificial intelligence and big data for advanced spatial analytics and decision support systems in geography. Malaysian Appl Geography 1(2):62–67

Hu C, Lee Z, Ma R, Yu K, Li D, Shang S (2010) Moderate resolution imaging spectroradiometer (MODIS) observations of cyanobacteria blooms in Taihu Lake, China. J Geophysical Res: Oceans 115(C4)

Huete AR (1988) A soil-adjusted vegetation index (SAVI). Remote Sens Environ 25(3):295–309

Jensen JR (2009) Remote sensing of the environment: an earth resource perspective. Pearson Education, India

Jernakoff P, Hick P, Ong C, Hosja W, Grigo S (1996) Mapping algal blooms using airborne digital multi-spectral video and the importance of bloom dynamics in the collection of in-water data. Mar Technol Soc J 30(4):36

Jorge VA, Granada R, Maidana RG, Jurak DA, Heck G, Negreiros AP, Dos Santos DH, Gonçalves LM, Amory AM (2019) A survey on unmanned surface vehicles for disaster robotics: main challenges and directions. Sensors 19(3):702

Karang IWGA, Pravitha NLPR, Nuarsa, IW, Basheer Ahammed KK, Wicaksono P (2024) High-resolution seagrass species mapping and propeller scars detection in Tanjung Benoa, Bali through UAV imagery. J Ecol Eng 25(1)

Karki S, Sultan M, Elkadiri R, Elbayoumi T (2018) Mapping and forecasting onsets of harmful algal blooms using MODIS data over coastal waters surrounding Charlotte County, Florida. Remote Sens 10(10):1656

Kattenborn T, Leitloff J, Schiefer F, Hinz S (2021) Review on convolutional neural networks (CNN) in vegetation remote sensing. ISPRS J Photogramm Remote Sens 173:24–49

Kennish MJ (2022) Management strategies to mitigate anthropogenic impacts in Estuarine and coastal marine environments: a review. Open J Ecol 12(10):667–688

Khan RM, Salehi B, Mahdianpari M, Mohammadimanesh F, Mountrakis G, Quackenbush LJ (2021) A meta-analysis on harmful algal bloom (HAB) detection and monitoring: a remote sensing perspective. Remote Sens 13(21):4347

Klemas VV (2009) Sensors and techniques for observing coastal ecosystems. In: Yang X (ed) Remote sensing and geospatial technologies for coastal ecosystem assessment and management, 1st edn. Springer, Berlin, pp 17–44

Klemas V (2011) Remote sensing techniques for studying coastal ecosystems: an overview. J Coast Res 27(1):2–17

Klemas V (2013) Airborne remote sensing of coastal features and processes: an overview. J Coast Res 29(2):239–255

Krishnamoorthi N (2016) Role of remote sensing and GIS in natural-disaster management cycle. Impl J Inter Res 2(3):144–154

Kumar S, Parida BR, Ahammed KB (2025) Flood risk assessment of the Kosi River Basin in North Bihar using synthetic aperture radar (SAR) data and AHP approach. Natural Hazards Research

Lakra D, Singh SK, Gupta SK, Kanga S (2024) Enhancing integrated resource management through remote sensing and GIS. J Geogr Cartogr 7(1):4265

Leidner AK, Turner W, Pettorelli N, Leimgruber P, Wegmann M (2012) Satellite remote sensing for biodiversity research and conservation applications: a committee on earth observation satellites (CEOS) workshop. http://remote-sensing-biodiversity.org/images/workshops/ceos/CEOS_SBA_Biodiversity_WorkshopReport_Oct2012_DLR_ Munich.pdf

Li M, Zhong L, Boicourt WC, Zhang S, Zhang DL (2007) Hurricane-induced destratification and restratification in a partially-mixed estuary. J Mar Res 65:(2)

Liu Z, Zhang Y, Yu X, Yuan C (2016) Unmanned surface vehicles: an overview of developments and challenges. Annu Rev Control 41:71–93

Liu C, Tao R, Li W, Zhang M, Sun W, Du Q (2020) Joint classification of hyperspectral and multispectral images for mapping coastal wetlands. IEEE J Sel Top Appl Earth Obs Remote Sens 14:982–996

Loza S, Danilov S, Schröter J, Nerger L, Maßmann S, Janssen (2013). Assimilating NOAA's SST data and in situ T, S profiles into the BSH operational circulation model for the North and Baltic Seas. In: Paper presented at the future of operational oceanography, congress center Hamburg, Hamburg, Germany, 8 October 2013–10 October 2013

Ma L, Liu Y, Zhang X, Ye Y, Yin G, Johnson BA (2019) Deep learning in remote sensing applications: a meta-analysis and review. ISPRS J Photogramm Remote Sens 152:166–177

Ma L, Li M, Tong L, Wang Y, Cheng L (2013) Using unmanned aerial vehicle for remote sensing application. In: Paper presented at the 21st international conference on geoinformatics, Kaifeng, China, 20–22 June 2013

Macusi ED, Katikiro RE, Deepananda KA, Jimenez LA, Conte AR, Fadli N (2011) Human induced degradation of coastal resources in Asia Pacific and implications on management and food security. J Nat Sci 9(10):13–28

Madkour H A (2015) Detection of damaged areas due to tourism development along the Egyptian Red Sea coast using GIS, remote sensing and foraminifera. Dissertation, National Institute of Oceanography and Fisheries

Mahabir R, Croitoru A, Crooks AT, Agouris P, Stefanidis A (2018) A critical review of high and very high-resolution remote sensing approaches for detecting and mapping slums: trends, challenges and emerging opportunities. Urban Sci 2(1):8

Mahrad BE, Newton A, Icely JD, Kacimi I, Abalansa S, Snoussi M (2020) Contribution of remote sensing technologies to a holistic coastal and marine environmental management framework: a review. Remote Sens 12(14):2313

Manley JE (2008) Unmanned surface vehicles, 15 years of development Paper presented at MTS/IEEE OCEANS Conference, Quebec city, California, 15–18 September 2008

Manno G, Re CL, Basile M, Ciraolo G (2022) A new shoreline change assessment approach for erosion management strategies. Ocean Coast Manage 225:106226

Martin S (2014) An introduction to ocean remote sensing. Cambridge University Press, United Kingdom

Maxwell SM, Hazen EL, Bograd SJ, Halpern BS, Breed GA, Nickel B, Teutschel NM, Crowder LB, Benson S, Dutton P, Bailey H (2013) Cumulative human impacts on marine predators. Nat Commun 4(1):2688

McCarthy MJ, Colna KE, El-Mezayen MM, Laureano-Rosario AE, Méndez-Lázaro P, Otis DB, Toro-Farmer G, Vega-Rodriguez M, Muller-Karger FE (2017) Satellite remote sensing for coastal management: a review of successful applications. Environ Manage 60:323–339

McLean RF, Tsyban A, Burkett V (2001) Coastal zones and marine ecosystems. In: McCarthy JJ, Canziani OF, Leary NA, Dokken DJ, White KS (eds) Climate change 2001: impacts, adaptation, and vulnerability. Contribution of working group II to the third assessment report of the intergovernmental panel on climate change. Cambridge University Press, Cambridge, pp 343–379

Mellin C, Andréfouët S, Kulbicki M, Dalleau M, Vigliola L (2009) Remote sensing and fish–habitat relationships in coral reef ecosystems: review and pathways for multi-scale hierarchical research. Mar Poll Bull 58(1):11–19

Meyer D (2016) Glider technology for ocean observations: a review. Ocean Sci Discuss 2016:1–26

Mishra D, Narumalani S, Rundquist D, Lawson M (2006) Benthic habitat mapping in tropical marine environments using QuickBird multispectral data. Photogram Eng Remote Sens 72(9):1037–1048

Muchiri N, Kimathi S (2016) A review of applications and potential applications of UAV. In: Paper presented at annual conference on sustainable research and innovation, May, 280–283

Nagendra H, Lucas R, Honrado JP, Jongman RH, Tarantino C, Adamo M, Mairota P (2013) Remote sensing for conservation monitoring: assessing protected areas, habitat extent, habitat condition, species diversity, and threats. Ecol Indic 33:45–59

NASA (2011) Landsat 7 Science data users handbook landsat project science office at NASA's Goddard space flight center in greenbelt, 186. https://landsat.gsfc.nasa.gov/wp-content/uploads/2016/08/Landsat7_Handbook.pdf. Accessed 10 Oct 2024

Noble MM, Harasti D, Pittock J, Doran B (2019) Linking the social to the ecological using GIS methods in marine spatial planning and management to support resilience: a review. Mar Pol 108:103657

Noble MM, Harasti D, Pittock J, Doran B (2021) Using GIS fuzzy-set modelling to integrate social-ecological data to support overall resilience in marine protected area spatial planning: a case study. Ocean Coast Manage 212:105745

Nobre AM, Ferreira JG (2009). Integration of ecosystem-based tools to support coastal zone management. J Coast Res SI(56):1676–1680

Nordin NH, Rahman AHA, Wahab NA, bin Tengku Sembok TM, bin Mohd Shukran MA, Yunus MSFM, Isnan S, Abdul MRAAB (2022). Monitoring of mangroves changes in Pulau Kukup using geographical information system (GIS). Earth Environ Sci 1064(1):012009

Oliver MJ, Glenn S, Kohut JT, Irwin AJ, Schofield OM, Moline MA, Bissett WP (2004) Bioinformatic approaches for objective detection of water masses on continental shelves. J Geophy Res Oceans 109(C7)

Ouellette W, Getinet W (2016) Remote sensing for marine spatial planning and integrated coastal areas management: achievements, challenges, opportunities and future prospects. Remote Sens Appl Soc Environ 4:138–157

Pajares G (2015) Overview and current status of remote sensing applications based on unmanned aerial vehicles (UAVs). Photogramm Eng Remote Sens 81(4):281–330

Paul TT, Dennis A, George G (2016) A review of remote sensing techniques for the visualization of mangroves, reefs, fishing grounds, and molluscan settling areas in tropical waters. In: Finkl C, Makowski C (eds) Seafloor mapping along continental shelves. Coastal Research Library, vol 13. Springer, Cham, p 105–123

Pelletier D (2020) Assessing the effectiveness of coastal marine protected area management: four learned lessons for science uptake and upscaling. Fron Mar Sci 7:545930

Pepe M, Fregonese L, Scaioni M (2018) Planning airborne photogrammetry and remote-sensing missions with modern platforms and sensors. Eur J Remote Sens 51(1):412–436

Petropoulos GP, Kalivas DP, Griffiths HM, Dimou PP (2015) Remote sensing and GIS analysis for mapping spatio-temporal changes of erosion and deposition of two Mediterranean river deltas: the case of the Axios and Aliakmonas rivers, Greece. Int J Appl Earth Obs Geoinf 35:217–228

Pettorelli N, Laurance WF, O'Brien TG, Wegmann M, Nagendra H, Turner W (2014a) Satellite remote sensing for applied ecologists: opportunities and challenges. J Appl Ecol 51(4):839–848

Pettorelli N, Safi K, Turner W (2014b) Satellite remote sensing, biodiversity research and conservation of the future. Philosophical Trans Royal Soc b: Biol Sci 369(1643):20130190

Porter DE, Field DW, Klemas VV, Jensen JR, Malhotra A, Field RT, Walker SP (2006) RESAAP final report: NOAA/NERRS remote sensing applications assessment project. University of South Carolina, Columbia, South Carolina

Rao M, Murthi KS (2006) Keeping up with remote sensing and GI advances—policy and legal perspectives. Space Pol 22(4):262–273

Roberts CM, Andelman S, Branch G, Bustamante RH, Carlos Castilla J, Dugan J, Halpern BS, Lafferty KD, Leslie H, Lubchenco J, Warner RR (2003) Ecological criteria for evaluating candidate sites for marine reserves. Ecol Appl 13:199–214

Roy S, Mahapatra M, Chakraborty A (2019) Mapping and monitoring of mangrove along the Odisha coast based on remote sensing and GIS techniques. Model Earth Sys Environ 5:217–226

Ryan JP, Johnson SB, Sherman A, Rajan K, Py F, Thomas H, Harvey JBJ, Bird L, Paduan JD, Vrijenhoek RC (2010) Mobile autonomous process sampling within coastal ocean observing systems. Lim Oceanog Methods 8(8):394–402

Santos CF, Domingos T, Ferreira MA, Orbach M, Andrade F (2014) How sustainable is sustainable marine spatial planning? Part I—linking the concepts. Mar Pol 49:59–65

Seitz RD, Wennhage H, Bergström U, Lipcius RN, Ysebaert T (2014) Ecological value of coastal habitats for commercially and ecologically important species. ICES J Mar Sci 71(3):648–665

Shahbazi M, Théau J, Ménard P (2014) Recent applications of unmanned aerial imagery in natural resource management. Giscience Remote Sens 51(4):339–365

Shivakrishna A, Ramteke K, Dhanya M, Charitha R, Aktar S, Singh R, Abidi ZJ (2020) Spatio-temporal distribution of water quality parameters in ramsar site-Kolleru Lake. Indian J Anim Res 54(6):753–760

Shucksmith RJ, Kelly C (2014) Data collection and mapping–Principles, processes and application in marine spatial planning. Mar Pol 50:27–33

Skoog AI (2010) The Alfred Nobel rocket camera. An early aerial photography attempt. Acta Astronautica 66(3–4):624–635

Stamoulis KA, Delevaux JM (2015) Data requirements and tools to operationalize marine spatial planning in the United States. Ocean Coast Manage 116:214–223

Stelzenmüller V, Lee J, South A, Foden J, Rogers SI (2013) Practical tools to support marine spatial planning: a review and some prototype tools. Mar Pol 38:214–227

Strittholt J, Miles L, Horning N, Fosnight E (2007) Sourcebook on remote sensing and biodiversity indicators. Tech Ser 32:203

Turner W (2013) Satellites: make data freely accessible. Nature 498(7452):37–37

Turner RK, Subak S, Adger WN (1996) Pressures, trends, and impacts in coastal zones: interactions between socioeconomic and natural systems. Environ Manage 20:159–173

Verfuss UK, Aniceto AS, Harris DV, Gillespie D, Fielding S, Jiménez G, Johnston P, Sinclair RR, Sivertsen A, Solbø SA, Storvold R (2019) A review of unmanned vehicles for the detection and monitoring of marine fauna. Mar Poll Bull 140:17–29

Walker ND, Wiseman WJ Jr, Rouse LJ Jr, Babin A (2005) Effects of river discharge, wind stress, and slope eddies on circulation and the satellite-observed structure of the Mississippi River plume. J Coast Res 21(6):1228–1244

Wei J, Lee Z, Shang S (2016) A system to measure the data quality of spectral remote-sensing reflectance of aquatic environments. J Geophys Res Oceans 121(11):8189–8207

White C, Halpern BS, Kappel CV (2012) Ecosystem service tradeoff analysis reveals the value of marine spatial planning for multiple ocean uses. Proc Natl Acad Sci 109(12):4696–4701

Yang J, Artigas FJ (2009) Mapping salt marsh vegetation by integrating hyperspectral and LiDAR remote sensing. CRC Press, Boca Raton, FL, USA, pp 173–187

Yang Z, Yu X, Dedman S, Rosso M, Zhu J, Yang J, Xia Y, Tian Y, Zhang G, Wang J (2022) UAV remote sensing applications in marine monitoring: knowledge visualization and review. Sci Total Environ 838:155939

Yue P, Jiang L (2014) BigGIS: how big data can shape next-generation GIS. In: Paper presented at the 3rd international conference on Agro-Geoinformatics, Beijing, China, 11–14 Aug 2014

Zheng G, DiGiacomo PM (2017) Uncertainties and applications of satellite-derived coastal water quality products. Prog Oceanog 159:45–72

A Review of the Blue Economy of the Coastal Bay of Bengal—Status, Potentials, and Challenges

Subba Rao Venkata Durvasula, Ram Mohan Rao Ponnapalli, and Suman Keerthi

Abstract The Bay of Bengal (BoB), one of the world's largest tropical marine ecosystems, comprises 29 distinct ecological regions and its hydrobiology is shaped by both natural and anthropogenic influences. Major natural drivers include substantial riverine inflow, monsoonal currents, cyclones, sea-level rise, and climate variability, while human pressures such as deforestation, industrial pollution, overexploitation, shipbuilding and aquaculture further impact the marine ecosystem. Monsoonal dynamics strongly regulate primary production, generating phytoplankton hotspots with chlorophyll-a concentrations up to 22.7 mg m^{-3} and carbon fixation rates of 3852 mg C m^{-3} d^{-1} in nearshore waters. These productive zones coincide with Potential Fishing Zones (PFZs), which significantly support the blue economy. Despite the presence of 485 recorded seaweed species, their economic potential remains largely untapped. Mangrove ecosystems spanning approximately 4992 km^2 across the Sundarbans, Odisha, Coringa, Pichavaram, and the Andaman–Nicobar Islands sustain diverse fauna but are increasingly degraded. Similarly, coral reefs—harboring 179 species in the Andaman–Nicobar Islands and 31 in the Gulf of Mannar—are threatened by recurrent bleaching. National research and development institutions, including CIBA, CMFRI, and MPEDA, have advanced aquaculture, oyster and finfish farming, and seaweed cultivation, with more than 400 hatcheries currently engaged in *Litopenaeus vannamei* culture in particular. To enhance blue economy initiatives, this study recommends (1) regionally relevant

S. R. Venkata Durvasula (✉)
Department of Biology, University of New Mexico, Albuquerque, NM, USA
e-mail: seshu35@gmail.com

Present Address:
R. M. R. Ponnapalli
Former Director of Fisheries Andhra Pradesh, Kakinada, India
e-mail: rammohanraokkd@gmail.com

S. Keerthi
Research and Development, BMR Industries Pvt. Ltd., Palavakkam, Chennai, Tamil Nadu, India
e-mail: keerthisuman@bmrgroups.org

K. K. Basheer Ahammed et al. (eds.), *Monitoring Coastal Marine Ecosystems Under Climate Change*, https://doi.org/10.1007/978-3-032-09137-6_15

environmental education, (2) integrative understanding of ecosystem–economy linkages, (3) regular environmental impact assessments, (4) adoption of integrated coastal zone management, and (5) community empowerment for participatory governance.

Keywords Phytoplankton production · Upwelling · Monsoons · Fishing zones

1 Bay of Bengal: The Environment

The concept of Blue Economy (BE) was first introduced in a book *"The Blue Economy: 10 Years, 100 Innovations, 100 Million Jobs"* (Pauli 2010). It gained global attention at the Rio + 20 Conference in 2012, which emphasized sustainable development and ocean-based economies (UNDESA 2014). The Blue Economy encompasses all activities related to oceans, seas, and coasts that promote sustainable use of marine ecosystems for economic growth, improved livelihoods, and job creation. Historically, ocean use was largely limited to coastal fishing, but today, the scope has expanded significantly with key components—sustainable fisheries, aquaculture, marine tourism, maritime transport, renewable ocean energy, marine biotechnology, habitat conservation, and climate change adaptation (see Fig. 1) that aim to balance economic development with environmental sustainability. In the recent years, there has been rapid growth in deep-sea mining and marine renewable energy development, particularly within countries' Exclusive Economic Zones (EEZs), whose legal jurisdictions have been clearly defined. The BE is projected to generate goods and services worth approximately US$24 trillion (Pouponneau et al. 2024). However, it faces significant challenges such as illegal, unreported, and unregulated (IUU) fishing, marine pollution, climate change, and habitat degradation (Davies et al. 2019; Martínez-Vázquez et al. 2021).

In this review, we present biological aspects of production, i.e., production by phytoplankton, seaweeds, mangroves, corals, and mariculture (coastal aquaculture) that could contribute to the Blue Economy of the Bay of Bengal. Primary production in the coastal waters of West Bengal, Odisha, Andhra Pradesh, Tamil Nadu, and the Andaman Nicobar Island complex is high, which in turn through scientific approach could ultimately sustain commercially important mariculture.

The Bay of Bengal (BoB), an offshoot of the Indian Ocean, is one of the world's largest tropical marine ecosystems (LME) and is located between latitude 5° and 24°N, and longitude 79° and 100°E. Spread across 2,173,000 km^2, it includes the coastal waters of Bangladesh, India, and Myanmar that have a coastline of 14,000 km and 4.3×10^6 km^2 total EEZ area. The habitats in the Bay of Bengal include estuaries, lagoons, coastal lakes, mudflats, rocky and sandy beaches, shoals, coral reefs, mangroves, marsh vegetation, flood prone areas, coastal dunes, reclaimed areas, and salt pans. Chilika, Pullicat, Kolleru and the wetlands in West Bengal are the main brackish water lakes. These habitats have an unparalleled biodiversity and provide ecosystem services to nearly 200 million coastal populations.

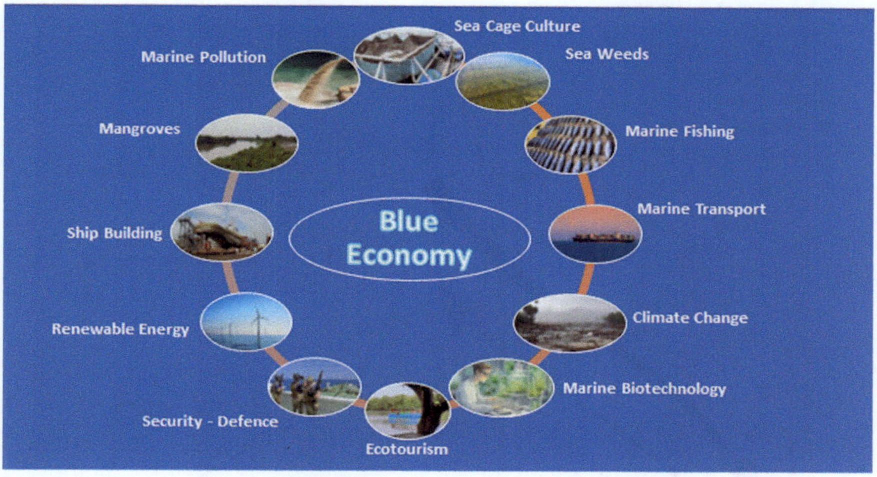

Fig. 1 The blue economy sectors

2 Bay of Bengal: Rivers and Monsoons

India is a land of rivers and monsoons. Major rivers such as the Ganges, Brahmaputra, and Meghna in the north; the Irrawaddy and Thanlwin in the east (originating from Myanmar); and the Mahanadi, Godavari, Krishna, and Cauvery in the west (from India) collectively discharge approximately 1.5×10^{12} cubic meters of water and 2000×10^6 tons of sediment into the Bay of Bengal each year (Mohanty et al. 2008).

During winter monsoon, (NEM Nov-Feb) winds blow from the northeast and large volumes of freshwater are driven along the east coast equator-ward, resulting in the East India Coastal Current (Fig. 2, EICC) (Rainville et al. 2022). During the warmest months (June–September) winds are reversed (SWM), and the southwesterly winds force high-salinity waters from the Indian Ocean and the Arabian Sea into the BOB resulting in a Summer Monsoon Current (Fig. 2, SMC).The monsoonal currents, upwelling of nutrient-rich sub-surface waters govern the biological production.

Along the east coast of India from Sunderbans to Gulf of Mannar, gradations are noticed in the low salinity and high silicate levels (Table 1) which suggest decreasing fluvial inflow. Salinity (PSU) ranged between 0.5 and 30.8 in the Sunderbans, Paradeep 1.1–29.8, Visakhapatnam 27.22–33.60, Kalpakam 23.56–34.15 and 31–36 in the Gulf of Mannar. Suffice to say that the nutrients showed a gradation, i.e., maximum silicates (μm) corresponded to 63.31–86.84 at Sunderbans, 16–153.2 (Paradeep), 12.03–40.0 (Visakhapatnam), 1.58–23.10 (Kalpakam), and 0.08–1.75 in the Gulf of Mannar. Gradations in the phosphate and nitrate levels exist with high values in the Sunderbans and the lowest in the Gulf of Mannar (Table 1).

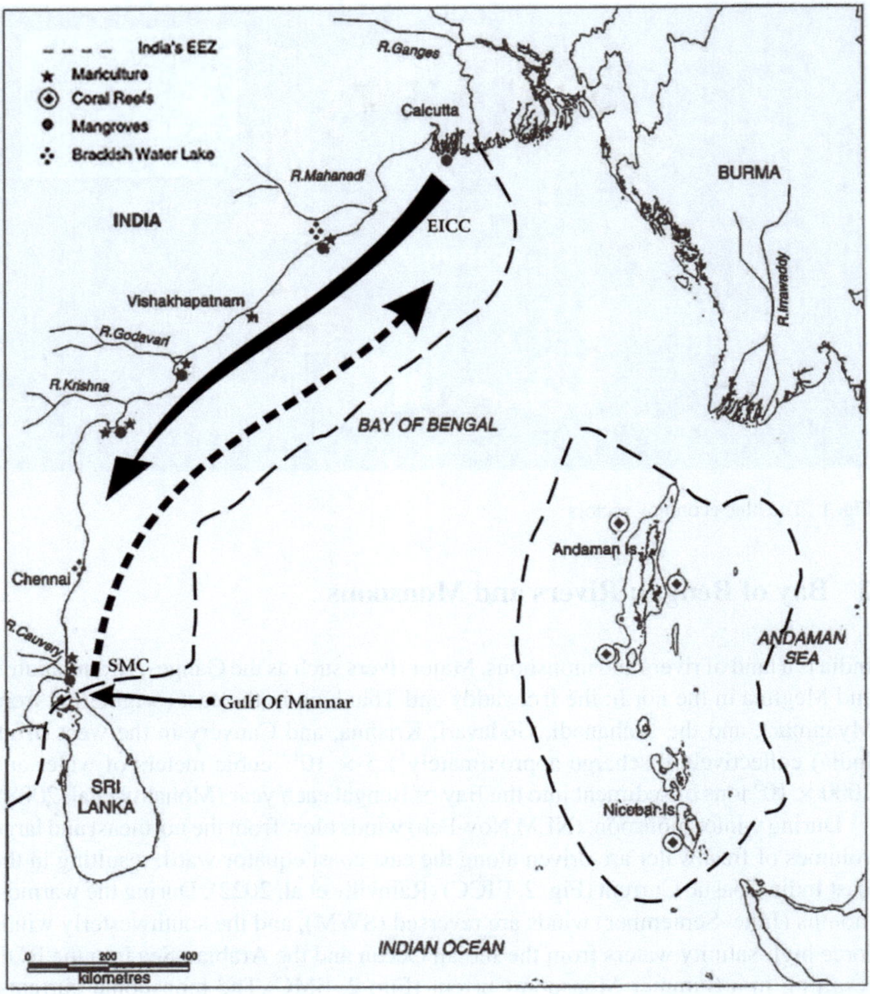

Fig. 2 The Bay of Bengal with its EEZ, ecosystems, East India Coastal Current (EICC) and the Summer monsoon Current (SMC)

3 Phytoplankton Production, Upwelling, monsoons, Fishing Zones

In the marine environment, primary production through photosynthesis serves as the foundation for sustaining all forms of life. Coastal waters exhibit high levels of primary productivity, contributing approximately 11.7 ± 0.53 gig tons of carbon per year (range: 10.9–12.4 Gt C y^{-1}), compared to 50.7 ± 1.38 Gt C y^{-1} (range: 48.7–52.5) produced across the global oceans (Kulk et al. 2020). The Bay of Bengal (BoB) hosts 29 distinct ecosystems—including estuaries, brackish water lakes and lagoons,

Table 1 Range of hydrobiological variables in the fringe waters of Bay of Bengal

Region	Temp ° C	Salinity PSU	Oxygen mg l^{-1}	PO_4 µm	SiO_2 µm	NO_3 µm	Chl a µg l^{-1}	References
Sunderbans Bhagirathi estuary	13.9–30.0	0-5–30.8	4.25–7.88	2.54–7.45	63.31–86.84	36.31–95.42	0.22–40.3*	Choudhury and Pal 2012
Chilika	24.9-30.1	9.0–20.7	6.7–7.7	0.5–1.7	50.1–86.7	1.4–4.9	10.1–23.3	Muduli and Pattnaik 2020
Paradeep	20.6–31.9	1.1–29.8	4.7–7.3	0.7–58.1	16–153.2	1.3–16.5	0.8–4.23	Sangeeta et al. 2018
Visakhapatnam	27.20–31.50	27.22–33.60	3.84–6.1	0.19–2.0	12.03–40.0	4.44–23.32	3.15–5.52	Nageswara Rao et al. 2016
Kakinada	24.3–29.4	20.3–34.5	5.29–7.08	0.19–1.14	6.94–42.4	0.74–9.02	0.5–7.1@	Ayajuddin et al. 2014
Nizampatnam Bay	27.95–29.15	32.5–33.1	5.38–8.43	0.26–0.72	3.60–6.50	0.66–1.40	0.24–2.68	Pandiyarajan et al. 2014
Kalpakam	28.50–30.60	23.56–34.15	5.08–6.33	0.05–1.96	1.58- 23.10	0–48.43	0.57–4.25	Satpathy et al. 2010
Coleroon	24.5–28.5	9.6–31.4 **	3.1- 7.5	0.2–3.1	55–125	10.1–23.4	2–7.5	Rajasekharan 2012
Parangipettai	27.15–32.4	27.11–34.33	4.23–5.5	1.24–6.89	2.29–11.10	1.24–6.89	0.35–3.72	Vajravelu et al. 2018
Gulf of Mannar	25–30	31–36	4.27–6.37	0.11 to 0.95	0.08 to 1.75	0–0.03	0.01–2.11	Deepika et al. 2019

* Manna et al. 2010. ** Silambarasan et al. 2015. @Latha et al. 2014

shallow shelf regions, continental slopes, deep abyssal zones, and offshore ridges—all of which contribute significantly to the region's overall primary productivity.

A high biodiversity exists with 276 phytoplankton species (Wang et al. 2022), 485 seaweeds (Ganesan et al. 2018), 48 species mangrove forests, 300 coral reefs, and 15 species sea grass meadows that provide economic support through fishing and shrimp farming. About 26 million depend on artisanal capture fisheries and aquaculture, 0.2 million on non-fish mangrove products and ~ 40 million associated with coastal tourism (BOBLME2015). The BOB is cyclone prone with 5–6 in a year. Due to the lack of large mechanized seaworthy fishing vessels, fishing activity in the BOB is mostly artisanal and limited to the fringe coastal waters. Because of their adjacency and ease of access, 66 million coastal population depend on the brackish water in the Sunderbans, Chilika Lake, Godavari, Krishna estuaries extensively.

During the SWM period, the BOB is rendered oceanic and also experiences coastal upwelling on the east coast of India and in the south Sri Lanka as well (Vinayachandran et al. 2004; Rao et al. 2006). Thus, these reversing monsoons and the upwelling of nutrient-rich bottom water act as forcing functions on the BOB phytoplankton ecology. There have been several seasonal planktological studies along the east coast: Hoogly estuary, Chilika Lake, Puri, Visakhapatnam, Pitchavaram, Kalpakam, Muthupettai, Porto Novo, Lake Pullicat, and Palk Bay (Table 2). A general feature is a bimodal peak distribution with a major peak during SWM and a minor one coinciding with the NEM. The qualitative and quantitative constituents and production characteristics of the major bloom during SWM are marine and contrast the more estuarine during the NEM. Of the coastal phytoplankton species, 20 occur in the BoB round the year. Heavy blooms of phytoplankton were reported mostly during pre-monsoon season (D'Silva et al. 2012; Oyeku and Mandal 2021; Padmakumar et al. 2012) and are attributed to the enrichment caused by riverine discharge or upwelling of nutrient-rich sub-surface waters.

Energy transfer in marine ecosystems begins with phytoplankton, forming the base of the food web. The flow of energy from phytoplankton to fish—and the length of the connecting food chains—can significantly influence tropho-dynamics and the sustainability of fisheries (Ryther 1969; Stocka et al. 2017). Assessing the suitability of the Bay of Bengal (BoB) coastal waters for primary production is therefore essential and requires spatial and temporal data. One avenue is the utilization of remote sensing data of photosynthetic pigments (Padmanabhan et al. 2000). Chlorophyll a (chl a), a key indicator of phytoplankton biomass, has been measured in the Indian Exclusive Economic Zone (EEZ). Data from 430 stations sampled between 1962 and 1988 show chl a concentrations during the southwest monsoon ranging from 1.1 to 93.2 mg Cm^{-2} (Sarupriya and Bhargava 1998), comparable to 1.1 to 50.7 mg m^{-2} observed in the Arabian Sea (Bhattathiri et al. 1980). These values varied with monsoon patterns, and hotspots of high chl *a* were typically observed near estuaries. In the Andaman Sea, chl a concentrations ranged from 0.4 to 34 mg m^{-2}.

Table 2 Phytoplankton biomass (chl *a*) and primary production in fringe waters of the Bay of Bengal

Fringe waters	Chl a mg m^3	References	Primary Pro. mgC m^3 d^1	Reference
Hoogly estuary	2.55–3.16	De et al. 2021	187.-500	Bhunia and Choudhury 1982
Chilika Lake	0.94–12.53	Raman et al. 1990	41.46–515.59	Benerjee et al. 1998
Gopalpur	0.9–17.22	Gouda and Panigrahi 1996	134.24–988.54	Panigrahy and Mahapatro 1984
Visakhapatnam	0.077–5.15	Vijayakumaran et al. 1996	92–3380 0.72–226.5	Subba Rao 1973 Vijayakumaran et al. 1996
Chennai	0.6–9.2	Nair 1990	1229	Madhu et al. 2002
Kalpakam	< 22.7	Satpathy and Nair 1990	36–3852	Satpathy and Nair 1980
Palk Bay	0.4–2.0	Sarangi 2008	1.8–2342	Nair 1970
Viziangam Bay			114–672	Jacob 1984
Vellar estuary	3.4–12.8	Arokiasundaram et al. 2015	550–1190, 2940	Arokiasundaram et al. 2015, Kawabata et al. 1993
Tuticorin	1.2–11.8	Gopinathan et al. 1994	114–1600	Gopinathan et al. 1994
Andaman	0.02–0.085*	Nair 1970	230–2700	Nair 1970
Nicobar	–	Nair 1970	170–2350	Nair 1970

*Devassy and Bhattathiri 1981

Significant spatial and seasonal variations were also observed in chlorophyll biomass (ranging from 0.077 to 22.7 mg m^{-3}) and primary production, which spanned from 0.72 to 3852 mg C m^{-3} d^{-1} (Table 2). A comparison between coastal and offshore waters clearly shows higher productivity in the former. Coastal waters exhibited chl *a* concentrations ranging from 0.6 to 55 µg L^{-1}, while primary production from 1.8 to 3852 mg C m^{-3} d^{-1} (Subba Rao 2000). In contrast, offshore waters had much lower productivity, with primary production ranging between 0.04 and 295.7 mg C m^{-3} d^{-1} (Table 2).

A more recent compilation by Subba Rao et al. (2023) reported chl *a* values between 1.1 and 42.9 mg m^{-2} and primary production between 187 and 5590 mg C m^{-2} d^{-1}. Thus a clear productivity gradient exists across the BoB, with higher values in the nearshore and shelf waters than in the oceanic province.

Löscher (2021) summarized primary production rates in the BoB as follows: 0.3–936 mg C m^{-2} d^{-1}, with peaks of 2200 mg C m^{-2} d^{-1} in nearshore zones, ~ 500 mg C m^{-2} d^{-1} on the shelf, and ~ 300 mg C m^{-2} d^{-1} in open ocean regions. Integrated primary production in oceanic offshore waters ranged from 89 to 221 mg C m^{-2} d^{-1}, often limited by nitrate depletion despite high silicate levels (Jyothibabu et al. 2004).

These patterns suggest the presence of "hotspots" of biomass and productivity in the BoB, particularly in regions influenced by freshwater input, nutrient enrichment, and estuarine processes. Some fringe or estuarine brackish waters, which are critical for mariculture, may exhibit even higher productivity, especially during ephemeral phytoplankton blooms.

Historical data (1908–2017) show that high nutrient concentrations—particularly nitrate (NO_3-N), phosphate (PO_4-P), and silicate (SiO_4-Si)—combined with low salinity, have triggered blooms of various phytoplankton groups such as prymnesiophyceae, raphidophyceae, bacillariophyceae, and dinophyceae around the Indian peninsula (Oyeku and Mandal2021).

Of interest, is the positive response of primary production to environmental perturbations. Sarma et al. (2019) reported that primary production in the upper 25 m of the Bay of Bengal (BoB) water column is comparable to that of the Arabian Sea. This productivity was linked to high concentrations of dissolved organic nitrogen (DON) and dissolved organic phosphorus (DOP), which together contribute between 70 and 99% of the total dissolved nutrients in the region. One hypothesis attributes enhanced primary production in the BoB to the presence of cyclonic (CE) and anticyclonic eddies (ACE), which influence nutrient availability through vertical mixing associated with divergence. These eddies can alter the nutrient composition and may have a positive effect on phytoplankton community structure (Sarma et al. 2020).

Additional studies suggest that episodic events—such as tropical cyclones (Maneesha et al. 2011) and mesoscale eddies (Vinayachandran and Mathew 2003)—can disrupt the stratified surface layer, promoting the upward transport of nutrient-rich deep waters and enhancing primary productivity. Beyond upwelling, biological production in the BoB also shows interannual variability linked to monsoon conditions. Jayalakshmi et al. (2022) found that during deficit monsoon (DM) years, the basin tends to be warmer, yet supports relatively higher chlorophyll a levels and primary production compared to normal monsoon (NM) years. Each year, more than 50 mesoscale eddies form in the Bay of Bengal (Chen et al. 2012). However, the specific impacts of these eddies on nutrient dynamics and primary productivity remain insufficiently quantified and warrant further investigation.

Thus, these data suggest that enrichment of the BOB fringe waters responded positively to environmental perturbations such as upwelling and sustained high levels of algal biomass and production. During the summer months, the disposal of dredged spoil from the entrance channel of Visakhapatnam harbor into the sea led to natural enrichment of the waters, resulting in blooms of *Thalassiosira subtilis* and *Chaetoceros curvisetus*. This enrichment caused a 3- to 13-fold increase in gross primary production (Subba Rao1973).

Analysis of data for BOB between latitude 5° and 24°N, and longitude 79° and 100°E. (Sarker et al. 2020) showed an increase in sea surface temperature from ~ 28 (during 1950) to ~ 29.2°C (during 2010) and the regression analysis yielded (Y = 0.01x + 7.42, R^2 0.603). During 1995 to 2020, primary productivity (mg C m^3) decreased from ~ 3200 to ~ 1500 (by 2018 and the regression was (y = + 73.73x

+ 15,044 R^2 0.455). This negative trend in primary productivity is in line with a global decrease in primary productivity and related to global warming (Vichi and Lévy 2016). However, a similar comparative analysis for the coastal fringe waters of the BOB would be interesting as our primary concern is more with the blue economy in the coastal waters than in the offshore.

Potential Fishing Zones (PFZ) can be identified based on factors such as sea surface temperature (SST), water transparency, sunlight penetration depth, and biological indicators like phytoplankton and chlorophyll (Solanki et al. 2000). These PFZs help fishermen easily locate fish shoals, thereby reducing the costs of commercial fishing operations. PFZs also influence the migration and recruitment patterns of various species (Nyitrai et al. 2012). The Catch Per Unit Effort (CPUE) is higher in PFZs compared to non-PFZs, further lowering fishing expenses. In the northern Andhra region (Visakhapatnam and Kalingapatnam), PFZs exhibit a cyclic pattern, with PFZs being closer to the coastal waters during summer and shifting to deeper waters in the pre-monsoon season (Nyitrai et al. 2012).

4 Seaweeds

Seaweeds are a valuable source of industrially important thickening agents and gels, such as agar, carrageenan, and alginates, which have wide applications in food, confectionery, pharmaceuticals, dairy, textiles, paper, paints, and varnishes (Kiling et al. 2013). In addition to their commercial value, seaweeds also provide natural chemicals, including pigments, mannitol, iodine, and products with cosmetic and therapeutic properties. Chemical analyses of *Sargassum wightii*, *Kappaphycus alvarezii*, and *Ulva fasciata* have revealed the presence of carbohydrates, proteins, tannins, phenols, flavonoids, anthocyanins, coumarins, steroids, saponins, glycosides, cardiac glycosides, terpenoids, triterpenoids, quinones, alkaloids, and acids (Selvaraj 2016). Furthermore, there is potential for utilizing seaweeds in the production of bioethanol, biofuels, fertilizers, bio stimulants, and other value-added bio-products.

Economic potential of seaweeds in India has been underused. A great diversity of seaweeds exists along the coast of Pulicat in Tamil Nadu, Andhra Pradesh, and Chilika in Odisha and also the Andaman Nicobar complex. Ganesan et al. (2018) reported 485 species from various coastal regions, with 133 species from the Andaman and Nicobar Islands (Muthuvelan et al. 2001), 14 species from the West Bengal coast (Mukhopadhyay and Pal 2002), and 14 species from the Odisha coast (Sahoo et al. 2003). Kaliaperumal et al. (2004) reported 80 species of Chlorophyta, 56 species of Phaeophyceae, and 146 species of Rhodophyta in the Gulf of Mannar. Notably, species such as *Halimeda* spp. and *Caulerpa* spp. were restricted to Palk Bay and the Gulf of Mannar. Additionally, *Ulva reticulata* and *Kappaphycus alvarezii* were not found in the reefs of Palk Bay, specifically near Mandapam and Rameswaram Island (Ayog 2024).

Seaweeds are less abundant in the lower salinity waters of West Bengal and Odisha, with only 14 species, recorded, compared to 65 species in Andhra Pradesh, 455 species in Tamil Nadu, and 133 in the Andaman and Nicobar Islands (Pati et al. 2016). Regarding annual yield, Odisha reported a relatively low harvest of 269,705 tons (wet weight), while Tamil Nadu saw significantly higher yields (Pati et al. 2016).

Compared to a global production of 35 million tons seaweed, worth about $16.5 billion, India, produces nearly 33,345 tons *Sargassum, Turbinaria, Gracilaria* and *Gelidiella*) (CMFRI, FRAD 2022) fetching an annual revenue of about $25 million. India has the potential to produce around 9.7 million tons of seaweed per year, according to the Central Marine Fisheries Research Institute (CMFRI 2012). The Indian government has come up with a package of $80 million to promote seaweed cultivation of up to 1.12 million tons by 2025. Among the global seaweed production through farming, *Kappaphycus alvarezii* and *Eucheuma denticulatum* contribute to 27.8% of the total production (FAO 2022). Seaweed cultivation does not require land, freshwater, or fertilizers and can thrive in varied temperatures and require minimal freshwater inputs. Advantages of seaweed cultivation include oxygenation of the water, mitigation of adverse impact of pollution, heavy elements, and carbon sequestration. Coastal communities are gainfully employed in seaweed cultivation. *Hydropuntia edulis*, *Gelidiella acerosa*, and *Hypnea musciformis* are cultivated in Krusadai Island in the Gulf of Mannar. In 1984, a fragment of the algae *K. alvarezii* (then known as *K. striatum*), originally from the Philippines, was acquired by CSIR-CSMCRI from Japan and is now farmed on the Tamil Nadu coast.

In the Gulf of Mannar, approximately 1555 fisherfolk from 14 coastal villages are involved in seaweed harvesting, including 1270 women and 285 men (Fig. 3). Along the Palk Bay coast, 670 fisherfolk from 24 villages harvest seaweeds, engaging 460 women. Around 1,500 households earn approximately $300 per month from seaweed harvesting (Ganesan et al. 2018).

There have been studies on culturing seaweeds (Chennubhotla et al. 1978; Kaliaperumal 1998; Kaladharan and Kaliaperumal 1999; Oza et al. 1994; Patel et al. 1979). Two methods are used for seaweed cultivation: vegetative propagation in

Fig. 3 Seaweed cultivation in the Gulf of Mannar

inshore waters is by inserting fragments of seaweed in braids of a network of nylon, polypropylene ropes or in nets tied to open bamboo rafts (Fig. 3 and Table 3). Vegetative propagation is employed for species such as *Ulva, Gelidiella acerosa, Gracilaria edulis, Gracilaria* spp., *Corchata, Gelidiopsis variabilis, Gelidium pusillum, Hypnea musiformis, Acanthophora spicifera, Hormophyta triqueta, Cystoseira trinodis*, and various species of *Sargassum* (Chennubhotla 1977; Chennubhotla et al. 1978). To avoid impact of sedimentation and grazing by fish, culturing was done in slightly deeper waters; 122 seaweed species are grown in 2–6-m depth waters with a pH (6.5–8.5) and transparency (2–6 m). Based on their suitability for seaweed cultivation, three zones—Green, Amber, and Blue—have been identified along the entire coast of India. Zone A sites, located more than 1 km from CRZ-IA, are considered most suitable for farming, as they are distanced from sensitive ecosystems. Zones B and C are also suitable for cultivation (Aayog 2024). The rope culture method for *Gelidiella acerosa* yielded 421 g/m (wet) per year (FAO 2022).

Another method of seaweed cultivation involves using spores, zoospores, monospores, tetraspores, and carpospores (Kaliaperumal 1998), which are collected on suitable substrates such as nets, bamboo splints, straw, or polypropylene. These spores are then cultured in petri dishes to grow germlings before being transplanted into natural seawater. *Sargassum plagiophyllum, Enteromorpha flexuosa,*

Table 3 Seaweed cultivation in the Bay of Bengal and associated commercial products (based on Ganesan et al. 2018)

Taxa	Method	Harvest (wet)	Commercial product
Natural			
Kappaphycus alvarezii	Bamboo rafts (3 × 3 m) with braided ropes, Polythene bags, fish net bags, and 50–60 kg planting material are used	45 days 240–1260 kg	κ-carrageenan
Hypnea musciformis	Monoline, coir ropes	25 m kg	kappa-carrageenan
Sarconema filiforme	Ropes and bamboo rafts	25 kg	α-carrageenan
Cultured			
Gelidiella acerosa	Long-line ropes, single rope floating technique, coral stone culture method and concrete stone method	421 g/m wet y-1	Pharmaceutical and bacteriological grade agar
Gracilaria edulis	Long-line rope single raft (3 × 3 m) Floating technique 5–6 kg planting material	45–60 days 40–50 kg	Food-grade agar
G. dura	Bamboo raft	45–60 kg	Agarose
Gracilaria salicornia	Tube-net	30–40 kg	Agar
Gracilaria crassa	Tube-net	45 kg	Agar

Ulva fasciata, and *Gracilaria edulis* are grown using spores (Kaliaperumal 1998). In Mandapam, *Gracilaria edulis,* and *Gelidiella acerosa* were successfully grown in seawater aquaria (Umamaheswara Rao 1973). However, cultivating *Gracilaria corticata* in seawater aquaria takes longer time. Defined media, such as the Provasoli culture medium or seawater enriched with agar–agar, can also be used to grow germlings. Additionally, rather than in open coastal waters, seaweeds can be cultured under controlled conditions in covered open raceways.

The seaweed industry in India is primarily a cottage industry, lacking sophisticated machinery (Coppen and Nambiar 1991). It relies on the natural stock of agar-yielding red seaweeds such as *Gelidiella acerosa*, *Gracilaria edulis*, and *G. crassa*, as well as algin-yielding brown seaweeds like *Sargassum* and *Turbinaria* (Table 1). There are 22 agar-producing units, mostly located in Tamil Nadu, and 15 units that produce algin. However, there are no units in India currently producing carrageenan (Kaladharan and Jayasankar 2015).

The Indian Council of Agricultural Research (ICAR), CMFRI, Central Salt and Marine Chemicals Research Institute, National Center for Sustainable Coastal Management, and the National Institute of Ocean Technology conduct research on seaweed. The Gulf of Mannar Marine Biosphere Reserve (GoMBR) was the first of its kind in South and Southeast Asia and was a precursor to the initiation of the Man and Biosphere (MAB) program by UNESCO in 1971. In 1986, the Government of Tamil Nadu declared it a Marine National Park, and in 1989, the Government of India designated it as the first MAB Marine Biosphere Reserve of India.

Significant progress is being made in the introduction and acclimatization of subtropical species that offer improved yields and high-quality colloids and carrageenan production. High-yielding strains of agarophytes, such as *Gracilaria NBR-10* and various species of *Gelidium* and *Pterocladia,* have also been successfully introduced. However, it is important to note that estuarine waters in the Bay of Bengal, particularly those with high tidal amplitudes, are unsuitable for seaweed cultivation. These waters, often turbid with high temperatures, are not conducive to seaweed growth. Additionally, short production cycles, lack of technical expertise, insufficient investment, absence of clear guidelines and regulations pose significant challenges to the expansion of seaweed production.

5 Mangroves

Mangroves grow in the intertidal zones of sea coasts and play a critical role in maintaining the health of coastal environments by supporting a wide range of biodiversity of flora and fauna. Healthy mangrove forests provide natural protection against hazards such as cyclones, coastal erosion, and salinity intrusion. Additionally, they serve as effective carbon sinks, sequestering significant amounts of carbon and contributing to climate change mitigation. Mangroves, safeguard coastal regions, and sustain local livelihoods through fishing, forestry, and tourism. According to the Global Mangrove Alliance's report "*The State of the World's Mangroves, 2024*",

(GMA) the global coverage of natural mangrove forests is approximately 0.147×10^6 km^2.

In India, the total mangrove cover is estimated at 4992 km^2, with 2114 km^2 classified as mangrove forests. The Sundarbans in West Bengal hold the largest share, covering 140,000 hectares (1400 km^2). Other significant mangrove areas in India include Andhra Pradesh (404 km^2), the Andaman and Nicobar Islands (404 km^2), and Puducherry (2 km^2), as reported in the *India State of Forest Report, 2021.*

6 The Sunderbans

Near the Hooghly River, the region supports 59 species of mangrove plants, with *Rhizophora mucronata, R. apiculata, Avicennia officinalis, A. marina, Ceriops* species, *Excoecaria agallocha, Acanthus ilicifolius*, and *Acristuchus* species are most dominant. Near the water edge *Rhizophora, Bruguiera, Sonneratia*, and *Avicennia* were dominant and in the middle were *Ceriops decandra, Aegiceras* species, and *Clerodendrum* species. Further inland, more salt-tolerant plants like *Suaeda* and *Lumnitzera* grow landward. *Avicennia marina* contributes approximately 1603 g m^2 y^{-1} (Ghosh et al. 1990).

Detritus and litter from mangrove vegetation form the primary food source for a variety of organisms, including crustaceans, fish, and shellfish, many of which use mangrove ecosystems as spawning and nursery grounds (Ghosh et al. 1990). Mangroves also serve multiple utilitarian purposes: they are used as firewood, construction materials, and sources of tannins for the leather industry; their leaves are fed to cattle, and various parts are used in traditional herbal medicine. The dense network roots stabilize coastline preventing soil erosion. However, this crucial ecosystem is under going threat from climate change, sea-level rise, pollution, and deforestation. Development of shrimp aquaculture expansion, oil palm plantations, and rice cultivation—and natural threats like tropical storms and rising temperatures contributed to 43% of global mangrove loss between 2000 and 2020 (GMA REPORT *The Indian Express*, July 27, 2024). At the Mangrove Ecology, Functioning and Management Conference, Siddique (2012) discussed the destruction of the Chakaria Sundarbans mangrove forest. This area, protected by the government since 1926, once supported 90% of the local population and had an estimated annual economic value ranging from \$200,000 to \$900,000 per hectare. Over the years, the entire mangrove region was cleared, mainly for shrimp farming, with little salt bed along human settlement.

Mangroves of the Sundarbans Biosphere Reserve (SBR) are facing continuous degradation. For the last few decades several species have been at varying levels of extinction. Historically, Indian mangroves covered nearly 6000 km^2 (FAO 2003; Gnanappazham and Selvam 2014), but were reduced to 4046 km^2 in 1987. During 1987 and 2011, their extent seems to have stabilized (Bhatt and Kathiresan 2011).

On a global scale more than 16% of species are in the state of becoming extinct, another 10% under degradation and if this continues the mangrove biome may vanish by the next 100 years (Polidoro et al. 2010). In India, the "Green India Mission" under the National Action Plan on Climate Change (2008) with a projected extension of 1000 km^2 mangrove forests, the Community-based Mangrove Management (CBMM) and the Community-based Forest Management (CBFM) developed into a Joint Mangrove Management (JMM). Deforestation of mangroves for shrimp cultivation will affect the carbon footprint. Saving the mangroves would help in improving the lives of the coastal populations. Restoration of mangroves would help in protecting the coastal and marine biodiversity by strengthening negative and climate adaptation and increase in biodiversity. Further it would also help in long-term economic support.

A recent study, using 36 environmental criteria assessed the habitat suitability of 18 species within the Sundarbans Biosphere, Sahana et al. (2022) studied using 10 machine learning algorithms and concluded that the slope and temperature were the most significant factors influencing the habitat distribution of the selected species. Other major drivers are deforestation, over-exploitation, natural hazards, extreme weather events, salinity intrusion, sea-level rise, coastal erosion, embankment construction, shrimp aquaculture, increasing population pressure, pollution, industrial waste, poverty, underdevelopment, and lack of public awareness.

Mahmooda et al. (2021) reviewed the key issues and challenges in mangrove management across the period from 1780 to 2020, identifying a wide range of threats to the Sundarbans ecosystem. Besides those given by Sahana et al. (2022), additional challenges include uncontrolled tourism, pollution, etc., are important. Mahmooda et al. (2021) noted a four-phased paradigm shift in the region's management objectives: from conversion for agriculture (1780–1875), to timber production for revenue collection (1876–1951), followed by inventory-based management (1952–1992), and finally, a move toward integrated and co-management strategies (1993–2020). This historical progression reflects the evolving priorities in mangrove conservation and the growing recognition of the need for inclusive, ecosystem-based management approaches.

7 The Mangroves of Odisha

In the state of Odisha, mangroves are located in a diverse range of plant habitats. These mangroves are distributed in the 3 major zones (Srivastava 2015; Nayak et al. 2016). Mangrove cover in the districts of Balasore, Bhadrak, Jagatsinghpur, Kendrapara, and Puri is approximately 2 km^2, 30 km^2, 8 km^2, 190 km^2, and 1 km^2, respectively (India State of Forest Report 2015). The galloping growth into aquaculture ponds in Mahanadi delta over the past two decades, and exploitation of mangroves for firewood, livestock fodder, aquaculture, and industrial development has resulted in severe damage environmental damage to the environment.

Nayak et al. (2016) suggested local educational campaigns on conservation of mangroves. About 90 plant species found in Odisha's mangrove ecosystems hold potential as future sources of potent bioactive compounds with medicinal and industrial applications (Das et al. 2020).

Observations of Agarwal et al. (2017) indicated that the clearance of just 1 hectare of mangrove forest in this region could result in the release of 26.94 tons of CO_2 into the atmosphere, contributing to global warming and temperature rise. Given the significant role mangroves play in carbon sequestration, they stressed that the conservation of these ecosystems should be prioritized within the Integrated Coastal Zone Management Plan (ICZMP) for Odisha.

8 The Mangroves of Coringa

The Coringa mangrove reserve forest is the second largest (124 km^2) mangrove ecosystem in the Indian sub-continent. It is located between 16^044' to 16^053N and long. 82^014' to82^022'E (Fig. 4). It has a reserve 3156 hectares, and 9442 hectares of Coringa extension forest. Coringa and its mangroves are the lifeline of over 40 villages. Spread over 320 km^2, of which 235 km^2 was protected under the Coringa Wild Life Service (CWLS) and bird watching, wildlife spotting, boating and fishing and nature trail and eco-tourism are popular activities. Fishing for the mud crabs- *Scylla serrata* is active in this mangrove (Fig. 4).

There are 35 species of mangroves represented by 17 families; and 19 are mangrove species. Avicenniaceae is the most abundant, followed by Rhizophoracea and Euphorbiaceae. *Avicennia marina*, *A. officinalis*, *Excoecaria agallocha*, *Aegiceras corniculatum*, *Sonneratia apetala*, *Ceriops decandra*, *Rhizophora apiculata,* and *R. mucronata* were the important mangrove species (Satyanarayana et al. 2002). Based on the species composition, 5 zonation groups were recognized and

Fig. 4 Coringa Mangroves and activities. **a**- destruction, **b**. Urbanization **c**. Shrimp farms **d**, **e**, **f**- Eco-Tourism, and **g**, **h** Mud Crab (*Scylla serrata*) fishing

the tidal level and salinity are the determining factors. *E. agallocha* was typical of landward locations, *A. marina* (along with *A. officinalis* and *A. alba*) were at the bed level and neap high tide level and a large group of sites inhabited by diverse species at spring high tide level conditions. Similar to Sunderbans, anthropogenic activities like conversion to rice cultivation and industrial development are the major drivers.

Between 1988 and 1998, mangrove cover in the region significantly declined (Table 3). Microalgae, flagellates, and ciliates contributed to the macrobenthos, while nematodes, copepods, foraminifera, ostracods, and polychaetes were part of the meiobenthos, and crustaceans, polychaetes, bivalves, and gastropods made up the macrobenthos. In the East Godavari River Estuarine Ecosystem (EGREE Final Report 2017), there were 34 molluscan species, 382 ichthyofaunal species, 245 avian species, and 14 mammal species. Approximately, 0.128 million fisherfolk from 44 villages rely on the mangroves as a primary source of firewood, construction timber, and fodder for livestock. Apart from cyclones, other significant threats to this environment include the over-exploitation of 3600 tons of mollusks for lime production, the conversion of mangrove areas into aquaculture ponds (Table 4), and the structural changes in the ecosystem due to sustained anthropogenic activities (Satyanarayana et al. 2002). Furthermore, the carbon content of the mangroves has decreased from 174.41 to 74.66 MT/ha, largely due to rising salinity levels and the impacts of climate change. To address these challenges, a polyculture plantation involving species such as *Aegiceras corniculatum*, *Avicennia officinalis*, *Bruguiera gymnorrhiza*, *Ceriops decandra*, *Sonneratia apetala*, and *Avicennia marina* is recommended. The Ministry of Environment, Forests and Climate Change and the Government of Andhra Pradesh, with support from UNDP and GEF, launched the project "Mainstreaming Coastal and Marine Biodiversity into Production Sectors in the East Godavari River Estuarine Ecosystem". The project's Final Report (2017) on the Establishment of a Knowledge Management System for EGREE recommended educating low-income fishermen about the importance of mangroves, encouraging various sectors to participate in mangrove plantation programs, promoting eco-tourism in mangrove areas, and developing integrated management planning with periodic environmental impact assessments and audits (Table 4).

The M. S. Swaminathan Research Foundation (MSSRF) launched the "Food for Work" program in eight villages across the Godavari and Krishna mangroves along the eastern coast of Andhra Pradesh (Ravishankar et al. 2001). The program emphasized the importance of community involvement in the management of mangrove resources and focused on restoration activities. As part of the initiative, the MSSRF trained women from fishing villages in tasks such as raising mangrove saplings, selecting degraded mangrove areas, canal alignment, and digging and planting. Through these efforts, 520 hectares of mangrove area were restored. Additionally, the program contributed to socio-economic development by addressing poverty and improving livelihoods in the local communities.

Table 4 Changes in land use—landcover in Coringa area during 1988–1999 (based on Dept. Ocean Development, Chennai, 2001)

	Area ha
Mangrove no change	11,567
Degraded to mangrove	248
Other land to mangrove	37
Mangrove to aquaculture	639
Mangrove to degraded	534
Mangrove to other land	213
Mangrove to mudflat	174
Aquaculture no change	625
Salt pan to aquaculture	418
Degraded area to aquaculture	159

9 The Mangroves of Pichavaram

The Pichavaram Mangroves, spanning approximately 1100 hectares, are located between the Vellar and Coleroon estuaries in the Cuddalore district of Tamil Nadu, India (Ghosh et al. 2022). These mangroves experience heavy rainfall during the northeast monsoon (October–December), which renders the water brackish, while the salinity remains high (35–45 PSU) throughout the rest of the year (Selvam et al. 2010; Sathyanathan et al. 2014). Key mangrove species in the area include *Avicennia, Rhizophora, Excoecaria*, and *Acanthus*, which provide vital habitats for a variety of wildlife such as birds, fish, crustaceans, and mollusks. The mangroves are home to various bird species, including herons, egrets, kingfishers, and pelicans, and serve as an important stopover for migratory birds. In addition to supporting biodiversity, the Pichavaram mangroves act as a natural barrier against coastal erosion, protecting the shoreline from storm surges and tsunamis. Local communities are actively engaged in conservation initiatives, with ongoing efforts in reforestation and sustainable resource management to safeguard the mangrove ecosystem from over-exploitation.

Afforestation efforts for the conversion of mangroves to aquaculture have led to an increase in the constituents of phytoplankton, macrobenthos, and meiobenthos (Venkanna, 1990). The Mangrove Ecology, Functioning, and Management Conference (2012) significantly advanced our understanding of the structure and functioning of mangrove ecosystems. It recommended using the Delphi System to evaluate and develop a comprehensive understanding of the entire mangrove ecosystem, rather than merely focusing on individual components (Mukherjee et al. 2012). The economic contribution of the Pichavaram mangroves was assessed using market pricing, surrogate pricing, and Willingness to Pay (WTP) methods (Deb Roy et al. 2014). The total economic value of the mangroves was estimated at approximately rupees 4.4 million. The indirect benefits, including offshore fisheries, carbon sequestration, and the storm protection function of the mangroves, accounted for 95% of this value. In contrast, direct fishery contributions and eco-tourism contributed 4.7%, while WTP accounted for just 0.03%.

On the southeast coast of India, the socio-economic conditions of mangrove-dependent communities have been negatively impacted by declining fish catches (Kathiresan and Rajendran 2002). There is greater concern about mangrove degradation among individuals in the 20–40 age group than in the older. Local communities attribute the recent decline in mangrove health to several factors: (1) untreated discharges from aquaculture industries; (2) reduced influx of fresh and tidal waters into the mangrove estuaries; (3) shallowness of mangrove water bodies due to siltation and sedimentation; and (4) increased fishing pressure resulting from the growing participation of non-traditional fishermen.

The need to raise awareness in coastal villages about preventing the degradation and depletion of mangrove ecosystems, as well as promoting their rejuvenation and restoration, was highlighted for the Andhra coastal region (Pradeep 2012). To address this, sociologists proposed the establishment of eco-clubs for 12–15-year-old school children, aiming to educate them on disaster preparedness, mitigation activities, and the importance of safeguarding endangered mangrove species such as *Rhizophora* and *Avicennia.*

From the 1980s, the "slash and burn industry", and conversion of the mangroves to aquaculture, and for oil palm plantations have damaged both mangroves and the traditional coastal communities. For Research and Development, National Mangrove Genetic Resource Centre was established in Bhitarkanika (Odisha) by the Environment Ministry. To augment mangrove cover in India, the Union government of India had launched the Mangrove Initiative for Shoreline Habitats and Tangible Incomes (MISHTI) program last June to plant mangroves in 540 sq.km on the coastline of 11 states and two Union territories (The Indian Express July 27, 2024). It is commendable that under the Green Tamil Nadu Mission restoration plan, The Tamil Nadu Department of Environment and Climate Change (2011–2012), with the involvement of the villagers, restored 375 hectares of mangroves in 2023–2024. About 0.4 $\times$ 10^6 *Avicennia*, *Rhizophora*, *Excoecaria* and *Agallocha* saplings are planted and an increase of 6,780 hectares by 2026 is planned.

10 Andaman and Nicobar

The Andaman and Nicobar Islands, located between 6°00'N and 92°94'E in the Bay of Bengal, encompass an Exclusive Economic Zone (EEZ) of 0.6 million km^2 and are renowned for their rich marine biodiversity. The archipelago covers an area of 663,629 km^2 and is known for its oceanic islands, which have a diverse array of mangrove, coastal, and marine ecosystems that remain relatively unexplored. The islands' biodiversity, coupled with their rich cultural heritage, tribal traditions, evergreen forests, wildlife sanctuaries, river tributaries, and pristine beaches, offers significant potential for both national and international eco-tourism. Mangroves cover 7.5% of the land area. The Zoological Survey of India (ZSI) and the Botanical Survey of India (BSI) have recorded 48 species of mangroves across these islands. A comprehensive review of literature on Andaman and Nicobar mangroves, dating back to Klos (1902), includes species such as *Rhizophora mucronata, R. apiculate, Bruguiera gymnorrhiza, B. cylindrica, Lumnitzera littorea, Avicennia officinalis, A. marina, Ceriops tagal, Sonneratia caseolaris, S. alba, Excoecaria agallocha, Aegiceras corniculatum*, and *Nypa fruticans* (Asokan et al. 2024). An overview of the mangroves is provided by the Middle Andaman Division, Department of Environment and Forests, Andaman and Nicobar Islands 2011–2012 reported 244 seaweed species which seem to resemble those from the west coast of Thailand (Karthik et al. 2021).

Land development and the extraction of mangroves for fuel have caused significant damage to the mangrove ecosystem. However, in the Middle Andamans, a Mangrove Interpretation Centre (MIC) has been established at Yeratta, where various species of mangrove nurseries and plantations are displayed. Given the region's significant potential for aquaculture, alongside the Zoological Survey of India (ZSI) and Botanical Survey of India (BSI), the Marine Products Export Development Authority (MPEDA) and Central Marine Fisheries Research Institute (CMFRI) are promoting seaweed and fish farming. Following the 2004 tsunami, approximately 1300 hectares were permanently inundated, making the area highly suitable for brackish water aquaculture (Chaturvedi et al. 2014). The Rajiv Gandhi Centre for Aquaculture (RGCA), a part of MPEDA, plays a crucial role in advancing aquaculture in this region. The project focuses on the domestication of tiger shrimp, genetic improvement, training, and collaborations with national and international organizations to foster the socio-economic development of the Bay Islands. However, the tsunami also severely impacted the mangroves of the Nicobar Islands (Nehru and Balasubramanian 2012), with 30% of the mangrove species previously recorded being absent during the post-tsunami survey. The abundance of various species at the tsunami-impacted sites was notably lower compared to the pre-tsunami situation.

11 Coral Reefs

Coral reefs play a significant role in the Blue Economy of the Bay of Bengal, contributing both directly and indirectly to the economic, ecological, and social wellbeing of the region. Coral reef ecosystems support a high biodiversity and fisheries and also play a crucial role acting as natural barriers against erosion and storm surges. Also, they hold untapped potential in emerging sectors like marine biotechnology and pharmaceuticals. Tourists are attracted to fishing, diving, and snorkeling and are thus a source of livelihood for coastal communities. However, their role in the blue economy in the Indian seas is not appreciated.

Coral reefs in the Bay of Bengal are in Palk Bay, Gulf of Mannar, and around Andaman and Nicobar Islands. In the Palk Bay, the coral ecosystem is between 09 17 14.8 N; 079 11 16.4 E at the East near Pamban to 09 17 34.4 N; 079 06 57.8 E and essentially fringing reefs with Scleractinia (Venkataraman and Rajan 2013). About 31 species of live corals belonging to 9 families and 18 genera are recorded (Marimuthu et al. 2016). Macroalgae are abundant in these coral reefs with 28 species of green algae, 13 brown algae, and 20 Red algae; and 20 species of boring sponges cause destruction to the reef system (Ramesh and Kannupamdi 1997). The coral reefs in the Andaman and Nicobar Islands have about 948.8 km^2 coral reefs (Jeyabaskaran et al. 2007) and are considered the richest in coral diversity with 179 species belonging to 61 genera (Kumar 1997). The reefs consist of fringing, channel, and patch types, and support rich marine life. Due to the Tsunami in December 2004, corals in these waters suffered a huge loss.

Growth rates of corals are low and highly variable between corals, between species, within reefs, and over wider spatial scales (Lough and Cantin 2014). Growth rate of the Indo-Pacific massive coral *Porites* spp. is about 4 $mm \cdot y^{-1}$ in the marginal reef environment of Hong Kong and about 25–30 $mm \cdot y^{-1}$ at Lihir Island in the Western Pacific Warm Pool (Lough and Cantin (2014).

However, coral reefs in the Bay of Bengal face several significant threats, sea surface temperature (SST) anomalies (31.80–32.30 °C in certain regions), coupled with the El Niño Southern Oscillation (ENSO) phenomena in 2010 resulted in bleaching of almost 70% of live coral (Majumdar et al. 2018). Also, heavy sedimentation from major river systems, pollution from agricultural runoff, plastic waste, and industrial discharge, siltation, unsustainable poaching, destructive fishing practices like blast and cyanide fishing impacted the corals. Rapid coastal development and unregulated tourism also contribute to habitat loss.

Measures to restore corals should take into consideration their low growth rates. Several suggestions were made to overcome the threats in coral reef diversity of Andaman Islands: intense systematic research activities (Jeyabaskaran et al. 2007), artificial reef restoration and silviculture (Majumdar et al. 2018), and adapting holistic management strategies (Mandal and Banerjee 2025).

12 Mariculture (Coastal Aquaculture)

In 2017, the global population reached 7.6 billion and is expected to reach 8.6 billion by 2030 (United Nations 2017). Food security will be a daunting challenge and global aquaculture production could play an important positive role (FAO 2018). Aquaculture, although includes finfish, mollusks, and crustaceans, shrimp culture is mainly practiced in tropical and subtropical regions. Among aquaculture-producing nations, China is ranked first (61.5% of global production) followed by India. Global aquaculture production reached 80MT food fish in 2016, of which inland aquaculture produced 51.4 million tons while both coastal aquaculture and mariculture (i.e., aquaculture in marine environment) produced 28.7 million tons (FAO 2018). However, the blue revolution of aquaculture has been under intense criticism because of the environmental impacts including habitat destruction, water pollution, eutrophication, biotic depletion, ecological effects, and diseases.

The CMFRI Mandapam was the first to attempt culturing milkfish (*Chanos chanos*) in 1958–1959 (FAO 2005–2024). Subsequently, the Central Marine Fisheries Research Institute (CMFRI) extended similar studies to other species, including oysters, mussels, and clams. Between 1972 and 1984, CMFRI developed techniques for breeding the Japanese pearl oyster (*Pinctada fucata*) and the blacklip pearl oyster (*Pinctada margaritifera*), and culture of the Indian backwater oyster (*Crassostrea madrasensis*). Although monocultures- of *Lethrinus, Epinephelus, Mugil cephalus, Chanos chanos, Etroplus suratensis, Epinephelus tauvina, Lates calcarifer* and *M. cephalushas* been tried culturing of *L. calcarifer* has been commercialized and around 2006–2009 at Visakhapatnam.

Only in the recent years, coastal Bay of Bengal became the platform for mariculture, particularly the shrimp. It has presented an unprecedented opportunity to meet food demands and improving living standards of coastal communities and emerged as a major producer from 20 MT in 1970 to 0.747 MT in 2020. To secure food security, eradicate poverty, and to maximize the opportunities, in 2021, FAO provided a clear roadmap and launched the Blue Transformation vision, consistent with a road map of Llewellyn et al. (2016) and Upadhyay and Mishra (2020).

The Central Institute of Brackish water Aquaculture (CIBA) and the Central Marine Fisheries Research Institute (CMFRI) have made pioneering contributions to mariculture studies. Research centers have been established across key coastal regions, including Barrackpur, Contai Srikakulam, Visakhapatnam, Narsapur, Ongole, Chennai, Cuddalore, Nagapattinam, Pattukkotai, Mandapam Camp, Tuticorin, and Kanyakumari. The Marine Products Export Development Authority (MPEDA) established the Andhra Pradesh Shrimp Seed Production Supply and Research Centre (TASPARC), the Orissa Shrimp Seed Production Supply and Research Centre (OSPARC) in Orissa which provided assistance and paved the way for the establishment of a number of private hatcheries.

Among the various advancements in aquaculture, the most prominent in India are marine finfish farming, mussel and oyster cultivation, and seaweed farming. The brackish water aquaculture sector was largely based on farming essentially *Penaeus monodon* and other shrimp species such as *P. indicus, P. merguiensis*, and *P. semisulcatus*, which were once considered promising candidates for commercial cultivation. However, the introduction of *Litopenaeus vannamei* from Taiwan around 2001 has significantly transformed the sector. Currently, Coastal Aquaculture Authority (CAA) has given permission for farming vannamei in 22,715 hectares and allowed 135 hatcheries to import broodstock for the production and supply of high-quality seed to farmers. *L. vannamei* is able to survive high temperatures (Albalat et al. 2022) and salinities from 0.5 to 45 PSU; the recommended standard is 10–20 PSU (Millard et al. 2021a, b). To ensure healthy growth and minimize aggressive behavior, recommended stocking density is maintained at 6–15 shrimps m^2 as overcrowding is likely to lead to aggressive behavior (Abdussamad and Thampy 1994; Pijl 2024).

The BMR Hatchery at Puducherry, India (Fig. 5 1a–1f) is to be credited for obtaining Government of India approved pilot-scale culture of *L. vannamei* in 2003, and started commercial production in 2009 following a risk assessment study by the National Bureau for Fish Genetic Resources. This Hatchery utilizes standard recommended aquaculture technological strategies. Sea water pumped through ozone treatment is filtered through a bag filter to remove 40–20 μm particles and then through a cartridge filter to remove particles > 0.2 μm. Water is further processed

Fig. 5 1-f Filtration System in BMR Shrimp Hatchery

through a carbon filter and followed by UV treatment. This water is stored in a cement reservoir and used for raising axenic local culture strains of marine diatoms such as *Thallasiosira weissflogii* (Fig. 6 2a–2f) More than 70 indigenous algal strains and 26 species were Isolated and cultured in a defined f/2 medium (Keerthi et al. 2014, 2017). *T. weissflogii* cultures were scaled up to obtain heavy algal biomass and used as feed for the nauplii and post-larvae (Fig. 5 2f). BMR group hatchery produced products ensuring high survival and yielding hi-quality post-larvae for fast growth

Fig. 6 a–f culture laboratory with 70 strains of indigenous 26 diatom species

in ponds. BMR Group hatcheries use extensively probiotic products like microbial water conditioner (Fig. 5 1a–1f) and feeds for specific health boost for enhanced stress and disease resistance in hatchery production. Post-larvae are now available year-round for supply to farmers throughout the year. Since 1995, the culture of *P. monodon* is affected by White Spot Syndrome Virus (WSSV), resulted in stagnation in the development of shrimp farming.

According to the MPEDA, and The Society of Aquaculture Professionals, India, *L. vannamei* accounts for around 95% of shrimp production. Although the investment costs are high, *L. vannamei* is more adaptable than *P. monodon*; it is tolerance for low salinity (PSU 10–25) compared to 29–35 of *P. monodon*, and post larval stocking density m^{-2} is 20–40 compared to 10–15 for *P. monodon* (Manilal 2013).The shrimp aquaculture has gained prominence evident from data (Table 5 and Fig. 6). Andhra Pradesh has the maximum brackish water area and cultivates shrimp in 50,000 ha. It produces 912,439 Mt of *L. vannamei* the maximum. The hatcheries, feed mills, farms, and processing plants related to shrimp culture have increased, and so did the sale values. There are 450 hatcheries, 50 feed mills, 100,000 farms, and 243 processing plants (Table 6) and 12 top players in the mariculture (Fig. 7).

Shrimp farming in India is export oriented. Primarily geared toward export markets, driven by its high profitability (Nguyen et al. 2020). In Nellore district, the most active in shrimp farming—the impact of stocking density on the growth, survival, and yield of *Penaeus vannamei* has been extensively studied (Thriveni et al. 2022). Stocking densities of post-larvae ranged between 107, 167 and a maximum of 960,000 per ha and on the average feed consumption 8050 kg, with a conversion efficiency between 75 and 94% (Thriveni et al. 2022). It depended on variation in climatic conditions such as rainfall, temperature, salinity, and pH. Shrimp culture has

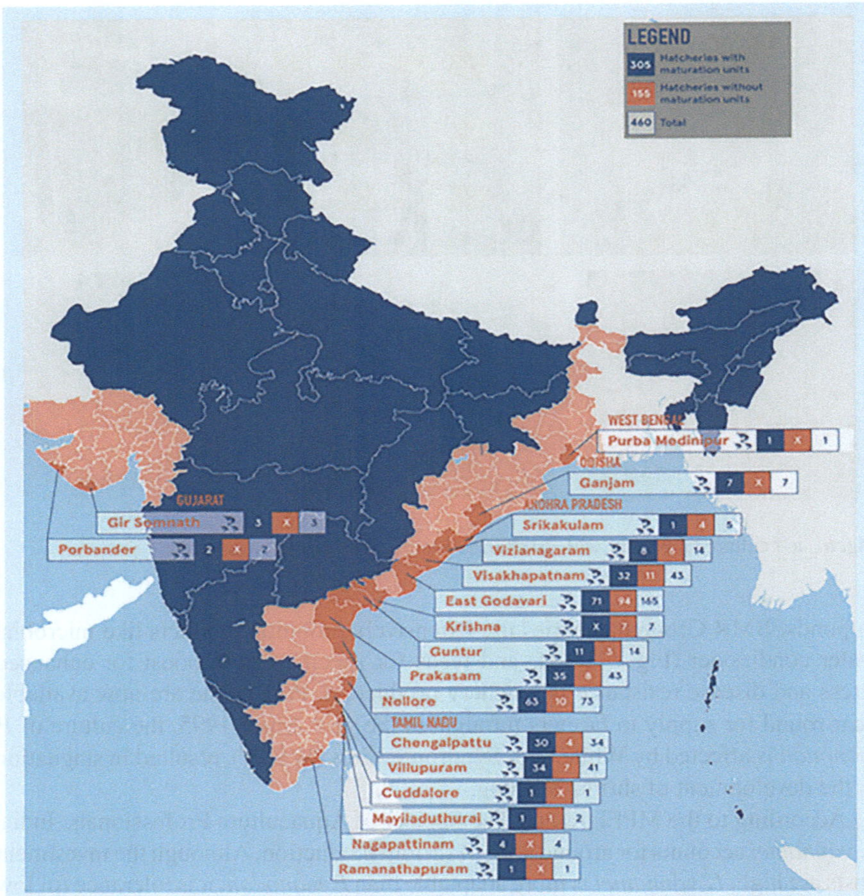

Fig. 7 Location of hatcheries (total 460; blue denotes 305 maturation and red 135 without maturation units) of monodon and vannameiin the Bay of Bengal. (*Source*: Pijl 2024)

also its major infectious disease problems: continues to face substantial challenges from infectious diseases. WSSV; White Spot Syndrome Virus, EMS/AHPND: *Vibrio parahaemolyticus*, EHP: *Enterocytozoon hepatopenaei* (EHP), White feces disease: Vibrios and (EHP), and Muscle Necrosis caused by *Vibrio harveyi.*

13 Endangered Animals Unique to the Study Area

1. There are several animals unique to the east coast of India and are of tourist interest (Fig. 8). The Horseshoe crabs known as Limulus, *Tachypleus gigas* and *Carcinoscorpius rotundicauda,* considered living fossil are distributed along the

Table 5 Number of processing facilities (Based on MPEDA's database Shrimp insights)

Region	West Bengal	Odisha	Andhra Pradesh	Tamil Nadu
Population millions	98	42	53	76
Coastline km	158	450	974	907
Farming	62,000	73,000	600,000	23,000
Estimated brackish water area (Ha)	4,05,000	31,600	1,50,000	56,800@
Cultivation area (Ha)	34,660	11,000	50,000	2907@
Hatcheries	2	3	351	4
Feed mills	1	7	28	83
Feed mills	1	7	28	83
Processing	38	25	93	30
Farmed shrimp production MT	46,021	44,460	912,439	39,558
L. vannamei				
P. monodon	24,345	2288	17,617	65
Total MT *	70,366	48,748	930,056	39,623

* MPEDA financial year 2022–2023 (MT), @ includes Puducherry

Table 6 Number of establishments associated with mariculture in Andhra Pradesh, India

	Hatcheries	Feed mills	Farms	Processing plants
Number of establishments	450	50	100,000	243
Sale values$ Bn	$0.18–0.24	$1.15–1.25	$4	$5
Top players *	Sapthagiri Group Vaisakhi Bios Golden Marine BMR Group, CP Group	Avanti Feeds, CP Group Deepak Nexgen, Devi Sea Foods Sandhya Marines	Not available	Devi Fisheries, Devi Sea Foods Sandhya Aqua, Kader Exports Nekkanti Sea Foods

* The top 5 for hatcheries is based on broodstock procurement in 2023; for feed it is based on an estimation of sales in FY 2022–2023

northeast coast of India extending from the extreme north of West Bengal through Orissa to the northern coast of Andhra Pradesh. Horseshoe crabs are often caught for their blood, which contains Limulus amebocyte lysate (LAL), a chemical used to detect bacterial endotoxins (Maloney et al. 2018) and their ecological conservation is necessary.

2. At multiple points in Odisha particularly Gahirmatha, Devio River mouth and Rushikulya ~ 0.64 million Olive Ridley Turtles *Lepidochelys olivaxcea* nest and lay their eggs (Fig. 8). Being world's largest mass nesting sites (rookeries) for olive ridley this is a spectacle. The major threats for their survival are construction of ports, incidental capture in trawl nets, and killing for meat. The Wildlife Institute of India (WII), Rushikulya Sea Turtle Protection Committee (RSTPC),

Fig. 8 Endangered animals unique to study area 1. Limulus ,2: Olive Ridley Turtles, 3. Whale shark, 4. Dugong

United Artists Association (UAA), and Green Life Rural Association (GLRA) are active in monitoring and protecting these turtles.

3. The Whale shark *Rhincodon typus* is a filter feeder sighted off Odisha, Kakinada, and Chennai (Fig. 8). They are massive and attain 14–21.9 m in length with a long lifespan and late maturation. The species is considered endangered by the IUCN.
4. Dugongs (*Dugong dugon*), commonly known as sea cows, are found in Indian waters, particularly in Gulf of Mannar, Palk Bay, and the Gulf of Kutch (see Figs. 4 and 8). They are the only existing species of herbivorous mammal that live in the sea and their population has declined to 200 individuals. A Task Force for Conservation of Dugongs in India (2007) is set up by the Ministry of Environment and Forests, Government of India.

Coringa is a reserve of the salt water crocodile, Olive Ridley turtles and the Indian Otter. Nearly 120 species of birds seek this area for nesting and breeding.

14 Limitations, Discussions, Challenges, Strategies, and Issues

There are a large number of publications and reviews that offer plenty of guidelines, mostly repetitive, but less on practical advice on the challenges, strategies, and issues. The blue economy being the bold nascent vision of future, policies are matched at the national level by India, Bangladesh, Indonesia, Australia, Mauritius, Seychelles, East Africa, and South Africa. Sustainable blue revolution, aqua business development in particular faces several challenges, ranging from regulatory issues to ecological impacts (Fig. 9). These challenges often act synergistically and natural disasters like

Fig. 9 Blue economy challenges

cyclones, heavy flood, and drought due to climate change are a major concern. Bay of Bengal experiences on average five to 6 cyclones annually (Sahoo and Bhaskaran, 2017) and prediction for the twenty-first century is an increase as much as 45% (Emanuel 2013). Impact of heavy floods, storm surges and drought due to climate change is a major concern. Although there is a wide consensus that Blue Economy can fulfill the requirements, environmental issues need to be considered (Fig. 9).

India is aiming at utilizing extensive coastline and Exclusive Economic Zone (EEZ) for economic benefits and has made strides in developing a comprehensive Blue economy policy aiming at food security, economic growth generating livelihood. However, Blue Economy faces multiple challenges such as Unusable Source Extraction (URE), habitat destruction, marine pollution. Climate Change Effects (CCE), lack of infrastructure, piracy, threats from climate change. Addressing these issues requires robust government initiatives and strategic measures. Additionally, interaction between escaped farmed stock and wild species, development of antibiotic-resistant microbial communities and usage of hormones and growth promoters are important considerations in mariculture. Inadequate supply of larval seed is the major problem and infection of the shrimp larvae due to 5 diseases compounds it further. The emergence of antibiotic resistance, and the use of hormones and growth promoters are important considerations in mariculture. Poaching and vandalism, predation by birds, poor transport facilities and cold chain logistics, i.e., cold storage and rapid transportation to maintain seafood quality are crucial.

15 Government of India Projects

India launched the Sagarmala project, the Pradhan Mantri Matsya Sampada Yojana (PMMSY), and the Fisheries and Aquaculture Infrastructure Development Fund (FIDF) aimed to enhance sustainable development in the Blue Economy sectors. The Gazette of India (CG-DL-E-21032024–253,271), in 2024 notified guidelines in great detail on the Aqua zonation, Aqua mapping, water quality for regulating coastal aquaculture selection, construction of aqua farm, health management monitoring, site selection, construction of aquaculture farms, health and waste management, periodic environment impact assessment, and integrated coastal zone management.

The Parliament enacted the Coastal Aquaculture Authority (CAA) Act to regulate the activities connected with aquaculture in coastal areas including Andhra Pradesh, Tamil Nadu, Karnataka, Goa, Puducherry, Odisha, West Bengal, Gujarat, Kerala, Diu and Daman, and the Andaman and Nicobar Islands. The CAA prescribes guidelines to ensure coastal aquaculture follows regulations in coastal environment and the livelihood of various sections of people living in the coastal areas. The Aquaculture Certification Council (ACC) advises on farming, hatcheries, and aqua culture quarantine procedures. Andhra Pradesh State Aquaculture Development Authority (APSDA ACT) established under APSADA Act, 2020 on Seed (Quality Control) Rules 2020 and Set up Minimum Standards to be maintained in shrimp and fish seed hatcheries.

The objective of the APSDA ACT is to: create an organizational setup for governing, formulate, execute, implement monitoring, regulate aquaculture and ancillary activities, ensure quality and certified aquaculture inputs like seed, feed, feed supplements, aqua chemicals, aqua products, medicines, etc., create enabling environment for trade and export of aqua products, promote establishment of infrastructure; markets facilities to promote skill up gradation and to promote Transfer of Technology.

The Government of India under the Marine Products Export Development Authority (MPEDA) is a coordinating agency, MPEDA works in collaboration with various central and state government bodies. It also promotes brackish water aquaculture for the production of prawns for export. India is also promoting ocean research and early warning systems, integrated coastal zone management, and the delineation of coastal economic zones, to support sustainable coastal and marine development. But to realize the goal consistent with UN Sustainable Development Goals (SDGs) a significant investment-estimated US$ 175 billion is required annually by 2030 (Chaturvedi 2014; Lee et al. 2020).

India is in its infancy in Blue economy. As it involves the centrally important rural fish farmer communities, educating them from the existing skill sets to meet the increasing demand is of utmost importance. Novel methodology, government policies, practical recommendations economic models, climate change, and comparison with neighboring countries are viewed as corporate activities and as such would be unfamiliar to the rural folks that are interested in earning their livelihood. As "Persuading the people to act is not the same thing as deciding what action to take"

(Colgan 2018), we point out that empowering local communities, particularly Indigenous groups, is essential for a holistic understanding of the marine environment and as inputs for policy-making.

16 Sea Cage Farming

Sea cage farming has emerged as a significant option boosting seafood production and is gaining momentum globally. It enables the large-scale cultivation of commercially valuable finfish and is recognized as one of the most efficient and economical methods for raising fish. The first open sea cage was launched in 2007 off Visakhapatnam coast in the Bay of Bengal by CMFRI (see Fig. 10). Cage mariculture produces organic matter that could also increase nutrient loads up to six-fold by 2050 (Islam 2005) which may promote an increase in harmful algal blooms (Bouman et al. 2013) which in turn could result in the development of fish kills. To maintain water quality, cage farming has to be promoted away from human settlements and discharge points of industrial and municipal waste.

The latest version is a cost-effective galvanized iron (GI) cage design suitable for low-investment farming operations and has proven successful in various Indian maritime states (CMFRI 2012). Compared to many maritime countries India is still in the early stages of developing cage culture. Currently, there are approximately 1500 cages were in inshore and brackish water areas, producing about 1500 tons of fish (NFDB 2018).There is considerable potential for further expansion of cages were culture farming in India. The ICAR and CMFRI estimate that utilizing just 1% of India's inshore waters could accommodate up to 820,000 cages and yield approximately 3.2 million tons of fish annually. This highlights the urgent need to advance sea cage farming in India. CMFRI has conducted a range of culture experiments using various cage and net materials. High-density polyethylene (HDPE) frames,

Fig. 10 Cage cultures at Nagayalanka, Kistna District, Andhra Pradesh

which are durable but expensive, are suitable for entrepreneus with substantial investments. For smaller groups and fishermen, cost-effective epoxy-coated galvanized iron (GI) frames are available, though they have a shorter lifespan than HDPE frames. CMFRI has tested various net materials including braided and twisted HDPE nets—which are durable for multiple seasons, and nylon nets, which are economical but require additional ballast due to their light weight. Long-lasting—and nylon nets, which are cost-effective but lightweight, requiring additional ballast for stability. The candidates fish/ shellfish species that are considered for rearing in sea cages include Seabass (*Lates calcarifer*), Silver Pompano (*Trachinotus blochii*), Cobia (*Rachycentron canadum*), Groupers (*Epinephelus* spp.), Snappers (*Lutjanus* spp.), and Spiny Lobster (*Panulirus* spp.). NFDB (2018) formulated guidelines for the sustainable growth of sea cage farming in India has published guidelines for sea cage farming in India. Open sea cage culture has the potential to improve coastal livelihoods, generate employment, and contribute to national food security. However, its success hinges on a comprehensive strategy and clear policy directive. Kiruba-Sankar et al. (2019) developed a policy brief that aims to tackle current issues, propose potential solutions, and introduce innovative approaches involving key stakeholders. It is intended to serve as a practical guide for policymakers, facilitating the growth and development of the open sea cage farms in Indian waters, viz., formulating a comprehensive sea cage farming policy, design and fabrication of cages and net material supporting structure facilities, breeding technology of the species to be cultured, developing proper cost-effective feed technology, disease management, and continuous R and D programs.

17 Artificial Intelligence

Similar to any other business, availability of research based technical knowledge, trained human capital, potential investments, access to high-potential innovations, scalability and publicity are necessary for blue economy. Policy and Institutional support for effective governance, technological advancements, and policy interventions are necessary to address gaps and enhance the sustainability of mariculture enterprises. Blue economy necessitates balancing investment, customers, Management, Innovation and measurement. Within the Blue Economy, Artificial Intelligence (AI) uses a set computer learning neural network that reviews data sources, publications, reports and all relevant materials. AI edits the bad data and helps optimizing data requirements for predictive modeling and resource optimization. Spalding (2024) observed that AI has the potential to move us forward to ocean-positive economic activities such as greener shipping, sustainable tourism, sustainable aquaculture, and new environmentally friendly technologies. A case study is the Kenyan Blue Economy which is in an alarming state due to threats from climate change, sea level rise, acidification and disrupted ocean currents (Gesami and Nunoo 2024). AI—a transformative tool—incorporated diverse stakeholder perspectives and provided a solid foundation into practical insights in decision making and management of their environment. Their Blue economy in 2018 was US$ 300 billion with a 49 million work force but AI predicted a substantial increase of work force and its economy by 2030.

In conclusion, the Bay of Bengal (BoB) is the largest unique tropical marine ecosystems (LME) and provides a rich diversity of habitats. The replacing South-west and North East monsoon govern its biology. Primary production in BoB coastal waters and the shelf waters ranged between 0.3 and 936 mg C m^2 d^{-1} with some hot spots (2200 mg C m^2 d^{-1}) in near shore, 500 in the shelf and 300 mg C m^2 d^{-1} in the open ocean. The rich mangroves along the coast line (Sunderbans, Andhra Pradesh, the Andaman and Nicobar Islands, and Puducherry) also contribute to high organic production. Coastal waters in Tamil Nadu sustain high seaweed growth. Mariculture shrimp yields in BoB are the highest and ranged between 200 and 500 kg per hectare per year and experienced rapid growth. BoB mariculture sector has significant potential to contribute to the blue economy. To meet the projected seafood demand, India needs to boost mariculture production significantly, potentially adding 4 to 5 million MT annually by 2030. These challenges could be effectively managed with mass education at all levels, technological improvements, policy, and strengthening of institutional frameworks.

For enhancement of India's Seafood Industry and the coastal mariculture we recommend: 1.Regulation of Spawning Period: Capture and protections during spawning periods to ensure sustainability, 2. Achieve Global Standards: Upgrade fishing harbors and landing centers to meet international standards, 3. Promote Marine Culture: Develop marine aquaculture with integrated management practices and aquaculture value, 4. Enhance cold storage and transportation to maintain seafood quality, and 5. Enhance Traceability: Improve traceability and chain-of-custody to meet global trade standards.

18 Addendum

Global seaweed production has tripled since 2000 to 32.4 million tons, led largely by Asia. As a climate-resilient and fast-growing bioresource, seaweed does not require freshwater, fertilizers, antibiotics, or land, while improving water quality, absorbing carbon, and mitigating ocean acidification. Seaweed being rich in protein, vitamin, mineral, and bioactive profile makes its usage is highly relevant for nutrition and public health programs.

Seaweed is increasingly used as food, feed, pharmaceuticals, cosmetics, biofuels, and bioplastics, and its India currently contributes less than 1 percent of global seaweed output, yet its sector—valued at $35–60 million—has a target of 9.7 million tons by 2030, supported by Government initiatives such as Pradhan Mantri Matsya Sampada Yoiana (PMMSY)MMSY, the Technology Information, Forecasting and Assessment Council (TIFAC) (Solanki et al. 2026) Seaweed Mission, and national research institutions. Techno-economic assessments show that a US$128 million investment could expand *Kappaphycus* farming to 26,000 hectares, generating US$96 million in revenue and over 108,000 jobs, with returns up to 63 percent—far higher than traditional aquaculture. Seaweed also aligns with the National Biofuel Policy, with large-scale cultivation offering significant bioethanol potential.

However, growth is constrained by fragmented governance, weak infrastructure, market volatility, and limited value-chain integration. Unlocking India's seaweed potential requires a cohesive national policy, climate-resilient infrastructure, stable markets with minimum procurement pricing, advanced technologies, diversified products, and strong community participation. With coordinated governance, technology-driven expansion, value-addition, and community-led models, seaweed can emerge as a key pillar of India's blue economy—supporting livelihoods, nutrition security, climate resilience, and sustainable economic growth.

Acknowledgements We are most grateful to Dr. Willem van der Pijl, for permission to use figure on Indian hatcheries, and to Dr. K.K. Basheer Ahammed for inviting us to contribute this chapter. We thank the reviewers and Dr. Doris Bleier, Editor Springer International for their constructive suggestions on the manuscript. We thank Dr. Gopi, Fisheries Development Officer, Tallarevu, Andhra Pradesh and Mr. Madhav Prabhas for assistance with the art work and Bala Durvasula for formatting the text.

References

Aayog N (2024) Strategy for the development of seaweed value chain. In: Strategy for the development of seaweed value chain fostering diversified livelihoods ICAR-CMFRI and NIOT. ISBN Number: 978-81-967183-2-9, ISBN Number: 978-81-967183-2-9

Abdussamad EM, Thampy DM (1994) Cannibalism in the tiger shrimp Penaeus monodon Fabricius in nursery rearing phase. J Aquac Tropics 9(1):67–75

Agarwal S et al (2017) Carbon sequestration by mangrove vegetations: a case study from Mahanadi mangrove wetland. J Environ Sci Comput Sci Eng Technol 7(1):016–029

Albalat A et al (2022) Welfare in farmed decapod crustaceans, with particular reference to Penaeus vanamei. Front Mar Sci. Sec Marine Fisheries, Aquaculture and Living Resources 9:2022. https://doi.org/10.3389/fmars.2022.886024

Arokiasundaram A et al (2015) Distribution of Chl.a in Vellar estuary and coastal waters of Parangipettai and Southeast coast of India. Int J Sci Inventions Today 2015:1–14

Banerjee RK et al (1998) Chilika lake—present and past. Central Inland Capture Fisheries Research Institute (Indian Council of Agricultural Research). Bulletin 80:103

Asokan HN et al (2024) An overview of Andaman and Nicobar Mangroves-Existing Day State of affairs. EIACP: Geodiversity and Impact on Environment 29 eISSN 0974-1356pISSN:0974-1364

Ayajuiddin MA et al (2014) Distribution of phytoplankton along an environmental gradient off Kakinada. Ind J Geo-Marine Sci 43:357–364

Bhatt JR, Kathiresan K (2011) Biodiversity of mangrove ecosystems in India. In: Bhatt JR, Macintosh DJ, Nayar TS, Pandey CN, Nilaratna BP (eds) Towards conservation and management of mangrove ecosystems in India. IUCN India. pp 1–34

Bhattathiri PMA, Devassy VP, Radhakrishna K (1980) Primary production in the Bay of Bengal during southwest monsoon of 1978. Mahasagar-Bull Natl Inst Ocean 13:315–323

Bhunia AB, Choudhury A (1982) Primary production of the estuarine water around Sagar Island. Sunderbans IJMS 11(1):87–89

BOBLME (2015) The Bay of Bengal: an overview of a large Marine Ecosystem. Phuket, Thailand: BOBLME. pp ii+36

Bouwman L et al (2013) Mariculture impacts include enrichment and eutrophication, interaction with food web. Environ Res Lett 8(044026):5. https://doi.org/10.1088/1748-9326/8/4/044026

Chaturvedi CS, Ram N, Pandey AK (2014) Aquaculture prospects in Indian and Nicobar Islands. India J Exp Zool India 17:183–187. ISSN 09/2-0030

Chen G et al (2012) The features and interannual variability mechanisms of mesoscale eddies in the Bay of Bengal. Cont Shelf Res 47:178–185. https://doi.org/10.1016/J.csr.2012.07.011

Chennubhotla VSK (1977) Food from the sea food from the sea weeds. Seaf Export J 9:21–24

Chennubhotla VSK, Kaliaperumal N, Kalimuthu S (1978) Culture of Gracilaria edulis in the inshore waters of Gulf of Mannar (Mandapam). Indian J Fish 25(1and2):228–229

Choudhury AK, Pal R (2012) Understanding the seasonal dynamics of primary productivity in relation to phytoplankton populations from the Bhagirathi—Hooghly estuary, eastern Indian coast. J Algal Biomass Utilization 3(4):80–88

CMFRI (2012) Handbook on open sea cage culture. In: Philipose Jayasree Loka KK, Krupesha Sharma Divu Damodaran SR, p 154

CMFRI FRAD (2022) Marine fish landings in India—2021. Technical report. ICAR-Central Marine Fisheries Research Institute, Kochi

Colgan C (2018) The blue economy handbook of the Indian Ocean Region: chapter two—the blue economy theory and strategy and Chapter 15 climate change and the blue economy of the Indian Ocean. Center for the Blue Economy, Middlebury Institute of International Studies

Coppen JJW, Nambiar P (1991) Bay of Bengal Programme, Post-Harvest Fisheries, Agar and Alginate Production from Seaweed in India. FAO Organization of the United Nations

Das A, Devi RS, Kumar S (2020) Mangrove plants of Odisha: an ecological indicators and sources of bioactive compounds. J Biodivers Conserv 4(1):298–298

Davies IP et al (2019) Governance of marine aquaculture: pitfalls, potential, and pathways forward. Mar Pol 104:29–36. https://doi.org/10.1016/j.marpol.2019.02.054

De TK et al (2021) Ecological assessment of Hoogly River considering a few of the more perturbed sites based on some relevant physico-chemical and biological variables—a part of AVIRIS-NG (NASA-ISRO) ground truth verification. Reg Stud Mar Sci 41(2021):101598. https://doi.org/10.1016/j.rsma.2020.101598

Deb Roy P et al (2014) Importance of mangrove conservation and valuation to women and men—a case study of Pichavaram Mangroves in India. Gender in aquaculture and fisheries: navigating change. Asian Fish Sci Special Issue 27S(2014):79–90

Deepika S et al (2019) Seasonal variation of phytoplankton diversity in seagrass ecosystem of Mandapam coast of Gulf of Mannar. J Exp Zool, India 21(1):499–506

Department of Environment and Forests 2011–2012. The Magnificent Mangroves of Andamans-An overview. Middle Andaman Division, Department of Environment and Forests, Andaman and Nicobar Islands 2011–2012

Devassy VP, Bhattathiri PMA (1981) Distribution of phytoplankton and chlorophyll *a* around Little Andaman Island. Ind J Mar Sci 10:248–252

D'Silva MS et al (2012) Algal blooms: a perspective from the coasts of India. Nat Hazards 63:1225–1253

EGREE Final Report (2017) Establishment of Knowledge Management System for East Godavari River Estuarine Ecosystem, Andhra Pradesh. Wildlife Institute of India Principal Investigator Dr. K. Sivakumar. p 449

Emanuel K (2013) Downscaling CMIP5 climate models shows increased tropical cyclone activity over the 21st century In: Santer D, Lawrence EO Livermore National 110 (30)12219–12224https://doi.org/10.1073/pnas.1301293110

FAO (1997) Mariculture Development: a new challenge. FAO Fisheries Circular No. 928 FIIU/C928. FAO, Rome. www.fao.org/docrep/w6602e/w6602e00.htm

FAO (2017) The future of food and agriculture: Trends and challenges. Food and Agriculture Organization of the United Nations, Rome
FAO (2018) The State of World Fisheries and Aquaculture 2018—Meeting the sustainable development goals. Rome. Licence: CC BY-NC-SA 3.0.
FAO (2022) The State of World Fisheries and Aquaculture 2022. Towards Blue Transformation. FAO, Rome. https://doi.org/10.4060/cc0461en
FAO (2005–2024) National Aquaculture Sector Overview. India. National Aquaculture Sector Overview. India Fact Sheets. In: Ayyappan S FAO Fisheries and Aquaculture Division [online]. Rome. Updated. [Cited 20Aug 2024]
FAO, IFAD, UNICEF, WFP and WHO (2018) The state of food security and nutrition in the world 2018. Building climate resilience for food security and nutrition. Rome, FAO
Ganesan M et al (2018) Seaweed resources in India —current status of diversity and cultivation: prospects and challenges. Botanica Marina 62. https://doi.org/10.1515/bot-2018-0056
Gesami BK, Nunoo J (2024) Artificial intelligence in marine ecosystem management: addressing climate threats to Kenya's blue economy. Front Mar Sci 11:1404104. https://doi.org/10.3389/fmars.2024.1404104
Ghosh PB et al (1990) Mangrove litter production in a tidal creek of Lothian Island of Sunderbans. India Indian Jour Mar Sci 19:292–293
Ghosh S et al (2022) Multiscale diagnosis of mangrove status in data-poor context using very high spatial resolution satellite images: a case study in Pichavaram Mangrove Forest, Tamil Nadu, India Remote Sens 14:2317. https://doi.org/10.3390/rs14102317
GMA Report (Global Mangrove Alliance) 27 July 2024. Indian Express
Gnanappazham L, Selvam V (2014) Response of mangroves to the change in tidal and fresh water flow—a case study in Pichavaram, South India Ocean and Coastal Management 102:131–138. https://doi.org/10.1016/j.ocecoaman.2014.09.004
Gopinathan C P et al (1994) Phytoplankton pigments in relation to primary production and nutrients in the inshore waters of Tuticorin, southeast coast of India. Indian J Mar Sci 23:209–212
Gouda R, Panigrahi RC (1996) Ecology of Phytoplankton in coastal waters off Gopalpur, East Coast of India. Indian J Mar Sci 25:81–84
Jacob RM, Kumar RV (1984) Primary productivity in the nearshore waters of Viziangam, Trivandrum. J Mar Biol Ass India 26(1 and 2):66–70
India State of Forest Report (2015) Forest Survey of India, Ministry of Environment, Forest and Climate Change, Kaulagarh Road, P.O-IPE, Dehradun, India
IGOFAO (2016) Sustainable intensification 984 of aquaculture in the Asia-Pacific region. In: Documentation of Successful Practices, eds W.985 Miao and K. K. Lal (Bangkok: Food and Agricultural Organization of the United Nations).
Islam MS (2005) Nitrogen and phosphorus budget in coastal and marine cage aquaculture and impacts of effluent loading on ecosystem: review and analysis towards model development. Mar Pollut Bull 50:48–61. https://doi.org/10.1016/j.marpolbul.2004.08.008
Jayalakshmi KJ et al (2022) Differential response of Bay of Bengal during deficit and normal monsoons. Environ Monit Assess 194(3):217. https://doi.org/10.1007/s10661-021-09743-2
Jeyabaskaran R Venkataraman K, Alfred JRB (2007) Implications for conservation of coral reefs in the Andaman and Nicobar Islands, India. https://www.Researchgate.net/publication/277983458
Jyothibabu R et al (2004) Differential response of winter cooling on biological production in the northeastern Arabian Sea and northwestern Bay of Bengal. Curr Sci 87:783–791
Kaladharan P, Kaliaperumal M (1999) Seaweed Industry in India. Naga 22(1):11–14
Kaladharan P, Jayasankar R (2015) Chapter 29 in 'Status of exploited marine fishery resources of India'. In: Mohan Joseph M, Jayaprakash AA (eds) Central Marine Fisheries research Institute, Kochi, pp 228–239
Kaliaperumal N (1998) Seaweed biotechnology. In: Proceeding of the first national seminar on marine biotechnology, pp 91–98
Kaliaperumal N, Kalimuthu S, Ramalingam JR (2004) Present scenario of seaweed exploitation and industry in India. Seaweed Res Util 26(1 and 2):47–53

Karthik P, Ramesh C, Mohanraju RA (2021) A checklist of seaweeds of the Andaman and Nicobar Islands, India: a way forward for seaweed cultivation, food, and drug applications. Environ Monit Assess 193:671. https://doi.org/10.10007/s10661-021-09458-4

Kathiresan K, Rajendran N (2002) Fishery resources and economic gain in three mangrove areas on the south-east coast of India. Fish Manage Ecol 9:277–283

Kawabata Z et al (1993) Phytoplankton biomass and productivity of different size fractions in the Vellar estuarine system, southeast coast of India. Indian J Mar Sci 22:294–296

Keerthi S et al (2014) A nutrient medium for development of cell dense inoculum in mixotrophic mode to seed mass culture units of Dunaliella salina. Algol Stud 147(1):7–28(22). https://doi.org/10.1127/algol_stud/2014/0177

Keerthi S (2017) Culture Studies on a Halophile Dunaliella salina from Tropical Solar Salterns, Bay of Bengal, India. Ch. 4, 65–86. In: Ravi Durvasula DV, Subba Rao (eds) Extremophiles from biology to biotechnology. CRC Press, New York

Kiling S, Turan S, Tekogul, Koru E (2013) Seaweeds for food and industrial applications. https://doi.org/10.5772/53172

Kiruba-Sankar R et al (2019) Expanding open sea cage culture in Andaman and Nicobar Islands: blue growth perspective. ICAR-Central Island Agricultural Research Institute, Port Blair, Andaman and Nicobar Islands, p 16

Kulk G et al (2020) Primary production, an index of climate change in the ocean: satellite-based estimates over two decades. Remote Sens 12(826). Remote Sens (2021), 13(17):3462. https://doi.org/10.3390/rs13173462

Kumar K (1997) The Coral Reef Ecosystem of the Andaman and Nicobar Islands: Problems and Prospects and the World-Wide Fund for Nature—India Initiatives for its Conservation. #9 Regional Workshop on the Conservation and Sustainable Management of Coral Reefs Dec. 1997 held at the M. S Swaminathan Research Foundation, Chennai, India, December 15–17, 1997. https://www.fao.org/4/x5627e/x5627e0f.htm#TopOfPage

Latha PT et al (2014) Retrieval and validation of chlrophyll-a concentrations in the coastal waters off Yanam and Kakinada (Godavari) Basin along East coast of India. J Indian Soc Remote Sensing 42. https://doi.org/10.10007/s12524-013-0282-3

Lee KH, Noh J, Jong SK (2020) The Blue Economy and the United Nations' sustainable development goals: challenges and opportunities. Environ Int 137:1–6

Llewellyn LE, English S, Barnwell S (2016) A road map to a sustainable Indian Ocean blue economy. J Indian Ocean Reg 12(1):52–66

LÖscher,C.R. 2021. Reviews and syntheses: trends in primary production in the Bay of Bengal—is it at a tipping point? Biogeosciences 18:4953–4963. https://doi.org/10.5194/bg-18-4953-2021 © Author(s) 2021

Lough JM, Cantin NE (2014) Perspectives on massive coral growth rates in a changing ocean. Biol Bull 226:187–202

Madhu NV, Maheswaran PA, Jyothibabu R (2002) Enhanced biological production off Chennai triggered by October 1999 super cyclone (Orissa). Curr Sci 82(12):1472–1479

Mahmooda H et al (2021) Paradigm shift in the management of the Sundarbans mangrove forest of Bangladesh: issues and challenges. Trees Forests and People 5(3–4) https://doi.org/10.1016/j.tfp.2021.100094

Majumdar SD (2018) Threats to coral reef diversity of Andaman Islands, India: a review. Reg Stud Mar Sci 24:237–250

Maloney T, Phelan R, Simmons N (2018) Saving the horseshoe crab: a synthetic alternative to horseshoe crab blood for endotoxin detection. PLoS Biol 16(10):e2006607. Published online 2018 Oct 12. https://doi.org/10.1371/journal.pbio.2006607

Mandal A, Banerjee M (2025) Understanding and improving sustainable use of coral reef resources. Acad Environmental Sci Sustain 2. https://doi.org/10.20935/AcadEnvSci7569

Maneesha K, Sarma V, Reddy NPC, Sadhuram Y, Murty TVR, Sarma VV (2011) Meso-scale atmospheric events promote phytoplankton blooms in the coastal Bay of Bengal. J Earth Syst Sci 120:773–782

Manilal A (2013) Effect of Stocking Density on the Survival Rate and Growth Performance in Penaeus monodon J Basic Appl Sci 10. https://doi.org/10.6000/1927-5129.2014.10.32
Manna S et al (2010) Dynamics of Sundarban estuarine ecosystem: eutrophication induced threat to mangroves Saline Syst 6:8. https://doi.org/10.1186/1746-1448-6-8
Marimuthu N (2016) Spatial variation in the health of coral reef communities of Palk Bay, southeast coast of India. Aquat Ecosyst Health Manage 19(4):360–367. https://doi.org/10.1080/14634988.2016.1258925
Martínez-Vázquez RM, Milán-García J, de Pablo Valenciano J (2021) Challenges of the Blue Economy: evidence and research trends. Environ Sci Eur 33:61. https://doi.org/10.1186/s12302-021-00502-1
Millard RS et al (2021) Global mRNA and miRNA analysis reveal key processes in the initial response to infection with WSSV in the Pacific Whiteleg Shrimp. MDPI. 13. https://doi.org/10.3390/v13061140
Millard R et al (2021) How do abiotic environmental conditions influence shrimp susceptibility to disease? A critical analysis focused on White Spot Disease. J Invertebr Pathol 186:107369. https://doi.org/10.1016/j.jip.2020.107369
Mohanty P et al (2008) Sediment dispersion in the Bay of Bengal. In Pratap K, Mohanty (eds) Monitoring and modelling Lakes and Coastal Environments. Springer Netherlands, pp 50–78. https://doi.org/10.1007/978-1-4020-6646-7_5
Muduli PR, Pattnaik AK (2020) Spatio-temporal variation in physicochemical parameters of water in the Chilika Lagoon. In Ecology, conservation, and restoration of Chilika Lagoon, India, pp 203–229
Mukherjee S, Chaudhuri A, Sen S, Homechaudhuri S (2012) Effect of cyclone Aila on estuarine fish assemblages in the Matla river of the Indian Sundarbans. J Trop Ecol 28:405–415. https://doi.org/10.1017/S026646741200020X
Mukhopadhyay A, Pal R (2002) A report on biodiversity of algae from coastal West Bengal (South and North 24-Parganas) and their cultural behavior in relation to mass cultivation program. Indian Hydrobiol 5:97–107
Muthuvelan B et al (2001) Standing crop biomass and comparative distribution of agarophytes, alginophytes and other algae in South Andaman. Indian Hydrobiol 4:130–138
Nageswara Rao L et al (2016) Variability of phytoplankton in the coastal waters of Visakhapatnam, Central East Coast of India (Bay of Bengal) with special reference to environmental parameters. J Biodivers Environ Sci (JBES) 9(4):73–86. ISSN: 2220-6663 (Print) 2222-3045 (Online). http://www.innspub.net
Nair PVR (1970) Primary productivity in the Indian Seas. Bull. Cent Mar Fish. Res Inst 22:56
National Conservation Strategy and Action Plan for the Dugongs and their Habitats in India (2012), UN (2007), Ministry of Environment and Forests Task Force, building on the UN Convention on Migratory Species (CMS) Dugong MoU signed in 2007
National Fisheries Development Board, Hyderabad (NFDB) (2018) Guidelines for Sea Cage Farming in India
Nayak PK et al (2016) Mangroves of Mahanadi delta in the state of Odisha and aspects of their conservation. Int J Conserv Sci 7(4):1095-1104
Nehru P, Balasubramanian P (2012) Mangrove species diversity and composition in the successional habitats of Nicobar Islands, India: a post-tsunami and subsidence scenario. Forest 427:70–77. https://doi.org/10.1016/j.foreco.2018.05.063
Nguyen KA et al (2020) Economic efficiency of extensive and intensive shrimp production under conditions of disease and natural disaster risks in Khanh Hoa and Tra Vinh provinces. Vietnam Sustainability 12(5):2140
Nyitrai D et al (2012) Trends in estuarine fish assemblages facing different environmental conditions: combining diversity with functional attributes. Aquat Ecol 46:201–214. https://doi.org/10.1007/s10452-012-9392-1
OECD (2017) Marine protected areas: economics, management and effective policy mixes, Paris: OECD Publishing, Paris. https://doi.org/10.1787/978982642/6208.en

Oyeku IG, Mandal K (2021) Historical occurrences of marine micro algal blooms in Indian peninsula: probable causes and implications. Oceanologia 63:51–70

Oza RM, Tiwari A, Rajyaguru MR, Goswamy S (1994) Laboratory and field culture of marine red alga Gracilaria verrucosa (Gracilariaceae, Rhodophyta). Indian J Mar Sci 23:157–161

Padmanabhan R, Silva JMN, Sudailamuthu K (2000) A remote sensing approach to monitor potential fishing zone associated with sea surface temperature and chlorophyll concentration Indian J Geo-Mar Sci 49(06):1025–1030

Pandiyarajan RS et al (2014) Distribution, abundance and diversity of phytoplankton in the inshore waters of Nizampatnam, South East Coast of India. Nopr.niscpr.res.in Corpus ID: 128773003

Panigrahy RC, Mahapatro KN (1984) Primary production studies in the surface waters of Gopalpur Sea. Orissa Coast Bulletin Environ Sci 1:1–4

Padmakumar KB, Menon NR, Sanjeevan VN (2012) Is occurrence of harmful algal blooms in the exclusive economic zone of India on the rise? Int J Ocean. Article ID 263946. https://doi.org/10.1155/2012/263946

Patel JB et al (1979) Experimental field cultivation of Gelidiella acerosa at Ervadi, India. In: Proceedingss International Symposium Marine Algae of the India Ocean Region. CSMCRI, Bhavanagar, pp 24–25 (Abstract)

Pati MP et al (2016) Uses of seaweed and its application to human welfare: a review. Int J Pharm Pharm Sci 8. ISSN-0975-1491

Pauli G (2010) Blue economy-10 years, 100 innovations, 100 million jobs. Paradigm Publications. ISBN-13978–091211190

van der Pijl W (2024) Shrimp insights guide to the Indian shrimp industry technical manuals and guides, 25 July 2024

Polidoro BA et al (2010) The loss of species: Mangrove extinction risk and geographic areas of global concern. PLoS ONE 5:e10095

Pouponneau A (2024) Riding the bluewave: unlocking finance for the sustainable blue economy. http://thecommonwealth.org

Pradeep KR (2012) Rejuvenation and restoration mangroves in coastal region. E-mail: sfird_org@yahoo.co.in /vroandhrasouth@gmail.com

Prasanna Kumar S et al (2002) Why Bay of Bengal is less productive during summer monsoon compared to the Arabian sea? Geophys Res Lett 29(881–884):2002. https://doi.org/10.1029/2002GL016013

Rainville L et al (2022) Water Mass exchanges between the Bay of Bengal and Arabian sea from multiyear sampling with autonomous gliders. J Phys Oceanogr 52(10):2377–2396. https://doi.org/10.1175/JPO-D-21-0279.1

Rajasekharan R (2012) Ecology of faunal communities on the Andaman and Nicobar Islands. In: Ecology of faunal communities on the Andaman and Nicobar Islands, pp 1–22. https://doi.org/10.1007/978-3-642-28335-2_1

Ramesh A, Kannupandi T (1997) Recent changes in the coral reef ecosystem of Palk Bay: a comparative status of previous reports and researches #19. Regional workshop on the conservation and sustainable management of coral reefs Dec 1997. Held at the M. S Swaminathan Research Foundation, Chennai, India, 15–17 Dec 1997. https://www.fao.org/4/x5627e/x5627e0p.htm

Rao RR et al (2006) Observed mini-cold pool off the southern tip of India and its intrusion into the south-central Bay of Bengal during summer monsoon season. Geophys Res Lett 33:L06607. https://doi.org/10.1029/2005GL025382

Ravishankar T et al (2001) Report for the period from April 2000-March 2000. Kakinada, M.S. Swaminathan Research Foundation

Raman AV et al (1990) Phytoplankton characteristic of Chilika Lake, a Brackish Water Lagoon along the East Coast of India. Indian J Mar Sci 19(1990):274–277

Ryther JH (1969) Photosynthesis and fish production in the sea. Science 166(3901):72–76

Sahana M, Areendran G, Sajjad H (2022) Assessment of suitable habitat of mangrove species for prioritizing restoration in coastal ecosystem of Sundarban Biosphere Reserve, India. Sci Rep 12:20997 (2022). https://doi.org/10.1038/s41598-022-24953-5

Sahoo D, Sahu N, Sahoo D (2003) A critical survey of seaweed biodiversity of Chilika Lake, India. Algae 18:1–12. https://doi.org/10.4490/ALGAE.2003.18.1.001
Sahoo,B. Bhaskaran.P. K. 2017. A comprehensive data set for tropical cyclone storm surge-induced inundation for the east coast of India. B. https://doi.org/10.1002/joc.5184
Sarangi K (2008, Sept) Monthly variability of chlorophyll and associated physical parameters in the southwest Bay of Bengal water using remote sensing data. Indian J Geo Mar Sci 37(3):256–266
Sarker S et al (2020) A bio-physicochemical perspective of the Bay of Bengal. J Mar Biol Assoc UK 100(4):517–528. https://doi.org/10.1017/S0025315420000442
Sarma VVSS et al (2019) Organic nutrients support high primary production in the Bay of Bengal. Geophy Res Lett 46:6706–6715a.pdf (743.4Kb). https://doi.org/10.1029/2019GL082262
Sarma VVSS et al (2020) Influence of eddies on phytoplankton composition in the Bay of Bengal. Cont Shelf Res 208. https://doi.org/10.1016/j.csr.2020.104241
Sarupriya JS, Bhargava RMS (1998) Seasonal distribution of Chlorophyll-a in the exclusive economic zone (EEZ) of India. Ind J Mar Sci 27:292–297
Satpathy KK (1990) Impact of power plant discharge on the physico-chemical characteristics of Kalpakkam coastal waters. Mahasagar (Bull Natl Inst Oceanogr) 23:117–125
Satpathy K et al (2010) Seasonal variation in physicochemical properties of coastal waters of Kalpakam, East coast of India with special emphasis on nutrients. Env Monit Assess 164:153–171. https://doi.org/10.1007/s1066-009-0882-0
Satyanarayana B et al (2002) Mangrove floristic and zonation patterns of Coringa, Kakinada Bay, East Coast of India Wetlands Ecology and Management, vol 10. © 2002 Kluwer Academic Publishers, the Netherlands, pp 25–39
Selvam V et al (2010) Pichavaram mangrove wetland: situation analysis. MS Swaminathan Research Foundation, Chennai, India, pp 37–39
Selvaraj S (2016) Phytochemical profiling of seaweeds collected from Pulicat Lake, Coramandal coast of South India. World J Pharm Pharm Sci 5(7):1292–1297. https://doi.org/10.20959/wjpps20167-7180
Shrivastava SS (2015) Mangroves of Odisha. A pictorial guide, p 03
Siddique MAM (2012) Destruction of hundred year's oldest mangrove Chakaria Sunderban forest: socio-economic impact on coastal communities. In: Proceedings international conference: meeting of Mangrove ecology, functioning and management (MMM3) 206 July 2012. Galle, Sri Lanka, p 174
Silambarasan A et al (2015) Monthly variations of physico-chemical characteristics in Coleroon estuary, Southeast Coast of India. Inven Rapid: Water Environ 2015(4). ISSN 2229-7766
Solanki HU, Dwivedi RM, Nayak R (2000) Generation of composite image using OCM chlorophyll and NOAA AVHRR SST for locating potential fishing grounds. In: Proceedings PORSEC 2000, vol 2. National Institute of Oceanography, Goa, pp 669–672
Solanki VK, Joshi NH, Baraiya AR (2026) A review on status of global and Indian Seaweed farming: Past and p. Int J Adv Biochem Res Resent Scen10:33–54. https://www.doi.org/10.33545/26174693.2026.v10.i1a.6864
Spalding MJ (2024) AI and improving the blue economy. J Ocean Tech 19(3024):28–35
Sreenivasan D, Srinivasan A, Padmavathy P (2007) Seasonal variation of phytoplankton diversity in seagrass ecosystem of Mandapam coast of Gulf of Mannar. https://www.researchgate.net/publication/329266878
Stocka CA et al (2017) Reconciling fisheries catch and ocean productivity 2017. PNAS, pp E1441–E1449
Subba Rao DV (1973) Effects of environmental perturbations on short-term phytoplankton production off Lawson's Bay, a tropical coastal embayment. Hydrobiologia 43:77–91. https://doi.org/10.1007/BF00014258
Subba Rao DV (2000) Bay of Bengal: Chapter 71. In: Sheppard C (ed) Seas at the Millennium. Elsevier Science Ltd, UK, pp 269–284
Subba Rao DV, Nageswara Rao CV, Keerthi S (2023) Bay of Bengal Monsoons; primary production and algal cultures. IO Bubble 18:6–12

Task Force Report Dugong in India (2012) First Draft 1 National conservation strategy and action plan for the dugongs and their habitats in India. In: Prepared by the task force for conservation of Dugongs in India Sept 2012. Ministry of Environment and Forests Government of India, p 45

Thriveni K et al (2022) Technical efficiency analysis of shrimp farms in Andhra Pradesh, India using stochastic frontier approach Indian. J Fish 69(2):119–127. https://doi.org/10.21077/ijf.2022.69.2.113857-14

Umamaheswara Rao M (1973) The seaweed potential of the seas around India. Proc. Symp. on Living Resources of the Seas Around India (1968), pp 687–692

UN 2007 World Population ageing (2007) Dept Economic and social Affairs. ST/ESA/SERA/260

UNDESA (2014) World urbanization prospects. United Nations, Department of Economic and Social Affairs, Population Division

Upadhyay DK, Mishra M (2020) Blue economy: emerging global trends and India's multilateral cooperation. Marit Aff: J Natl Marit Found India 16(1):30–45

Vajravelu M et al (2018) Seasonal influence of physico-chemical parameters on phytoplankton diversity, community structure and abundance at Parangipettai coastal waters, Bay of Bengal, South East Coast of India. Oceanologia 60: Issue 2, April–June 2018, Pages 114–127

Venkanna P (1990) Present Status of the estuarine flora of the Godavari and the Krishna. J Bombay Natural Hist Soc 88:47–54

Venkataraman K, Rajan R (2013) Status of coral reefs in Palk Bay. Rec Zool Surv India 113(Part-1):01-11-201

Vichi M, Levy M (2016) A reduction in marine primary productivity driven by rapid warming over the tropical. Indian Ocean Geophys Res Lett 43:826–833

Vijayakumaran KN, Rao B, Radhakrishna K (1996) Surface productivity and related hydrography off Visakhapatynam during premonsoon and winter months. Ind J Mar Sci 25:29–34

Vinayachandran PN, Mathew S (2003) Phytoplankton bloom in the Bay of Bengal during the northeast monsoon and its intensification by cyclones. Geophys Res Lett 30(11):1572. https://doi.org/10.1029/2002GL016717

Vinayachandran P et al (2004) Biological response of the sea around Sri Lanka to summer monsoon. Geophys Res Lett 31(1). https://doi.org/10.1029/2003GL018533

Wang X, Sun J, Yu H (2022) Distribution and environmental impact factors of phytoplankton in the Bay of Bengal during autumn. Diversity 14:361. https://doi.org/10.3390/d14050361

Harnessing Blue Carbon Ecosystems: A Global Strategy for Climate Mitigation and Coastal Sustainability

Manali Santra, Chandra Shekhar Dwivedi, and Arvind Chandra Pandey

Keywords Blue Carbon Ecosystems · Mangroves · Carbon sequestration · Climate change mitigation · Coastal sustainability · Geospatial technology

1 Introduction

With the increasing global climate change and environmental deterioration, the need of natural solutions for reducing carbon emissions and conserving biodiversity has become more and more important. Among these solutions, Blue Carbon Ecosystems (BCEs) stand out as a unique and powerful tool. BCEs such as mangroves, seagrasses, and tidal salt marshes are coastal and marine environments that possess the remarkable capacity to capture and retain carbon dioxide (CO_2) from the atmosphere at rates far higher than those of land-based forests.

The exceptional capability of BCEs not only establishes their crucial role in combating climate change but also emphasizes their wider ecological, social, and economic importance.

(Raihan 2023). Blue carbon is the carbon that is sequestered by different marine and coastal ecosystems of the Earth. By contrast to terrestrial ecosystems, which mostly store carbon in trees and soil, BCEs store carbon in both the above-ground plant biomass and the sediments below (Quevedo and Kohsaka 2024). This dual storage capacity makes BCEs particularly efficient carbon sinks. For example, while tropical forests are known for their high rates of carbon sequestration, BCEs can

M. Santra (✉) · C. S. Dwivedi · A. C. Pandey
Department of Geoinformatics, Central University of Jharkhand, Ranchi 835222, Jharkhand, India
e-mail: manali.1926020100[3]@cuj.ac.in; manali.santra.ms@gmail.com

C. S. Dwivedi
e-mail: chandra.dwivedi@cuj.ac.in

K. K. Basheer Ahammed et al. (eds.), *Monitoring Coastal Marine Ecosystems Under Climate Change*, https://doi.org/10.1007/978-3-032-09137-6_16

sequester up to five times more carbon per unit area compared to these and do so over much longer timeframes, often thousands of years (Lavery et al. 2013).

Furthermore, BCEs only encompass a minute portion of the Earth's surface—approximately 0.5% of the seafloor—yet they account for a substantial portion of all carbon stored in marine sediments (Schile et al. 2017). This disproportionate contribution to global carbon sequestration highlights the crucial role these ecosystems play in maintaining the planet's carbon balance (Stankovic et al. 2023). BCEs are diverse and varied, each playing a specific role in carbon sequestration and offering distinct ecological benefits. The three primary types of BCEs—mangroves, seagrass, and tidal salt marshes—are distributed across the globe, from tropical to temperate regions, and each contributes uniquely to the overall carbon budget (Basheer Ahammed et al. 2024; Kumar et al. 2024). Beyond their carbon sequestration capabilities, BCEs provide a multitude of ecosystem services that are vital for the health of the planet and the well-being of human populations, particularly those living in coastal areas.

Although they have great importance, BCEs are among the most endangered ecosystems on the planet. Ever since the mid-twentieth century, the global loss of salt marshes is estimated to be at 50%, mangroves at 35%, and seagrass meadows at 29%. These losses are driven by a combination of climate change and human activities, including coastal development, pollution, and unsustainable resource extraction.

Rising sea levels, increased storm intensity, and higher temperatures are exacerbating the vulnerability of BCEs. For instance, mangroves and salt marshes are being "squeezed" between rising seas and human infrastructure, limiting their ability to migrate landward and maintain their ecological functions (Fu et al. 2021). Coastal development for urbanization, agriculture, and aquaculture has led to the widespread destruction of BCEs. For example, the conversion of mangrove forests into shrimp farms has not only eliminated crucial carbon sinks but also released significant amounts of stored carbon back into the atmosphere (Sahana et al. 2022). Similarly, pollution from agricultural runoff and wastewater discharges has degraded water quality, affecting the health and productivity of seagrass meadows. Recognizing the urgent need to protect and restore BCEs, many countries and international organizations are developing and implementing strategies to conserve these vital ecosystems. These efforts are often integrated into broader climate mitigation and adaptation frameworks, highlighting the interconnectedness of environmental, social, and economic goals. The future of BCEs depends on the actions taken today to protect and restore these critical ecosystems. As the world faces the dual challenges of climate change and biodiversity loss, BCEs offer a natural solution that addresses both issues simultaneously (Alongi 2012). Nevertheless, the ongoing deterioration of these ecosystems poses a significant risk to the achievement of worldwide climate objectives and the well-being of millions of individuals who rely on them. In order to guarantee the enduring viability of BCE, it is imperative to include blue carbon strategies into both national and international climate policies, enhance the implementation of environmental legislation, and encourage community involvement in conservation initiatives. By harnessing the power of BCEs, the global community (Bertram et al. 2021; Peter et al. 2021) can make significant strides toward a more sustainable and resilient future, where both people and nature thrive.

2 Types of Blue Carbon Ecosystems (BCEs)

BCEs refer to coastal and marine habitats that possess the capacity to absorb and retain substantial quantities of carbon. These ecosystems play a crucial role in combating climate change by effectively absorbing carbon dioxide (CO_2) from the atmosphere and storing it in plant and sedimentary matter. There exist multiple distinct varieties of BCEs, each with distinctive attributes, distributions, and roles in carbon sequestration.

2.1 Mangrove Forests

These are one of the most well-known types of BCEs. Mangroves are salt-tolerant plants and shrubs, predominantly inhabit intertidal zones along tropical and subtropical coastlines.

They are highly efficient at sequestering carbon due to their dense root systems, which trap and store organic carbon in the soil.

Global Distribution and Coverage: Mangroves are found in over 120 countries, with significant concentrations in Southeast Asia, West Africa, and Central America. They cover an estimated 15 million hectares globally, making them one of the largest BCE.

Carbon Sequestration Potential: Mangroves have a higher carbon sequestration potential compared to other tropical forests with the ability to store up to four times more carbon per hectare. Carbon is stored in both the biomass of trees and in waterlogged soils, which have the ability to retain carbon for thousands of years.

Additional Ecosystem Services: In addition to storing carbon, mangroves offer vital services such as coastal defense, fish habitat, and water purification. They play a vital role in sustaining the livelihoods of several coastal communities.

2.2 Seagrass Meadows

Seagrass meadows are underwater plant communities found in shallow marine environments. These flowering plants form dense underwater meadows that are crucial for carbon sequestration and marine biodiversity.

Global Distribution and Coverage: Seagrass meadows are found along coastlines on every continent except Antarctica. They are particularly abundant in the shallow coastal waters of Europe, Australia, and Southeast Asia.

Carbon Sequestration Potential: Seagrass exhibits a remarkable capacity for carbon sequestration, mostly due to the extensive storage of carbon in the sediments within their root systems. Seagrass meadows have the capacity to trap carbon at a rate up to 35 times greater than that of tropical rainforests.

Ecological importance: Seagrass meadows have ecological significance as they provide habitat for a diverse array of marine organisms, such as fish, turtles, and crustaceans. In addition, they serve to stabilize sediments, mitigate coastal erosion, and enhance water quality by effectively filtering out contaminants.

2.3 Tidal Salt Marshes

Tidal salt marshes are coastal wetlands prone to regular inundation. These ecosystems are characterized by dense stands of salt-tolerant plants, which play a crucial part in carbon sequestration.

Global Distribution and Coverage: Tidal salt marshes are found in temperate regions around the world, particularly along the coastal areas of Europe, Australia, and China along with the United States.

Carbon Sequestration Potential: Tidal salt marshes are highly effective at sequestering carbon in their soils, which are usually saturated with water and lacking in oxygen. That inhibits the breakdown of organic material, therefore enabling the long-term storage of carbon.

Role in Coastal Protection: Tidal salt marshes act as natural buffers against storm surges and flooding. They also support biodiversity, provide habitat for wildlife, and filter nutrients and sediments from the water.

Other Potential BCEs:

While mangroves, seagrasses, and tidal marshes are the primary BCEs, other coastal and marine environments also have the potential to contribute to carbon sequestration.

2.4 Coral Reefs

Coral reefs, while traditionally not considered major carbon sinks, contribute to carbon cycling in coastal environments. Their role in blue carbon is less about direct carbon storage and more about supporting the broader ecosystem processes that facilitate carbon sequestration in adjacent BCEs like seagrasses and mangroves.

2.5 *Kelp Forests*

Kelp forests are large marine communities found in cold, nutrient-rich waters. These ecosystems are being increasingly recognized for their potential to sequester carbon, particularly in deep ocean sediments when detached kelp sinks to the seafloor.

Every type of BCE possesses distinct attributes that provide numerous benefits to the process of carbon sequestration and storage. For example, although mangroves excel at storing biomass above the land, seagrass meadows are more efficient at storing carbon below the ground. Tidal salt marshes, due to their sediment accumulation, offer effective long-term solutions for carbon storage. The range of functions performed by these ecosystems highlights the need for preserving the variety of BCE in order to optimize their combined effect on mitigating climate change. Understanding the different types of BCE is crucial for developing targeted conservation strategies. Each ecosystem plays a unique role in carbon sequestration and provides a range of other ecosystem services that are vital for coastal communities and global environmental health. Protecting and restoring these ecosystems is essential for mitigating climate change and ensuring the resilience of coastal areas in the face of environmental challenges.

3 Ecosystem Services Provided by Blue Carbon Ecosystems

BCEs provide a wide range of interconnected services that are crucial for preserving environmental health, enhancing human welfare, and controlling global warming (Choudhary et al. 2024). There are four main areas into which the services can be classified: providing, regulatory, cultural, and supporting services.

3.1 *Provisioning Services*

It refers to the direct products and resources that BCEs provide to humans. These include essential resources that support livelihoods, economies, and overall well-being.

Fisheries: BCEs have played an essential part as breeding grounds for numerous economically significant fish species, serving as environments that support the reproduction, growth, and prosperity of fish and shellfish. For example, mangroves play a vital role as essential breeding areas for young fish and crustaceans, which then move to different marine environments such as coral reefs or the open sea. These ecosystems provide essential support for both local and global fisheries, thereby making major contributions to the food security and economic stability of coastal populations.

Forest Products: Mangrove forests provide a range of resources, different kinds of forest products, in addition to their contribution to fisheries. These sources provide lumber for construction, fuelwood, and charcoal manufacture, especially in areas where there are limited options for alternate energy sources. In addition, mangroves offer non-timber resources such as honey, tannins, and traditional remedies obtained from the bark and leaves of mangrove trees. These resources are vital for the well-being of local populations.

Genetic Resources: BCEs contain an extensive diversity of species, some of which have distinct genetic characteristics that are significant for scientific research, agriculture, and biotechnology. For instance, the genetic variability present in seagrass meadows and mangroves presents an opportunity for the creation of crops that possess greater resistance to climate change, pests, and diseases. Exploring genetic resources from BCEs has the potential to result in findings that can have positive impacts on medicine, agriculture, and environmental sustainability.

3.2 Regulating Services

It describes the advantages derived from the control and management of ecosystem processes. BCE maintains the environmental stability and mitigates climate change through various regulating services.

Carbon Sequestration: Carbon sequestration is a crucial service supplied by BCEs, as they have the capacity to store carbon. They can sequester carbon dioxide from the atmosphere by absorbing it and storing it in their plant and sediment structures. This method aids in mitigating global climate change by diminishing the quantity of CO_2 in the atmosphere, thereby reducing the greenhouse effect.

Climate Regulation: In addition to carbon sequestration, BCEs have an impact on local and regional climates by regulating temperature, humidity, and patterns of precipitation. For instance, the thick coverings of mangrove forests lower temperatures and enhance humidity in coastal regions, promoting more consistent and pleasant microclimates.

Coastal Protection: BCEs serve as inherent buffers against coastal hazards, including storm surges, erosion, and the SLR. Mangroves, characterized by their complex root structures, disperse wave energy and capture sediment, thereby mitigating the effects of storms and averting coastal erosion. Tidal salt marshes and seagrass meadows contribute to the stabilization of shorelines and shield coastal infrastructure against potential harm.

Water Purification: BCEs help in the process of water purification by effectively removing contaminants and excessive nutrients from the water column through filtration. Seagrass meadows, for instance, capture suspended sediments and assimilate nutrients, enhancing water clarity and quality. This function is crucial for preserving the health of nearby ecosystems, such as coral reefs, which are highly susceptible to alterations in water quality.

3.3 Cultural Services

Cultural services refer to the intangible advantages that individuals derive from ecosystems, specifically encompassing recreational, spiritual, and aesthetic aspects.

Recreation and Tourism: BCEs are extremely popular locations for ecotourism and recreational activities. Tourists are drawn to coastal locations abundant in BCE due to activities like snorkeling, birding, and kayaking, which in turn boost local economy. Mangrove forests and seagrass meadows provide unparalleled opportunities to engage with nature and raise environmental consciousness, owing to their abundant biodiversity and picturesque landscapes.

Cultural Heritage and Spirituality: BCEs hold significant cultural and spiritual importance for numerous coastal communities. For example, specific mangrove forests hold religious significance in certain societies, functioning as locations for ceremonies and customary activities. These ecosystems play a crucial role in the cultural identity of indigenous and coastal inhabitants, offering a feeling of connection and belonging.

Aesthetic and Educational Value: BCEs are highly esteemed for their inherent aesthetic appeal and rich biodiversity. They offer chances for environmental education and scientific study, contributing to the promotion of awareness regarding the significance of conservation and the function of ecosystems in supporting life on Earth. The aesthetic appeal of these landscapes also serves as a source of inspiration for art, literature, and photography, bettering human civilization.

3.4 Supporting Services

Supporting services refer to the fundamental ecological processes that are essential for the operation of all supplementary ecosystem services. Maintenance of the health and production of BCEs necessitates the implementation of these processes.

Nutrient Cycling: BCEs are essential for the nutrient cycling of nitrogen and phosphorus, which are critical for the development and productivity of plants and other species. Mangroves and seagrass meadows sequester and redistribute organic material, hence enhancing the nutrient-dense sediments that sustain a wide variety of species.

Habitat Provision: BCEs offer essential habitats for a diverse range of species, encompassing fish, birds, invertebrates, and plants. The intricate root systems of mangroves provide refuge and reproductive habitats for a multitude of marine species, while seagrass meadows sustain a wide range of marine diversity. The preservation of these habitats is crucial for the sustenance of biodiversity and the robustness of marine ecosystems.

Biodiversity Maintenance: BCEs are hotspots of biodiversity, supporting a diverse array of species that are adapted to the unique conditions of coastal environments. The high biodiversity found in these ecosystems contributes to the overall stability and resilience of the marine environment, helping to sustain ecological processes and services.

4 Carbon Storage Potential of Blue Carbon Ecosystems

The carbon sequestration capacity of BCE varied greatly according on the ecosystem type, ecological health, and geographic location (Powlson et al. 2011; Costanza et al. 2014). Quantifying the carbon sequestration capacities of various Biochar Ecosystems (BCEs) is essential for maximizing their contribution to mitigating climate change.

4.1 *Mangroves*

Mangroves are among the most effective ecosystems for carbon sequestration, with a remarkable ability to store carbon both above and below ground.

Carbon Storage Capacity: Mangroves can store between 800 and 1200 metric tons of carbon per hectare, with some studies suggesting even higher values in certain regions. The carbon is stored in the biomass of the mangrove trees (trunks, branches, leaves, and roots) as well as in the soil. The below-ground carbon storage, in particular, is substantial because mangrove soils are rich in organic matter and are often waterlogged, which slows down the decomposition of organic material.

Time Frames: The carbon stored in mangrove ecosystems can remain sequestered for centuries or even millennia if the ecosystem remains undisturbed. This long-term carbon storage makes mangroves particularly valuable in the context of climate change mitigation.

4.2 *Seagrass Meadows*

Seagrass meadows are another critical BCE with significant carbon sequestration capabilities, particularly due to their ability to store carbon in sediments.

Carbon Sequestration Capabilities: Seagrass meadows sequester carbon at rates of approximately 83–370 g of carbon per square meter per year, depending on the species and environmental conditions. Unlike mangroves, the most of carbon in seagrass ecosystems is stored below ground. These sediments can accumulate large amounts of organic carbon over time, creating substantial carbon sinks.

Below-Ground Carbon Storage: The importance of below-ground carbon storage in seagrass meadows cannot be overstated. The thick root and rhizome systems of seagrass plants stabilize sediments and trap organic material, preventing its decomposition and release back into the atmosphere. The carbon stored in seagrass sediments can remain locked away for thousands of years, provided the meadows are not disturbed.

4.3 Tidal Salt Marshes

Tidal salt marshes are highly productive ecosystems that play a significant role in carbon storage, particularly in their dense, organic-rich soils.

Carbon Storage Capacity: Tidal salt marshes can store between 200 and 300 metric tons of carbon per hectare. While this is generally lower than the carbon storage capacity of mangroves, tidal salt marshes are still highly efficient at sequestering carbon, especially in temperate regions where they are most common.

Comparison with Other BCEs: Tidal salt marshes are often compared to mangroves and seagrass meadows in terms of their carbon sequestration potential. Although they may store less carbon per hectare than mangroves, their extensive coverage in some regions makes them significant contributors to global carbon sequestration. Additionally, like seagrass meadows, the majority of carbon in tidal salt marshes is stored in the soil, which can remain sequestered for long periods.

Each component of the Blue carbon ecosystem has a vital function in the process of carbon sequestration. Notably, mangroves typically provide the greatest carbon storage capacity per unit area, with tidal salt marshes and seagrass meadows following closely behind. Through understanding the benefits of each type of ecosystem, it is feasible to formulate conservation measures that are more accurate and effective, therefore maximizing the capacity of global BCE to sequester carbon.

5 Blue Carbon Wealthy Countries

Countries with vast marine carbon ecosystems significantly contribute to the global carbon cycle by sequestering and storing carbon. These ecosystems not only sequester and retain substantial amounts of carbon dioxide (CO_2) but also offer essential functions such as providing protection to coastlines, fostering marine life, and sustaining

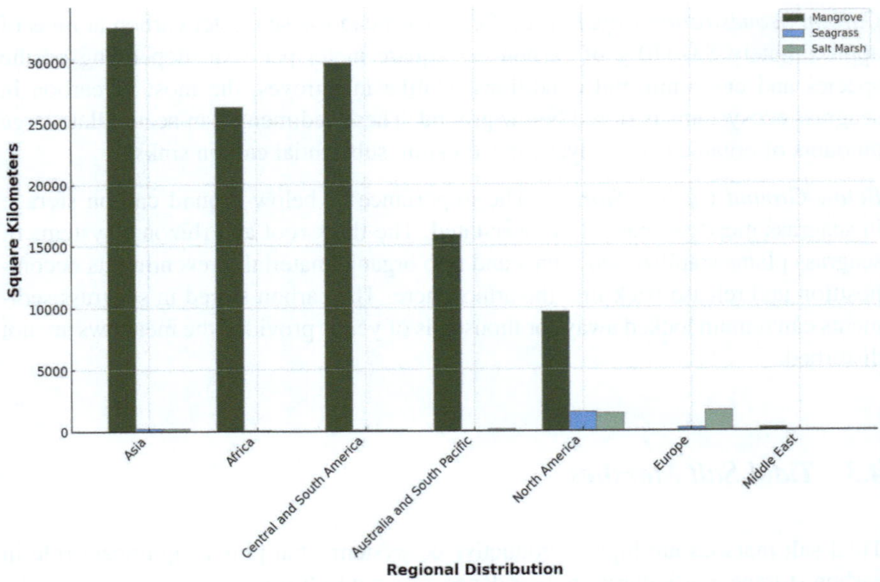

Fig. 1 Worldwide distribution of blue carbon ecosystems (Himes-Cornell et al. 2018)

local livelihoods. Continuous surveillance of the BCE through diverse conservation and management initiatives is crucial for reducing the impacts of climate change and safeguarding coastal biodiversity. In the face of rising sea levels and increased human activity along coastlines, effective management guarantees the continued vitality and provision of these crucial services by these ecosystems (Fig. 1).

The impacts of global warming are not uniformly distributed throughout the planet, and neither is the distribution of sinks for carbon sequestration (Moore et al. 2022; Harris et al. 2012). Some regions are more vulnerable to climate change, while others possess vast ecosystems capable of absorbing significant amounts of CO_2. This uneven distribution creates a complex dynamic in global climate policy, where the benefits of carbon sequestration are shared globally, but the responsibilities and capacities are unevenly distributed. Based on the economic value generated from carbon sequestration via BCEs, the social cost of carbon (SCC) is estimated. The SCC represents the economic damages associated with the emission of one additional ton of CO_2, making it a crucial tool for policymakers in evaluating the cost-effectiveness of climate mitigation strategies. The blue carbon wealth varies among countries based on the area distribution and health condition of the BCEs, as well as the Specific Cost of Carbon in the Country (CSCC). The CSCC reflects how much a nation values the avoided harm from carbon emissions, which depends on its economic system, climate vulnerability, and other socioeconomic factors (Huang et al. 2018).

An interesting concept in the context of blue carbon wealth is the concept of donor and receiver. Donor and receiver countries refer to the roles they play in the global distribution of benefits derived from carbon sequestration in BCEs. Donor nations sequester more carbon through their BCEs than the economic benefits they receive, as the majority of these benefits are realized in other countries with higher CSCCs. In contrast, receiver countries gain a lot from the donor countries' carbon sequestration because the avoided climate damages result in significant economic savings for them (Lal 2005; Milica et al. 2023). The size of the BCEs and the CSCCs controls the distribution of BC wealth. Donor countries play a key role in global climate adaptation and mitigation efforts, but the receiver countries with higher CSCCs are where the economic value of BCEs is most apparent.

Australia is at the top of the list of donor countries, followed by Indonesia and Cuba. These countries have vast and well-preserved blue carbon ecosystems that sequester large amounts of CO_2, making them crucial players in global climate regulation (Jiawen et al. 2024). A significant amount of blue wealth is also observed in Brazil and the United States, among other countries. However, the blue carbon wealth generated in these donor countries is largely transferred to nations with higher CSCCs, such as the United States, China, and India, which are major receiver countries (Vivek et al. 2011). This transfer of wealth highlights the interconnectedness of global climate efforts, where the actions taken by one country can have profound impacts on others (Fig. 2).

The concept of donor and receiver countries underscores the importance of international cooperation in blue carbon management. It is essential that global policies recognize the contributions of donor countries and provide them with the necessary support and incentives to maintain and expand their BCEs. Additionally, receiver countries, which benefit from global carbon sequestration efforts, should actively

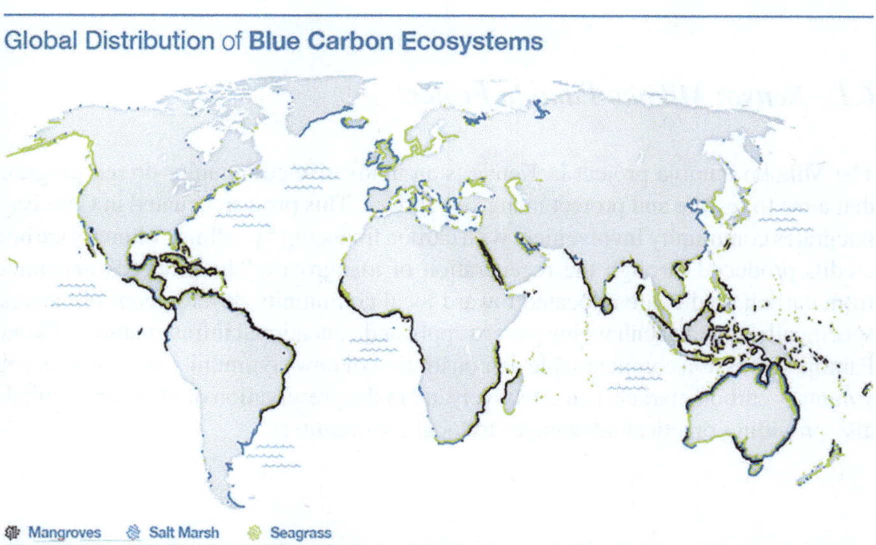

Fig. 2 Global distribution of blue carbon ecosystem. *Source:* bluecarboninitiative.org

participate in financing and supporting conservation initiatives in donor nations. This global partnership is crucial for ensuring that blue carbon ecosystems continue to thrive and contribute to climate change mitigation, ultimately benefiting all nations (Guri et al. 2018).

Moreover, there is a growing recognition of the need to integrate blue carbon strategies into national and international carbon markets. By allowing countries to trade carbon credits based on the sequestration provided by their BCEs, donor countries could receive direct financial compensation for their contributions to global climate goals. This approach not only provides economic incentives for conservation but also fosters a more equitable distribution of the benefits associated with blue carbon wealth (Pan et al. 2011; Carpenter et al. 2009).

In conclusion, the role of blue carbon ecosystems in global climate mitigation is undeniable. Countries with extensive BCEs hold a unique position acting as natural carbon sinks that provide significant economic and environmental benefits. The dynamics between donor and receiver countries highlight the importance of global collaboration in managing these vital ecosystems. Through cooperation, countries can guarantee the optimization and fair distribution of the advantages of blue carbon, so increasing the chances of a sustainable and resilient future for everyone.

6 Blue Carbon Conservation Initiatives Taken by Countries

Global nations are progressively acknowledging the value of blue carbon ecosystems (BCEs) in easing climate change and preserving coastal landscapes.

6.1 Kenya: Mikoko Pamoja Project

The Mikoko Pamoja project in Kenya is an innovative community-driven program that aims to restore and protect mangrove forests. This project, situated in Gazi Bay, integrates community involvement with carbon financing by selling voluntary carbon credits produced through the regeneration of mangroves. The proceeds generated from carbon credits are allocated toward local community development initiatives, specifically aimed at enhancing water supply and educational infrastructure. Mikoko Pamoja serves as a commendable demonstration of how community engagement and voluntary carbon markets can effectively aid in the preservation of blue carbon while also providing practical advantages to local communities.

6.2 Vietnam: Markets and Mangroves Project

The Markets and Mangroves project in Vietnam combines environmental preservation with economic incentives by promoting organic shrimp farming that is connected to the conservation of mangroves. This research showcases the capacity of integrating market-driven incentives, such as organic certification, with the conservation of blue carbon habitats. The project helps to carbon sequestration and improves local livelihoods by implementing sustainable shrimp farming practices that are compatible with mangrove conservation. This strategy emphasizes the significance of harmonizing economic activity with conservation objectives to guarantee the enduring viability of BCEs.

6.3 India: Sundarbans Mangrove Restoration

The Sundarbans in India is home to one of the largest mangrove ecosystems globally. The Sundarbans Mangrove Restoration Project is an extensive attempt designed to restore deteriorated mangrove regions in order to improve the absorption of carbon and offer advantages for climate adaptation. This project is a component of the Verified Carbon Standard (VCS) and specifically targets reforestation, soil carbon accounting, and community engagement. By incorporating soil carbon into its carbon accounting, the research accurately measures the complete carbon storage capacity of the mangrove ecosystem. In addition, the initiative promotes economic growth in local communities by generating employment opportunities and enhancing their ability to withstand the effects of climate change.

6.4 Madagascar: Blue Forests Project

The Blue Forests Project in Madagascar is an initiative to integrate the preservation of mangroves within the nation's REDD + policy, to reduce GHG emissions. This initiative is centered around the preservation and rehabilitation of mangrove ecosystems, which play a significant part in both capturing carbon and safeguarding coastal areas. The objective of the Blue Forests Project is to include mangrove forests into carbon financing systems, enabling Madagascar to utilize international carbon markets as a means to finance conservation initiatives. This effort emphasizes the possibilities of incorporating blue carbon solutions into wider national climate policy and REDD + frameworks.

Blue carbon conservation programs suffer from several obstacles, such as the complexity and expense of bringing projects into carbon markets that adhere to regulations, so restricting funding prospects. Precisely measuring soil carbon is a difficult task, but it is crucial in order to accurately assess the capacity to store carbon.

Community participation is essential for the success of these programs, but it can be hard to ensure regular involvement and fair distribution of benefits. Restoration and long-term sustainability efforts face considerable challenges due to the variety of local conditions and the extent of deterioration in BCEs.

The actions carried out by nations such as Kenya, Vietnam, India, and Madagascar demonstrate the capacity of blue carbon projects to contribute to both the mitigation of climate change and the promotion of sustainable development. These projects are utilizing carbon funding mechanisms, community engagement, and policy support to safeguard and rehabilitate crucial blue carbon habitats. The knowledge gained from these endeavors offers vital understanding for the future advancement of blue carbon programs and emphasizes the importance of international collaboration in preserving these crucial ecosystems.

6.5 Australia: Reducing GHG Using Floating Wetlands

Floating wetlands are an innovative solution designed by Blue Carbon Lab to curb down greenhouse gas (GHG) emissions and reduce nutrient loads in wastewater treatment plants (WWTPs). These systems consist of buoyant structures that support native wetland plants, which absorb nutrients directly from the water column, acting as biological filters. This not only helps in reducing nutrient concentrations but also lowers the emission of GHGs from wastewater lagoons, making floating wetlands a promising strategy for decreasing the carbon footprint of WWTPs (Fig. 3).

In collaboration with Westernport Water and Clarity Aquatic, new designs for floating wetlands are being developed and tested in Australia. The goal of these projects is to enhance the effectiveness of floating wetlands in nutrient removal and GHG reduction. Initial studies of the prototype floating wetlands have demonstrated significant environmental benefits: reductions of total nitrogen and total phosphorus by 58% and 49%, respectively, and a decrease in GHG emissions by up to 50%. Additionally, these systems have shown potential in reducing concentrations of toxic PFAS chemicals and other pollutants. This technology offers a scalable and sustainable solution for improving water quality and addressing climate change. As part of broader climate mitigation strategies, floating wetlands present a viable option for WWTPs worldwide, contributing to greener practices and engaging stakeholders in the adoption of innovative environmental technologies. By integrating these systems, countries can enhance their efforts to combat climate change while improving water management practices.

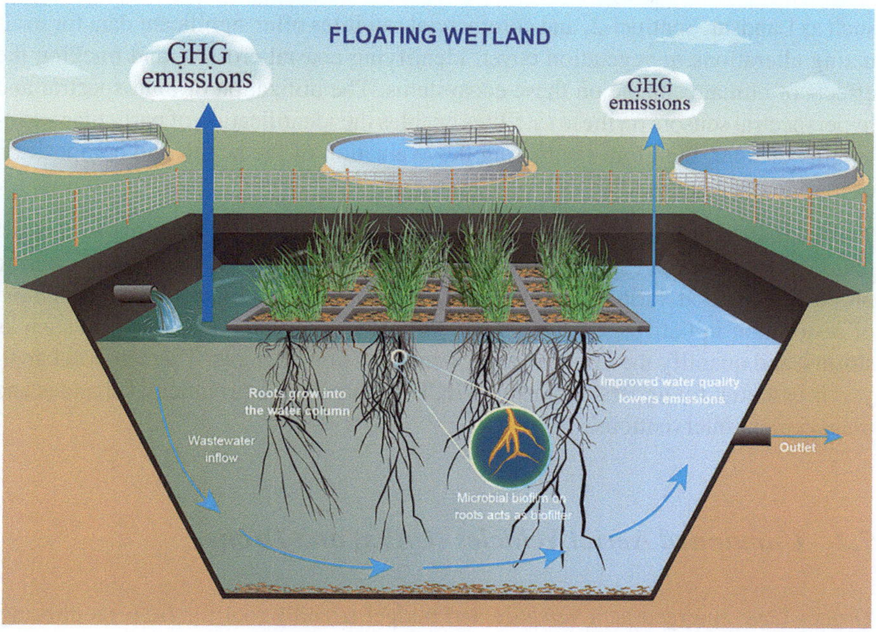

Fig. 3 Concept of floating wetlands. *source*: www.bluecarbonlab.org

7 Geospatial Technology to Monitor Blue Carbon Ecosystems

Geospatial technology has significantly enhanced the monitoring and management of BCEs, such as mangroves, seagrass meadows, and tidal salt marshes. These technologies provide efficient methods for tracking changes in BCEs across time, assessing their state, and quantifying their influence on carbon sequestration. Geospatial technologies, including remote sensing, Geographic Information Systems (GIS), and unmanned aerial vehicles (UAVs or drones), are actively involved in the monitoring of these vital environmental systems.

7.1 Remote Sensing for BCE Monitoring

Remote sensing is a crucial tool for monitoring BCEs because it allows for the collection of data over large and often inaccessible areas. This technology uses satellites and aerial platforms to capture imagery and other data that can be analyzed to detect changes in ecosystem structure, health, and carbon storage capacity.

Satellite Imagery: High-resolution satellite photography is utilized to monitor the scope and state of BCEs over vast geographical regions. Remote sensing satellites such as Landsat, Sentinel-2, and commercial satellites offer significant data for evaluating alterations in vegetation cover, identifying coastal erosion, and tracking the effects of climate change on these ecosystems. The utilization of multispectral and hyperspectral sensors on these satellites enables the identification of particular wavelengths of light, therefore facilitating the evaluation of the health and composition of marine ecosystems.

Temporal Monitoring: One of the strengths of remote sensing is its ability to conduct temporal monitoring, capturing changes in BCEs over time. By combining satellite photos captured at various time intervals, scientists may monitor the advancement of restoration initiatives, evaluate the consequences of natural phenomena such as storms, and quantify the effectiveness of conservation attempts. This temporal analysis is essential for understanding how BCEs respond to environmental changes and management interventions.

7.2 Unmanned Aerial Vehicles (UAVs) and Drones

Drones have emerged as a versatile and budget friendly tool for BCE monitoring. They can capture high-resolution images and generate detailed 3D models of ecosystems, providing data at a much finer scale than satellite imagery.

High-Resolution Data Collection: Drones equipped with multispectral and thermal sensors can collect data at resolutions as fine as a few centimeters. This high level of detail allows for precise monitoring of small-scale changes in vegetation structure, species composition, and carbon sequestration potential. For example, drones can be used to monitor the growth of newly planted mangroves or to assess the condition of seagrass meadows after restoration efforts.

Elevation and Biomass Mapping: Drones can also generate Digital Elevation Models (DEMs) and biomass maps of BCEs. These models are essential for understanding the topography of these ecosystems, which influences tidal flow, sediment deposition, and carbon storage. By combining elevation data with vegetation indices derived from multispectral imagery, researchers can estimate the biomass of BCEs, which is directly related to their carbon sequestration capacity.

Cost-Effectiveness and Accessibility: Drones are relatively inexpensive and easy to deploy, making them accessible tools for local communities, researchers, and conservation organizations. Their ability to cover difficult-to-access areas, such as remote coastlines or dense mangrove forests, makes them invaluable for monitoring projects where traditional methods may be impractical.

7.3 Geographic Information Systems (GIS)

GIS is a critical component of BCE monitoring, providing a platform for integrating and analyzing spatial data from various sources, including remote sensing, drone imagery, and field surveys.

Data Integration and Analysis: GIS allows for the integration of diverse datasets, enabling comprehensive analysis of BCEs. For example, satellite imagery can be overlaid with drone-derived maps and field data to assess the spatial distribution of carbon stocks, identify areas at risk of degradation, and prioritize regions for conservation or restoration.

Spatial Modeling and Prediction: GIS can be used to develop spatial models that predict changes in BCEs under different scenarios, such as sea-level rise or increased human activity. These models can inform management decisions by identifying areas most likely to be affected by environmental changes and guiding targeted conservation efforts.

Mapping and Visualization: GIS is also essential for visualizing BCE data, creating maps that communicate complex information to stakeholders, policymakers, and the public. These maps can illustrate the extent of BCEs, highlight areas of high carbon storage, and show the progress of restoration projects over time.

7.4 Machine Learning (ML) and Artificial Intelligence (AI)

ML and AI have enhanced the capabilities of remote sensing for BCE monitoring. These technologies enable more accurate and efficient analysis of large datasets.

Object-Based Image Analysis (OBIA): OBIA is a machine learning technique that segments satellite or drone imagery into meaningful objects rather than individual pixels, allowing for more accurate classification of land cover types. In BCE monitoring, OBIA can be used to differentiate between species of mangroves, track their spreading, and assess the success of restoration efforts at a species level.

Random Forest (RF) Classifier: The RF algorithm, a type of machine learning model, is frequently used to monitor BCE. It can classify different land cover types based on various features extracted from the imagery, such as spectral values, elevation, and texture. This method has been shown to achieve high accuracy in classifying coastal ecosystems, as demonstrated in the recent studies that used it to monitor salt marsh restoration.

8 Challenges and Future Directions

While geospatial technologies offer powerful tools for monitoring BCEs, several challenges remain. The high cost and limited availability of ultra-high-resolution satellite imagery, the need for standardized methodologies for carbon accounting, and the integration of diverse data sources are ongoing challenges in the field. However, continued advancements in remote sensing, drone technology, and machine learning hold promise for overcoming these challenges.

The future of BCE monitoring will likely involve the integration of multiple geospatial technologies to create comprehensive, multi-scale monitoring systems. These systems will enable real-time tracking of BCE health and carbon storage, providing the data needed to support global climate mitigation efforts and the sustainable management of coastal ecosystems.

9 Conclusion

Preserving and restoring Blue Carbon Ecosystems is crucial for tackling urgent environmental issues such as climate change, biodiversity decline, and coastal damage. Countries with substantial biodiversity conservation efforts (BCEs), such as Australia, Indonesia, and Kenya, are at the forefront of developing new techniques to maintain these ecosystems. These measures not only protect the BCEs but also improve their ability to store carbon and offer other essential ecosystem services. The incorporation of geospatial technologies has significantly enhanced the capacity to observe and oversee BCEs, guaranteeing that conservation efforts are both efficient and enduring. Moreover, the notion of donor and receiver countries highlights the necessity for international collaboration in the management of BCEs, where the advantages of carbon sequestration are shared worldwide, but the responsibilities must also be equitably shared. As nations worldwide increasingly adopt sustainable development methods, the significance of BCEs in regulating global temperature and maintaining environmental health cannot be emphasized enough. By persistently investing in these ecosystems and adopting cutting-edge management approaches, countries can guarantee the ongoing growth of BCEs and their valuable contribution to a sustainable and resilient future for everyone.

References

Alongi DM (2012) Carbon sequestration in mangrove forests. Carbon Manag 3(3):313–322. https://doi.org/10.4155/cmt.12.20

Basheer Ahammed KK, Karang IWGA, Nuarsa IW, Indrawan IGS, Hendrawan IG, Dewi NMNBS, Pandey AC (2024) Mapping of the spatio-spectral dynamics of Mangrove Chlorophyll Concentrations via Sentinel-2 Satellite Imagery. In: Forum geografi, vol 38(2). pp 244–256

Bertram C, Quaas M, Reusch TBH (2021) The blue carbon wealth of nations. Nat Clim Chang 11:704–709. https://doi.org/10.1038/s41558-021-01089-4

Carpenter SR, Mooney HA, Agard J, Capistrano D, DeFries RS, Díaz S, Duraiappah AK, Oteng-Yeboah A, Pereira HM, Perrings C, Reid WV, Sarukhan J, Scholes RJ, Whyte A (2009) Science for managing ecosystem services: beyond the millennium ecosystem assessment. Proc Natl Acad Sci 106(5):1305–1312. https://doi.org/10.1073/pnas.0808772106

Choudhary B, Dhar V, Pawase AS (2024) Blue carbon and the role of mangroves in carbon sequestration: its mechanisms, estimation, human impacts, and conservation strategies for economic incentives. J Sea Res 199:102504. https://doi.org/10.1016/j.seares.2024.102504

Costanza R, De Groot R, Sutton P, Van Der Ploeg S, Anderson SJ, Kubiszewski I, Turner RK (2014) Changes in the global value of ecosystem services. Glob Environ Chang 26:152–158. https://doi.org/10.1016/j.gloenvcha.2014.04.002

Fu XM, Tang HY, Liu Y (2021) Resource status and protection strategies of mangroves in China. J Coast Conserv 25:42. https://doi.org/10.1007/s11852-021-00800-z

Guri K et al (2018) Carbon emission of global construction sector. Renew Sustain Energy Rev 81:1906–1916. www.sciencedirect.com/science/article/pii/S1364032117309413

Harris NL, Brown S, Hagen SC, Saatchi SS, Petrova S, Salas W, Hansen MC, Potapov PV, Lotsch A (2012) Baseline map of carbon emissions from deforestation in tropical regions. Science 336(6088):1573–1576. https://doi.org/10.1126/science.1217962

Himes-Cornell A, Pendleton L, Atiyah P (2018) Valuing ecosystem services from blue forests: a systematic review of the valuation of salt marshes, sea grass beds, and mangrove forests. Ecosyst Serv 30:36–48. https://doi.org/10.1016/j.ecoser.2018.01.006

Huang L, Krigsvoll G, Johansen F, Liu Y, Zhang X (2018) Carbon emission of global construction sector. Renew Sustain Energy Rev 81:1906–1916. https://doi.org/10.1016/j.rser.2017.06.001

Jiawen L, Failler P, Ramrattan D (2024) Blue carbon accounting to monitor coastal blue carbon ecosystems. J Environ Manage 352:120008. https://doi.org/10.1016/j.jenvman.2023.120008

Kumar KCVN, Ahammed KB, Lele N, Nimmi N, Kubo S, Rao KN, Harikumar PS, Gopinath G, Bhattacharya BK (2024) Assessing the significance of leaf chlorophyll content for measuring the vegetation health in mangrove species along the Kerala Coast, India–A multi-proxy approach. Reg Stud Mar Sci 77:103702

Lal R (2005) Forest soils and carbon sequestration. For Ecol Manage 220(1–3):242–258. https://doi.org/10.1016/j.foreco.2005.08.015

Lavery PS, Mateo M, Serrano O, Rozaimi M (2013) Variability in the carbon storage of seagrass habitats and its implications for global estimates of blue carbon ecosystem service. PLoS ONE 8(9):e73748. https://doi.org/10.1371/journal.pone.0073748

Milica S, Janmanee P, Nattacha K, Prathep A (2023) National assessment and variability of blue carbon in seagrass ecosystems in Thailand. Mar Pollut Bull 197:115708. https://doi.org/10.1016/j.marpolbul.2023.115708

Moore AC, Hierro L, Mir N, Stewart T (2022) Mangrove cultural services and values: current status and knowledge gaps. People Nat 4(5):1083–1097. https://doi.org/10.1002/pan3.10375

Pan Y, Birdsey RA, Fang J, Houghton R, Kauppi PE, Kurz WA, Phillips OL, Shvidenko A, Lewis SL, Canadell JG, Ciais P, Jackson RB, Pacala SW, McGuire AD, Piao S, Rautiainen A, Sitch S, Hayes D (2011) A large and persistent carbon sink in the world's forests. Science 333(6045):988–993. https://doi.org/10.1126/science.1201609

Peter I et al (2021) Blue carbon as a natural climate solution. Nat Rev Earth Environ 2(4):20–30. https://doi.org/10.1038/s43017-021-00224-1

Powlson DS, Whitmore AP, Goulding KWT (2011) Soil carbon sequestration to mitigate climate change: a critical re-examination to identify the true and the false. Eur J Soil Sci 62(1):42–55. https://doi.org/10.1111/j.1365-2389.2010.01342.x

Quevedo JMD, Kohsaka R (2024) A systematic review of cultural ecosystem services of blue carbon ecosystems: trends, gaps, and challenges in Asia and beyond. Mar Policy 159:105898. https://doi.org/10.1016/j.marpol.2023.105898

Raihan A (2023) A review of tropical blue carbon ecosystems for climate change mitigation. J Environ Sci Econ 2(4):14–36. https://doi.org/10.56556/jescae.v2i4.602

Sahana M, Areendran G, Sajjad H (2022) Assessment of suitable habitat of mangrove species for prioritizing restoration in coastal ecosystem of Sundarban Biosphere Reserve, India. Sci Rep 12:20997. https://doi.org/10.1038/s41598-022-24953-5

Schile LM, Kauffman JB, Crooks S, Fourqurean JW, Glavan J, Megonigal JP (2017) Limits on carbon sequestration in arid blue carbon ecosystems. Ecol Appl 27(3):859–874. https://doi.org/10.1002/eap.1489

Stankovic M, Panyawai J, Khanthasimachalerm N, Prathep A (2023) National assessment and variability of blue carbon in seagrass ecosystems in Thailand. Mar Pollut Bull 197:115708. https://doi.org/10.1016/j.marpolbul.2023.115708

Vivek K et al (2011) Carbon emission limits required to satisfy future representative concentration pathways of greenhouse gases. Geophys Res Lett 38(5):23–29. https://doi.org/10.1029/2010gl046270

Community-Based Disaster Management and Mitigation Plan

Coastal Multi-hazard Vulnerability Assessment Along East Coast Deltaic Plain of India—A Geospatial Approach

P. P. Amisha, Sreelakshmi Prakash, Abin Varghese, Jyothi Vijayan, Ruksana Salim, S. Aravindh, and K. R. Baiju

Abstract Coastal zones have undergone significant changes in the twenty-first century due to population growth, expanding urban development, industrialization, and other human activities. The Krishna-Godavari deltaic region in India is highly exposed to the impacts of coastal hazards and extreme weather events, which disrupt the livelihoods of people living in coastal areas. Identifying priority areas with high vulnerability is crucial for effective disaster management and resilience efforts. A Coastal Vulnerability Index was created by integrating several sub-indices such as exposure, sensitivity, and adaptive capacity derived from 15 indicators. These indicators primarily reflect physical, climatic, and socio-economic factors, and were weighted using the analytical hierarchy process (AHP). The study identifies 857.319 km^2 of the area are highly vulnerable which includes Amalapuram and Machilipatanam mandals. Mummidivaram mandal in East Godavari district is highly vulnerable due to greater exposure to risks, while Amalapuram, Pithapuram, and Tuni mandals in the same district show high sensitivity to coastal hazards. In contrast, Narsapur mandal in West Godavari district has low adaptive capacity, which increases its vulnerability. It is anticipated that the vulnerability analysis will serve as valuable tools for land use planning, which is crucial for safeguarding local assets and formulating emergency measures tailored to coastal natural disasters, such as sea-level inundation and intense tropical storms.

Keywords Multi-hazard · Vulnerability · CVI · Krishna-Godavari · AHP · Sensitivity · Exposure · Adaptive capacity

P. P. Amisha · S. Prakash · A. Varghese (✉) · J. Vijayan · R. Salim · S. Aravindh · K. R. Baiju
Dr. R Satheesh Centre for Remote Sensing and GIS, School of Environmental Sciences, Mahatma Gandhi University, Kottayam, Kerala, India
e-mail: rsgismgu@gmail.com

K. K. Basheer Ahammed et al. (eds.), *Monitoring Coastal Marine Ecosystems Under Climate Change*, https://doi.org/10.1007/978-3-032-09137-6_17

1 Introduction

Coastal zones represent dynamic interfaces in which marine and fluvial processes interact to maintain shoreline integrity. These regions furnish both tangible ecosystem services and intangible cultural and socio-economic benefits critical to national development (Sankari et al., 2015). However, coastal environments are continually modified by natural hydro-geomorphic forces and compounded by anthropogenic activities, leading to progressive changes in morphology and increased environmental stress (Baig et al., 2021).

The IPCC's Fifth Assessment Report (2014) indicates that low-lying coastal areas will experience substantially heightened risk throughout the twenty-first century and beyond. Coastal zones frequently concentrate the economic gains derived from maritime transport, fisheries, tourism, and recreational activities, all of which support human settlements. According to India's 2011 Census, roughly 14.2% of the national population resides in coastal districts, and nearly 500 million people live in or near river deltas, with population densities continuing to rise. Although these regions are endowed with rich natural resources, they are simultaneously beset by widespread poverty, unsustainable resource exploitation, and recurrent natural disasters—such as cyclones and tsunamis—that pose serious threats to the lives and livelihoods of coastal communities.

The eastern coastline of India experiences an average of four to six tropical cyclones per year, originating in the southeastern Arabian Sea and the northern Bay of Bengal (Sahoo and Bhaskaran, 2018). Along this margin lie the major deltaic systems of the Mahanadi, Krishna-Godavari and Sundarbans, whose low-lying topography and unconsolidated muddy substrates confer high susceptibility to hydrodynamic and geomorphic hazards. This intrinsic vulnerability not only compromises the resilience of coastal infrastructure and communities but also constitutes a significant impediment to the pursuit of sustainable development objectives. Syvitski et al., 2009 reported that thirty-three major river deltas around the world, including those in India, are sinking primarily due to reduced aggradation and accelerated compaction. Krishna, Ganges, Brahmaputra, and Godavari delta show the high-risk trend among the 48 deltas of the world under the anthropogenic factors (Tessler et al., 2015). These natural hazards cannot be avoided, but their impact can be mitigated through the implementation of appropriate preparedness plans and mitigation measures (Mahendra et al., 2011).

Vulnerability, as a dynamic concept, refers to the extent to which individuals, communities, or systems are prone to experiencing harm when exposed to external stresses (IPCC, 2012). Coastal Vulnerability Index (CVI) is the primary quantitative framework used to assess coastal risk by integrating physical, environmental, social, and economic factors to evaluate impacts in the aftermath of hazard events (Sahoo and Bhaskaran, 2018). Remote sensing techniques play a vital role in vulnerability analysis by providing accurate and comprehensive data on various environmental and socio-economic factor (Ali et al., 2019). Multiple investigations have examined coastal vulnerability to distinct hazards, such as tsunamis, cyclones, storm events, shoreline erosion rates, and sea-level rise (Pal et al., 2022). To quantify this vulnerability, researchers have applied a range of assessment frameworks and indicator sets, including the Community Vulnerability Index (CoVI), the Physical Vulnerability Index (PVI), and Social Vulnerability measures (Basheer Ahammed and Pandey, 2019a, b; Baig et al., 2021). The Federal Emergency Management Agency (FEMA) was the first to initiate the Multi-hazard Mapping Initiative (MMI) to develop multi-hazard maps for coastal areas (Lowe, 2003). Nowadays remote sensing and GIS have become widely popular in the field of natural risk assessment and vulnerability analysis. Combining the statistical or analytical techniques with spatial data enhances the accuracy and acceptability of results (Ali et al., 2019). Some of the popular multi-criteria decision analysis (MCDA) methods along with GIS are analytical hierarchy process (AHP), frequency ratio method (FR), artificial neural network (ANN), fuzzy logic (Ali et al., 2019; Kia et al., 2012; Pourghasemi et al., 2012). Machine learning is a modern approach in vulnerability assessment, offering a data-driven alternative to traditional models. Integrating machine learning algorithms with these MCDA methods will improve accuracy and efficiency in identifying vulnerable areas (Tasnuva and Bari, 2024).

This study seeks to generate a spatially explicit, multi-hazard vulnerability map for the Krishna-Godavari delta by employing the analytical hierarchy process (AHP). The significance of this study stems from the growing urbanization of India's coastal areas which increases the vulnerability of expanding urban population to cyclones, coastal floods, and other coastal hazards. While climate change effects cannot be entirely averted, their severity can be mitigated through robust preparedness planning and strategic interventions—outcomes that this research is designed to facilitate.

2 Study Area

The Krishna-Godavari Basin, nestled along the eastern coast of India, spans approximately 15,000 km^2 on land and extends over 25,000 km^2 offshore. Geographical expansion of delta from 15°30′N to 17°N latitude and 80°00′E to 82°30′E and it covers an area of approximately 7370 km^2. The study area consists of 11 taluks (mandals), comes within four districts of Andhra Pradesh such as Guntur, Krishna, East and West Godavari (Fig. 1). The 336 km long Kirshna-Godavari (K-G) twin delta coast has mangrove forest mainly on both sides of three distributaries of Krishna River. This region is considered to be the rice bowl of the state. About thirty five wide variety of mangrove species are found in this region, of these 16 are true mangroves with 19 associates (Kubo et al., 2018).

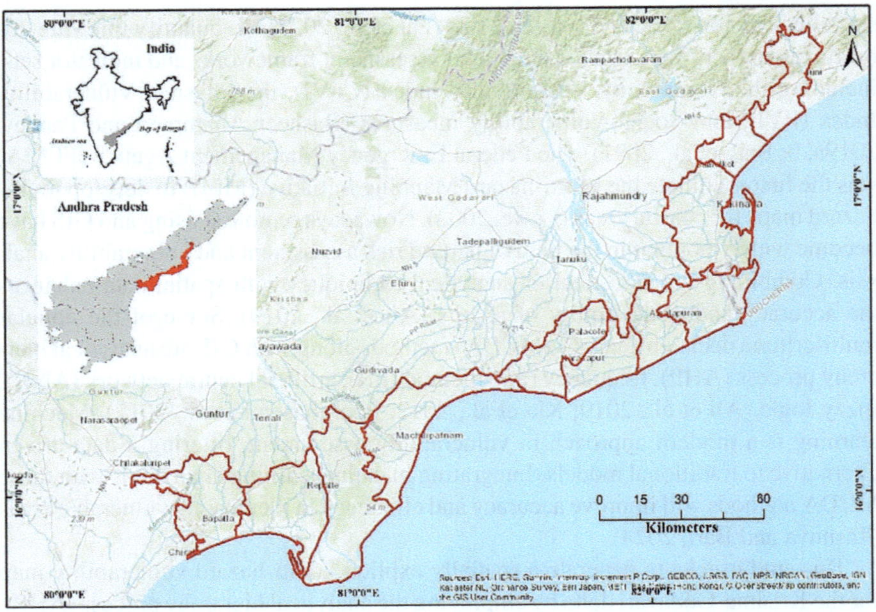

Fig. 1 Location map of study area

3 Methodology

Coastal vulnerability to multi-hazards of Krishna-Godavari delta was assessed using various sub-indices like exposure, sensitivity, and adaptive capacity (Sarkar et al., 2019; Gosh and Mistri, 2021). Vulnerability sub-indices were calculated based on 15 parameters primarily considering physical, climatic and socio-economic variables. The dataset used for the current study are given in Table 1.

Table 1 Datasets used for the study

Parameter	Data sources
Land use land cover	Landsat 8 OLI TIRS, USGS Earth explorer
Elevation, slope, stream	SRTM DEM, USGS Earth explorer
Roads, shoreline	Google Earth Pro
Cyclone track	Typhoon Joint Centre
Rainfall	GES DISC, Indian Meteorological Department
Storm surge height	India WIRS
Population density, household density, Literacy rate, work participation rate, % of families residing in the area	Census of India, 2011

Exposure parameters act as internal drivers of vulnerability caused by natural hazard (Gosh and Mistri, 2021). In the present study, land use land cover (LULC), slope, elevation, flood inundation risk, stream density, proximity to shoreline, cyclone track and road and rainfall deviation were considered as exposure parameters. Satellite images include Landsat 8 OLI/TIRS (path and row—141/48, 141/49, 142/48,142/49; date of acquisition 02/02/2023) and SRTM DEM with 30 m spatial resolution was downloaded from Earth explorer (www.earthexplore.usgs.gov). Topographical factors like land elevation, slope, streams, and flood inundation risk map were extracted from SRTM DEM. Hydrological processes including infiltration, evapotranspiration, and runoff generation are governed by land use types (Rahmati et al., 2016). LULC was digitized from Landsat 8 by visual interpretation. Shoreline and roads were demarcated from Google Earth Pro using on screen digitization techniques and proximity map was prepared. To identify the region most susceptible to cyclonic events, cyclone tracks were collected for last 10 years from Joint Typhoon Warning Centre (https://www.metoc.navy.mil/jtwc/jtwc.html) and proximity to cyclone track was prepared by using Euclidean distance tool in ArcGIS. Monthly rainfall data for the period of 2019–2023 were downloaded from NASA GES DISC (https://disc.gsfc.nasa.gov/) and annual average rainfall was calculated. The rainfall deviation was calculated using Equation below:

$$\text{Rainfall deviation} = \frac{\text{Recorded rainfall} - \text{Average rainfall}}{\text{Average rainfall}} \times 100 \tag{1}$$

The average rainfall of the region was taken as 968.238 mm from Indian Meteorological Department (https://apwrims.ap.gov.in/mis/rainfall/).

Storm surge records for the past century were obtained from the India Water Resource Information System portal (https://indiawris.gov.in/wris). For each taluk in the Krishna-Godavari delta, the maximum storm surge height corresponding to a 100 year return period was used to assign risk rating. Based on the storm surges heights, eleven taluks were grouped into four classes, with risk ratings from 1 (lowest) to 4 (highest) for vulnerability index calculations (Sahana and Sajjad, 2019). Flood inundation risk was evaluated by considering combined effects of storm surge, cyclonic activity, heavy rainfall, and embankment breaches producing water depths of < 1 m up to 5 m. Spatial inundation extents at these depth thresholds were delineated using SRTM DEM data (Poulter and Halpin, 2008; Sahana and Sajjad, 2019). Inundated pixels were reclassified by depth category, and the percentage of each taluk's area under each inundation level was computed to yield mean flood inundation risk. Finally, taluk-specific inundation risk scores were calculated using the square-root-of-product-mean algorithm as defined in Eq. 2 (Kumar et al., 2010; Sahana and Sajjad, 2019).

$$\text{Flood inundation risk} = \sqrt{\frac{(a \times b \times c \times d \times e)}{5}} \tag{2}$$

where *a*, *b*, *c*, *d*, and *e* represent the area percentage of the taluk under different level of flood inundation. Based on flood inundation risk, five classes of flood hazard were obtained using 20th, 40th, 60th, and 80th percentile accordingly.

In environmental hazard assessments, sensitivity denotes a system's susceptibility to natural disturbances through direct or indirect pathways (IPCC, 2007). In this study, three demographic indicators—population density, household density, and the proportion of the child population—are employed to represent sensitivity. Population density is defined as the number of individuals per square kilometer, while household density refers to the number of households per square kilometer (Ali et al., 2019).

Adaptive capacity is the ability of a system or community to withstand the negative consequences of hazards by utilizing its available resources, skills, and technology (IPCC, 2007). Both exposure and sensitivity can be influenced by that system's adaptive capacity. Literacy rate, work participation rate, percentage of families residing in the area are the adaptive capacity parameters used to calculate the adaptive capacity index. These demographic data were collected from primary census abstract of the year 2011 (https://www.censusindia2011.com/).

3.1 Coastal Vulnerability Index Analysis (CVI)

Coastal vulnerability to oceanogenic multi-hazards was assessed using analytical hierarchy process (AHP) introduced by Professor Thomas L. Saaty (Saaty, 1980, 1990) gives a better understanding of parameters into a hierarchical way. Several studies proved that AHP can be effectively used in coastal vulnerability assessments as it provides a structured and systematic way to evaluate and prioritize multiple contributing factors, offering more accurate estimations than the conventional methods. A pairwise comparison matrix was constructed to evaluate the relative importance of the fifteen selected coastal vulnerability factors. Each factor was assigned a rank from 1 to 9 according to its relative significance (Table 3). Subsequently, individual PCMs were generated for each sub-index to derive priority vectors, which quantify parameter weights and enable consistency assessment (Table 4). A consistency ratio (CR) of 0.10 or lower is generally accepted as indicating reliable judgments; values approaching zero reflect a very robust decision-support framework (Ghosh and Mistri, 2021; Ali et al., 2019). The CR defined as:

$$\text{Consistency Ratio(CR)} = \frac{\text{Consistency index(CI)}}{\text{Random index (RI)}} \tag{3}$$

The average random consistency index value was derived from Saaty's scale which is shown in Table 2.

The consistency index (CI) was calculated using Equation below;

$$\text{CI} = \frac{\lambda \max - n}{n - 1} \tag{4}$$

Table 2 Random consistency index (RI) (Saaty, 1980)

N	1	2	3	4	5	6	7	8	9	10	11	12	13	14	15
RI	0	0	0.58	0.9	1.12	1.24	1.32	1.41	1.45	1.49	1.51	1.53	1.56	1.57	1.5

Table 3 Saaty's (1990) numerical scale of importance for AHP preference

Scale	Numerical rating	Reciprocal
Extremely importance	9	1/9
Very to extremely strong importance	8	1/8
Very strongly importance	7	1/7
Strongly to very strongly importance	6	1/6
Strongly importance	5	1/5
Moderately to strongly importance	4	1/4
Moderately importance	3	1/3
Equally to moderately importance	2	1/2
Equally importance	1	1

where " λmax" is the largest Eigen value of the considered matrix and "n" is the total number of sub-indicators.

The total vulnerability index is derived by combining the sub-indices using an averaging approach. The (Eq. 5) was used for calculating coastal vulnerability of the region.

$$V = \frac{[E + S + (1 - AC)]}{3} \tag{5}$$

where V, E, S, and AC represent the vulnerability, exposure, sensitivity, and adaptive capacity receptively.

4 Results and Discussion

The coastlines of the Krishna-Godavari deltaic plain are highly dynamic, influenced by climatic variations and fluvial processes. Increasing frequency of tropical cyclones in the Bay of Bengal, underscores the need to assess coastal vulnerability to flooding amid climate change and sea-level rise.

Table 4 Pairwise comparison matrices for various parameters

(a) Exposure

Parameters	A	B	C	D	E	F	G	H	I	Normalized Eigen weight
Distance from road (A)	1	1/2	1/3	1/4	1/5	1/6	1/7	1/8	1/9	0.019
LULC (B)	2	1	1/2	1/3	1/4	1/5	1/6	1/7	1/8	0.026
Slope (C)	3	2	1	1/2	1/3	1/4	1/5	1/6	1/7	0.037
Elevation (D)	4	3	2	1	1/2	1/3	1/4	1/5	1/6	0.053
Stream density (E)	5	4	3	2	1	1/2	1/3	1/4	1/5	0.076
Proximity to shoreline (F)	6	5	4	3	2	1	1/2	1/3	1/4	0.109
Proximity to cyclone track (G)	7	6	5	4	3	2	1	1/2	1/3	0.154
Rainfall deviation (H)	8	7	6	5	4	3	2	1	1/2	0.218
Flood inundation risk (I)	9	8	7	6	5	4	3	2	1	0.307

Maximum Eigen value = 9.401, CI = 0.05, CR = 0.035

(b) Sensitivity

Parameters	J	K	L							Normalized Eigen weight
Population density (J)	1	3	5							0.633
Household density (K)	1/3	1	3							0.26
Percentage of child population (L)	1/5	1/3	1							0.106

Maximum Eigen value = 3.039, CI = 0.019, CR = 0.033

(c) Adaptive capacity

Parameters	M	N	O							Normalized Eigen weight
Percentage of families residing in the area (M)	1	3	5							0.633
Literacy rate (N)	1/3	1	3							0.26
Work participation rate (O)	1/5	1/3	1							0.106

Maximum Eigen value = 3.039, CI = 0.019, CR = 0.033

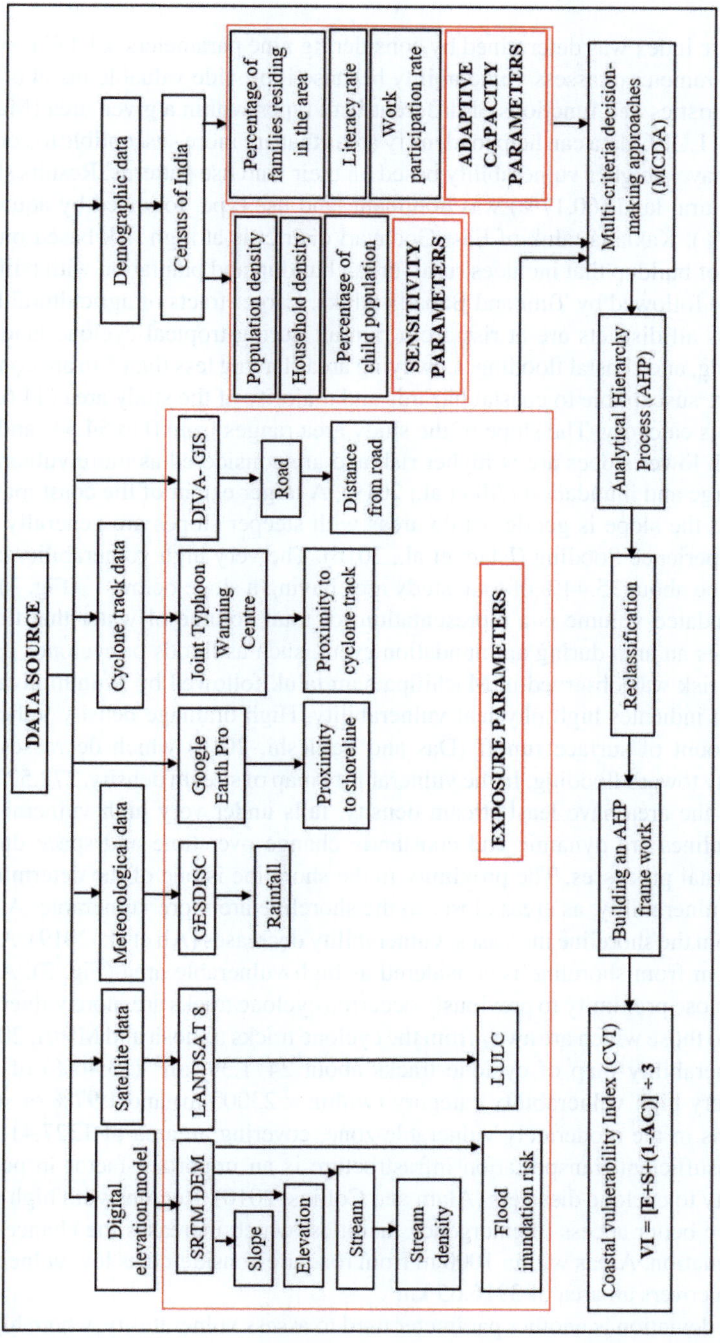

Fig. 2 Flowchart of methodology for vulnerability mapping

4.1 Exposure

The exposure index was determined by considering nine parameters. LULC is often used as a parameter to assess vulnerability because it provide valuable insights into the characteristics and functions of different land types within a given area (Murali et al., 2013). LULC data can help to identify areas that are more susceptible to certain hazards or have a higher vulnerability based on their land use patterns. Results show that agricultural land (60.17%) was dominant land use type, followed by aquaculture (12.34%). Kakiada taluk of East Godavari district is at high risk based on the percentage of buildup that includes rural–urban buildup and plantation with buildup area (Fig. 3) followed by Tuni and Bapatla taluks. Larger tracts of agricultural land cover across all districts are at risk from rainfall during tropical cyclone landfall, river flooding, and coastal flooding. Low lying areas having less than 5 m are considered as more susceptible to coastal hazards and majority of the study area (44.62%) comes in this category. The slope of the study area ranges from 0 to 54.34° and the regions with lower slopes are at higher risk and are considered as more vulnerable to storm surge and inundation (Ali et al., 2019). A larger extent of the coast may be inundated if the slope is gentle, while areas with steeper slopes are generally less likely to experience flooding (Islam et al., 2016). The very high vulnerability class is found to be about 35.44% of total study area having a slope below 1° (Fig. 3).

The inundated volume is a representation for total volume of water that floods or submerges an area during an inundation event, such as floods or cyclones. High inundation risk was observed in Machilipatnam taluk followed by Mummidivaram taluk which indicates high physical vulnerability. High drainage density indicates a great amount of surface runoff (Das and Pardeshi, 2018) which decreases the vulnerability toward flooding. In the vulnerability map of stream density, 272.52 km^2 (3.69%) of the area have least stream density, falls under very high vulnerability zone. Shorelines are dynamic and constantly change over time and space due to various coastal processes. The proximity to the shoreline is one of the determining factors in vulnerability, as areas closer to the shoreline are more vulnerable. As the distance from the shoreline increases, vulnerability decreases (Ali et al., 2019). Areas within 500 m from shoreline is considered as high vulnerable area (Fig. 3). Areas within the close proximity to previously occurred cyclone tracks are more vulnerable compared to those which are away from the cyclone tracks (Ghosh and Mistri, 2021). In the vulnerability map of cyclone tracks about 2471.39 km^2 (33.46%) of area falls into very high vulnerability category (within < 23000 m) and 1.97% of study area remains in the moderately vulnerable zone, covering an area of 1327.41 km^2 (Fig. 3). Insufficient transportation infrastructure is an important factor in people vulnerability to cyclone disasters (Alam and Collins, 2010). Regions with high road density have better access to emergency facilities, which increases the chances for timely evacuation. Areas within 1000 m from road are considered as low vulnerable zone which covers an area of 3916.65 km^2.

Rainfall deviation is another parameter used to assess vulnerability, where higher deviations indicate increased vulnerability, while lower deviations indicate lower

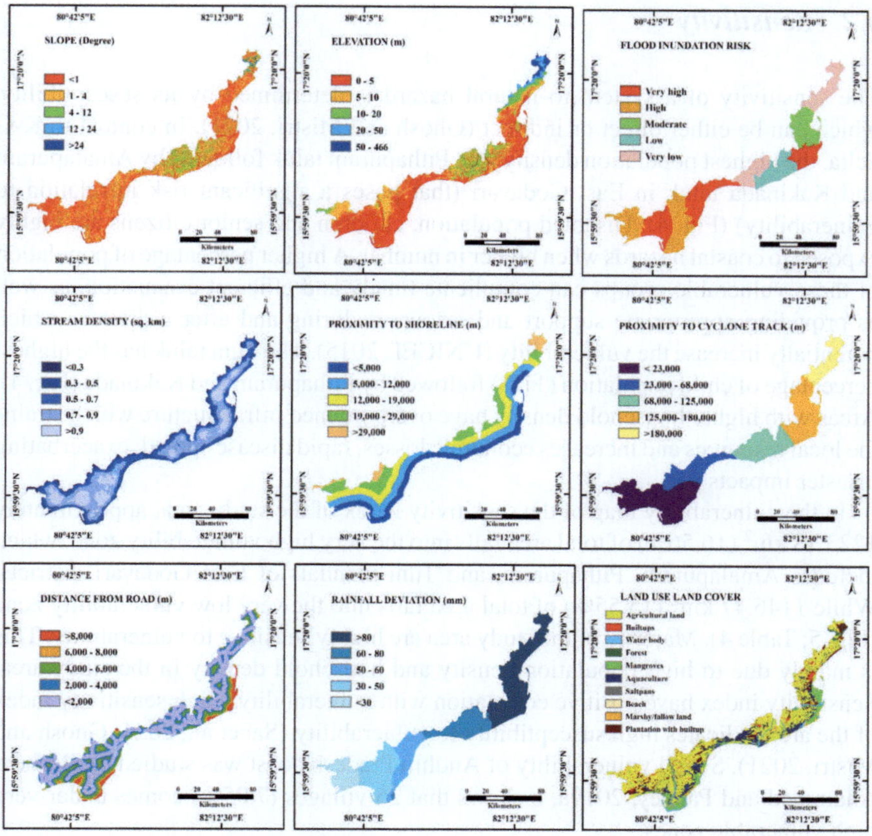

Fig. 3 Exposure parameters used for study

vulnerability (Ghosh and Mistri, 2021). Around 2668.61 km^2 (36.21%) of the study area has higher rainfall deviation, belongs to very highly vulnerable category. Conversely, an area of 820.65 km^2 (11.13%) falls in very low vulnerable zone due to low rainfall deviation (Fig. 3).

Exposure map shows that around 13.736% falls into the very high vulnerability zone, covering an area of 995.14 km^2 which includes Mummidivaram, Machilipatnam, and Repalle mandals (Fig. 5).Of these mandals Mummidivaram mandal of East Godavari district is more exposed to vulnerability. The high vulnerability of the area is due to lower elevation, gentle slope, proximity to shoreline and cyclone tracks, greater distance from road, land use patterns, and high flood inundation risk.

4.2 Sensitivity

The sensitivity of a system to natural hazard is determined by its susceptibility, which can be either direct or indirect (Ghosh and Mistri, 2021). In context to K-G delta, the highest population density is at Pithapuram taluk followed by Amalapuram and Kakinada taluk in East Godavari (that poses a significant risk in relation to vulnerability) (Fig. 4). Disabled population, children and senior citizens are highly exposed to coastal hazards when higher in number. A higher percentage of population of these vulnerable groups can complicate timely and efficient evacuation, as well as providing appropriate support and resources during and after a disaster which potentially increase the vulnerability (UNICEF, 2015). The Tuni taluk has the highest percentage of child population (11%) followed by Pithapuram and Kakinada (Fig. 4). Areas with higher household density have overburdened infrastructure which strains the local resources and increases economic losses, rapid disease spread, exacerbating disaster impacts.

In the vulnerability map of the sensitivity index of the study area, approximately 1223.18 km^2 (16.56%) of total area falls into the very high vulnerability zone, which includes Amalapuram, Pithapuram, and Tuni mandals of East Godavari districts. While 1146.47 km^2 (15.55%) of total area falls into the very low vulnerability zone (Fig. 5; Table 4). Majority of the study area are highly sensitive to vulnerability. This is mainly due to high population density and household density in the study area. Sensitivity index have positive correlation with vulnerability. High sensitivity index of the area indicates high susceptibility to vulnerability (Sai et al., 2021; Ghosh and Mistri, 2021). Social vulnerability of Andhra Pradesh coast was studied by Basheer Ahammed and Pandey, 2019a, b shows that 26 villages (7.05%) comes under very high vulnerable zones.

4.3 Adaptive Capacity

Adaptive capacity refers to the inherent potential of a system to withstand the adverse impacts of hazards using available resources, skills and technology (IPCC, 2007). Percentage of families residing in the area, literacy rate, and work participation rate were the three key factors for preparedness and evacuation planning operations (Sahoo and Bhaskaran, 2018). Communities with higher literacy rate are more likely to understand the warning messages, evacuation instructions and other critical information which can help those informed decisions and take appropriate actions to mitigate risks. High adaptive capacity index was found in Narsapur taluk in West Godavari district due to low families residing in the area, high literacy rate and work participation rate (Fig. 4). Individuals and households with lower participation rates have limited financial resources making Pithapuram and Avanigadda taluks more vulnerable in terms of coping capacity to coastal hazards. Adaptive capacity of

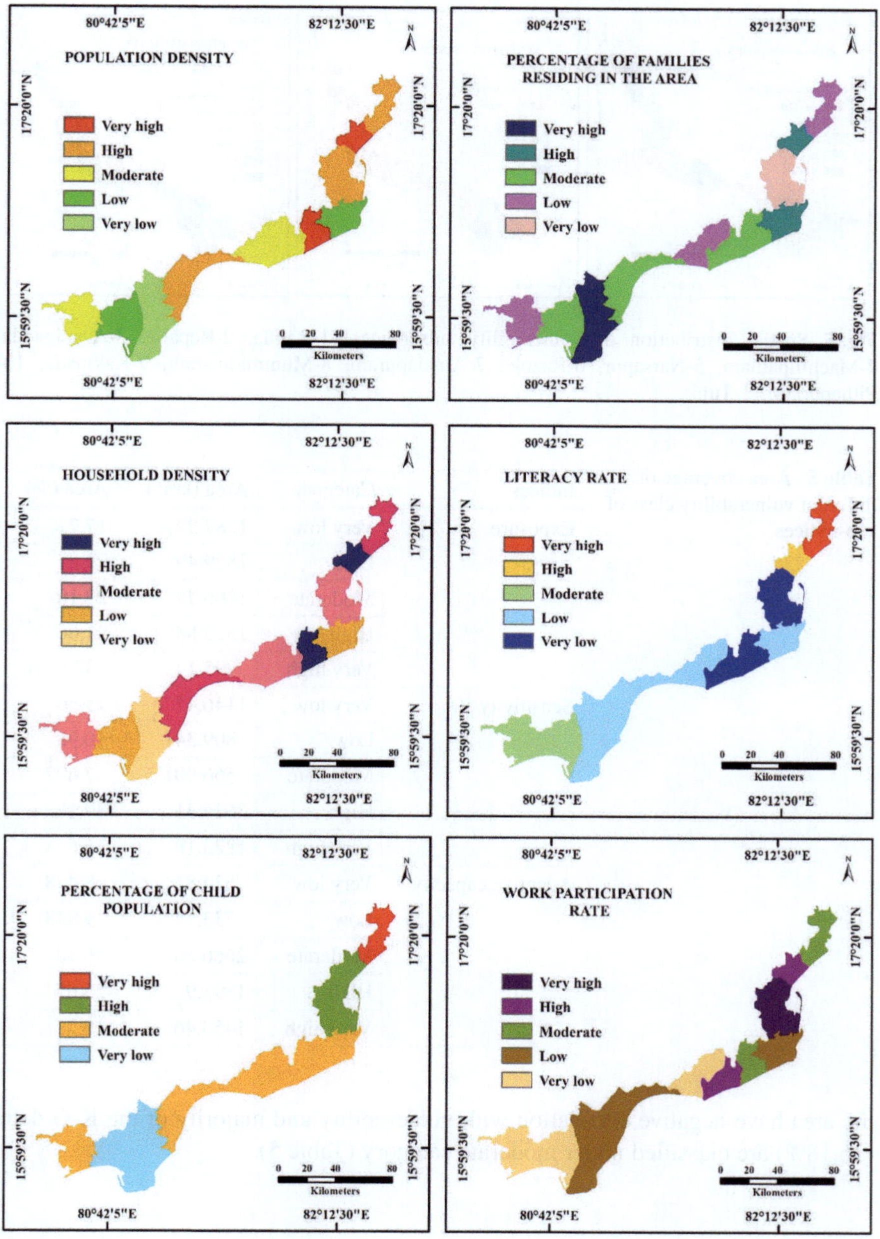

Fig. 4 Sensitivity and adaptive capacity parameters used for study

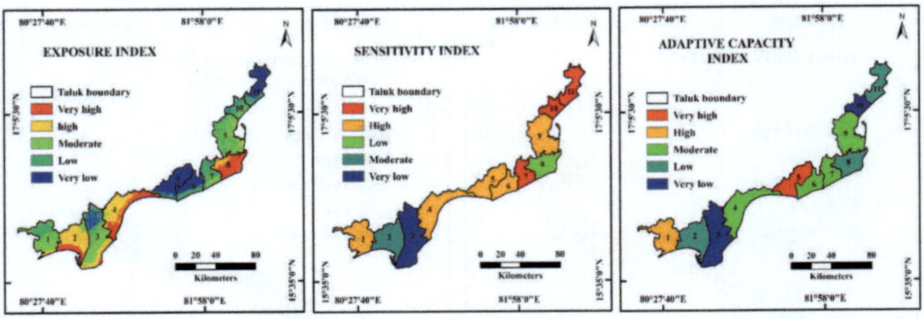

Fig. 5 Spatial distribution of vulnerability sub-indices (1-Baptla, 2-Repalle, 3-Avanigadda, 4-Machilipatnam, 5-Narsapur, 6-Razole, 7-Amalapuram, 8-Mummidivaram, 9-Kakinada, 10-Pithapuram, 11-Tuni)

Table 5 Area coverage of different vulnerability class of sub-indices

Indices	Category	Area (km^2)	Area (%)
Exposure	Very low	1287.29	17.77
	Low	1829.49	25.25
	Moderate	1606.79	22.18
	High	1525.88	21.06
	Very high	995.14	13.74
Sensitivity	Very low	1146.47	15.56
	Low	809.341	10.98
	Moderate	566.901	7.692
	High	3624.31	49.18
	Very high	1223.18	16.6
Adaptive capacity	Very low	624.81	8.478
	Low	733.177	9.948
	Moderate	2666.86	36.18
	High	1891.9	25.67
	Very high	1453.46	19.72

the area have negative correlation with vulnerability and majority of the K-G delta (36.18%) are classified under moderate category (Table 5).

4.4 Coastal Vulnerability Index

Coastal multi-hazard vulnerability is analyzed based on 15 indicators primarily considering physical, climatic, and socio-economic variables through AHP. The vulnerability maps of the study area were categorized into five levels of vulnerability:

very low, low, moderate, high, and very high (Table 5). The present study revealed that the majority of the K-G delta are vulnerable to coastal multi-hazard events and about 857.31 km^2 (11.83% of the study area) are considered as highly vulnerable zones which includes Amalapuram and Machilipatanam mandals (Fig. 6). Lower elevation, gentle slope, high flood inundation risk, proximity to cyclone track, high sensitivity, and low adaptive capacity makes these taluks more susceptible to coastal hazards. Mangroves play a crucial role in coastal protection by acting as a natural barrier against storm surges, coastal erosion, and sea-level rise. Loss of mangroves leads to decrease of the natural protection provided by these ecosystem and makes coastal communities especially vulnerable to threats posed by storms, flooding, and erosion (Kubo et al, 2018). The degradation of mangroves also effects the wildlife in the Coringa Wildlife Sancturay (Ramesh et al., 2021; Nunna, 2023). The study shows that Coringa Wildlife Sanctuary in East Godavari district is belonging to highly vulnerable category. Less vulnerability was found in Avanigadda mandal in Krishna district due to high adaptive capacity and fewer sensitivity factors. Similar study was conducted by Basheer Ahammed and Pandey, 2019a, b in East Coast of India using sea-level rise, historical surge height, and shore line change rate. It was observed that Krishna district is most vulnerable and West Godavari considered as least vulnerable (Table 6).

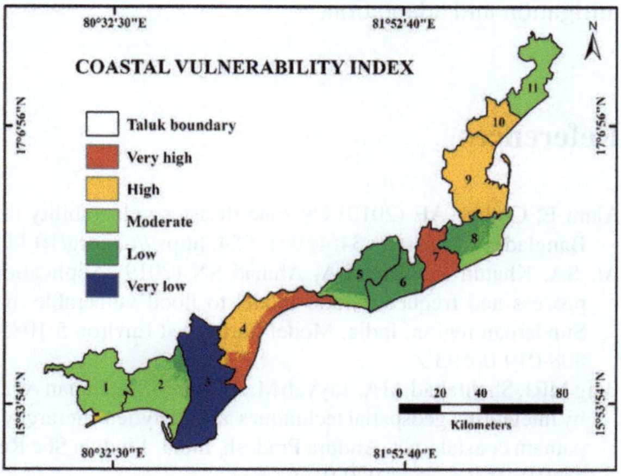

Fig. 6 Coastal vulnerability status of Krishna-Godavari delta

Table 6 Geographical coverage of coastal vulnerability classes of K-G delta

Sl. no.	Category	Area (km^2)	Area (%)
1	Very low	1135.98	15.68
2	Low	1346.62	18.59
3	Moderate	2238.65	30.9
4	High	1666.04	23
5	Very high	857.319	11.83

5 Conclusion

The densely populated coastal region along the India's coast is highly vulnerable to natural hazards like tropical cyclones. On an average 4–6 cyclones are formed in the Bay of Bengal yearly and therefore, a greater risk exists for storm surges flooding, inland flooding from rainfall heavy winds as compared to Arabian Sea. Keeping in this view understanding of such risk is essential to understand the future vulnerability and risk scenarios in the delta. The coastal vulnerability of Krishna-Godavari delta region was studied within a geographic information system by making use of 15 parameters considering exposure, sensitivity, and adaptation capacity index by applying AHP techniques. The composite maps of three indices signify that Avanigadda taluk in Krishna district is the least exposed, whereas Amalapuram taluk in East Godavari districts is highly vulnerable to natural hazards due to high sensitivity and low adaptive capacity of the region. Coastal vulnerability has been categorized into five zones with a significant portion (33%) of the K-G delta falling within high to very high vulnerability areas with least mitigation building capacity to handle such catastrophic impacts. The study make use of GIS-based multi-criteria approach a most reliable and coast effective approach for monitoring and management of multi-hazard. The findings of this study can provide foundational information for local authorities to develop and implement effective measures for cyclone hazard mitigation and adaptation.

References

Alam E, Collins AE (2010) Cyclone disaster vulnerability and response experiences in coastal Bangladesh. Disasters 34(4):931–954. https://doi.org/10.1111/j.1467-7717.2010.01176.x

Ali SA, Khatun R, Ahmad A, Ahmad SN (2019) Application of GIS-based analytic hierarchy process and frequency ratio model to flood vulnerable mapping and risk area estimation at Sundarban region, India. Model Earth Syst Environ 5:1083–1102. https://doi.org/10.1007/s40808-019-00593-z

Baig MRI, Shahfahad AIA, Tayyab M, Asgher MS, Rahman A (2021) Coastal vulnerability mapping by integrating geospatial techniques and analytical hierarchy process (AHP) along the Vishakhapatnam coastal tract, Andhra Pradesh, India. J Indian Soc Remote Sens 49:215–231. https://doi.org/10.1007/s12524-020-01204-6

Basheer Ahammed KK, Pandey AC (2019a)Coastal social vulnerability and risk analysis for cyclone hazard along the Andhra Pradesh, East Coast of India.KN-J Cartogr Geogr Inf 69(4):285–303.https://doi.org/10.1007/s42489-019-00029-9

Basheer Ahammed KK, Pandey AC (2019b) Geoinformatics based assessment of coastal multi-hazard vulnerability along the East Coast of India. Spat Inf Res 27(3):295–307.https://doi.org/10.1007/s41324-018-00236-y

Das S, Pardeshi SD (2018) Integration of different influencing factors in GIS to delineate groundwater potential areas using IF and FR techniques: a study of Pravara basin, Maharashtra India. Appl Water Sci 8(7):197. https://doi.org/10.1007/s13201-018-0848-x

Ghosh S, Mistri B (2021) Assessing coastal vulnerability to environmental hazards of Indian Sundarban delta using multi-criteria decision-making approaches. Ocean Coast Manage 209:105641. https://doi.org/10.1016/j.ocecoaman.2021.105641

IPCC (2007) Climate change 2007: impacts adaptation and vulnerability. In: Parry ML, Canziani OF, Palutikof, JP Linden van der PJ, Hanson CE (eds) Contribution of working group II to the fourth assessment report of the intergovernmental panel on climate change. Cambridge University Press Cambridge, UK, p 976

IPCC (2012) Managing the risks of extreme events and disasters to advance climate change adaptation. In: Field CB, Barros V, Stocker et al TF (eds) A special report of working groups I and II of the intergovernmental panel on climate change. Cambridge, UK and New York, NY, USA, p 582

IPCC, Climate Change (2014) Synthesis report. In: Contribution of working groups I, II and III to the fifth assessment report of the intergovernmental panel on climate change. Geneva, Switzerland, p 151

Islam MA, Debashis M, Ashraf D, Syed HA (2016) Coastal multi-hazard vulnerability assessment along the Ganges deltaic coast of Bangladesh–a geospatial approach. Ocean Coast Manage 127:1–15. https://doi.org/10.1016/j.ocecoaman.2016.03.012

Kia MB, Pirasteh S, Pradhan B, Mahmud AR, Sulaiman WNA, Moradi A (2012) An artificial neural network model for flood simulation using GIS: Johor River Basin, Malaysia. Environ Earth Sci 67(1):251–264

Kubo S, Kumar KCVN, Demudu G, Malini BH, Rao KN, Agrawal R, Ramakrishnan R, Rajawat AS (2018) Monitoring of eco-restoration of mangrove wetlands through time series satellite images: a study on Krishna-Godavari Delta region, east coast of India. Geogr Rev Jpn Ser B 90(2):66–75. https://doi.org/10.4157/geogrevjapanb.90.66

Kumar TS, Mahendra RS, Nayak S, Radhakrishnan K, Sahu KC (2010) Coastal vulnerability assessment for Orissa State, east coast of India. J Coastal Res 26(3):523–534. https://doi.org/10.2307/40605480

Lowe AS (2003) The federal emergency management agency's multi-hazard flood map modernization and the national map. Photogramm Eng Remote Sens 69(10):1133–1135. https://doi.org/10.14358/PERS.69.10.1133

Mahendra RS, Mohanty PC, Bisoyi H, Kumar TS, Nayak S (2011) Assessment and management of coastal multi-hazard vulnerability along the Cuddalore-Villupuram, east coast of India using geospatial techniques. Ocean Coast Manage 54(4):302–311. https://doi.org/10.1016/j.ocecoaman.2010.12.008

Mani Murali R, Ankita M, Amrita S, Vethamony P (2013) Coastal vulnerability assessment of Puducherry coast, India, using the analytical hierarchical process. Nat Hazard 13(12):3291–3311. https://doi.org/10.5194/nhessd-1-509-2013

Nunna VK (2023) Degradation of mangroves along coastline of Andhra Pradesh and strategies for conservation. In: Fifth world congress on disaster management: volume V, pp 145–153

Pal S, Yadav VK, Sharma A, Ananthan PS, Qureshi NW, Dey S, Jana P, Karmakar S, Ojha SN (2022) Coastal multi-hazard vulnerability mapping: a case study along the coast of South 24 Parganas District, East Coast of India. J Indian Soc Remote Sens 50(9):1701–1712. https://doi.org/10.1007/s12524-022-01558-z(

Poulter B, Halpin PN (2008) Raster modelling of coastal flooding from sea-level rise. Int J Geogr Inf Sci 22(2):167–182. https://doi.org/10.1080/13658810701371858

Pourghasemi HR, Pradhan B, Gokceoglu C (2012) Application of fuzzy logic and analytical hierarchy process (AHP) to landslide susceptibility mapping at Haraz watershed, Iran. Nat Hazards 63(2):965–996. https://doi.org/10.1007/s11069-012-0217-2

Rahmati O, Zeinivand H, Besharat M (2016) Flood hazard zoning in Yasooj region, Iran, using GIS and multi-criteria decision analysis. Geomat Nat Haz Risk 7(3):1000–1017. https://doi.org/10.1080/19475705.2015.1045043

Ramesh DA, Muthukrishnan L, Balamurugan S, Karthi N, Dhivya S (2021) Ecosystem assessment for sustainable management: a case study of Coringa Mangroves, Andhra Pradesh, India. Arthshastra Indian J Econ Res 10(2–3):8–26. https://doi.org/10.17010/aijer/2021/v10i2-3/167171

Saaty TL (1980) The analytic hierarchy process (AHP). J Oper Res Soc 41(11):1073–1076

Saaty TL (1990) How to make a decision: the analytic hierarchy process. Eur J Oper Res 48(1):9–26. https://doi.org/10.13128/Aestimum-7138

Sahana M, Sajjad H (2019) Vulnerability to storm surge flood using remote sensing and GIS techniques: a study on Sundarban Biosphere Reserve, India. Remote Sens Appl: Soc Environ 13:106–120. https://doi.org/10.1016/j.rsase.2018.10.008

Sahoo B, Bhaskaran PK (2018) Multi-hazard risk assessment of coastal vulnerability from tropical cyclones–a GIS based approach for the Odisha coast. J Environ Manage 206:1166–1178. https://doi.org/10.1016/j.jenvman.2017.10.075

Sai KS, Manjunath A, Krishnaiah S (2021) Assessment of flood vulnerability index for Krishna river basin in Krishna, Guntur and Prakasam Districts of Andhra Pradesh. Int J Syst Assur Eng Manage 1–14

Sankari TS, Chandramouli AR, Gokul K, Surya SM, Saravanavel J (2015) Coastal vulnerability mapping using geospatial technologies in Cuddalore-Pichavaram coastal tract, Tamil Nadu, India. Aquat Procedia 4:412–418

Sarker MNI, Wu M, Alam GM, Shouse RC (2019) Livelihood vulnerability of riverine-island dwellers in the face of natural disasters in Bangladesh. Sustainability 11(6):1623. https://doi.org/10.3390/SU11061623

Syvitski JP, Kettner AJ, Overeem I, Hutton EW, Hannon MT, Brakenridge GR, Day J, Vörösmarty C, Saito Y, Giosan L, Nicholls RJ (2009) Sinking deltas due to human activities. Nat Geosci 2(10):681–686. https://doi.org/10.1038/ngeo629

Tasnuva A, Bari QH (2024) Integrating machine learning algorithms and fuzzy AHP for assessing livelihood vulnerability in southwestern coastal Bangladesh. Regional Stud Mar Sci 79:103825. https://doi.org/10.1016/j.rsma.2024.103825

Tessler ZD, Vörösmarty CJ, Grossberg M, Gladkova I, Aizenman H, Syvitski JP, Foufoula-Georgiou E (2015) Profiling risk and sustainability in coastal deltas of the world. Science 349(6248):638–643. https://doi.org/10.1126/science.aab3574

UNICEF (2015) Unless we act now: the impact of climate change on children

Preserving the Coast of Merauke, South Papua: Integrating Local Wisdom and Artificial Intelligence for Sustainable Eco-marine Tourism

Yohanes Hendro Pranyoto and Dedimus Berangka

Abstract Sand exploitation in the coastal region of Merauke, South Papua, Indonesia, has caused severe environmental damage, including coastal abrasion, groundwater pollution, and infrastructure damage. Poverty, low environmental awareness, and weak law enforcement are the leading causes of this problem. Eco-marine tourism uses the cultural heritage of the Marind tribe as an alternative way to handle sustainable coastal resources. It enhances environmental awareness and assists local people in increasing their income. The indigenous knowledge or local wisdom of the Marind people, which includes traditional ecological insights and customary values, plays a vital role in establishing sustainable marine tourism. The Kali Maro Festival is a successful example of integrating local wisdom values into tourism. Even though there are possibilities, problems like lousy infrastructure and possible conflicts with other economic activities still need to be fixed. AI and ML can help protect coastal natural resources and encourage eco-marine tourism that is good for the environment. By judiciously and ethically utilizing these technologies, AI and ML can offer governments and communities more profound and accurate insights into various aspects, including natural resource mapping, environmental and social impact analysis, and more informed and data-driven decision-making.

Keywords Sand exploitation · Eco-marine tourism · Ecological awareness · Local wisdom · Marind tribe · Merauke · South Papua · Artificial intelligence (AI) · Machine learning (ML)

Y. H. Pranyoto (✉) · D. Berangka
Saint James Catholic College of Merauke, Merauke, South Papua, Indonesia
e-mail: yohaneshenz@stkyakobus.ac.id

D. Berangka
e-mail: dedimus@stkyakobus.ac.id

K. K. Basheer Ahammed et al. (eds.), *Monitoring Coastal Marine Ecosystems Under Climate Change*, https://doi.org/10.1007/978-3-032-09137-6_18

1 Introduction

Merauke is a regency located in the South Papua Province of Indonesia. This district is the outermost and foremost region in Eastern Indonesia and borders directly with the country of Papua New Guinea, with an area of approximately 46,791.63 km. Compared to other regencies, Merauke has the longest coastline at 930.2 miles (Hapsara, 2024, p 12). Merauke Regency consists entirely of lowlands. Approximately 80% of the area in Merauke Regency is wetlands, such as swamps, rivers, mangrove forests, peatlands, rice fields, river deltas, reservoirs, irrigation canals, and floodplains (Sunarni et al., 2019). Agricultural and fishery commodities are the primary commodities in this region, especially rice, tubers, shrimp, and fish, although land cultivation and fishing methods generally still use traditional methods.

As a coastal area, Merauke has much potential for marine natural resources and beach tourism. However, the government and community need to manage this potential effectively. In 2023, the Central Bureau of Statistics data indicated that 10.01% (24,030 people) of Merauke people were impoverished (Hapsara, 2024, p 13). This figure will likely increase in the coming years unless the local government does not address this. Therefore, it is clear that the local government must empower the economy of the community, especially the Marind people, who are the Indigenous tribe in the Merauke Regency so that they can prosper in their land.

The Marind tribe, as an indigenous tribe of South Papua, especially Merauke Regency, consists of several sub-tribes, including Marind Pantai, who lives on the coast (Gamu & Pranyoto, 2023). The Marind Pantai people live on the coast in Merauke Regency and are generally economically poor. They do not have sufficient skills, capital, and knowledge to compete with migrants from various regions (Risamasu and Gebze, 2020). Some people use this to their advantage, offering a quick way to earn money by selling their land. The land purchased is generally around the coast at a cheap price. This land is then massively exploited for sand to be sold in the city at high prices as building materials (Hallatu et al., 2021).

Infrastructure development in Merauke Regency has surged with the division of Papua Province into South Papua Province in 2022. Sand, one of the basic building materials, is one commodity that is becoming increasingly needed. The seashore in Semangga District and Naukenjerai District is the source of more than 90% of the sand requirements. The beach is the primary source of sand, given that no volcanoes can release volcanic material for sand mining in South Papua. The local government has a spatial plan for this, although it often needs to be revised in practice. The Merauke Regency Spatial Plan for 2010–2030 delineates the following land use allocations: land designated for non-cultivation is 2,015,279 ha, while land designated for cultivation is 2,491,821 ha. Of this area, the land designated as a sand mining area is 2,161.12 (0.05%), concentrated in the Malind and Okaba sub-districts (Djamali et al., 2016).

The practice of illegal sand mining around the coast has been rampant for years, and there is no law enforcement action from the government, either preventively, repressively, or curatively. Limited natural resources and low-quality human resources, both from an economic and ecological standpoint, have caused the sale of land around the coast to be quite massive since 2000 and has increased since 2014 after the issuance of Law Number 23 of 2014 concerning regional government, which regulates the licensing and supervision of class C excavation materials (including sand) to become the authority of the province, limiting the space for law enforcement (Hallatu et al., 2021).

This illegal sand mining practice devastates the coastal ecology. Combined with rising sea levels every year, it results in alarming coastal abrasion, with Indonesia's highest level of coastal abrasion in the Merauke region, South Papua (Lekatompessy and Maturbongs, 2021). In the last ten years, Merauke's coastal ecosystems, particularly mangrove habitats, have suffered significant damage. Due to rising seawater intrusion, tidal flooding, and a decline in the quality of freshwater supplies to satisfy the needs of the Merauke people, most of Merauke's coastal communities are now almost unable to use groundwater for consumption because seawater has become brackish due to pollution (Widiastuti et al., 2016).

Coastal degradation has many ecological, economic, socio-cultural, and ethical consequences (Sarang et al., 2024). It is essential to find a solution to this problem so that Marind people living along the coast of Merauke do not have to move. As a result, the threat of loss of cultural identity for younger generations will increase (Wijaya et al., 2022a, b). In response to this situation, some religious and community leaders in Merauke have tried to implement an eco-marine tourism program since 2015 to increase people's ecological awareness of the importance of environmental protection (Kardi and Wiasta, 2020). The program was initiated in 2009 by religious leaders in Wendu Parish, Archdiocese of Merauke, who collaborated with traditional leaders in several villages such as Wendu, Urumb, Nasai, and surrounding areas. They are attempting to engage the community in ceasing the sale of coastal land, enhancing agricultural productivity, and developing a beach tourism zone. At first, the eco-marine tourism program encountered problems because the people in the area were used to getting straight money from selling sand.

In contrast, eco-marine tourism requires a longer compensation process, more complex management patterns, marketing, and careful planning (Gustiarini et al., 2023). This project aims to raise the standard of living in nearby communities by giving them new ways to make money by taking care of tourist attractions. It also puts protecting nature and the long-term health of Merauke's coastal ecosystems at the top of its priorities. Instead of selling land, the community will have a steady income stream from running tourist attractions. Furthermore, they receive instruction regarding environment (Lakshmi et al., 2024).

Problems such as sand exploitation on the coast of South Papua and efforts to develop eco-marine tourism still poorly need to be better planned and managed. In managing coastal natural resources and developing sustainable eco-marine tourism, the government and the community must combine local wisdom approaches with the latest technology (Louati et al., 2024). Artificial intelligence (AI) and machine learning (ML) technologies have significant potential for addressing these challenges. AI and ML can provide governments and communities, especially eco-marine tourism businesses, with more profound and accurate insights into various aspects, from natural resource mapping to environmental and social impact analysis, ultimately supporting more informed and data-driven decision-making (Gesami and Nunoo, 2024).

For the government, using AI, for example, with satellite imagery, can help monitor the impact of sand excavation on the coast of South Papua and law enforcement efforts by applying local regulations, mitigating disasters, and planning appropriate coastal area development. As for the community, especially eco-marine tourism businesses, AI is beneficial for developing sustainable eco-marine tourism. AI and ML can help analyze tourist data, preferences, and market trends to design attractive and relevant tour packages. In addition, AI can also be used to predict the number of tourist visits so businesses can optimize available capacity and facilities. AI and ML can also help customize the traveler experience by suggesting activities, accommodations, and food that align with each person's hobbies and needs. Tourism activities might use AI to lessen their damage to ecosystems. By constantly using AI to watch over coastal ecosystems, businesses can quickly spot possible environmental damage and take the necessary steps to stop it. The Marind indigenous people in South Papua attempt to balance economic development, environmental preservation, and community welfare through AI and ML technology's ethical and practical use.

2 Theoretical Framework

Theotonio Dos Santos explains that dependency is an imbalance between developed and developing countries (Siregar, 2019). Dependence can also mean that the progress in developing countries is only due to the economic development of developed countries and capitalism (Chandra et al., 2023). In this context, Merauke Regency, as part of South Papua Province as a developing region, is highly dependent on other regions. In terms of human development, Merauke falls somewhere in the middle of the pack in 2021, according to the HDI indicator.

According to Paul Baran, the process of utilizing human or natural resources by other parties causes dependence on developing countries or regions due to unfair economic relations with other regions (Sahrul and Muhammad, 2020). Other more developed regions use natural resources from a developing region. It makes developing regions even more dependent on more developed regions in the economic field, and finally, it spreads to social, cultural, and political fields as well.

Merauke Regency in 2021 ranked 9th in the dependency ratio of districts/cities in Papua Province (Fig. 1). This shows that Merauke is still highly dependent on other districts or regions regarding economy and development. As a result, the exploitation of natural resources and the lack of sustainable development measures in Merauke are consistent with the predictions of dependency theory and ground realities. To solve this problem, we need a good government that protects the rights of Indigenous people and puts community-based natural resource management into action (CBNRM) (Shereni and Saarinen, 2020).

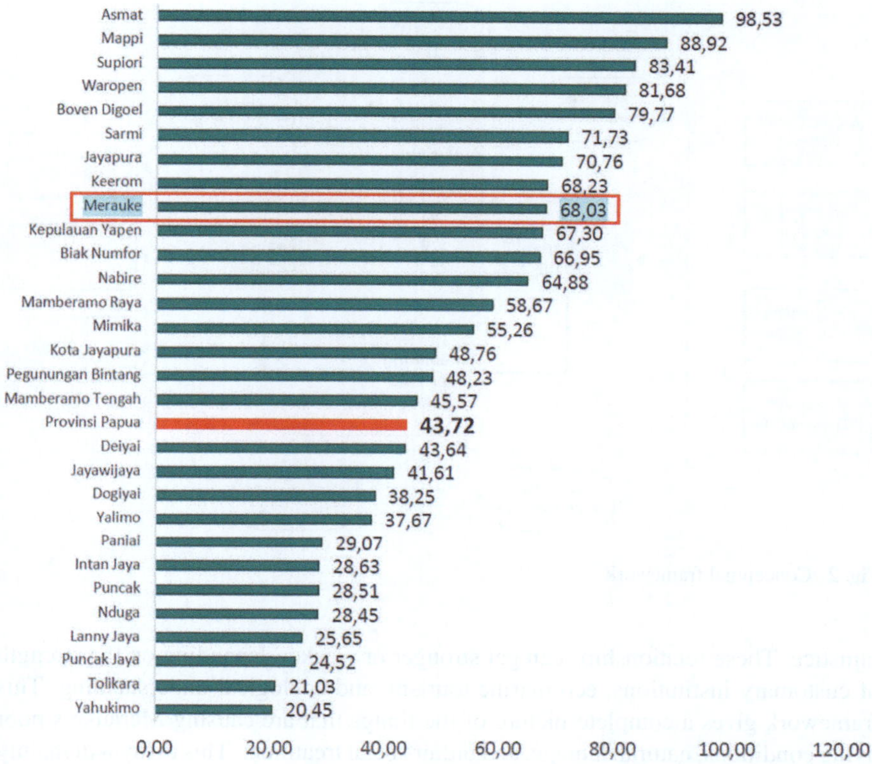

Fig. 1 Dependency ratio of regency/city in Papua Province 2017–2021 (Walilo, 2021, p 41)

The concept of Indigenous peoples is also fundamental in this study. Recognizing Indigenous peoples' rights to their lands, natural resources, and traditional practices (Reumi, 2018) is essential for places like Merauke (Oja et al., 2018). It is morally necessary for development to recognize the rights of indigenous people to their lands, natural resources, and cultural practices (Thontowi and Wartini, 2023).

Through the community-based natural resource management (CBNRM) model, local communities can become the main actors in managing their environment (Katikiro et al., 2024). The community's long-term goals, local knowledge, and well-established traditions can all affect decisions. For instance, CBNRM along the coast of Merauke helps protect shoreline areas and make the most of marine resources. Native American groups are in charge of managing natural resources. The government respects ethnic diversity by promoting fair, sustainable, and inclusive growth.

The framework shows that the role of companies, environmental conditions, government regulations, and social and cultural factors directly influence natural resource exploitation and social conflict (Fig. 2). These variables mediate the independent variables' relationships with environmental degradation, poverty and social

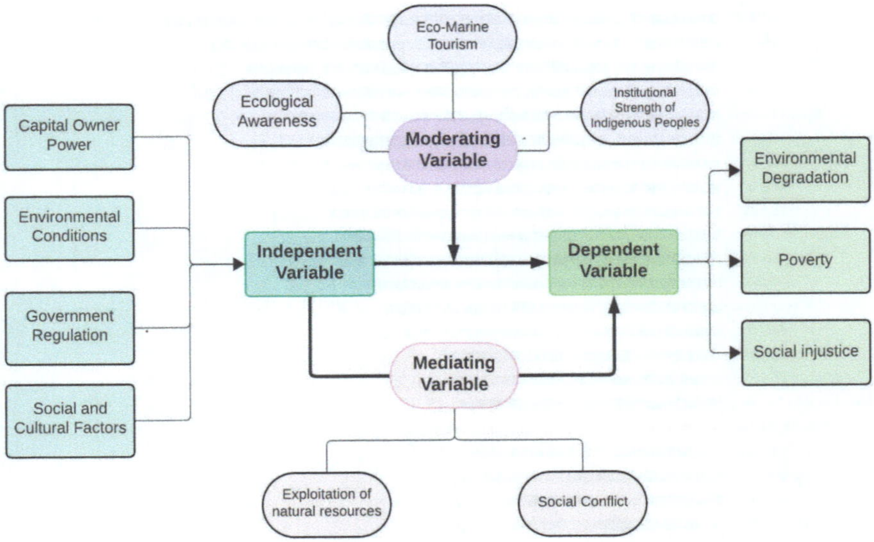

Fig. 2 Conceptual framework

injustice. These relationships can get stronger or weaker depending on the strength of customary institutions, eco-marine tourism, and ecological understanding. This framework gives a complete picture of the things that are causing Merauke's poor living conditions, natural damage, and unfair social treatment. This essay will mainly discuss the effects of natural degradation, specifically how careless use of natural resources can hurt coastal environments. To solve this issue, we need a method that simultaneously considers the natural, social, economic, and cultural (Fatima et al., 2024).

3 Methodology

3.1 Study Area

Merauke, a regency in South Papua Province, Indonesia, is located in the south-east of Papua Island, bordering the country of Papua New Guinea to the east. It is geographically located between 5°15'-8°45' (South latitude) and 137°40'-141°00' (East longitude). With a total area of 46,791.63 km^2, it is one of Indonesia's largest districts (Sanadi and Rumboirusi, 2024). Most of Merauke is lowland with extensive swamps and mangrove forests, especially in the south. These swamps are important carbon stores and habitats for diverse flora and fauna (Ririhena et al., 2017). In the north, tropical rainforests are still relatively well preserved. These forests are home to many endemic Papuan species and are critical in maintaining ecosystem balance. Merauke also has a long coastline of 930.2 miles, bordering the Arafura Sea (Hapsara, 2024, p 13).

Demographically, the population of Merauke Regency reached approximately 338,773 people in 2024 (Hapsara, 2024, p 13). There are many groups, but the Marind Tribe, which lives in wetlands and along the coast, is the native Merauke tribe. In addition, there are other tribes, such as Muyu, Mandobo, and Auyu, as well as various immigrant tribes from various regions in Indonesia. Most of Merauke's population relies on agriculture, plantations, fisheries, and utilization of forest products. Indigenous people still maintain local wisdom when interacting with nature (Hallatu and Palittin, 2023). Even though Indonesian is the legal language, many people still speak local languages every day, such as Marind. Most of the people in Merauke are Catholic, Islam, and Protestant Christianity (Hapsara, 2024, p 16).

The Dutch colonial government established Merauke Regency on February 12, 1902 (Papua, 2010). Initially, Merauke became the center of government and trade in the southern region of Papua. Over time, Merauke developed into an important port city, especially after the opening of a shipping route between Merauke and Australia in the early twentieth century (Jeminta and Mbaraka, 2005). After Indonesia's independence, Merauke became part of Papua Province. Merauke was designated as the capital of the newly formed South Papua Province on July 25, 2022, following a lengthy procedure under the Law of the Republic of Indonesia No. 14 of 2022 concerning the Establishment of South Papua Province (Betaubun et al., 2023). In recent decades, Merauke has undergone significant social and economic changes, mainly due to infrastructure development, investment in the plantation and mining sectors, and population migration from outside Papua (Fig. 3).

This study concentrates on two districts in Merauke: Naukenjerai and Semangga (Fig. 4), specifically in several villages in the coastal area where the Marind Pantai tribe lives. Previous research found prevalent and concerning sand mining activity in both areas (Djamali et al., 2016). This study's locations were Ndalir, Tomer, Wendu, Bahor, and Nasai beaches. Some beaches, such as Ndalir Beach and Tomer Beach in Naukenjerai district, are active sand mining sites. In Semangga District, sand mining has changed some beaches, such as Wendu Beach, Bahor Beach, and Nasai Beach. These beaches are now slowly becoming eco-marine tourism spots or beaches.

Fig. 3 Marind tribe past and present (Letsoin, 2015)

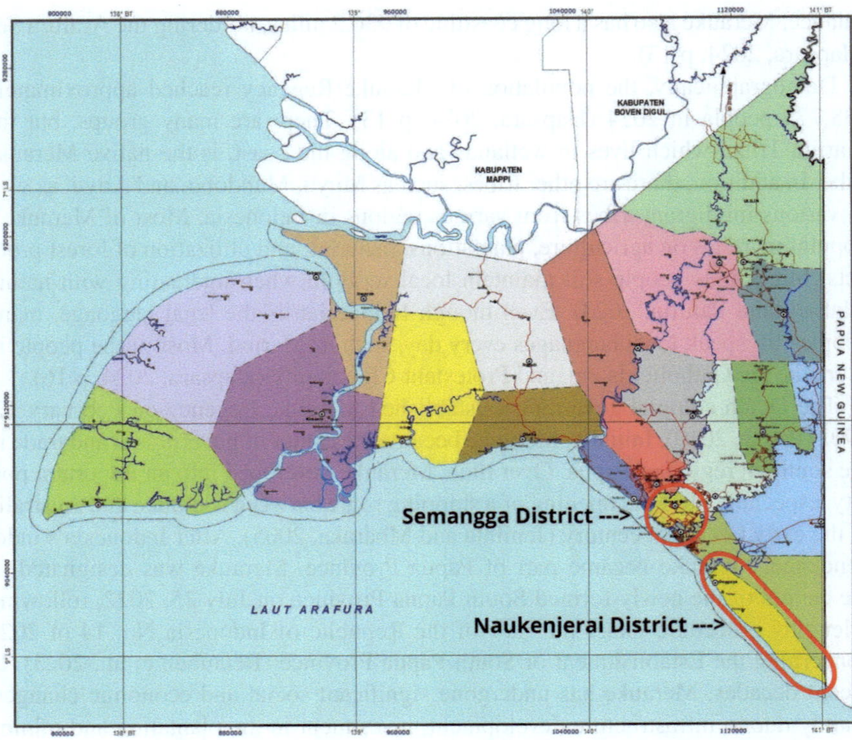

Fig. 4 Research location in Semangga and Naukenjerai districts, Merauke Regency, South Papua Province (Walilo, 2021)

3.2 *Method*

This paper is a research study using a qualitative-descriptive approach. It aims to describe the role of local wisdom-based eco-marine tourism in increasing community ecological awareness, reducing sand mining activities, and preserving coastal ecosystems amid the threat of seawater abrasion. The method employed in this research is ethnoscience, the study of scientific knowledge through a culture tied to nature and evolving in society (Jannah, 2022).

The research was conducted over six months, from February to July 2024, at the designated research sites (Fig. 4). Methods used to collect data included observing, interviewing, and studying records. Methods for analyzing data focused on looking at firsthand and second-hand data sources. Field observations and interviews with religious, communal, and traditional leaders and sand mining actors were primary data sources. This study used a systematic approach to identify essential informants, ensuring they met our requirements for knowledge of sand mining, coastal ecosystem degradation, customary authority, and eco-marine tourist management, among other topics. Public documents, strategies, policies about geographical and territorial planning, local ordinances, and scholarly studies served as secondary sources of information. The researcher used an inductive-qualitative approach to assess the data, which focuses on extracting individual meanings from passages and quotations that enable subsequent processes of open coding, triangulation, and interpretation of results that adequately capture the situation's complexity regarding perspectives and impacts (Creswell and Creswell, 2017; Chambliss and Schutt, 2013).

4 Discussion

4.1 *Mapping of Natural Resources and Impacts of Sand Exploitation in Coastal Areas*

South Papua and Merauke Regency governments should work together to collect accurate, complete, and current information about coastal natural resource mapping and the effects of sand extraction in South Papua so that they may make decisions based on data. Artificial intelligence (AI) and machine learning (ML) are promising for better using coastal natural resources and reducing the negative impacts of sand mining in South Papua. These tools can help us find new ways to solve challenging problems. For example, we can watch changes in the shoreline in real time, catch illegal mining early, and guess how these activities will affect the environment in the long run.

So far, the data obtained by the author mostly comes from research results from previous researchers and journalists. Therefore, the opportunity to use AI and ML in natural resource mapping and impact assessment of sand exploitation in the coastal areas of southern Papua has advantages over traditional methods (Pandey et al., 2023). AI and ML can improve monitoring and evaluation by processing and analyzing vast amounts of data quickly and accurately, covering a larger area (like the coast of South Papua), and being objective (AI and ML can provide objective and unbiased analysis, helping with transparent and evidence-based decision-making) (Chen et al., 2023).

In natural resource mapping, AI can look at satellite pictures and lidar data to find and group different coastal ecosystems, like mangrove forests, seagrass beds, river deltas, and coral reefs (Basheer Ahammed et al., 2024; Wasim et al., 2024). ML models map more accurately and faster than hand-drawn methods because researchers train them to recognize patterns and distinctive characteristics of each ecosystem. Satellite images, field data, and machine learning models enable artificial intelligence (AI) to assess natural resource stocks in specific areas, including fish populations, mangrove biomass, and sand mining quantities. Participating in eco-marine tourism and conserving resources necessitates comprehension of fundamental principles (Table 1).

Local governments can use relevant agencies such as provincial or district Energy and Mineral Resources Offices to improve the accuracy and efficiency of natural

Table 1 Potential beach sand mining areas in Merauke Regency (Walilo, 2021)

No.	Location	East longitude (E)	Southern latitude (S)	Metric ton	Description
1	Samkai Village	$140^{0}39'24,17''$	$8^{0}52'10,28''$	2,000,000	The beach sand is brownish and medium-grained; sedimentation is still ongoing
2	Nasem Village	$140^{0}51'86,11''$	$8^{0}65'06,11''$	1,000,000	
3	Kuler Village	$140^{0}54'28,61''$	$8^{0}68'50,00''$	5,000,000	The beach sand is brownish, contains shells, is medium-coarse-grained, and spreads over 300–400 Ha with an average thickness of 1.5 m
4	Onggaya Village	$140^{0}55'63,06''$	$8^{0}70'60,75''$	7,000,000	
5	Tomer Village	$140^{0}57'20,72''$	$8^{0}73'19,80''$	8,000,000	
6	Domande Village	$139^{0}98'19,80''$	$8^{0}16'66,67''$	7,500,000	Beach sand near the mouth of the river is gray-brown and contains shells. It is medium-coarse grained
7	Urumb Village	$140^{0}31'41,66''$	$8^{0}45'00,33''$	5,000,000	
8	Wendu Village	$140^{0}29'35,66''$	$8^{0}43'10,30''$	4,500,000	

resource mapping, mainly sand, by integrating artificial intelligence (AI) technology and satellite imagery (Wasim et al., 2024). The application of AI to high-resolution satellite imagery enables the identification and analysis of sand deposits accurately, quickly, and efficiently compared to conventional methods. AI can be trained to recognize patterns and characteristics of sand deposits in satellite images, even in hard-to-reach areas (Kapatamoyo, 2024). The result is a complete, *up-to-date* picture of sand's location, amount, and quality. This knowledge is essential for planning land use, managing mines in a way that does not harm the environment, and catching people who steal sand. Adding AI and satellite images to data from drones and *ground s*tudies can help make maps more accurate and complete with more information.

The utilization of AI and ML technology can also be a solution to strengthen law enforcement efforts against illegal sand mining. AI and ML can be integrated to monitor and analyze the impact of sand exploitation, especially on shoreline changes (Dang et al., 2022). Multi-temporal satellite images, which record shoreline conditions at various periods, can be processed using AI to accurately detect changes in beach morphology (Figs. 5 and 6). AI can identify erosion and accretion patterns indicative of illegal sand mining activities (Kumar et al., 2020).

To make ML algorithms, it can also combine data from different sources, like satellite pictures, drone photos taken from above, and measurements taken in the field. These computers learn to spot trends in how the shoreline changes and guess the long-term effects of illegal sand mining on coastal ecosystems. AI and ML analysis give law enforcement officers much helpful information that helps them

Fig. 5 One of the soil or sand mining activities using heavy equipment in Urumb Village

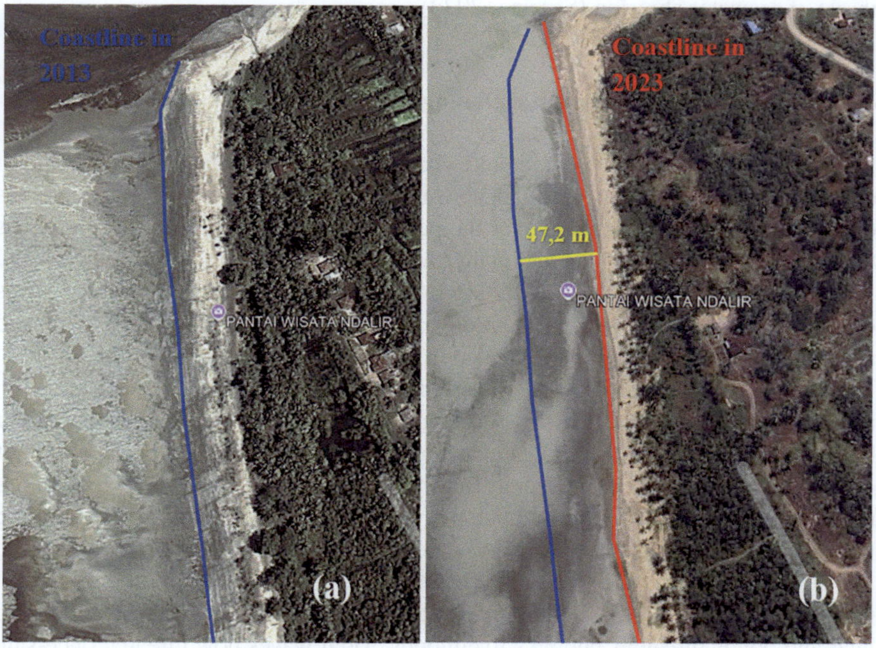

Fig. 6 Comparison of the coastline at Ndalir Beach in 2013 (**a**) and 2023 (**b**)

find illegal mining sites, determine how much money they have lost, and develop focused enforcement strategies.

Law enforcement efforts on illegal sand mining are urgently needed because the process of extracting sand and soil on the coast of Merauke, especially in Semangga and Naukenjerai Districts, is still carried out massively. Heavy machinery, such as excavators, has made the once clandestine excavation process much more visible.

Evidence from the researcher's interviews and observations points to the excavation on vacant lots once owned by locals but now owned by shady businesspeople who profited from the city's sand. Generally, for one truckload of sand, landowners get around Rp 100,000–200,000 for beach sand and around Rp 200,000–400,000 for land sand. The sand is sold for between Rp 1,500,000 and Rp 2,500,000 when it gets to the customer, which depends on how far away the consumer is.

Law enforcement officials and traditional authorities should have acted to prevent digging during the research process. Land dug up for sand and soil is left unattended until it becomes a pool of water. It causes massive seawater abrasion in coastal areas. Since 1990, the average global sea level has increased by 3.4 mm per year due to climate change. The rate has risen to 3.6 mm per year since the mid-2000s and is projected to attain 4.5 mm per year by 2030 (Gehrels and Garrett, 2021; Yincan, 2017). Merauke itself is geographically a lowland with an altitude of approximately 3.5 m above sea level (meters above sea level) with a composition of 50% alluvial soil, making the Merauke region very vulnerable to coastline deterioration due to the threat of seawater abrasion (Sergino, 2022).

4.1.1 Impact of Environmental Damage

Sand mining activities on the Merauke coast have increased the risk of damage to coastal ecosystems due to seawater abrasion. Abrasion of the Merauke coast has caused severe physical damage over the past few decades. The coastline continues to shrink, swallowing up land and mangrove forests once home to various flora and fauna. Large palm trees are now falling over one by one because the rough waves of the Arafura Sea are wearing them down. Aside from that, things like roads and residential places near the coast are also in danger.

The coastline in these two districts continues to shrink significantly, although the shrinkage rate differs. Ndalir Beach in the Naukenjerai district is a particular highlight. In the last ten years, its coastline has retreated 47.2 m or an average of 4.7 m yearly. Meanwhile, at Urumb Beach in Semangga district, the coastline has retreated 39.3 m or an average of 3.9 m per year in the past ten years. These alarming figures illustrate how quickly the waves are eroding the land in this region. Coastal abrasion in Merauke is not just a change in landscape. Abrasion has real impacts that threaten the lives of coastal communities and ecosystems. The loss of agricultural land and ponds threatens people's livelihoods. Tidal waves now threaten once-safe homes. Coastal erosion also affects species. Mangrove trees, which naturally protect the coast, are shrinking, putting many fish and wildlife species at risk of extinction.

The Merauke Regency Central Bureau of Statistics reported that mangrove forests in Merauke covered an area of 322,952.96 ha in 2023 (Sanadi and Rumboirusi, 2024). Merauke is one of Indonesia's districts with the most extensive mangrove forests (Sunarni et al., 2019). The function of mangroves itself is essential to protect the coast from abrasion, provide habitat for various types of animals and plants, provide carbon storage, and provide a source of livelihood for the community, especially the Marind Pantai tribe, most of whom still apply hunting and gathering lifestyle

Fig. 7 Comparison of mangrove forest in the Maro River delta in 2013 (**a**) and 2023 (**b**)

(Suharno and Kadir, 2023). However, with the current conditions amid development and the threat of abrasion, the area of mangrove forests in Merauke continues to shrink by an average of 6.71%/year (Widiastuti et al., 2016) (Fig. 7).

Rising sea levels and seawater intrusion into the land also cause freshwater pollution caused by increasing the salinity level of water for community consumption. Coastal communities dominated by the Marind Pantai tribe still rely on traditional wells as a water source for consumption and daily needs. However, with the declining quality of groundwater, they have switched to bottled or refillable drinking water.

Coastal communities in Semangga and Naukenjerai districts generally no longer use traditional wells as spring water sources for consumption due to high salinity levels. Generally, the salinity level of water in Merauke affected by seawater abrasion can range from 0.5 to 3 ppt (Merauke, 2023). Water with salinity below 0.5 ppt is considered freshwater that is safe for consumption. In comparison, water with salinity levels above three ppt is considered brackish and unfit for consumption without prior treatment. As groundwater quality declined, the community turned to refilled drinking water for cooking and consumption (Table 2). It directly impacts the community's economy, resulting in additional expenditure for low-income families to purchase refill drinking water.

Groundwater pollution also impacts the livelihoods of coastal residents. Communities living as farmers must bear the risk of reduced crop quality as groundwater salinity levels significantly impact crop fertility and reduce yields (Nicolas et al., 2023). In addition, fishermen also experience negative impacts. Seawater can change where fish and other marine life live in river estuaries and brackish waters. It can

Table 2 Percentage of respondents based on main water source for drinking and cooking

Water source	Total	Percentage (%)
Refill or packaging	33	66
Rainwater	8	16
Well protected	4	8
Borehole well	3	6
Pipe well	2	4
Total	50	100

reduce the number of fish in the area and the size hunters can catch (Love et al., 2008). It can threaten the sustainability of their livelihoods, which depend heavily on fisheries resources.

Merauke's coastal communities in Naukenjerai and Semangga districts are generally farmers and fishermen. Therefore, they are the first to feel the impact of coastal sand exploitation, which has resulted in a decline in the quality of agricultural land due to high levels of groundwater salinity and declining fish populations in river estuaries, brackish waters, and waters around the coast.

4.1.2 Impact of Infrastructure Damage

Sea scouring has damaged critical infrastructure in the Semangga and Naukenjerai districts, especially roads and bridges close to the coast. Roads that used to be smooth are now worn down and have holes in them. The waves have even washed away some roads. It makes it hard to get transportation between areas. Bridges connecting coastal villages are also in danger of collapsing. Some existing bridges were built independently and use semi-permanent materials. The condition of these transportation facilities hampers the mobility of the population and the distribution of goods. Abrasion poses a risk to coastal communities. The impending waves have rendered previously susceptible dwellings unstable, compelling some individuals to vacate their residences since they might no longer remain there. The afflicted communities faced significant hardship in losing their houses and possessions.

Damage to the roads that connect coastal villages in the Semangga district, like Urumb, Yatom, Bahor, Wendu, Nasai, and Matara, affects many parts of community life, especially business. Damaged or blocked road access from wear and tear has made getting goods and services harder for people who need them. It has made it harder to sell food and fish to people in the area. It could make goods and services more expensive in towns nearby. Tourist sites in this area are also highly dependent on road access because the distance from Merauke city to the coastal villages in Semangga district is quite far, around 23 km. An area will be hampered in its development if transportation access is still constrained because road access is very influential on development in an area (Prapti et al., 2015).

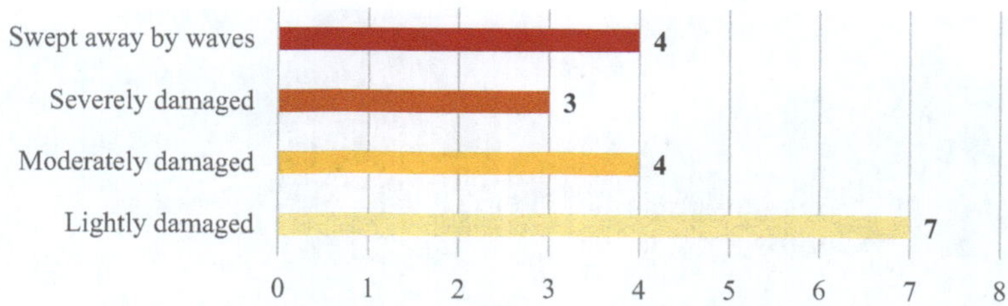

Fig. 8 Condition of Residents' houses affected by abrasion in Wendu Village and Urumb Village in 2023–2024 (Observation results)

According to the author's observations during approximately eight years of living in Merauke, road access to and from coastal villages in Semangga district, such as Urumb, Yatom, Bahor, Wendu, Nasai, and Matara villages, has no meaningful development such as the addition of alternative routes or the construction of permanent drainage channels around the road. It messes up the community's social and economic life and makes it hard for people to get health and education services. These coastal areas have much promise for natural resources and tourism but need better infrastructure to make it happen. The local government should immediately deal with this widespread worry to encourage fair growth (Fig. 8).

Field observations show that several houses living on the coast are affected by sea abrasion. Waves lightly damaged seven dwellings, moderately damaged four, severely damaged three, and destroyed four between 2023 and 2024. This figure excludes properties that abrasion destroyed in prior years and will continue rising in the following years unless authorities take genuine preventative action. Residents recount that they cannot sleep well at night because they hear the sound of the waves coming closer to their homes. The coastline is shrinking every year, and the waves threaten their homes. Some informants from the Marind tribal elders said that in their childhood, Urumb and Wendu villages were monumental, and the boundaries of their settlements reached far out to sea. The vicious waves, however, had already engulfed them. The loss of residential land owing to increasingly severe sea abrasion will eventually force the population to relocate.

Abrasion also happens in public places like schools, health centers, and places of worship. These buildings are getting more easily damaged, which could stop public services and neighborhood events. Many problems make it hard for kids to attend school, get medical care, and attend religious events. One of the most severely affected places of worship is the Catholic Church in the Sacred Heart Parish of Wendu, Archdiocese of Merauke, located on the coast of Kampung Wendu. This historic church complex is of great value to Catholics because Catholic missionaries first preached in Wendu Village (Rawung and Willem, 2005). Abrasion has rendered the first Dutch mission church at Wendu Village inaccessible. Unfortunately, a building with such high historical value should be damaged and lost due to abrasion (Fig. 9).

Due to sea abrasion in Semangga and Naukenjerai Districts, Merauke, infrastructure damage has impacted people's lives. Comprehensive and sustainable handling

Fig. 9 Photos **a** and **b** of *the impact of abrasion on the damage to the church building* (personal documentation). Photo **c** is from Santi Andriani (2024), the *impact of tidal flooding* on *people's houses*. Photo **d** taken by Agapitus Batbual (2019)*abrasion damages roads and bridges in Ndalir Village*

efforts are needed to protect vital infrastructure and ensure the welfare of coastal communities. In addition to short-term handling efforts, long-term mitigation and adaptation are also needed. Mitigation can be in the form of efforts to reduce the impact of abrasion, such as mangrove planting and coastal protection infrastructure development. Meanwhile, adaptation is an effort to adjust people's way of life to changing environmental conditions, for example, by relocating settlements to safer areas. However, the most urgent action is to stop illegal sand mining activities that are still rampant because any preventive measures will not significantly impact sand mining if they continue (Hallatu et al., 2021).

4.2 *Analysis of Factors Causing Sand Exploitation in Coastal Areas*

Many things affect beach sand mining in the Semangga and Naukenjerai Districts of Merauke Regency. Each of these causes directly affects the use of sand in coastal places. Here is a look at some of the main reasons for this:

Table 3 Percentage of respondents based on livelihoods

People's livelihoods	Total	Percentage (%)
Farmers	24	48
Fisherman	20	40
Labor	3	6
Self-employed	2	4
Public servant	1	2
Total	50	100

4.2.1 Economic Factors

The high poverty rate and limited formal employment opportunities in the region have encouraged people to seek alternative sources of income, including illegal sand sales. Marind communities on the coast generally rely on marine and agricultural products to make ends meet (Table 3). Although Marind people have an entrepreneurial spirit, vocational skills training is needed to compete in the world of work (Kaok et al., 2022). In addition, the presence of companies and contractors also increases the high demand for sand for infrastructure development in Merauke and surrounding areas.

It creates an economic incentive for communities to engage in illegal sand extraction and sale. Illegal sand sales offer attractive short-term financial benefits for people with low or no regular income who need cash quickly to fulfill their daily needs (Djamali et al., 2016). The presence of companies and infrastructure development projects, while potentially bringing economic benefits to the region, also poses a dilemma for local communities. The high demand for sand creates short-term economic opportunities and encourages the exploitation of non-renewable natural resources that damage the environment. It is in line with the results of an interview in the field with one of the residents who sold his land for exploitation. He stated that he sold his farm to finance his wife's medical treatment. He took this action because he lacked a stable income and the money for her treatment. Facing the challenge of generating income, some residents sold their land to meet their daily needs. Some residents sold their land to fulfill their daily needs because of the difficulty of earning an income.

4.2.2 Social and Cultural Factors

Lack of knowledge about the detrimental impacts of sand mining on the environment and the sustainability of natural resources is one of the social factors pushing many coastal villages in Merauke to sell their land for sand extraction. The Marind people living in the southern region of Papua consider forests and land mothers who guard, care for and raise them (Wijaya et al., 2022a, b, p 280). Even the Marind tribe only considers forests and land as the only inheritance that can be passed on to their children and grandchildren, which means that only land and forests have economic value and sacredness for the Marind tribe. However, in this modern era, the sacredness of the Marind tribe living in the Merauke Regency area began to fade when indigenous people began to recognize money as another commodity that replaced land (Wijaya et al., 2022a, b, p 284).

The low level of education also affects the low level of ecological understanding. The results showed that most of the people who sold their land had, at most, a primary school education (70%), and some did not go to school (25%). The Parish Priest of Wendu, Archdiocese of Merauke, asserts that the poor educational level has led the residents of Wendu, Urumb, and neighboring villages to hastily sell their land without comprehending the long-term ramifications (Fanumbi, 2024). The lack of formal education makes them vulnerable to tempting offers from outsiders who want to utilize their land without considering the ecological and social impacts that will be caused (Djamali et al., 2016).

Lack of access to knowledge on sustainable natural resource management further exacerbates this situation. Less educated communities rely on traditional knowledge that may no longer be relevant to today's environmental challenges (Gamu and Pranyoto, 2023). Because of this, they need help to make intelligent choices about how to use the land. Enabling more people to get information and help manage natural resources in a way that does not harm them is very important. The government, customary institutions, non-governmental groups, and other interested parties should work together to give communities, especially those in areas that are likely to be exploited for their land, the right kind of education and training.

4.2.3 Geographical Factors

Merauke, the capital of the newly formed South Papua province in 2022, is experiencing rapid development. The demand for building materials, mainly sand, increases yearly due to various ongoing infrastructure development projects, such as constructing roads, bridges, government buildings, and other public facilities. Sand is one of the main ingredients in mortar and concrete manufacturing, so its need is vital in the construction sector (Pasalli, 2022).

The geographical condition of Merauke, a lowland with no volcanoes, means that the Merauke region does not have a natural source of sand sedimentation. As an alternative, the community takes beach sand as a building material despite its lower quality. The Merauke Regency Government, since 2011, has tried to import sand from other areas such as Timika and Makassar, but the more expensive price factor makes consumers prefer local sand (Kompas.com, 2011). Moreover, one could contend that the beaches in Merauke include considerable sand reserves, rendering them a readily accessible resource for illicit and indiscriminate extraction. There is much sand on coastal beaches, but the region's sand supply is limited because there are no natural sources of sedimentation. It makes the area subject to overexploitation. This situation also makes the harmful effects of illegal sand mining worse. Taking out much sand without letting it grow back naturally can hurt the earth even more.

Also, some sand mining sites in the Naukenjerai area are far away and hard to get to, which makes it hard to keep an eye on and stop illegal sand mining.

4.2.4 Institutional or Policy Factors

Different government departments and related organizations must collaborate and coordinate their efforts to enforce the law and enhance environmental protection, contributing to the ongoing expectation of sand quarrying along Merauke's coast. In managing natural resources, the central government, the provinces, and the district governments have some overlapping power. It can lead to uncertainty and legal loopholes that the perpetrators of illegal sand excavation exploit.

In 2011, the local administration of Merauke Regency released Merauke Regent Regulation No. 14/2011 on the Management of Non-Metallic Mineral and Rock Mining Business to enforce the legislation on sand excavation. The regulation states: "All non-metal mineral and rock mining enterprises must secure an IUP or IPR (non-metal mineral and rock mining license) from the Regent before operation." However, this regulation lasted only two years because 2014 Law No. 23/2014 on Regional Government was issued, which withdrew the authority to license and supervise class C minerals to the province through the relevant agencies (Government MR, 2011). Sand excavation included in class C excavation materials must obtain a permit from the Papua Mining Service. At the same time, the Merauke government can only supervise public order using the Regional Regulation (Batbual, 2018).

Customary institutions in Merauke have also demonstrated weak supervision and regulatory factors. Because sand excavation often occurs in areas under customary rights, this sometimes prevents law enforcement officials from acting against the perpetrators. Law Number 21 of 2001 about Special Autonomy for Papua Province recognizes Indigenous Papuans' customary rights, as does Papua Province Regional Regulation Number 5 of 2022 concerning Recognition and Protection of Customary Law Communities in Papua Province (Batbual, 2018). The Marind Indigenous People's Organization (LMA) also finds itself largely powerless because the perpetrators selling sand are generally Indigenous Papuans driven by economic needs. Furthermore, the lack of strict and measurable enforcement of customary law against violators tempts hamlet owners and customary rights owners to sell sand (Widiastuti et al., 2018).

4.3 The Role of Eco-marine Tourism

Taking sand from the coast of Merauke has hurt the environment. This behavior hurts ecosystems along the coast and puts people who depend on marine resources at risk. In this case, eco-marine tourism is an option that can be good for both the earth and the bottom line.

4.3.1 Concept and Benefits of Eco-marine Tourism

Eco-marine tourism is a sustainable tourism approach that focuses on the exploration and appreciation of marine ecosystems in a responsible and low-impact manner (Kardi and Wiasta, 2020). The ultimate goal is to minimize detrimental impacts on the maritime environment while offering local people significant tourism experiences and economic advantages (Alkhudri et al., 2021). Eco-marine tourism combines the principles of ecology and sustainable tourism (Crockett and Stow, 2021). Some basic principles of eco-marine tourism include conservation, education, community participation (CBNRM), and sustainability (Kardi and Wiasta, 2020).

Conservation principles in eco-marine tourism encourage several actions, such as protecting and preserving marine ecosystems, coral reefs, mangroves, and marine animals. Preserving marine biota and ecosystems is critical to making coastal tourism attractive to tourists while protecting coastal nature. Furthermore, eco-marine tourism should promote educational initiatives among tourism site managers, local communities, and tourists to raise awareness of the importance of marine ecosystems and urge responsible behavior. It can increase collective ecological awareness that communities may need to acquire in formal school education. Eco-marine tourism promotes community empowerment via community-based natural resource management (CBNRM). Managing coastal resources as tourist attractions can involve sustainable planning and management and benefits to local communities from tourism activities. It is essential to ensure that tourism activities can occur long-term without damaging the environment or harming communities. Eco-marine tourism can slowly change the community's mindset to earn money instantly by selling sand because it offers a sustainable and environmentally friendly alternative livelihood based on the local knowledge and local wisdom of the Marind Community.

On the coast of the Naukenjerai district, the local community manages several eco-marine tourism destinations, such as Ndalir Beach, Onggaya Beach, and Katmap Beach. Some of the most famous beaches are Onggaya Beach and Ndalir Beach because they offer white sand and unspoiled natural scenery. Semangga District has several beaches, such as Urumb Beach, Bahor Beach, Wendu Beach, and Anasai Beach. Among these beaches, Anasai Beach is quite famous (Fig. 10) because it is well-managed and equipped with adequate public facilities for tourists. Beaches in Merauke typically derive their names from nearby villages; for instance, Urumb Beach derives its name from Urumb village.

Eco-marine tourism increases community income economically. The average monthly income from coastal tourism management increases from year to year, IDR 2 million in 2021, IDR 2.6 million in 2022, and IDR 3.1 million in 2023 (Fig. 11). The potential income from beach tourism can certainly still be increased with various innovations, one of which is by utilizing artificial intelligence (AI) technology. Developing eco-marine tourism in coastal areas aims to generate economic benefits while preserving the coastal environment.

The development of coastal tourism that fosters ecological awareness is needed, for example, by planting mangroves. Various parties, such as the Papua River Basin Agency, have initiated a mangrove planting program at Urumb Beach in November

Fig. 10 Anasai Beach tourism managed by Indigenous people

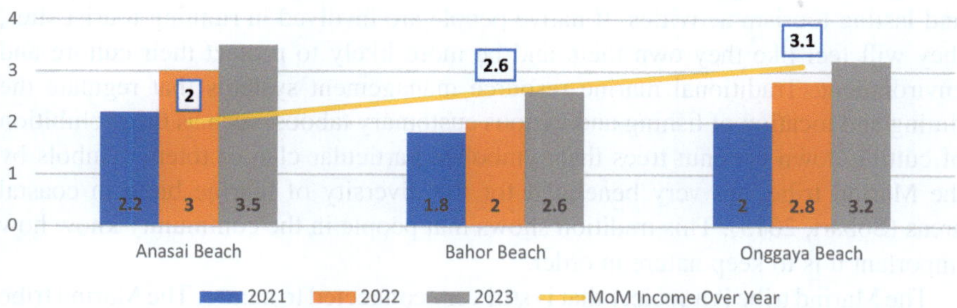

Fig. 11 Comparison of eco-marine tourism revenue from month to month for a year (in million rupiah)

2019 (Asry, 2019). The Main Navy Base (Lantamal XI) Merauke and the community planted 5800 mangrove trees at Urumb Beach in October 2019 (Yonavilbia, 2019). In addition, SMA Negeri Plus Urumb students planted 100 coconut seedlings and dozens of mangrove seedlings at Notif Beach in October 2022 (Kobun, 2022). This beach reforestation activity involves the surrounding community and shows them the importance of maintaining the beach ecosystem and the sustainability of beach tourist attractions and generations.

Some community elements also routinely have access to coastal tourism management and marketing training. Groups that receive training include tourism site managers, tourism activist groups, creative economic actors, and tourism awareness groups. Previously supported by relevant agencies, community groups independently carry out beach management, with local religious and community leaders supervising its implementation. Religious and traditional leaders play an essential role in the supervisory function because they have moral and spiritual authority that the community respects (Fanumbi, 2024). Their presence directs beach management activities to focus on economic benefits while maintaining local wisdom and promoting environmental sustainability. Their presence directs beach management activities to focus on economic benefits while maintaining local wisdom and promoting environmental sustainability.

4.3.2 Local Wisdom of the Marind Tribe in Eco-marine Tourism on the Coast of Merauke

The development of eco-marine tourism based on local wisdom is very appropriate to be implemented on the coast of Merauke because it not only provides economic benefits for the community but can also preserve the rich culture and traditions of the Marind tribe. Many Marind community traditions are essential for the preservation of coastal environments. The Marind people's local knowledge, like their traditional ecological knowledge and morals, can be a strong foundation for creating responsible and lasting tourism activities. If native people are involved in running tourist sites, they will feel like they own them and be more likely to protect their culture and environment. Traditional marine resource management systems that regulate the timing and location of fishing and various customary taboos, such as the prohibition of cutting down coconut trees that symbolize particular clan or totem symbols by the Marind tribe, are very beneficial for the diversity of marine biota in coastal areas (Sobari, 2015). This tradition shows that people in the community know how important it is to keep nature in order.

The Marind tribe is a society that is still very connected to nature. The Marind tribe does not see nature the way most people do today. They value nature and the earth a lot. Protecting and preserving nature and the environment is a moral responsibility of humans, which is as important as respecting other people's lives (Hendrawati, 2011, p 9). Modern society generally has an anthropocentric perspective, which considers humans the center of the universe. Only humans have value and worth; nature and everything in it is a means or tool to fulfill human needs. This perspective tends to lead to exploitative and indifferent attitudes and behaviors toward nature, which leads to environmental damage and destruction (Hendrawati, 2011, p 10). On the other hand, the Marind Tribe considers the natural environment as their mother, meaning that the Marind Tribe's cosmological insight provides concepts, principles, and views to preserve, protect, and make the natural environment sustainable with an inward-looking philosophy. In a sense, nature is more than just a source of life for the Marind people; it is life itself (Kalilago, 1992, p 6).

The Marind tribe's traditional belief system actively influences the management of eco-marine tourism, as the Marind people value nature and strive to prevent its destruction by human actions. In addition, the Marind community still upholds traditional values related to tourist attractions, such as *wambat*, which means cooperation between residents to overcome difficulties (Fanumbi, 2024). This principle of cooperation is still held firmly by coastal communities and is used to manage coastal tourist attractions with the principle of cooperation. The implementation is to create a coastal tourism business group of administrators and members. They are responsible for managing tourist attractions on behalf of the village community, planning, developing tourist sites, marketing, and reporting accountability to the community. The community jointly builds public facilities around the beach using local materials and traditional designs typical of the Marind coastal tribe, such as *honai* (temporary houses in the forest). It makes picket and guard duties in shifts at tourist sites. The fundamental principle of *wambat* or cooperation is thus manifested in their joint efforts to build infrastructure, manage daily operations, and ensure the security of the tourist sites (Fanumbi, 2024).

Some religious and community leaders in the Semangga district are trying to incorporate traditional and religious ceremonies into available tour packages to increase the potential of eco-marine tourism in coastal areas. Thus, tourists can enjoy the natural beauty of the beach and sea and experience the local community's rich culture and spirituality. This partnership attracts tourists looking for a more profound and critical vacation experience while helping local economies grow long-term (Fig. 12).

One form of integrating traditional events into tourism activities is the Kali Maro Festival tour package launch in 2023, held every October. The Kali Maro Festival combines the Marind tribe's harvest ceremony, known as the *ndambu* ceremony, with other tourism activities. The *ndambu* ceremony is an essential celebration after

Fig. 12 Kali Maro Festival in 2023 (Fanumbi, 2024)

a successful harvest season (Hendrawati, 2011). The ceremony aims to express gratitude and thanks to God and ancestors for the bountiful harvest and invoke blessings for the upcoming harvest.

The Kali Maro Festival spans nearly a month and features a variety of religious and cultural activities organized by the Marind tribe. The festival ends with a thanksgiving ceremony at the mouth of the Kali Maro and continues to the surrounding beaches. Traditional dances and songs are essential parts of the Kali Maro Festival. They will dance and sing their songs and dances while traditional instruments like the tifa and flute play in the background. There are stories in these songs and dances about farming, nature, and how people connect with God. Then, the religious and traditional leaders will pray and give thanks while giving different crops, like tubers, fruits, and fish, to God and the ancestors. The beach is the center of this action, making it the most essential part of the event. There will be an art show and a group meal at the end of the event. The community members will gather to share a meal, reinforcing their brotherhood and unity bonds (Fanumbi, 2024).

People celebrate their spirituality at the Kali Maro Festival, a necessary time to keep ancestral traditions alive and ensure that people and nature can live together happily. In addition, the event can be a draw for cultural tourists, showing them the rich culture of the Marind people and helping the local economy. The number of tourists from various regions will increase during the festival. Communities can also sell handicrafts, produce, and sell services such as local tour guides, homestays, equipment rentals, and transportation services. The Kali Maro Festival is integral to the local economy because it helps it grow. As the number of tourists rises, so does the need for places to stay, food, and other services that help them. It opens up new business opportunities for folks in the surrounding area. The attendees of the event can experience this positive effect. Eco-marine tourism in Merauke is becoming increasingly famous as a place to visit for cultural and natural reasons, and it can happen any time of the year.

4.3.3 Challenges of Eco-marine Tourism

Implementing eco-marine tourism in Merauke is undoubtedly open to various challenges. It is crucial to fix problems like insufficient infrastructure, people who need help understanding conservation, communities not good at handling tourist spots, and possible conflicts with other economic activities like fishing and sand quarrying. Basic infrastructure such as roads, transportation, accommodation, and other supporting facilities still need to be improved in many locations in Merauke. It can hinder tourists' accessibility and reduce the quality of their tourism experience (Alkhudri et al., 2021).

The need for public awareness of the importance of conservation is also a challenge in developing eco-marine tourism on the Merauke coast. Not all people realize the importance of conservation efforts. Some people oppose conservation efforts in coastal areas (Table 4). Not all communities in coastal areas realize the need to make conservation efforts such as planting mangroves or coconut trees. With a deep understanding of the importance of protecting marine ecosystems, it will be easier to establish and maintain responsible and sustainable tourism practices. Therefore, education and socialization of the community about the importance of conservation are critical in developing eco-marine tourism in Merauke.

Another problem is that the community needs to learn how to run tourist sites. The local Marind community still runs most of the eco-marine tourism sites along the coast of Merauke. It is especially true in the Semangga and Naukenjerai areas. Although there is training from the local Tourism Office, not all levels of society can access this training (Ino, 2024). In the spirit of preserving local wisdom, the Marind community faces challenges in managing tourist attractions. Skills needed to welcome and serve tourists, such as foreign language proficiency, knowledge of tourism management, and marketing, still need to be improved. Effective communication with tourists necessitates a profound comprehension of their cultural values. All of this is a challenge for those who used to live in harmony with nature but now have to adapt to the demands of the modern tourism industry.

A further challenge in developing eco-marine tourism is the potential for conflict with other economic activities, such as sand mining. Some people depend on sand mining for their income, but eco-marine tourism tries to protect the environment, which conflicts with these activities. Sand mining still leaves enormous holes in the natural scenery, which is a sight that tourists who want to enjoy the natural beauty of eco-marine tourism find interesting. The need for local communities to make money and to protect nature for a healthy future are at odds with each other. People in the neighborhood have been making quick cash for a long time by selling sand to collectors.

It makes it difficult for them to shift to more sustainable livelihoods, such as eco-marine tourism, which requires time and investment in skills development and infrastructure (Gustiarini et al., 2023). This challenge requires a holistic approach involving education, economic empowerment, and strict law enforcement to ensure local communities benefit from eco-marine tourism without compromising environmental sustainability (Lakshmi et al., 2024).

Table 4 Respondents' initiative in reforestation in coastal areas of their desire

Village name		Very often	Often	Rare	Never	Total
Ndalir	Total	0	0	1	9	10
	Percentage (%)	0	0	10	90	100
Tomer	Total	0	0	2	8	10
	Percentage (%)	0	0	20	80	100
Urumb	Total	0	1	7	7	15
	Percentage (%)	0	6	47	47	100
Wendu	Total	0	1	4	10	15
	Percentage (%)	0	6	27	67	100
Total		0	2	14	34	50
Percentage (%)		0	0	28	68	100

Despite numerous challenges, the Merauke coast presents various opportunities for developing eco-marine tourism. The natural beauty, unspoiled biodiversity, and unique local wisdom of the Marind people present significant opportunities for development into sustainable tourist attractions. This year, the Merauke Regency Tourism Office has designated the Serapu area in Anasai village as a focal point for government efforts in developing tourist destinations. In 2024, the authorities will designate the coastal area of Semangga district, stretching from Serapu to Anasai villages, as a tourist village (Keo, 2024). It is an essential step toward maximizing the benefits of eco-marine tourism in the area. With help from the government and the community, the Serapu to Anasai area aims to become an excellent example of eco-marine tourism growth based on local knowledge. This project aims to help local economies while protecting culture and natural resources for future generations.

4.4 Artificial Intelligence and Machine Learning for Sustainable Eco-marine Tourism: Potential, Challenges, and Limitations

Artificial intelligence (AI) and machine learning (ML) are becoming more popular as disruptive technologies with enormous potential to revolutionize several sectors, including marine tourism (Ku and Chen, 2024). It can be essential to use AI and ML to grow coastal tourism, especially *eco-marine* tourism in South Papua, which has much coastal wildlife and local knowledge. AI and ML can help *eco-marine tourism* bring in more money and support long-term management by monitoring ecosystem health, finding threats, and giving decision-makers correct data. AI and machine learning can also help guests and locals care more about the environment by giving them fun and interactive information. So, AI and ML can help reduce harmful effects on South Papua's shoreline environment, like illegal sand mining, coastal pollution, habitat loss, and overfishing. It will help ensure that natural resources will be around for future generations.

4.4.1 Potential of AI and ML in Eco-marine Tourism

Eco-marine tourism is a growing idea that combines the natural beauty of beaches, marine, and coastal areas with ideas about being environmentally friendly. However, it is getting harder and harder to balance the needs of tourists with the need to protect the sea environment. The tools of artificial intelligence (AI) and machine learning (ML) show promise for new and creative ways to solve this problem (Bulchand-Gidumal, 2020) (Fig. 13).

First, AI and ML enable innovative approaches to environmental monitoring. Marine environments are the foundation for eco-marine tourism (Supriyanto, 2022). AI and machine learning can analyze data from various sources, including satellite

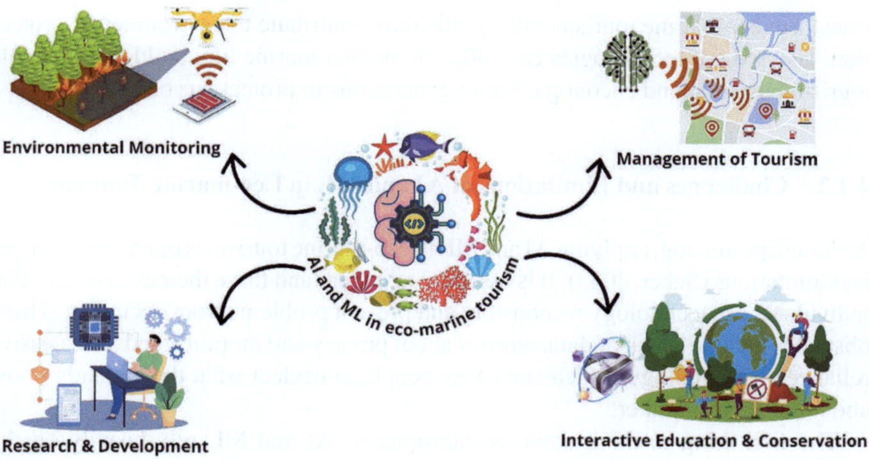

Fig. 13 Application of artificial intelligence (AI) and machine learning (ML) in eco-marine tourism management

photos, underwater sensors, and drones, to monitor the condition of coral reefs, coastlines, and mangrove vegetation, measure water quality, and detect illegal sand excavation activities. An AI-powered early warning system can notify the authorities during illegal sand excavation, thus preserving the coastal nature (Lamsal and Kumar, 2020).

Second, AI and ML can assist in optimal tourism management by predicting the number of tourist visits based on historical data, social media trends, and other external factors (Pabilando et al., 2022). This information allows managers to optimize the capacity of each tourist site, avoiding over-tourism that can damage the marine ecosystem. AI can also be used to plan tourist routes that minimize disturbance to marine habitats by avoiding sensitive fish spawning areas or coral reefs.

Third, AI can be used to develop interactive education and conservation, such as interactive applications and *virtual reality* (VR) platforms that educate tourists about the importance of marine conservation (Koh et al., 2023). Through such apps, tourists can learn about marine biodiversity, the impacts of pollution, and ways to contribute to conservation efforts. The immersive experience through VR allows them to virtually "dive," interact with marine life, and experience underwater beauty firsthand.

Fourth, ML can help in conducting more in-depth scientific research. ML can assist scientists in analyzing complex research data, such as genetic, acoustic, and oceanographic data, to understand marine ecosystems in greater depth (Xu et al., 2021). Researchers utilize ML to identify migration patterns of marine life, predict the spread of invasive species and beach sand excavation patterns, and develop predictive models for the conservation of both natural resources and species. AI and ML transform eco-marine tourism and significantly contribute to environmental protection. Using these technologies carefully can protect marine life, build a sustainable tourism business, and encourage future generations to protect our oceans.

4.4.2 Challenges and Limitations of AI and ML in Eco-marine Tourism

Although promising, applying AI and ML in eco-marine tourism is open to challenges and limitations (Seker, 2023). It is essential to understand these themes to ensure that individuals use technology responsibly and prevent problems from occurring. These obstacles include a lack of data, worries about privacy and morality, and an excessive reliance on technology, which can cause people to neglect what they already know about the subject matter.

The first thing to think about is data quality. AI and ML rely heavily on the quality of the data being used (Nivedhaa, 2024). Poor decisions and misleading analysis can happen when data is missing, wrong, or biased. Getting representative and high-quality statistics can take much work in the complex and changing marine environment. For instance, damage to underwater sensors can cause wrong data, and weather or atmospheric conditions can change the quality of satellite images and lead to wrong conclusions.

Second, problems related to data privacy to keep private data safe. AI and ML are used in tourism to get unique information from travelers, like where they are, what they like, and where they have been on vacation. It is essential to keep this kind of information safe and protect people's privacy from misuse or illegal access (Devineni, 2024). Examples of problems include data leakage and misuse of personal data for commercial purposes or profiling without travelers' consent. Third, issues related to ethical considerations in order to use technology responsibly. Considering ethical issues such as fairness, transparency, and responsibility is essential when utilizing AI and ML. Individuals should exercise caution in utilizing these tools to prevent any occurrence of unfair or discriminatory treatment. Algorithm bias presents challenges, as AI programs trained on biased data make choices or suggestions that unfairly impact certain groups (Nivedhaa, 2024). In addition, AI decision-making is sometimes complex for humans to understand, potentially raising questions about accountability (Pandey et al., 2023).

Fourth, the reliance on technology can ignore the human factor. Too much reliance on AI and ML can reduce human interaction and ignore local knowledge or wisdom in managing eco-marine tourism (Bulchand-Gidumal, 2020). Putting together local knowledge and the experiences of local groups with technology is a huge chance to make progress. For instance, fishing estimates based only on sensor data may not consider what local fishermen know about how fish move through the area. In addition, automated tourist ticket booking applications can reduce opportunities for interaction between tourists and local communities. These issues and limits show that AI and ML are not the only perfect ways to fix all issues in eco-marine tourism. Developing and utilizing these technologies judiciously is imperative, considering social, ethical, and environmental considerations (Pabilando et al., 2022). To make an eco-marine tourism business that lasts and benefits everyone, technology, people, and local knowledge must all work together.

5 Conclusions and Recommendations

This study shows that using sand for development along the coast of Merauke, especially in the Semangga and Naukenjerai Districts, has had nasty effects on the environment and people's lives. Some apparent effects of this activity are damage to coastal environments, wear and tear on the coast, groundwater pollution, and infrastructure damage. Exploitation of sand is ongoing because of poverty, lack of information about the environment, high demand for sand, geographical factors, and weak institutional functions.

Because of these problems, eco-marine tourism built on local knowledge is starting to look like a possible answer. This method gives coastal villages in Merauke a long-term way to make a living. It urges protecting the nature and the culture of the Marind tribe, which lives in the coastal area and is the largest. The ecological knowledge, traditional practices, and artistic expressions of the Marind people can serve as a foundation for an ethical and sustainable marine tourism initiative. The Kali Maro Festival, for instance, does an excellent job of combining native ceremonies with tourist activities. It makes the festival stand out and helps the community's economy. However, there are problems with putting eco-marine tourism into action, such as needing more facilities, people needing to care about the environment, and people not knowing how to run tourist sites. There may also be problems with other economic activities. Machine learning (ML) and artificial intelligence (AI) can help us overcome these problems and make eco-marine tourism even more successful in the long term. People from different groups must work together to solve these problems and make eco-marine tourism in Merauke a healthy industry. Religious leaders, communities, the government, customary institutions, and non-governmental organizations comprise these groups.

This study presents several suggestions. Initially, enhancing institutions and regulations: The government should facilitate collaboration and coordination among essential institutions. Local authorities must review and update regulations concerning tourism and the management of natural resources. There should also be stricter execution of the law against illegal sand mining. Second, it is essential to develop critical infrastructure like roads, transportation, lodging, and other supporting facilities to make Merauke easier to get to and improve the quality of the tourist experience. Third, AI and ML technology facilitates environmental monitoring, enhances tourism management, promotes conservation education, and enables the creation of an integrated database that encompasses sand exploitation data, environmental data, socio-economic data, and tourism data to bolster research and program development.

Fourth, empower local communities by providing training and mentoring in eco-marine tourism management, which encompasses language skills, service, marketing, financial management, and the use of AI. Fifth, developing tourism goods based on local knowledge: To make unique and appealing tourism goods for businesses and tourists, the Marind tribe's local knowledge, like traditional ceremonies, ecological knowledge, and handicrafts, needs to be mixed and expanded. Sixth, collaboration and partnership: The government, local communities, businesses, and institutions both at home and abroad need to build mutually beneficial partnerships to accelerate the transfer of knowledge and technology and apply research results in development programs for coastal management and sustainable eco-marine tourism development in Merauke. Following these suggestions should help eco-marine tourism grow in Merauke, which would also help the local economy, protect the environment and culture, and set a good model for other places that want to grow eco-marine tourism.

6 Future Research

Although there have been studies examining the application of artificial intelligence (AI) and machine learning (ML) in the study of sand exploitation and eco-marine tourism (Kumaragurupharanet al., 2022; Roy, 2024), there is still room for further exploration. Some recommendations for further research topics that are considered crucial include the impact of sand exploitation on coastal damage, the economic and social impacts of eco-marine tourism, and the development of AI and ML for adaptive management of eco-marine tourism. In-depth research is needed to quantitatively analyze the impact of sand exploitation on coastal degradation, particularly the rate of coastal erosion and damage to coastal vegetation such as mangroves. This research can utilize AI and ML techniques, such as multi-temporal satellite image analysis with *deep learning,* to detect shoreline and coastal vegetation extent changes over time (McAllister et al., 2022). Moreover, researchers can conduct predictive modeling of coastal erosion using *machine learning* to forecast erosion rates influenced by the intensity of sand exploitation and various environmental factors. Research can also help map erosion-prone areas with GIS spatial analysis to identify no-sand-exploitation zones to protect vulnerable areas (Seale et al., 2022).

Further research on the economic and social impacts of eco-marine tourism is also needed to comprehensively assess the economic and social impacts of eco-marine tourism programs for local communities. This research can take the form of surveys and econometric analysis to assess the contribution of eco-marine tourism to community income, job creation, and local economic growth or social network analysis with AI to understand better the structure and dynamics of stakeholder relationships in the development of eco-marine tourism (Monoarfa et al., 2019).

Meanwhile, AI and ML development research for adaptive eco-marine tourism management is needed to improve AI and ML capabilities in supporting adaptive and sustainable eco-marine tourism management in the South Papua region. Some potential development directions include the development of an AI-based tourist destination recommendation system that considers capacity, tourist preferences, and environmental conditions, the use of *computer vision* to automatically monitor marine biodiversity and detect illegal activities in conservation areas, and the development of predictive models to optimize marketing and promotion strategies for eco-marine tourism (Pabilando et al., 2022). By carrying out this comprehensive and innovative follow-up research, the utilization of AI and ML can significantly contribute to preserving the coastal environment and developing sustainable eco-marine tourism on the coast of South Papua.

References

Alkhudri AT, Heriyanto N, Zulfikar F, Sukiman, Nugroho TT (2021) From eco marine mart to ecotourism: development of local wisdom and marine creative economy as tourism excellence. In: Proceedings of the 1st annual international conference on natural and social science education (ICNSSE 2020), vol 547, pp 307–314. Atlantis Press. https://doi.org/10.2991/assehr.k.210430.046

Andriani S (2024) Swept away by rob flood, two houses lost trace. Retrieved 20 Aug 2024, from https://infopublik.id/kategori/nusantara/836179/tersapu-banjir-rob-dua-rumah-warga-hilang-jejak

Asry (2019) BWS papua plants mangroves on urumb beach to prevent abrasion. Retrieved 21 Aug 2024, from https://portal.merauke.go.id/news/4848/bws-papua-tanam-mangrove-di-pantai-urumb-guna-cegah-abrasi.html

Basheer Ahammed KK, Karang IWGA, Nuarsa IW, Indrawan IGS (2024) Mapping of the spatio-spectral dynamics of mangrove chlorophyll concentrations via Sentinel-2 Satellite Imagery. Forum Geogr 38(2):244–256

Batbual A (2018) Sand mining craze in merauke, residents fear disaster. Retrieved 21 Aug 2024, from https://www.mongabay.co.id/2018/12/12/tambang-pasir-menggila-di-merauke-warga-khawatir-bencana/

Batbual A (2019) Sand mining rampant, mangroves gone, Merauke Beach Abrasion alarming. Retrieved 20 Aug 2024, from https://www.mongabay.co.id/2014/04/29/penambangan-pasir-marak-mangrove-hilang-abrasi-pantai-merauke-mengkhawatirkan/

Betaubun BY, Waas RF, Pelu HD (2023) Juridical analysis of regional apparatus governance of South Papua Province through law number 14 of 2022. J Restor Justice, 97–113. https://doi.org/10.35724/jrj.v7i1.5266

Bulchand-Gidumal J (2020) Impact of artificial intelligence in travel, tourism, and hospitality. In: Handbook of e-tourism. Springer, pp 1–20. https://doi.org/10.1007/978-3-030-05324-6_110-1

Chambliss DF, Schutt RK (2013) Making sense of the social world: methods of investigation. SAGE Publications, London

Chandra NA, Hudaefah, Hasanah SN (2023) Dependency theory creates injustice for countries like Indonesia. In: Nugroho KS (ed) The pros and cons of development issues. CV. Agung Ilmu, Bandung, pp 20–25

Chen L, Chen Z, Zhang Y, Liu Y, Osman AI, Farghali M, Hua J, Ihara I, Rooney DW, Yap PS, Alfatesh A (2023) Artificial intelligence-based solutions for climate change: a review. Environ Chem Lett 21(5). https://doi.org/10.1007/s10311-023-01617-y

Creswell JW, Creswell JD (2017) Research design: qualitative, quantitative, and mixed methods approaches. SAGE Publications Inc., London

Crockett D, Stow DA (2021) Ecotourism, marine science and the environment. King Abdulaziz University, Jeddah

Dang KB, Dang VB, Ngo VL, Vu KC, Nguyen H, Nguyen DA, Nguyen TDL, Pham TPN, Giang TL, Nguyen HD, Do TH (2022) Application of deep learning models to detect coastlines and shorelines. J Environ Manage 320:115732. https://doi.org/10.1016/j.jenvman.2022.115732
Devineni SK (2024) AI in data privacy and security. Int J Artif Intell Mach Learn (IJAIML) 3(1):35–49. https://doi.org/10.17605/OSF.IO/WCN8A
Djamali RA, Betaubun P, Hermanuadi D, Syaban RA (2016) Cognitive mapping of the causes and impacts of sand explotation along the coastline in Merauke Regency. In: National seminar on research and community service results. Jember State Polytechnic, Jember, pp 114–119
Fanumbi A (2024) The role of the Catholic church for the preservation of the Merauke coastal environment. (Pranyoto YH, Interviewer) Merauke, Papua Selatan
Fatima S, Abbas S, Rebi A, Ying Z (2024) Sustainable forestry and environmental impacts: assessing the economic, environmental, and social benefits of adopting sustainable agricultural practices. Ecol Front 44:1119–1127. https://doi.org/10.1016/j.ecofro.2024.05.009
Gamu FA, Pranyoto YH (2023) The concept of religious moderation in the context of local wisdom totemism of the Marind-Anim community in Yaba Maru village, Tanah Miring district, Merauke Regency. J Pastor Issues XI(2):47–69. https://doi.org/10.60011/jumpa.v11i2.136
Gehrels R, Garrett E (2021) Rising sea levels as an indicator of global change. In: Letcher TM climate change (Third Edition). Elsevier, pp 205–217. https://doi.org/10.1016/B978-0-12-821575-3.00011-6
Gesami BK, Nunoo J (2024) Artificial intelligence in marine ecosystem management: addressing climate threats to Kenya's blue economy. Front Mar Sci 11:1–11. https://doi.org/10.3389/fmars.2024.1404104
Government MR (2011) Merauke regent regulation number 14 year 2011 on the management of non-metallic mineral and rock mining businesses. In: Regional regulation of Merauke regency. Merauke Regency Government, Merauke
Gustiarini A, Tampubolon N, Saranga N, Mustiah D, Kurniawan (2023) Sustainable tourism eco resort criteria for the South Waigeo waters of the Raja Ampat Islands. Indopacific J Aquat Resour VII(3):273–282. https://doi.org/10.46252/jsai-fpik-unipa
Hallatu TG, Wisadirana D, Mu'adi S, Chawa A (2021) Illegal sand mining and sar local wisdom: a case study in Merauke. Int J Soc Sci Humanities V(1):45–56. https://doi.org/10.29332/ijssh.v5n1.827
Hallatu TG, Palittin ID (2023) Muhu holy forest, protected area of Mahuze clan as a natural and cultural conservation. Tech Soc Sci J XLIX:428–433. https://doi.org/10.47577/tssj.v49i1.9837
Hapsara CM (2024) Merauke regency in figures 2024. BPS Merauke Regency, Merauke. Retrieved Aug 2024, from https://meraukekab.bps.go.id/id
Hendrawati Y (2011) Environmental ethics in the Dema-Totem-Klen Myth of The Marind-Anim Tribe of Papua: its relevance for environmental preservation in Indonesia. Faculty of Philosophy University of Gadjah Mada Yogyakarta, Yogyakarta
Ino (2024) These four components participate in tourism digital marketing training at Merauke Regency. Retrieved 22 Aug 2024, from https://portal.merauke.go.id/news/7300/empat-unsur-ini-ikuti-pelatihan-pemasaran-digital-pariwisata-di-kabupaten-merauke.html
Jannah F (2022) Ethnoscience studies based on local wisdom. Mualimin Publisher, Jayapura
Jeminta FS, Mbaraka R (2005) Merauke in history: from JH Kroesen-Drs. Johanes Gluba Gebze, Toward Agropolis. Sinar Terang, Merauke
Kalilago R (1992) Wasur national park: the life of the Marind-Anim people. Faculty of Social and Political Sciences, University of Cenderawasih, Jayapura
Kaok M, Inggrid P, Risamassu M (2022) Analysis of the entrepreneurial spirit of the marind community on the Coast of Lampu Satu Merauke. J Econ Soc Sci XIII(1):59–68. https://doi.org/10.35724/jies.v13i1.4378
Kapatamoyo M (2024) Artificial intelligence in natural resources management. In: 24th Biennial conference of the international telecommunications society (ITS): "new bottles for new wine: digital transformation demands new policies and strategies". International Telecommunications Society (ITS), Seoul, pp 1–12. Retrieved from https://hdl.handle.net/10419/302527

Kardi C, Wiasta IW (2020) Eco-marine-tourism development towards the renewal of coastal communities in Gerokgak District, Bali. E-J Tour VII(1):73–87

Katikiro R, Mahajan SL, Medard M (2024) The establishment, persistence, and scaling of marine community-based natural resource management in four coastal districts in Tanzania. Mar Policy 169:106347. https://doi.org/10.1016/j.marpol.2024.106341

Keo GS (2024) Merauke's serapu to Anasai area is to be launched as a tourism village. Retrieved 23 Aug 2024, from https://teropongnews.com/2024/07/kawasan-serapu-sampai-anasai-merauke-akan-dilaunching-jadi-kampung-wisata/

Kobun F (2022) Students at SMAN Plus urumb plant coconut and mangrove saplings. Retrieved 21 Aug 2024, from https://suryapapua.com/siswa-siswi-di-sman-plus-urumb-tanam-anakan-pohon-kelapa-dan-mangrove/

Koh LY, Wu M, Wang X, Yuen KF (2023) Willingness to participate in virtual reality technologies: public adoption and policy perspectives for marine conservation. J Environ Manage 334:117480. https://doi.org/10.1016/j.jenvman.2023.117480

Kompas.com. (2011) Merauke imports sand from outside. Retrieved 20 Aug 2024, from https://nasional.kompas.com/read/2011/06/13/03055269/merauke-datangkan-pasir-dari-luar?newnavbar=1

Ku EC, Chen C-D (2024) Artificial intelligence innovation of tourism businesses: from satisfied tourists to continued service usage intention. Int J Inf Manage 76:102757. https://doi.org/10.1016/j.ijinfomgt.2024.102757

Kumar L, Afzal MS, Afzal MM (2020) Mapping shoreline change using machine learning: a case study from the eastern Indian coast. Acta Geophys 17:1127–1143. https://doi.org/10.1007/s11600-020-00454-9

Kumaragurupharan MD, Ahmed IJGM, CV (2022) Controlling of illegal sand mining using machine learning and Ocr. Int J Innov Res Elect Electron Instrum Control Eng 10(6):37. https://doi.org/10.17148/IJIREEICE.2022.10607

Lakshmi VV, Srivalli B, Srivinas T, Ugandhar T (2024) Exploring eco-tourism perspectives for sustainable tourism in Telangana. Int J Humanities Soc Sci XI(3):8–13. https://doi.org/10.14445/23942703/IJHSS-V11I3P102

Lamsal R, Kumar V (2020) Artificial intelligence and early warning systems. In: AI and robotics in disaster studies. Springer Nature Singapore, Singapore. https://doi.org/10.1007/978-981-15-4291-6_2

Lekatompessy RL, Maturbongs EE (2021) Factors in efforts to overcome abrasion on the coast in the Merauke regency area. Dialogue: J Public Adm III(1):1–13

Letsoin Y (2015) Cultural and tourism (Merauke Papua Indonesia). Retrieved 20 Aug 2024, from https://yennyletsoin.blogspot.com/

Louati A, Louati H, Alharbi M, Kariri E, Khawaji T, Almubaddil Y, Aldwsary S (2024) Machine learning and artificial intelligence for a sustainable tourism: a case study on Saudi Arabia. Int Soc Inf Stud (IS4SI), 15(9):516. https://doi.org/10.3390/info15090516

Love JW, Gill J, Newhard JJ (2008) Saltwater intrusion impacts fish diversity and distribution in the Blackwater river drainage (Chesapeake Bay watershed). Wetlands 28:967–974. https://doi.org/10.1672/07-238.1

McAllister E, Payo A, Novellino A, Dolphin T, Medina-Lopez E (2022) Multispectral satellite imagery and machine learning for the extraction of shoreline indicators. Coast Eng 174:104102. https://doi.org/10.1016/j.coastaleng.2022.104102

Merauke DE (2023) Merauke regency environmental status report (SLH) year 2023. Merauke Regency Government, Merauke

Monoarfa SF, Hamzah SN, Yapanto LM (2019) Economic impact analysis of marine tourism to community revenue. Sci J Fish Mar 7(3):67–72

Nicolas F, Kisekka I, Daccache A, Kamai T, Ben-Gal A, Ochoa J, Ogunmokun F (2023) Assessing salinity impacts on crop yield and economic returns in the Central Valley. Agric Water Manage 287(9). https://doi.org/10.1016/J.AGWAT.2023.108463

Nivedhaa (2024) A comprehensive review of AI's dependence on data. Int J Artif Intell Data Sci (IJADS) 1(1):1–11

Oja H, Fitriani SG, Laode IC, Maturan AY (2018) The role of indigenous peoples (LMA) in the control of forest and land fires in Merauke. In: 1st International conference on global issue for infrastructure, environment & socio-economic development. IOP Publishing, Makassar, pp 1–6. https://doi.org/10.1088/1755-1315/235/1/012061

Pabilando RS, Tapaya TL, Carreon JS, Gupta AD, Naseem S (2022). Artificial intelligence in tourism management. Edurights Global, Tamil Nadu

Pandey DK, Hunjra AI, Bhaskar R, Al-Faryan MA (2023) Artificial intelligence, machine learning and big data in natural resources management: a comprehensive bibliometric review of literature spanning 1975–2022. Resour Policy 86(A). https://doi.org/10.1016/j.resourpol.2023.104250

Papua PD (2010) History of Papua past and present. Papua Provincial Government, Jayapura

Pasalli DA (2022) Physical characteristics and durability of concrete with sand extracted from Merauke Regency, Papua. Faculty of Engineering, Diponegoro University, Semarang. Retrieved from https://eprints2.undip.ac.id/id/eprint/12464/

Prapti L, Suryawardana E, Triyani D (2015) Analysis of the impact of road infrastructure development on the growth of people's economic enterprises in Semarang City. J Socio-Cult Dyn XVII(2):82–103

Rawung, Willem H (2005) Tracing the missionary trail in South Papua. Archdiocese of Merauke, Merauke

Reumi F (2018) Perception of rights to control the land over natural resources in indigenous people of Papua, Indonesia. J Law Policy Glob LXXIX(2):121–130

Ririhena SW, Nahumury M, Simatupang DO (2017) Strategy for biodiversity conservation efforts in Wasur national park of Merauke in Merauke Regency. E3S Web Conf 73(2):1–6. https://doi.org/10.1051/e3sconf/20187304001

Risamasu PI, Gebze EP (2020) Entrepreneurship in the perspective of Marind's young generation. J Econ Soc Sci XI(1):39–47. https://doi.org/10.35724/jies.v11i1.2679

Roy BK (2024) Role of artificial intelligence and technologies in improving ocean health in promoting tourism. In: Artificial intelligence and edge computing for sustainable ocean health. Springer Nature Switzerland. https://doi.org/10.1007/978-3-031-64642-3_2

Sahrul P, Muhammad A (2020) Analysis of rice import dependency in Indonesia. J Darussalam Econ Perspect VI(2):212–222

Sanadi EY, Rumboirusi WC (2024) South Papua Province in figures 2024. BPS South Papua Province, Merauke

Sarang RK, Wea D, Gaol EL, Pudjiarti ES (2024) Marind indigenous ecological justice: laudato Si' and ecological resilience facing industrial expansion in Merauke, South Papua. Int J Sci Soc VI(3):424–437. https://doi.org/10.54783/ijsoc.v6i3.1255

Seale C, Redfern T, Chatfield P, Luo C, Dempsey K (2022) Coastline detection in satellite imagery: a deep learning approach on new benchmark data. Remote Sens Environ 278:113044. https://doi.org/10.1016/j.rse.2022.113044

Seker F (2023) Evolution of machine learning in tourism: a comprehensive review of seminal research. J Artif Intell Data Sci 3(2):54–79

Sergino (2022) Merauke's disappearance from the world map. In: Novita E, (ed) Retrieved 10 Aug 2024, from https://portal.merauke.go.id/news/782/hilangnya-kota-merauke-dari-peta-dunia.html

Shereni NC, Saarinen J (2020) Community perceptions on the benefits and challenges of community-based natural resources management in Zimbabwe. Dev Southern Africa XXXVIII(6):879–895. https://doi.org/10.1080/0376835X.2020.1796599

Siregar H (2019) Interdependence and inequality in the world political economy. Communitarian J II(1):182–200. https://doi.org/10.56985/jc.v2i1.97

Sobari H (2015) Human philosophy of Marind Anim Tribe in Dema Myths. Faculty of Ushuluddin, Syarif Hidayatullah State University, Jakarta. Retrieved 21 Aug 2024

Suharno, Kadir A (2023) Adaptation and local knowledge of the Marind Anim Tribes in the utilization of Paperbark trees (Melaleuca leucadendron) in Merauke, South Papua, Indonesia. Biodiversitas J Biol Divers 24(11). https://doi.org/10.13057/biodiv/d241157

Sunarni, Maturbong MR, Rahmania R (2019) Zoning and structure of mangrove communities on the Coast of Merauke Regency. Natl Mar J XIV(3):165–178. https://doi.org/10.15578/jkn.v14i3.7961

Supriyanto EE (2022) Blue tourism: treating marine ecosystems and increasing the potential of maritime tourism in Indonesia. Indones J Tour 138–148

Thontowi J, Wartini S (2023) Effectuating food sovereignty without violating the rights of indigenous people: a case study of the merauke integrated food and energy estate project. Int Conf Adv Sci Innov (ICASI). KnE Social Sciences, Tripoli, pp 616–624. https://doi.org/10.18502/kss.v8i9.13376

Walilo Y (2021) Analysis of the development framework of Papua Province in 2021. Regional Government of Papua Province, Jayapura

Wasim M, Pandey AC, Basheer Ahammed KK, Dwivedi CS (2024) Two decades of coral bleaching in selected islands of Pacific Ocean: a holistic impact assessment. Regnal Stud Mar Sci 72:103648

Widiastuti MM, Ruata NN, Arifin T (2016) Economic valuation of mangrove ecosystem in the coastal of Merauke district. J Mar Fish Socio-Econ 11:147–159. https://doi.org/10.15578/jsekp.v11i2.3856

Widiastuti MM, Samderubun G, Arifin T (2018) Policy strategy for combating beach sand quarrying at Nasai Beach—Merauke regency. J Soc Policy VIII(1):27–38

Wijaya NW, Pelu HD, Samderubun F (2022a) Socio-legal analysis of the role of customary institutions in resolving land conflicts in customary forests in the Merauke regency area. J Leg Commun VIII(2):282–291. https://doi.org/10.23887/jkh.v8i2.50873

Wijaya N, Pelu HD, Samderubun F (2022b) Socio-legal analysis of the role of customary institutions in resolving land conflicts in customary forests in the Merauke district area. J Leg Commun VIII(2):283–291

Xu Y, Liu X, Cao X, Huang C, Liu E, Qian S et al. (2021) Artificial intelligence: a powerful paradigm for scientific research. Innov 2(4). https://doi.org/10.1016/j.xinn.2021.100179

Yincan Y (2017) Sea level change, sea water intrusion, and coastal land subsidence. In: Marine geo-hazards in China. Elsevier, pp 587–656. https://doi.org/10.1016/B978-0-12-812726-1.00014-0

Yonavilbia E (2019) Lantamal XI Merauke plants 5800 mangrove trees. Retrieved 21 Aug 2024, from https://infopublik.id/kategori/nusantara/378615/lantamal-xi-merauke-tanam-5800-pohon-mangrove?video=

An Integrated Coastal Vulnerability Analysis of Selected Coastal Areas of Kerala Using Bayesian Belief Network and GIS

Girish Gopinath, A. L. Achu, Sanju Eldhose, and N. P. Jessiya

Abstract Coastal vulnerability encompasses the susceptibility of populations and environments to disruptions caused by coastal hazards, including storms, flooding, and shoreline erosion. These hazards pose considerable risks to human life and property. In recent years, the Malabar Coast of India (especially in Kerala state) is become prone to various coastal hazards and affects millions of people. In this context, an integrated study of the coast of Kozhikode, Ernakulam, and Thiruvananthapuram districts of Kerala is conducted to identify the coastal vulnerable areas. Physical factors such as slope, lithology, geomorphology, and land use/land cover, along with socio-economic indicators including household density, population density, literacy rate, and the proportion of marginalized populations, are taken into account. Thereafter, physical vulnerability, socio-economic vulnerability, and integrated coastal vulnerability is modeled using the Bayesian belief network. The results indicate that a significant portion of coastal areas exhibit high levels of both physical and socio-economic vulnerability. Furthermore, the integrated coastal vulnerability assessment reveals that approximately 14% of Kozhikode, 18% of Ernakulam, and 20% of Thiruvananthapuram districts are classified within very high vulnerability zones. The findings of the present study are useful for long-term sustainable development of Kerala coast.

Keywords Coastal vulnerability · Bayesian Belief Network · GIS · Kerala

G. Gopinath (✉) · A. L. Achu · S. Eldhose
Department of Climate Variability and Aquatic Ecosystems, Kerala University of Fisheries and Ocean Studies, Kochi, India
e-mail: gg@kufos.ac.in

N. P. Jessiya
Department of Civil Engineering, National Institute of Technology, Kozhikode, Kerala, India

K. K. Basheer Ahammed et al. (eds.), *Monitoring Coastal Marine Ecosystems Under Climate Change*, https://doi.org/10.1007/978-3-032-09137-6_19

1 Introduction

Coastal regions are especially vulnerable to climate change impacts due to their close proximity to water bodies and the complex interplay between human activities and natural systems (Ummenhofer and Meehl 2017). The extent of this vulnerability is influenced by a coastal system's susceptibility, resilience, and capacity to withstand hazards (Holling et al. 1973). As reported by UNESCO, more than 40% of the global population lives within 100 km of a coastline (UN). As more people relocate to densely populated urban regions close to the coast, this percentage is predicted to rise in the ensuing decades. Coastal resilience is a top concern since these areas are severely threatened by increasing sea levels, stronger storms, and other environmental hazards. Numerous human activities have a substantial impact on coastal habitats, frequently having a negative impact on ecosystems and the communities that depend on them. Along coastlines, rapid development, urbanization, and tourism cause habitat degradation (mangroves, coral reefs, and dunes, etc.), increasing pollution, and modifications to natural coastal processes (Gopinath 2009). Sea level rise, driven by the melting of ice sheets and the thermal expansion of seawater due to global warming, exacerbates coastal flooding, accelerates shoreline erosion, and promotes saltwater intrusion into freshwater systems. These processes pose significant threats to infrastructure, displace populations, and degrade natural ecosystems. Since 1993, global sea levels have increased by more than 10 cm as a consequence of climate change. According to the World Economic Forum (2024), by the year 2100, up to 410 million people may be at risk from coastal flooding due to continued sea level rise.

India's coastline extends approximately 7500 km, comprising about 5400 km along the mainland and the remainder distributed among three island groups—Lakshadweep in the southern Arabian Sea and the Andaman and Nicobar Islands in the eastern Bay of Bengal. Nearly 250 million people reside within 50 km of the Indian coastline (Integrated Coastal Zone Management Project ICZMP 2010). Consequently, the ecological services provided by India's marine and coastal ecosystems are crucial to the nation's economic development. Given that a substantial proportion of natural ecosystems, industrial hubs, and ports are located along the coast, India's overall economic growth is significantly dependent on its coastal regions. Due to a combination of natural hazards, human activities, and climate change, Indian coastal areas face significant vulnerabilities (Naga et al. 2022). Erosion affects over 45% of Indian coastline. The worst affected states include West Bengal, Odisha, Kerala, and Tamil Nadu, which have lost significant territory and forced residents to relocate. Human activities like sand mining, port expansion, and the building of coastal structures that obstruct natural sediment flows, frequently make this erosion severe (Dinesh Kumar et al. 2007; Girish and Seralathan 2005). Low-lying coastal areas in India are seriously threatened by rising sea levels, especially in the Sundarbans, which are home to largest mangrove forest and millions of people in the world. The degradation of critical coastal habitats weakens natural protective barriers, increasing the vulnerability of coastal populations. Major coastal cities such as Mumbai, Chennai, and Kolkata are facing heightened flood risks due to rising sea levels, intensified storm surges, and extreme rainfall events (Chakrabortty et al. 2025). Additionally, the Bay of Bengal has witnessed a rise in both the frequency and intensity of cyclones, with severe events like Cyclone Amphan (2020) causing extensive damage to homes, agricultural lands, and infrastructure along the coast (Basheer Ahammed et al. 2023; Medha et al. 2023; Chettiyam et al. 2023). Coastal communities, especially small-scale fishermen, are especially susceptible because of diminishing fish stocks, habitat degradation, and growing industrial fisheries competition.

Kerala is situated along the southwest coast of India and has a long seashore of 600 km. The coast is characterized by asymmetrical topography dominated by undulating subdued hills and steep slopes and altitude ranges from below mean sea level (MSL) to 2695 m above MSL (Sreelekshmi et al. 2021; Kumar et al. 2014). Kerala has the highest proportion of residents who live along the coast. Kerala faces significant challenges related to coastal erosion, the degradation of coastal vegetation, the management of coastal resources, and the pressures of intense commercial activities, all of which require urgent attention. Coastal erosion is particularly severe, affecting approximately 63% of the state's 590 km-long shoreline. Key natural drivers of erosion in Kerala include intense rainfall, unstable sandy beaches, the disruption of offshore mud deposits, and the discharge of sediment-pore waters. Districts such as Alappuzha, Ernakulam, and Kozhikode are especially vulnerable, experiencing notable land loss that threatens coastal communities and infrastructure. Furthermore, Kerala has witnessed a dramatic decline in its mangrove cover, with an estimated 95% loss over the past three decades (Shaji 2014; Sreelekshmi et al. 2021).

Despite challenges such as overexploitation and unsustainable resource use, researchers continue to assess the rates of coastal change and associated vulnerabilities. Investigations into coastal erosion typically employ a combination of field surveys, remote sensing technologies, modeling approaches, and laboratory analyses. Interest in coastal process research, particularly in the southernmost regions of India, intensified following the 2004 tsunami, with the application of remote sensing and GIS techniques becoming more widespread. Studies often incorporate physical parameters such as geomorphology, lithology, land use/land cover, and shoreline change (Kumar and Kunte 2012). The Digital Shoreline Analysis System (DSAS), a free extension tool provided by the USGS, operates within the ArcGIS platform (Pereira et al. 2020). Socio-economic vulnerability at the micro-administrative unit level for three districts was assessed (Karuppusamy et al. 2021). Key socio-economic factors such as population density, household density, literacy rate (Ghosh and Mistri 2021), child population ratio, marginalized groups, and primary workers were used to evaluate coastal vulnerability (Karuppusamy et al. 2021). The study employed the Bayesian Belief Network (BBN), a probabilistic machine learning technique, to assess coastal vulnerability. The Bayesian network is a directed acyclic graph (DAG), where nodes represent domain variables (e.g., attributes of a dataset), and arcs between nodes denote probabilistic dependencies, quantified through conditional probability distributions (CP tables) for each node (Cheng and Greiner 2001). This method integrates both qualitative and quantitative data into a comprehensive probabilistic model, while GIS is used to manage spatial data for input, storage, analysis, and mapping. The fusion of Bayesian Network and GIS via a cell-by-cell comparison technique offers a novel approach to conducting probabilistic spatial analysis (Sahin et al. 2019). In, BBN, each node defines physical and socio-economic parameters and dependencies of these are illustrated using arcs. The data, along with the BBN developed in "Netica," were analyzed and visualized using GBay, an online platform that allows users to link their BNs with spatial data. This platform enables iterative network runs to integrate dynamics and feedback, as well as the addition of geo-processing calculations to capture spatial interactions (Stritih et al. 2020). In this present study, coastal vulnerability of three major coastal centers of Kerala, southern of India viz. Kozhikode, Ernakulam, and Thiruvananthapuram have been estimated and analyzed. This zone, which includes numerous industries, mining sites, and other development projects, is one of the primary socio-economic zones of the country. A significant population lives in the coastal districts.

2 Study Area

A significant diversity in coastal geomorphology, exposures, and ecosystems leads the coastal region of Kerala experiences a variety of environmental conditions. Kozhikode is geographically situated between latitudes 11°08′16.39″ N and 11°42′2.34″ N, and longitudes 75°31′41.22″ E and 76°09′1.22″ E. The district features an 80 km-long coastline along the Arabian Sea (Fig. 1a), extending from Mahe in the north, a coastal town near the northern border of the district, to the Kadalundi River in the south.

Extreme waves, coastal flooding, and coastal erosion are the major issue that is facing by the study area. Saline water intrusion is another threat for the coastal villages such as Payyoli and Thikkodi. 44 coastal villages in Kozhikode that are less than 5 km from the shoreline were taken into consideration for the study (Fig. 1). The Ernakulam district is situated in the central region of Kerala, India, at a Northern Latitude10°00′ N to 10°6′53.38″ and Eastern Latitude 76° 23′ 24.47″ to 76°33′ E (Fig. 1). It is having the district has a 46 km-long coastline. A major portion of the study area lies on low elevation. Sea level rise is the major threat that is faced by the study area. Shoreline retreat due to rising sea levels presents a significant threat to coastal communities. Villages such as Paravoor, Elooor, and Edakochi are experiencing saltwater intrusion. The study focused on 28 coastal villages located within 5 km of the shoreline. Thiruvananthapuram district, situated on the southwest coast of India, lies between latitudes 8°17′27″ N and 8°46′57″ N, and longitudes 76°40′25″ E and 77°05′56″ E (Fig. 1). The district has a 78 km-long coastline, with coastal erosion being one of the primary threats in the study area. Other than coastal erosion coastal flooding and extreme waves are the major threats. 46 coastal villages under 5 km from the shoreline were considered for the study (Fig. 1).

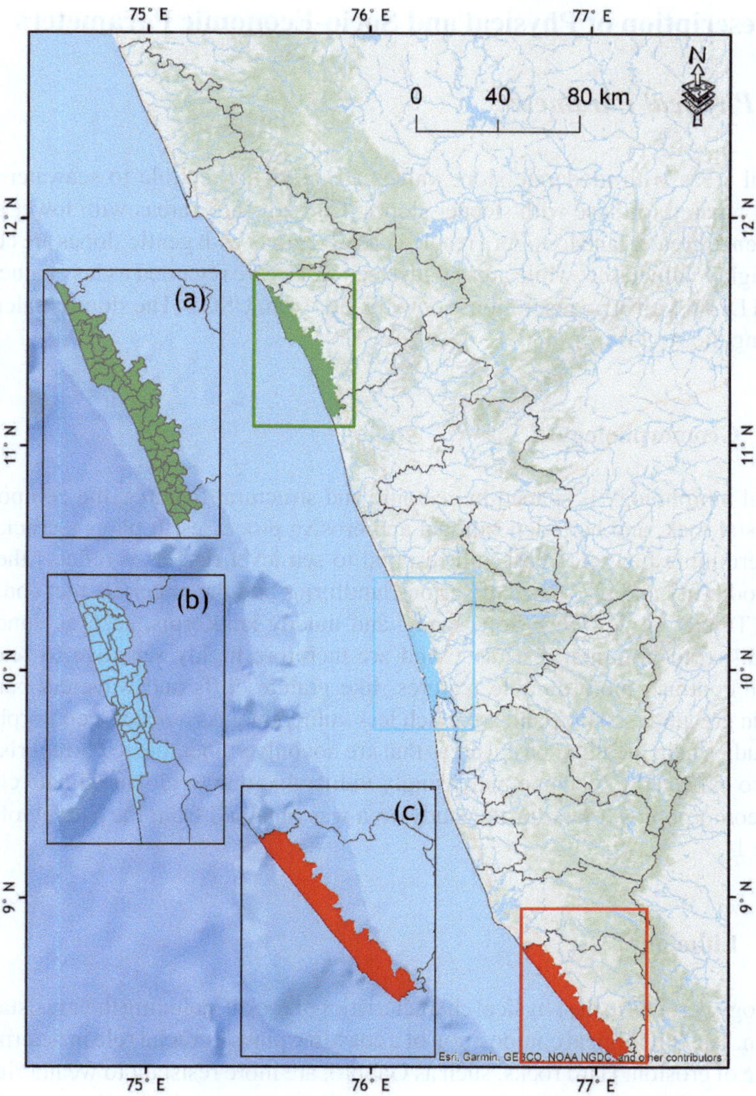

Fig. 1 Location map of the study area

3 Description of Physical and Socio-Economic Parameters

3.1 Physical Parameter

Coastal areas with moderate slope values are more susceptible to seawater intrusion compared to those with steeper slopes. Consequently, areas with lower slopes experience greater land loss. As a result, coastal regions with gentle slopes are considered highly vulnerable, while areas with steep slopes are regarded as less vulnerable. SRTM DEM with 30 m resolution is downloaded from USGS. The slope is calculated by using the ArcGIS.

3.1.1 Geomorphology

Coastal morphology is shaped by tectonic and structural features, the composition of coastal rock, and both depositional and erosive processes. It plays a crucial role in determining how the coast will respond to sea level rise, as it reflects the relative erodibility and resistance of various landforms and the materials that constitute them (Thieler et al. 1999). Soft, sandy, and muddy landforms, such as dunes and mudflats, offer minimal resistance and are therefore highly sensitive to sea level rise. In contrast, more durable features, like granite cliffs and wave-cut benches, provide greater resistance and are much less vulnerable. As a result, geomorphological studies help identify coastal areas that are not only vulnerable to current risks but are also expected to become increasingly vulnerable due to global climate change. The geomorphology data for the study area was obtained from the Survey of India (Table 1).

3.1.2 Lithology

Lithology pertains to the physical characteristics of rocks, including their resistance to erosion. The lithological composition of a coastline plays a crucial role in determining the rate of erosion. Hard rocks, such as Gabbro, are more resistant to weathering and erosion, leading to slower changes along coastlines composed of such materials, like granite (e.g., Land's End). In contrast, softer rocks, such as limestone, are more prone to weathering and erosion, resulting in faster coastal changes, as seen in coastlines made of chalk (e.g., Dorset) (Table 2).

3.1.3 Land Use/ Land Cover (LU/LC)

Understanding land use and land cover is crucial for effective coastal management. Land use refers to the ways in which land is developed and utilized, whereas land

Table 1 Geomorphological features of the study area

District	Geomorphology	Total area (%)
Kozhikode	Lower plateau (Lateritic)	50.13
	Coastal plain	30.58
	Pedi plain Weathered/buried	13.08
	Piedmont zone	0.15
	Flood plain	0.15
	Rock exposure	0.03
	Marshy	0.59
	Water body	5.28
Ernakulam	Coastal plain	72.07
	Flood plain	0.2
	Lower plateau (Lateritic)	0.23
	Marshy	0.38
	Water body	27.12
Thiruvananthapuram	Flood plain	0.01
	Lower plateau (Lateritic)	69.2
	Pedi plain Weathered/buried	11.24
	Residual hill	0.31
	Rock exposure	0.01
	Water body	3.55
	Coastal plain	15.69

Table 2 Lithological features

District	Lithology	Total area (%)
Kozhikode	Basic rocks	4.43
	Charnockite group of rocks	30.14
	Migmatite complex	45.52
	Sand and silt	17.79
	Tank/WB/River	2.11
Ernakulam	Sand and silt	75.49
	Tank/WB/River	24.51
Thiruvananthapuram	Charnockite group of rocks	3.16
	Khondalite group of rocks	34.81
	Laterite	1.37
	Migmatite complex	6.56
	Sand and silt	20.58
	Sandstone and clay with lignite intercalations	33.17
	Tank/WB/River	0.36

cover describes the physical characteristics of the land surface, such as forests, vegetation, agricultural areas, and water bodies. Rapid urbanization, industrial expansion, and population growth along coastlines have significantly altered land use and land cover patterns. These changes exacerbate the vulnerability of coastal regions to natural hazards, including flooding, storm surges, and inundation. Moreover, the development of infrastructure, increased waste generation, and urban sprawl disrupt ecological systems and biodiversity. Remote sensing has emerged as the most efficient and reliable method for monitoring and mapping these dynamic land use/land cover changes (Sudha Rani et al. 2015) (Table 3).

Table 3 Land use/ land cover of the study area

District	Land use/ land cover	Total area (%)
Kozhikode	Villages	7.83
	Mixed crop	73.54
	Agricultural	0.24
	Town/cities	1.07
	Crop land	9.07
	Scrub land	1.62
	Degraded	0.13
	Water body	6.28
	Barren rocky/sheet rocks	0.22
Ernakulam	Villages	1.85
	Mixed crop	40.39
	Agricultural	18.19
	Town/cities	12.76
	Crop land	0.88
	Water body	25.93
Thiruvananthapuram	Villages	7.69
	Mixed crop	73.95
	Town/cities	7.37
	Crop land	5.54
	Scrub land	0.12
	Degraded	0.17
	Water body	5.14
	Barren rocky/sheet rocks	0.02

3.2 Socio-Economic Parameters

Not only the natural hazards cause vulnerability, socio-economic situation of the community also adds to the vulnerability. The community with low socio-economic conditions will be affected more than the community that has a higher standard of living. This study evaluates six socio-economic parameters to assess the vulnerability of coastal communities in the study area.

3.2.1 Household Density

Household density is the number of houses per sq.km area of villages. Household density is directly proportional to the vulnerability. Household density is calculated using the Eq. 1 (Karuppusamy et al. 2021)

$$HD = \frac{Hn}{TA} \tag{1}$$

where HD is the household density, Hn is the number of households and TA represents total geographical area

3.2.2 Population Density

Population density of a village can directly affect the vulnerability of that area. A thickly populated area is highly exposed to the coastal hazards. The population increases higher it will affect the standard of the living people. Population density is calculated using the Eq. 2

$$PD = \frac{TP}{TA} \tag{2}$$

where PD stands for population density, TP is the total population and TA is the total geographical area.

3.2.3 Child Population Ratio

The no. of children in a population has a vital role when calculating the vulnerability of a community. Children are highly affected by vulnerable conditions physically and mentally. Extra care should be given to them during a vulnerable situation. The child population ratio is calculated using the Eq. 3

$$CPR = \frac{CP}{TP} \times 100 \tag{3}$$

where CPR is the child population ratio, CP stands for child population, and TP is the total population.

3.2.4 Literacy Rate

Literacy rate is an important parameter when considering the coastal vulnerability, the literate people can handle and analyze the vulnerable situation more effectively than the illiterates. Literacy is connected to the std of living also, high literacy rate of a village means the standard of living in that village is high. Literacy rate is calculated using the Eq. 4 (Ghosh and Mistri 2021).

$$\mathrm{LR} = \frac{\mathrm{TL}}{\mathrm{TP}} \times 100 \tag{4}$$

where LR stands for literacy rate, TL is the number of literates, and TP is the total population.

3.2.5 Marginalized Population Ratio

Keralites are a highly civilized population but some people are not having the privileges as others. These communities will be more exposed to vulnerable situations hence even if that is a smaller community, it should be considered. The marginal population ratio is calculated using the Eq. 5.

$$\mathrm{MPR} = \frac{\mathrm{TS}}{\mathrm{TP}} \times 100 \tag{5}$$

where MPR is the *Marginalised population* ratio, and TS is the sum of marginalized population ($population\ of\ SC + population\ of\ ST$) and TP is the total population.

4 Methodology

4.1 Bayesian Belief Network (BBN)

Bayesian Networks are probabilistic models that utilize directed acyclic graphs to represent the relationships between various variables. These models are grounded in Bayes' probability theory, which provides a framework for understanding and calculating conditional probabilities (Sanuy et al. 2020). Bayes' Theorem is a fundamental mathematical formula used to compute these conditional probabilities. It figures prominently in subjectivist or Bayesian approaches to epistemology, statistics, and inductive logic can be explained using Eq. 6.

$$P(\text{Pi})- = \frac{p(\text{Oi}) * p(\text{Oi})}{p(\text{Pi})} \tag{6}$$

where Oi is the probability of child node and Pi is the probability of Parent node. Probability of occurring an event (Oi) based on an influential event (Pi) can be understood using bayes theorem. Bayes theorem investigates every probability of occurring an event based on another event

A Bayesian network is composed of nodes and arcs, where each node represents a random variable, which can be either continuous or discrete. The arcs signify a direct influence between nodes, with the absence of a directed edge indicating independence between nodes. For example, in Fig. 2, nodes A, B, C, and D represent random variables. If node B is connected to node A by a directed arrow, node A is referred to as the parent of node B. Conversely, node C is independent of node A. Each node in the network has a conditional probability distribution, P(Xi | Parent(Xi)), which expresses the effect of the parent node on the child node. Bayesian networks are built on joint probability distributions and conditional probabilities. They compute the likelihood of an event by considering both dependent and independent variables. Dependent variables are those whose probability is affected by the changes in other variables, whereas independent variables remain unaffected by others. As a probabilistic graphical model, a Bayesian network offers a systematic approach to modeling complex problems by defining conditional independence assumptions for the known variables, while also accommodating unknown variables.

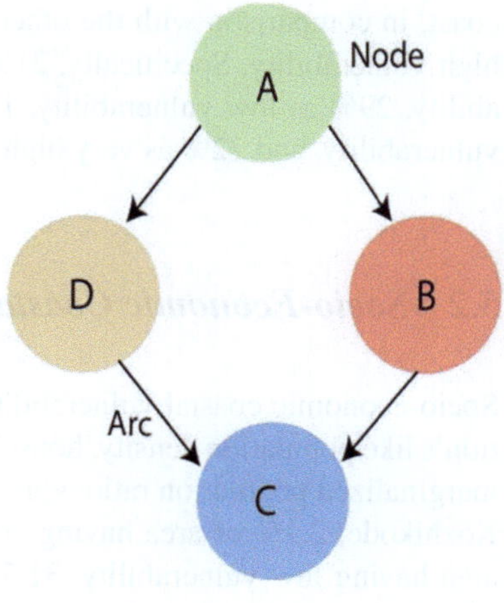

Fig. 2 Bayesian belief network

5 Result and Discussion

An integrated coastal vulnerability analysis using Bayesian Belief Networks (BBN) and Geographic Information Systems (GIS) combines both socio-economic and physical factors to evaluate the overall vulnerability of coastal regions. This method helps in understanding the complex interrelationships between different vulnerability factors and provides a probabilistic framework for decision-making.

5.1 *Physical Coastal Vulnerability Index (PCVI)*

Physical coastal vulnerability index of coasts of Thiruvananthapuram, Ernakulam, and Kozhikode are calculated using physical parameters like slope, geomorphology, lithology, and land use land cover. Kozhikode shows fluctuating vulnerability states such as low, moderate, and high indices. About 7.3% of the area is very low vulnerable, 24.3% of the area are low vulnerable, 31.3% of the area are under moderate vulnerable conditions, 22.7% of the area are highly vulnerable, and 14.4% of the area are very highly vulnerable (Fig. 3a). Ernakulam shows less physical vulnerability than considering other two districts. most of the regions face less vulnerable conditions. In the study area, 29.3% of the region is classified as very low vulnerability, 22.8% as low vulnerability, 23.5% as moderate vulnerability, 20.9% as high vulnerability, and 3.5% as very high vulnerability (Fig. 3b). Thiruvananthapuram coast, in comparison with the other two districts, exhibits a larger area under very high vulnerability. Specifically, 21% of the area is categorized as very low vulnerability, 29% as low vulnerability, 15.9% as moderate vulnerability, 12.1% as high vulnerability, and 22% as very high vulnerability (Fig. 3c).

5.2 *Socio-Economic Coastal Vulnerability Index (SECVI)*

Socio-economic coastal vulnerability index is calculated by accessing the information's like population density, household density, literacy rate, child population ratio, marginalized population ratio which that defines the standard of living of people. In Kozhikode, 2.1% of area having very low socio-economic vulnerability, 48.2% of area having low vulnerability, 31.5% of area having moderate vulnerability, 9.8% of area having highly vulnerable conditions, and 8.4% of area having very high vulnerability (Fig. 3d). In Ernakulam, 1.1% area having very low socio-economic vulnerability, 25.3% of area having low vulnerability, 49.3% of area having moderate vulnerability, 15.1% of area having highly vulnerable conditions, and 9.2% of area having very high vulnerability (Fig. 3e). 8.7% of area of Thiruvananthapuram having very low socio-economic vulnerability, 21.1% of area having low vulnerability,

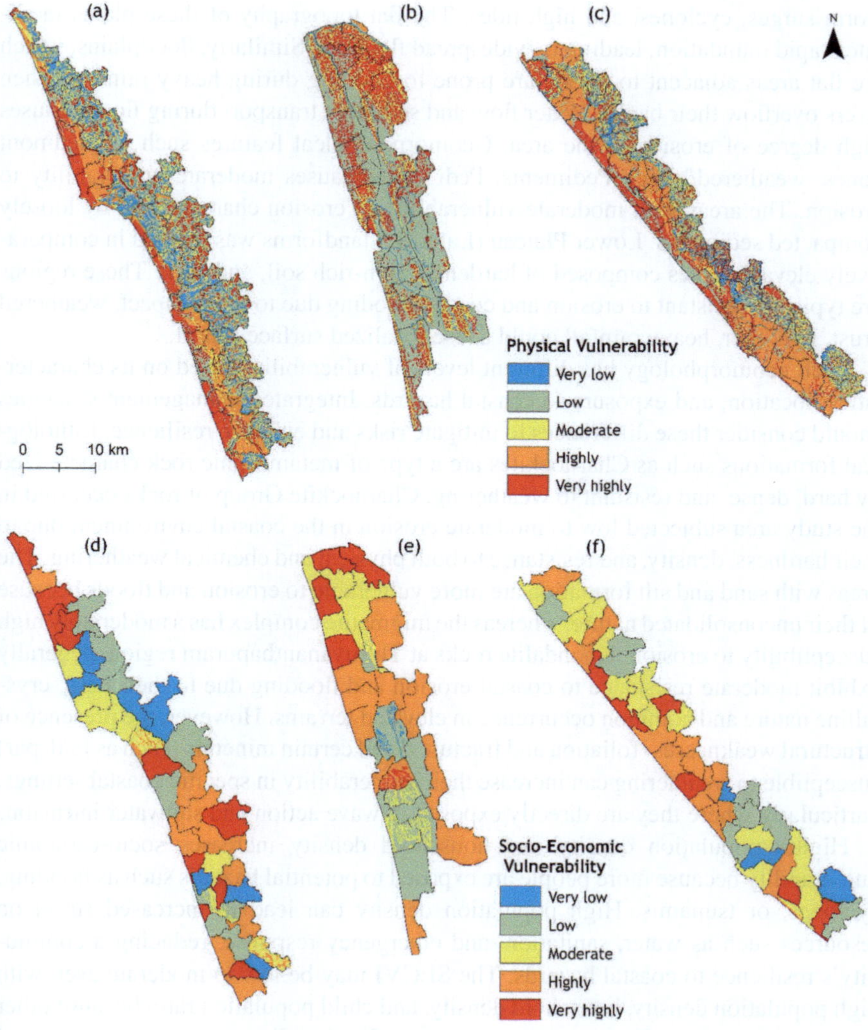

Fig. 3 Physical and socio-economic vulnerability of the study area

42.9% of area having moderate vulnerability, 20.9% of area having highly vulnerable conditions, and 6.4% of area having very high vulnerability (Fig. 3f).

Coastal plains, marshy lands, floodplains, and water bodies significantly contribute to the high coastal vulnerability in the study area. Coastal plains, which are low-lying, flat areas adjacent to the sea, are highly susceptible to coastal erosion. These areas are often composed of alluvial deposits such as sand, silt, and clay, making them particularly vulnerable to tidal influences and wave action. The loose, unconsolidated nature of the sediments exacerbates the risk of erosion, while their proximity to the ocean increases the potential for coastal flooding, especially during storm surges, cyclones, and high tides. The flat topography of these plains facilitates rapid inundation, leading to widespread flooding. Similarly, floodplains, which are flat areas adjacent to rivers, are prone to flooding during heavy rains or when rivers overflow their banks. Water flow and sediment transport during floods causes high degree of erosion in the area. Geomorphological features such as piedmont zones, weathered/buried Pediments, Pedi plains causes moderate vulnerability to erosion. The areas with moderate vulnerability to erosion characterized by loosely compacted sediments. Lower Plateau (Lateritic) landforms was located in comparatively elevated areas composed of hardened, iron-rich soil, and rock. These regions are typically resistant to erosion and coastal flooding due to the compact, weathered crust. However, heavy rainfall could cause localized surface runoff.

Each geomorphology has different levels of vulnerability based on its characteristics, location, and exposure to coastal hazards. Integrated management strategies should consider these differences to mitigate risks and enhance resilience. Lithological formations such as Charnockites are a type of metamorphic rock characterized by hard, dense, and resistant to weathering. Charnockite Group of rocks occurred in the study area subjected low to moderate erosion in the coastal environment due to their hardness, density, and resistance to both physical and chemical weathering. The areas with sand and silt formation are more vulnerable to erosion and floods because of their unconsolidated nature, whereas the migmatite complex has a moderately high susceptibility to erosion. Khondalite rocks at Thiruvananthapuram region, generally exhibit moderate resistance to coastal erosion and flooding due to their hard, crystalline nature and common occurrence in elevated terrains. However, the presence of structural weaknesses (foliation and fractures) and certain minerals (such as feldspar) susceptible to weathering can increase their vulnerability in specific coastal settings, particularly where they are directly exposed to wave action and saltwater intrusion.

Higher population density and household density increases socio-economic vulnerability because more people are exposed to potential hazards such as flooding, cyclones, or tsunamis. High population density can lead to increased strain on resources such as water, sanitation, and emergency response, reducing a community's resilience to coastal hazards. The SECVI may be low to moderate even with high population density, household density, and child population ratio because other factors such as high literacy rate, economy, effective disaster management, good infrastructure, low marginalized population ratio, natural barriers, and effective governance. These factors significantly enhancing the community's overall resilience to coastal hazards. This balance demonstrates the importance of a comprehensive approach to vulnerability assessment, where multiple social, economic, and environmental factors are considered together.

5.3 Integrated Coastal Vulnerability

The Integrated Coastal Vulnerability Index (ICVI) for coastal villages located within 5 km of the coastline in Kozhikode, Ernakulam, and Thiruvananthapuram districts

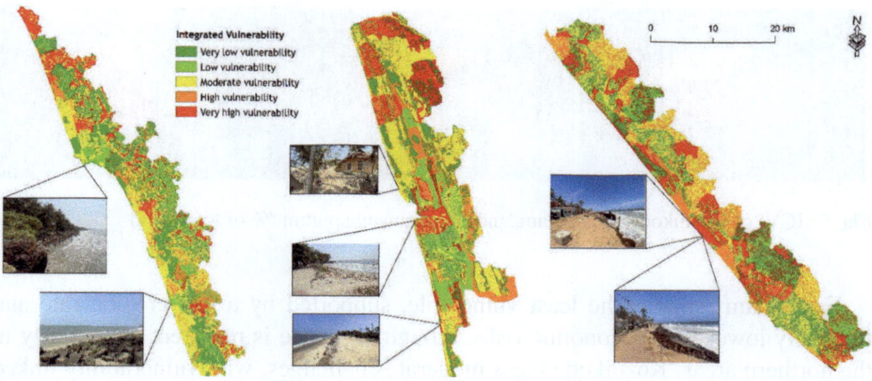

Fig. 4 Integrated coastal vulnerability map

was determined by combining the physical vulnerability index and the socio-economic vulnerability index using the Bayesian Belief Network (BBN) approach (Fig. 4). The results indicate that most areas within these districts face significant vulnerability, with Ernakulam exhibiting the lowest and Thiruvananthapuram the highest levels of vulnerability among the three. In Kozhikode, approximately 29% of the coastal region falls under very low vulnerability, 11% under low vulnerability, 37% under moderate vulnerability, 9% under high vulnerability, and 14% under very high vulnerability. Notably, the northern parts of Kozhikode display higher vulnerability levels.

In the Ernakulam coastal region, 19% of the area falls under very low vulnerability, 16% under low vulnerability, 29% under moderate vulnerability, 18% under high vulnerability, and another 18% under very high vulnerability. The northern part of the district predominantly exhibits higher vulnerability levels (Fig. 4). In the Thiruvananthapuram coastal region, 22% of the area is classified as very low vulnerability, 14% as low vulnerability, 25% as moderate vulnerability, 19% as high vulnerability, and 20% as very high vulnerability. Highly vulnerable zones are dispersed across the district. Thiruvananthapuram records the greatest proportion of land under high and very high vulnerability categories (39%), suggesting extensive vulnerability throughout the district (Fig. 5). Ernakulam has the lowest proportion of highly vulnerable areas (36% under high and very high vulnerability), making it the least vulnerable among the three districts. In Kozhikode, vulnerability is more varied, with most areas showing moderate vulnerability (37%), and the northern part being more vulnerable (Fig. 4). The results indicate that while all three districts face significant vulnerability, Thiruvananthapuram has more widespread and scattered areas of high vulnerability, suggesting a need for focused disaster risk reduction efforts across the entire district. In contrast, Ernakulam shows relatively less vulnerability, with risk concentrated in specific areas, particularly in the northern part. Kozhikode has a mix of vulnerability levels, with the northern region being the most at risk. Low ICVI of Ernakulam is benefits from lower physical vulnerability, allowing for more stable land conditions and reduced exposure to direct coastal hazards.

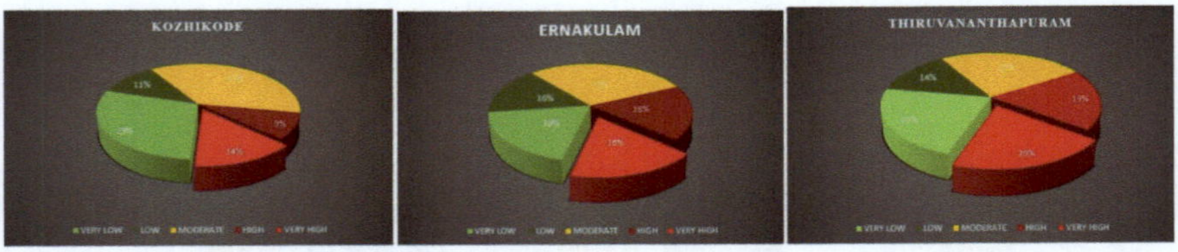

Fig. 5 ICVI of Kozhikode, Ernakulam, and Thiruvananthapuram (% of total area)

Ernakulam remains the least vulnerable, supported by a low erosion rate and relatively lower socio-economic risks, though vigilance is required, particularly in the northern areas. Kozhikode faces moderate challenges, with vulnerability linked to localized erosion and socio-economic factors, particularly in its northern part. Efforts to manage and reduce erosion, along with targeted socio-economic support, could help mitigate risks.

6 Conclusion

Integrated coastal vulnerability index of three major coastal centers of Kerala was assessed in the present work using Bayesian belief network, a machine learning technique. About 14%, 18%, and 20% of areas of Kozhikode, Ernakulam, and Thiruvananthapuram are under high risk of coastal vulnerability. The northern regions of Kozhikode and Ernakulam predominantly encompass the highly vulnerable areas, whereas the central part of Thiruvananthapuram contains the majority of the vulnerable zones. While looking at the physical vulnerability 14.4, 3.5, and 22% of the study areas of Kozhikode, Ernakulam, and Thiruvananthapuram are under highly vulnerable situations. The socio-economic vulnerability index reveals that 8.4, 9.2, and 6.4% of the study areas of Kozhikode, Ernakulam and Thiruvananthapuram are highly vulnerable. This study highlights the effectiveness of remote sensing and machine learning techniques in assessing coastal vulnerability. The study recommended the enhancement of coastal protection measures, such as erosion control structures and ecosystem-based approaches (like mangrove restoration) in highly vulnerable areas like Thiruvananthapuram. Improvement of community adaptability and readiness, particularly required in extremely susceptible places. By maintaining ongoing surveillance and making infrastructural investments, particularly in the low vulnerable areas, should preserve the current low vulnerability. Communities with moderate to high levels of susceptibility must properly concentrate erosion control efforts on vulnerable areas and enhance socio-economic resilience.

Acknowledgements The first author gratefully acknowledges the financial support provided by the Kerala University of Fisheries and Ocean Studies (KUFOS) through the KUFOS Aided Research Project (KARP), project number DoR (1)/5587/2021.

References

Arun Kumar A, Kunte PD (2012) Coastal vulnerability assessment for Chennai, east coast of India using geospatial techniques. Nat Hazards 64:853–872

Basheer Ahammed KK, Pandey AC, Parida BR, Wasim, Dwivedi CS (2023) Impact assessment of tropical cyclones amphan and nisarga in 2020 in the northern Indian ocean. Sustainability 15(5):3992

Bhattacharyya P, Dash PK, Padhy SR, Roy KS, Neogi S, Nayak AK (2020) Soil carbon dynamics in different land-use and management systems in tropical coastal regions of India. In: Carbon management in tropical and sub-tropical terrestrial systems, pp 89–102

Chakrabortty R, Ali T, Atabay S, Roy P, Pande CB (2025) Impact of climate change scenario on sea level rise and future coastal flooding in major coastal cities of India. Sci Rep 15(1):28689. https://doi.org/10.1038/s41598-025-13394-5

Cheng J, Greiner R (2001) Learning bayesian belief network classifiers: Algorithms and system. In: Advances in artificial intelligence: 14th biennial conference of the canadian society for computational studies of intelligence, AI 2001 Ottawa, Canada, Proceedings vol 14. Springer Berlin Heidelberg, 7–9 June 2001, pp 141–151

Chettiyam Thodi MF, Gopinath G, Surendran UP, Prem P, Al-Ansari N, Mattar MA (2023) Using RS and GIS techniques to assess and monitor coastal changes of coastal islands in the marine environment of a humid tropical region. Water 15(21):3819

Dinesh Kumar PK, Gopinath G, Laluraj CM, Seralathan P, Mitra D (2007) Change detection studies of Sagar Island, West Bengal (India) using IRS IC LISS III DATA. J Coast Res 23(6):1498–1502. https://doi.org/10.2112/05-0599.1

Ghosh S, Mistri B (2021) Assessing coastal vulnerability to environmental hazards of Indian Sundarban delta using multi-criteria decision-making approaches. Ocean Coast Manag 209:105641

Gopinath G (2009) Critical coastal issues of Sagar Island, East Coast of India. Environ Monit Assess 160:555–561

Gopinath G, Seralathan P (2005) Rapid coastal erosion of the Sagar Island, West Bengal. Environ Geol 48 (8):1058–1067. https://doi.org/10.1007/s00254-005-0044-9

Gopinath G, Thodi MFC, Surendran UP, Prem P, Parambil JN, Alataway A, Al-Othman AA, Dewidar AZ, Mattar MA (2023) Long-term shoreline and islands change detection with digital shoreline analysis using RS Data and GIS in humid tropical Kerala, India. Water 15(2), 244. https://doi.org/10.3390/w15020244

Gutierrez BT, Plant NG, Thieler ER (2011) A Bayesian network to predict coastal vulnerability to sea level rise. J Geophys Res: Earth Surf 116(F2)

Holling CS (1973) '1.3', Annu Rev Ecol Syst 4(1):1–23. Available at: https://doi.org/10.1146/annurev.es.04.110173.000245

Integrated Coastal Zone Management Project (2010) ICZMP Report. Government of India

Jay S (2010) Planners to the rescue: Spatial planning facilitating the development of offshore wind energy. Mar Pollut Bull 60(4):493–499. Available at: https://doi.org/10.1016/j.marpolbul.2009.11.010

Karuppusamy B et al (2021) Revealing the socio-economic vulnerability and multi- hazard risks at micro-administrative units in the coastal plains of Tamil Nadu, India. Geomat, Nat Hazards Risk 12(1):605–630. Available at: https://doi.org/10.1080/19475705.2021.1886183

Kumar PKD, Gopinath G, Murali RM, Muraleedharan KR (2014) Geospatial analysis of long-term morphological changes in Cochin estuary, SW Coast of India. J Coast Res 30 (6):1315–1320. https://doi.org/10.2112/JCOASTRES-D-12-00244.1

Medha, Mondal B, Dolui G, Tafsirul Islam SM, Bera MM (2023) Impact of land inundation caused by cyclone 'Amphan' across Bangladesh and India using spatial damage assessment framework. In: Environmental management and sustainability in India: case studies from West Bengal, Cham: Springer International Publishing, pp 187–214

Naga Kumar KCV, Deepak PM, Basheer Ahammed KK, Rao KN, Gopinath G, Dinesan VP (2022) Coastal vulnerability assessment using geospatial technologies and a multi-criteria decision making approach–a case study of Kozhikode District coast, Kerala State, India. J Coast Conserv 26(3):16

Pereira GF et al (2020) Assessing the rate of shoreline changes of Rameswaram Island, Tamilnadu, India, using MATLAB Component Runtime Utility & Digital

Sahin O et al (2019) Spatial Bayesian Network for predicting sea level rise induced coastal erosion in a small Pacific Island. J Environ Manag 238:341351. Availableat:https://doi.org/10.1016/j.jenvman.2019.03.008

Sanuy M et al (2020) A Bayesian network methodology for coastal hazard assessments on a regional scale: the BN-CRAF. Coast Eng 157:1–10. Available at: https://doi.org/10.1016/j.coastaleng.2019.103627

Shaji J (2014) Coastal sensitivity assessment for Thiruvananthapuram, west coast of India. Nat Hazards 73(3):1369–1392. Available at: https://doi.org/10.1007/s11069-014-1139-y

Shoreline Analysis System Int J Eng Adv Technol, 9(3):3318–3321. Available at: https://doi.org/10.35940/ijeat.c5675.029320

Sreelekshmi S, Veettil BK, Nandan SB, Harikrishnan M (2021) Mangrove forests along the coastline of Kerala, southern India: current status and future prospects. Reg Stud Mar Sci 41:101573

Stritih A et al (2020) An online platform for spatial and iterative modelling with Bayesian Networks. Environ Model Softw 127:104658. Available at: https://doi.org/10.1016/j.envsoft.2020.104658

Sudha Rani NNV, Satyanarayana ANV, Bhaskaran PK (2015) Coastal vulnerability assessment studies over India: a review. Nat Hazards 77:405–428

Thieler ER, Gayes PT, Schwab WC, Harris MS (1999) Tracing sediment dispersal on nourished beaches: two case studies. In: Proceedings of the international conference on coastal engineering and science of coastal sediment processes, vol 4(3), 2118–2136

Ummenhofer CC, Meehl GA (2017) '1.1' Philos Trans R Soc B: Biol Sci 372(1723). Available at: https://doi.org/10.1098/rstb.2016.0135

UN. https://www.un.org/esa/sustdev/natlinfo/indicators/methodology_sheets/oceans_seas_coasts/pop_coastal_areas.pdf

WEF 2024 World Economic Forum (2024). https://www.weforum.org/stories/2024/09/rising-sea-levels-global-threat/#:~:text=This%20article%20was%20first%20published,10cm%20over%20the%20last%20decade

Harnessing Community Strengths: Disaster Management and Mitigation in Coastal Areas

Mohd Akhter Ali and M. Kamraju

Abstract Coastal regions are increasingly vulnerable to natural disasters, exacerbated by climate change and rising sea levels. Effective disaster management relies heavily on the involvement of local communities, yet this potential is often underutilized in disaster mitigation policies. This paper investigates the role of community strengths—local knowledge, self-organization, and traditional practices—in enhancing disaster resilience in coastal areas. A comprehensive literature review reveals that community-based disaster risk management (CBDRM) initiatives, such as mangrove restoration in the Sundarbans, early warning systems in the Philippines, and the utilization of indigenous knowledge in Pacific Island nations, demonstrate significant positive outcomes in disaster preparedness and response. However, several barriers hinder effective community engagement, including funding limitations, infrastructural constraints, lack of technical knowledge, socio-political issues, and gender inequality. By examining successful case studies and identifying the barriers to community engagement, this paper provides insights into how local strengths can be integrated into formal disaster management frameworks. The findings have important policy implications for local, regional, and national governments, suggesting that greater emphasis on community involvement can lead to more resilient coastal communities.

Keywords Community-based disaster management (CBDRM) · Coastal vulnerability · Disaster resilience · Local knowledge · Early warning systems · Indigenous knowledge · Socio-political barriers · Gender inequality

M. A. Ali (✉)
Department of Geography, Directorate of Admissions, Osmania University, Hyderabad, India
e-mail: drmohdakhterali@gmail.com

M. Kamraju
Department of Geography, APSG, Hyderabad, India

K. K. Basheer Ahammed et al. (eds.), *Monitoring Coastal Marine Ecosystems Under Climate Change*, https://doi.org/10.1007/978-3-032-09137-6_20

1 Introduction

Coastal areas are some of the most hazard-prone regions in the world due to their proximity to large water bodies, which exposes them to a range of natural disasters. These areas are increasingly threatened by the adverse effects of climate change, including rising sea levels, intensified storm activity, and more frequent and severe flooding. According to the Intergovernmental Panel on Climate Change (IPCC), global sea levels have risen by approximately 20 cm since 1900, with the rate accelerating in recent decades due to melting polar ice and thermal expansion of the oceans (IPCC 2021). These environmental changes make coastal regions particularly susceptible to natural disasters such as cyclones, storm surges, tsunamis, and coastal erosion, which can lead to widespread destruction of infrastructure, livelihoods, and ecosystems (Basheer Ahammed and Pandey 2022).

The vulnerability of coastal areas is compounded by the fact that many of these regions are densely populated and economically reliant on activities that depend on proximity to the sea, such as fishing, tourism, and shipping. The United Nations estimates that more than 600 million people—roughly 10% of the world's population—live in coastal areas that are less than 10 m above sea level, making them vulnerable to the impacts of climate change and natural disasters (United Nations 2020). As these areas grow more vulnerable, the need for effective disaster management and mitigation strategies becomes ever more urgent.

While significant efforts have been made by governments, international organizations, and non-governmental organizations (NGOs) to address the risks associated with natural disasters in coastal areas, community involvement remains an underutilized resource in official disaster management frameworks. Communities living in these high-risk areas often possess valuable local knowledge and traditional practices that have been passed down through generations, allowing them to adapt to their environment and manage the risks associated with natural disasters. These community-driven efforts include indigenous early warning systems, resilient building practices, and the protection of natural buffers such as mangroves and coral reefs, which help reduce the impact of storm surges and flooding (Adger et al. 2005). However, disaster management policies frequently overlook or marginalize community knowledge in favor of top-down, technocratic approaches that may not be as effective or sustainable in the local context.

Research by Gaillard and Mercer (2013) emphasizes the importance of integrating local and indigenous knowledge into formal disaster risk reduction (DRR) strategies. They argue that communities are often the first responders to disasters and can offer critical insights into vulnerabilities and coping mechanisms that external agencies may miss. Similarly, Shaw et al. (2014) highlight the role of local leadership in enhancing community resilience, noting that communities with strong local governance and participatory disaster management structures tend to recover more quickly and effectively from disasters. Despite these insights, there remains a significant gap in the integration of community strengths into national and international disaster mitigation policies.

This paper seeks to examine the role of community strengths, including local knowledge, self-organization, and traditional practices, in disaster management and mitigation in coastal areas. By investigating a range of community-led disaster management practices, this study aims to show how these initiatives contribute to reducing vulnerability and building resilience in disaster-prone regions. The research focuses on understanding how these communities harness their strengths to prepare for, respond to, and recover from natural disasters, as well as how they collaborate with governmental and non-governmental organizations to enhance disaster mitigation efforts.

This study also aims to contribute to the growing body of knowledge on community-based disaster risk management (CBDRM), which emphasizes the importance of community participation in all phases of disaster management—from preparedness to response and recovery. The paper will provide case studies from various coastal regions where community-led initiatives have proven successful and will propose strategies for scaling these practices to other vulnerable areas.

2 Literature Review

Disaster Vulnerabilities in Coastal Areas:

Coastal areas are particularly vulnerable to a variety of natural hazards due to their proximity to the sea and exposure to dynamic weather patterns. **Storm surges**, **coastal flooding**, and **coastal erosion** are the most frequent and destructive events, with devastating effects on both human populations and ecosystems. Nicholls and Cazenave (2010) highlight the increased risk of coastal flooding due to rising sea levels caused by climate change. Their research shows that even small increases in sea levels can lead to severe flooding, particularly in low-lying coastal areas, which house a significant portion of the global population.

Coastal erosion, another major threat, is exacerbated by human activities such as **sand mining**, **deforestation of mangrove forests**, and **urban development**, which weaken natural coastal defenses. Studies have shown that **mangroves, coral reefs**, and **dunes** serve as natural buffers, reducing the impact of waves and storm surges. Research by Das and Crépin (2013) demonstrates that communities with intact mangrove ecosystems experience significantly less damage from storm surges compared to those without, underscoring the ecological role of these habitats in disaster mitigation.

Moreover, coastal areas are particularly vulnerable to tropical cyclones and hurricanes. Emanuel (2005) found that the intensity of tropical cyclones has increased over the past few decades, a trend linked to rising ocean temperatures caused by climate change. These storms result in widespread destruction, often displacing millions of people and leading to massive economic losses. The IPCC (2014) projects that the frequency and severity of extreme weather events, including cyclones and typhoons, will continue to increase, putting coastal regions at heightened risk of disaster.

While modern technologies and infrastructure improvements can reduce some vulnerabilities, Adger et al. (2005) argue that communities in coastal areas often lack the necessary resources and support to implement such measures, making them highly susceptible to disasters. Vulnerability in coastal regions is also closely linked to socio-economic factors, with poorer communities often living in the most vulnerable areas and having limited access to disaster preparedness and recovery resources.

Community-Based Disaster Risk Management (CBDRM):

Community-based disaster risk management (CBDRM) has emerged as a key strategy for addressing disaster vulnerabilities in coastal areas. CBDRM emphasizes the active participation of local communities in all stages of disaster management, from preparedness and risk reduction to recovery and rehabilitation. The approach is grounded in the belief that **local communities are the best positioned to understand their specific vulnerabilities and resources**, making them critical players in reducing disaster risks.

Shaw and Okazaki (2004) describe CBDRM as an effective strategy for empowering communities to take ownership of disaster risk reduction efforts. They point to case studies in **Bangladesh**, where community-driven initiatives, such as **early warning systems** and the construction of **cyclone shelters**, have significantly reduced fatalities during cyclones. Similarly, research by Paton and Johnston (2001) demonstrates the importance of community participation in building disaster resilience, particularly in regions where government intervention is limited or delayed. These studies highlight how communities that are well-organized and informed about disaster risks can effectively minimize the impact of natural hazards.

The success of CBDRM often hinges on the use of **local knowledge** and **traditional practices**. For instance, indigenous coastal communities in the **Philippines** have long relied on **natural indicators**—such as changes in wind patterns or animal behavior—to predict storms and tsunamis. In a study by Gaillard et al. (2008)**,** it was found that these local knowledge systems, when integrated with modern disaster management techniques, can significantly enhance the accuracy and effectiveness of early warning systems.

However, despite its advantages, CBDRM is not without challenges. Twigg (2015) identifies several gaps in the existing literature on CBDRM, particularly the lack of robust frameworks for evaluating its long-term effectiveness. While there are numerous success stories of CBDRM reducing disaster risks in the short term, questions remain about its sustainability, especially in the face of rapidly changing environmental and socio-economic conditions. Moreover, CBDRM initiatives often lack sufficient financial and technical support, which limits their scalability and broader application.

Challenges in Community Engagement:

Despite the proven benefits of community engagement in disaster management, there are significant challenges that hinder the full realization of its potential. One major barrier is the **lack of resources** available to local communities. Many coastal communities, particularly in developing countries, face chronic poverty, inadequate infrastructure, and limited access to technology, all of which reduce their capacity to implement effective disaster risk reduction measures (Cutter et al. 2003). These resource limitations make it difficult for communities to invest in disaster preparedness or recovery efforts, leaving them highly vulnerable when disasters strike.

Additionally, there is often a **gap in technical knowledge** between communities and formal disaster management agencies. While communities possess rich traditional knowledge, they may lack access to modern disaster management tools such as **geographic information systems (GIS)**, **satellite-based early warning systems**, or advanced **risk modeling** techniques. Gaillard and Mercer (2013) argue that effective disaster management requires a merging of local and scientific knowledge systems, but achieving this integration is often challenging due to **communication barriers**, lack of trust, or insufficient training.

Socio-political factors also play a crucial role in limiting community engagement in disaster management.

In many cases, marginalized groups—such as **women**, **indigenous peoples**, and **ethnic minorities**—are excluded from decision-making processes related to disaster risk management, even though they are among the most vulnerable to disasters. Fordham (2011) notes that gender inequality in disaster management is a persistent issue, with women often having limited access to resources and decision-making platforms, despite their critical role in community resilience.

Similarly, Pelling (2007) argues that political power dynamics within communities can hinder effective engagement, as local elites may dominate decision-making processes, sidelining the needs and voices of the most vulnerable members of the community.

Moreover, there are institutional barriers to community engagement. Governments and NGOs may not always recognize the value of community participation, or they may prefer top-down approaches that prioritize technical solutions over local input.

Bankoff (2003) discusses how disaster management policies are often influenced by external donors and international agencies, leading to the imposition of technocratic solutions that may not align with the realities on the ground. This disconnect between external agencies and local communities can reduce the effectiveness of disaster risk reduction efforts, as externally driven solutions may fail to account for local contexts and needs.

Finally, there is the issue of **sustainability**. While community-based initiatives may be effective in the short term, maintaining them over the long term can be difficult due to a lack of continuous funding, changing environmental conditions, or shifts in community priorities. Twigg (2015) notes that without long-term institutional support, many community-driven disaster risk reduction initiatives struggle to remain viable over time.

Hypothesis:

The central hypothesis of this study is that **communities with robust traditional practices, strong local leadership, and established self-organization mechanisms**

demonstrate higher resilience in disaster-prone coastal areas. The integration of local knowledge and leadership into disaster management strategies helps reduce vulnerability and improves recovery outcomes.

Shaw et al. (2014) found that when communities are empowered to take ownership of their disaster management processes, they are better able to adapt to changing environmental conditions and manage risks effectively. This paper posits that the incorporation of these community strengths into broader disaster management frameworks will lead to more sustainable and effective disaster mitigation strategies in coastal areas.

Research Questions:

i. **What are the most effective community-led disaster management practices in coastal areas?** This question seeks to identify the key practices that communities in coastal regions use to manage disaster risks. It focuses on traditional knowledge, early warning systems, and community-based infrastructure that mitigates the impact of disasters.
ii. **How can coastal communities improve collaboration with governments and NGOs for disaster mitigation?** While communities play a critical role in disaster management, they often face challenges in aligning their efforts with those of governments and NGOs. This question explores how better coordination and resource-sharing can enhance disaster mitigation outcomes.
iii. **What role does local knowledge and leadership play in enhancing the resilience of coastal communities?** Local knowledge and leadership are often crucial in disaster-prone areas, where external aid may be delayed. This question aims to investigate how these factors contribute to a community's ability to recover and rebuild after a disaster.

3 Methodology

This research employs a comprehensive approach using secondary data to analyze community-led disaster management and mitigation strategies in coastal regions. The focus is on three distinct areas: The Sundarbans (India and Bangladesh), coastal regions of the Philippines, and Pacific Island nations. Each of these regions presents unique challenges and opportunities for studying the role of community strengths in disaster resilience.

For the **Sundarbans**, secondary data sources include government reports from the Indian National Disaster Management Authority (NDMA) and Bangladesh's Ministry of Disaster Management and Relief. These reports provide insights into disaster preparedness plans, risk assessments, and the effectiveness of community-based interventions. Additionally, academic studies from journals such as *Global Environmental Change* and *Disasters* are reviewed to understand the local adaptation strategies employed by communities in response to cyclones, flooding, and coastal erosion. Historical records of past disasters, including data on their impacts and recovery efforts, further illuminate the region's disaster management practices.

In the **Philippines**, the research draws on publications from the National Disaster Risk Reduction and Management Council (NDRRMC) and reports by international NGOs like the Red Cross and Oxfam. These sources offer a detailed account of community-based disaster risk management practices, including the integration of local knowledge and traditional methods. Academic research from sources such as *International Journal of Disaster Risk Reduction* provides additional context on how coastal communities in Leyte and Samar have developed resilience against frequent typhoons and storm surges. Historical disaster data, including records of major typhoons and their impacts, are also analyzed to understand the evolution of disaster management strategies over time.

For **Pacific Island nations**, secondary data includes reports from the Pacific Islands Forum and documents from regional organizations like the Secretariat of the Pacific Community. These sources detail the challenges faced by small island nations in managing risks associated with sea level rise, cyclones, and tsunamis. Academic articles from journals such as *Environmental Science & Policy* and *Climate and Development* are reviewed to examine the role of indigenous knowledge and local practices in enhancing disaster resilience. Historical records and geospatial data on disaster events and community responses provide further insights into the effectiveness of various mitigation strategies.

The methodology involves several key analytical tools. **Thematic analysis** is used to identify and interpret recurring themes and patterns across the secondary data. This includes analyzing how traditional knowledge and local practices are integrated into disaster management strategies and the effectiveness of these approaches. **Statistical analysis** is applied to quantitative data, such as disaster impact statistics and resilience indices, to summarize findings and identify trends. Descriptive statistics, including frequencies and percentages, are used to present data on community preparedness and response measures. **GIS mapping** is employed to visualize vulnerable zones, community resilience measures, and the spatial distribution of disaster impacts. This helps to map out areas prone to specific hazards and assess the spatial effectiveness of community-led initiatives.

Finally, a **comparative analysis** across the three study areas is conducted to draw conclusions about commonalities and differences in community-led disaster management practices. This analysis highlights successful strategies that could be adapted or replicated in other regions, as well as context-specific challenges and opportunities. By synthesizing findings from secondary sources, the research provides a robust understanding of how community strengths contribute to disaster resilience and identifies best practices for future disaster management efforts.

4 Discussion

The analysis of secondary data across the Sundarbans, the Philippines, and Pacific Island nations reveals several crucial insights into the effectiveness of community-driven disaster management strategies. These strategies significantly contribute to resilience in coastal areas by leveraging local knowledge, traditional practices, and grassroots organization.

In the **Sundarbans**, the integration of mangrove restoration with local fishing practices exemplifies a successful community-led approach. Mangrove forests act as natural barriers against storm surges and coastal erosion, while local communities' involvement in their conservation helps in sustaining these protective benefits (Wickramasinghe and Stoll 2021). The traditional knowledge of local communities about the ecological importance of mangroves has led to successful collaborative efforts in restoration, which enhances both ecological stability and disaster resilience (Das and Vincent 2009).

In the **Philippines**, the implementation of early warning systems by community-based organizations has proven effective in mitigating the impacts of typhoons and floods. For instance, community networks have played a crucial role in disseminating early warnings and organizing evacuations, which has significantly reduced casualties and property damage during typhoon events (Santos and van den Berg 2020). The Philippines' experience underscores the value of local engagement in disaster preparedness and response, showing how community-led initiatives can complement formal disaster management systems (Bautista 2019).

For **Pacific Island nations**, the use of indigenous knowledge for disaster prediction and response demonstrates the effectiveness of integrating traditional practices with modern disaster management strategies. Indigenous knowledge, such as observing natural indicators like wind patterns and animal behavior, has been instrumental in anticipating and responding to environmental threats (Nunn 2015). This approach highlights the resilience of Pacific Island communities and the potential for incorporating traditional practices into broader disaster management frameworks (Barnett and Webber 2012).

These insights emphasize the effectiveness of community-driven strategies and highlight innovative approaches that could be adapted and replicated in other coastal regions to enhance disaster resilience.

Table 1 provides a detailed overview of community-based disaster management initiatives across three distinct regions: The Sundarbans in India and Bangladesh, the Philippines, and selected Pacific Island nations such as Tuvalu and Fiji. Each row of the table highlights a different initiative, focusing on its type, key outcomes, level of community involvement, and the supporting organizations involved.

In the **Sundarbans**, the primary initiative is **mangrove restoration**. This effort aims to restore and conserve mangrove forests, which are crucial for protecting coastal areas from storm surges and erosion. The restoration projects have led to several significant outcomes, including enhanced coastal protection, reduced damage from storm surges, and improved biodiversity in the region. Local communities are

Table 1 Overview of community-based disaster management initiatives

Region/ Country	Type of initiative	Key outcomes	Community involvement	Supporting organizations
Sundarbans (India and Bangladesh)	Mangrove restoration	• Enhanced coastal protection • Reduced storm surge damage • Improved biodiversity	• Local communities involved in planting and maintenance • Use of traditional knowledge	• Sundarbans Conservation Foundation • Local NGOs and government agencies
Philippines	Community-based early warning systems	• Reduced casualties during typhoons • Improved evacuation times • Increased awareness	• Local radio stations and community networks • Training programs for residents	• Philippine Red Cross • Local government units (LGUs) • NGOs
Pacific Island Nations (e.g., Tuvalu, Fiji)	Indigenous knowledge utilization	• Improved disaster prediction • Enhanced community preparedness • Strengthened coastal defenses	• Application of traditional knowledge in forecasting • Community-led adaptation projects	• Regional environmental organizations • Local traditional leaders

actively involved in these initiatives, participating in the planting and maintenance of mangroves and contributing their traditional ecological knowledge to the projects. Supporting organizations such as the Sundarbans Conservation Foundation, along with various local NGOs and government agencies, provide the necessary resources and support to sustain these efforts.

In the **Philippines**, the focus is on **community-based early warning systems**. These systems are designed to alert communities about impending typhoons and floods, thereby improving disaster preparedness and response. The key outcomes of these initiatives include a reduction in casualties, improved evacuation times, and increased awareness about disaster risks. Local networks, including community radio stations, play a vital role in disseminating warnings, while training programs help residents prepare for disasters. The Philippine Red Cross, local government units (LGUs), and various NGOs are instrumental in supporting and implementing these early warning systems, ensuring effective communication and coordination during emergencies.

For the **Pacific Island nations** like Tuvalu and Fiji, the initiative revolves around **utilizing indigenous knowledge**. This approach leverages traditional practices and knowledge to predict and respond to environmental changes and natural disasters.

Key outcomes include improved disaster prediction, enhanced community preparedness, and strengthened coastal defenses. Communities actively apply traditional knowledge to disaster management, with local projects focusing on adapting these practices for modern use. Regional environmental organizations and local traditional leaders support these efforts by facilitating the integration of indigenous knowledge into disaster management strategies.

Table 1 illustrates how community-based disaster management initiatives vary across different regions, each tailored to local needs and contexts. The table highlights the importance of community involvement and the role of supporting organizations in enhancing disaster resilience and preparedness. By summarizing these initiatives, Table 1 provides a clear understanding of how local knowledge and practices can be effectively integrated into disaster management frameworks to achieve significant outcomes in different coastal regions.

Mangrove Restoration is the primary disaster management practice in the Sundarbans. This strategy involves large-scale reforestation projects aimed at restoring mangrove forests to enhance coastal protection against storm surges and erosion. The benefits of mangrove restoration are significant, including improved coastal defense, increased biodiversity, and a natural barrier that mitigates the effects of extreme weather events. In the Philippines and Pacific Island nations, this practice is not applicable as their primary focus lies elsewhere.

In the **Philippines**, **community-based early warning systems** are a key disaster management practice. These systems involve the use of local radio stations and community networks to disseminate early warnings about typhoons and floods, combined with training programs to educate residents on disaster preparedness. The key benefits of this practice include a reduction in casualties, improved evacuation times, and heightened disaster awareness among community members. Mangrove restoration and indigenous knowledge utilization are not primary practices in the Philippines.

For **Pacific Island nations** like Tuvalu and Fiji, the practice of **indigenous knowledge utilization** is central to disaster management. This involves applying traditional forecasting methods and adaptation strategies to predict and respond to environmental changes and natural disasters. The benefits of using indigenous knowledge include improved disaster prediction, enhanced community preparedness, and strengthened coastal defenses. Mangrove restoration and early warning systems are not emphasized in these regions.

Table 2 provides a comparative overview of these disaster management practices, illustrating how different regions implement various strategies to address their unique vulnerabilities and enhance disaster resilience. It highlights the effectiveness of community-based approaches and the diverse benefits they bring to disaster management efforts in different coastal contexts.

Funding Limitations Present a significant barrier across all regions. In the Sundarbans, the lack of financial resources affects the scale and sustainability of mangrove restoration projects. Similarly, in the Philippines, inadequate funding impacts the effectiveness of early warning systems and training programs. Pacific Island nations

Table 2 Comparison of disaster management practices

Practice/Strategy	Sundarbans	Philippines	Pacific Island nations	Key benefits
Mangrove restoration	• Large-scale reforestation projects • Protection against storm surges and erosion	• Not applicable	• Not applicable	• Enhanced coastal protection • Biodiversity improvement • Natural barrier against extreme weather
Community-based early warning systems	• Not applicable	• Local radio stations and networks • Training programs for residents	• Not applicable	• Reduced casualties • Improved evacuation times • Increased disaster awareness
Indigenous knowledge utilization	• Not applicable	• Not applicable	• Traditional forecasting and adaptation • Community-led disaster preparedness	• Improved disaster prediction • Enhanced community preparedness • Strengthened coastal defenses

face challenges in funding projects that utilize indigenous knowledge. To mitigate these issues, regions collaborate with NGOs and international organizations for financial support, seek government grants, and leverage subsidies to enhance their disaster management efforts.

Infrastructural Constraints also pose challenges. The Sundarbans struggles with poor infrastructure in remote areas, making it difficult to access and restore mangrove sites. In the Philippines, inadequate infrastructure hampers the dissemination of early warnings and affects transportation during disasters. Pacific Island nations face limitations in infrastructure for modernizing traditional practices. Strategies to address these constraints include investing in infrastructure improvements, developing local networks, and enhancing communication channels to support effective disaster management.

Lack of Technical Knowledge is another barrier. In the Sundarbans, there is a need for advanced technical expertise in mangrove restoration. The Philippines requires more technical training to optimize early warning systems. Pacific Island nations experience gaps in integrating traditional and modern practices. Mitigation strategies include capacity-building workshops, technical training programs, and collaboration with experts to improve technical knowledge and application in disaster management.

Socio-Political Issues affect community engagement in different ways. In the Sundarbans, political instability and conflicts over resource use hinder project implementation. The Philippines faces political challenges in coordinating disaster responses and managing resource distribution. In Pacific Island nations, social and political dynamics influence the application of traditional knowledge. Strengthening local governance, engaging local leaders in decision-making, and addressing political and social challenges are essential strategies for overcoming these barriers.

Gender Inequality impacts participation in disaster management activities. In the Sundarbans, gender biases affect involvement in restoration projects. In the Philippines, gender disparities influence disaster preparedness and response roles. In Pacific Island nations, traditional gender roles affect participation in disaster management practices. Promoting gender-inclusive policies and encouraging equal participation in disaster management are key strategies to address gender inequality.

Table 3 provides a clear overview of the various barriers to effective community engagement in disaster management and the strategies employed to overcome these challenges in different regions. By highlighting these issues and solutions, the table underscores the importance of addressing barriers to enhance community involvement and improve disaster resilience.

Comparison with Existing Literature:

The findings of this study align with existing literature that underscores the importance of community involvement in disaster management. Research by Wilhite and Glantz (1985) highlights that local knowledge and community participation are crucial for effective disaster risk reduction. This study confirms these findings by illustrating successful community-led strategies, such as mangrove restoration in the Sundarbans and early warning systems in the Philippines, which echo the conclusions of previous research (Lewis 1999; Pelling 2011).

However, this study extends prior knowledge by providing new insights into innovative practices that have not been extensively documented. For example, the integration of indigenous knowledge in Pacific Island nations, as revealed in this study, adds depth to the understanding of how traditional practices can be systematically integrated into modern disaster management frameworks (Nunn 2015). This finding extends the work of Pelling (2011) by providing practical examples of community-driven resilience measures that can be adapted to different contexts.

Policy Implications:

The findings of this study have several important policy implications:

i. **Integration of Local Knowledge**: Governments should actively involve communities in disaster planning and response efforts, leveraging their local knowledge and traditional practices. This could be achieved through the establishment of participatory planning processes and the inclusion of community representatives in disaster management committees (Wilkin et al., 2017).

Table 3 Barriers to effective community engagement

Barrier	Sundarbans	Philippines	Pacific Island nations	Mitigation strategies
Funding limitations	• Limited financial resources for large-scale projects	• Insufficient funds for early warning systems and training	• Limited funding for indigenous knowledge projects	• Collaboration with NGOs and international organizations • Government grants and subsidies
Infrastructural constraints	• Poor infrastructure in remote areas • Difficulty in accessing mangrove restoration sites	• Inadequate infrastructure for early warning dissemination • Poor transport networks in some areas	• Limited infrastructure for implementing traditional practices in modern contexts	• Investment in infrastructure improvements • Development of local networks and communication channels
Lack of technical knowledge	• Limited technical expertise for advanced restoration techniques	• Need for more technical training for early warning systems	• Gaps in technical knowledge about integrating traditional and modern practices	• Capacity-building workshops • Technical training programs
Socio-political issues	• Political instability affecting project implementation • Conflicts over resource use	• Political challenges in coordinating disaster response • Inequities in resource distribution	• Social and political dynamics affecting traditional practices • Influence of external policies	• Strengthening local governance • Engaging local leaders in decision-making
Gender inequality	• Gender biases affecting participation in restoration activities	• Gender disparities in disaster preparedness and response roles	• Gender roles influencing involvement in traditional practices	• Promoting gender-inclusive policies • Encouraging equal participation in disaster management

ii. **Support for Community-Based Initiatives**: Policies should provide financial and technical support for community-led initiatives such as mangrove restoration and early warning systems. Creating grant programs and fostering partnerships with NGOs can help support and scale successful local disaster resilience projects (Schrader et al. 2020).

iii. **Strengthening Capacity Building**: Developing training programs for community members to enhance their skills in disaster management is crucial. These programs should focus on risk assessment, emergency response, and the use of technology for early warning systems, thereby strengthening community capacity (O'Brien et al. 2019).
iv. **Promoting Collaboration**: Effective disaster management requires coordination between communities, government agencies, and NGOs. Policies should encourage such partnerships to ensure a unified approach to disaster risk reduction and response (Mastrorillo et al. 2016).
v. **Adapting Policies to Local Contexts**: Disaster management policies should be tailored to the specific needs and contexts of different coastal regions. This involves customizing strategies based on local hazards, socio-economic conditions, and cultural practices (Kuriakose et al. 2018).

Challenges and Limitations:

Several challenges and limitations affect the effectiveness of community-driven disaster management:

i. **Funding and Resource Constraints**: Many communities face significant financial limitations that hinder their ability to implement and sustain disaster management initiatives. Lack of funding can restrict the scope and effectiveness of community-based projects (Foster and Ritchie 2018).
ii. **Infrastructural Limitations**: Inadequate infrastructure, such as insufficient cyclone shelters or early warning systems, can impede the effectiveness of disaster management efforts (Adger et al. 2005). Addressing these infrastructural gaps is essential for improving disaster resilience.
iii. **Social Factors**: Social issues, including gender inequality and exclusion of marginalized groups, can affect the inclusiveness and effectiveness of community-driven disaster management. Addressing these social factors is crucial to ensure that all community members benefit from disaster resilience measures (Enarson and Morrow 1998).
iv. **Data Reliability and Scope**: The reliance on secondary data limits the findings to the availability and quality of existing sources. Variations in data accuracy, coverage, and timeliness can affect the reliability of the results. Additionally, the geographical scope of the study may limit the generalizability of the findings to other coastal regions (Huq et al. 2004).

Addressing these challenges and incorporating the study's findings into policy frameworks, governments and organizations can enhance their disaster management strategies and better harness community strengths to improve resilience in coastal areas.

5 Conclusion

This paper has explored the crucial role of community strengths—such as local knowledge, self-organization, and traditional practices—in disaster management and mitigation, focusing on coastal areas through case studies in the Sundarbans, the Philippines, and Pacific Island nations. The analysis reveals that community-driven approaches significantly enhance resilience to natural disasters by integrating local expertise and practices into disaster management strategies.

In the Sundarbans, community-led mangrove restoration projects and the application of traditional ecological knowledge have proven effective in bolstering disaster resilience. The restoration of mangroves acts as a natural barrier against storm surges and erosion, while local knowledge about tidal patterns and storm behaviors enhances disaster preparedness and response (Das and Vincent 2009; Wickramasinghe and Stoll 2021). Similarly, in the Pacific Island nations, the use of indigenous knowledge for predicting and responding to environmental changes has been instrumental in improving disaster readiness and response (Nunn2015). These examples demonstrate the value of integrating traditional practices into modern disaster management frameworks.

In the Philippines, community-based early warning systems and disaster preparedness training have played a vital role in reducing casualties and property damage during typhoons and floods. These initiatives show how local involvement in disaster management can improve the effectiveness of early warning systems and ensure timely and efficient responses to natural hazards (Santos and Berg 2020; Bautista 2019). The experience of the Philippines underscores the importance of community engagement in disaster preparedness and highlights the benefits of training and resource allocation for local populations.

The findings from these case studies underscore the broader implications for disaster management policies. To build more resilient coastal communities, it is essential to support and integrate local knowledge and practices into disaster management frameworks. Policymakers should focus on enhancing community involvement, providing financial and technical support for community-led initiatives, and building local capacity through targeted training programs. Additionally, fostering collaboration among communities, government agencies, and NGOs can create a unified and effective approach to disaster risk reduction.

Despite the benefits of community-driven strategies, several challenges remain, such as funding limitations, infrastructural constraints, and social inequalities. Addressing these challenges is crucial for maximizing the effectiveness of community-based approaches. Future research should aim to explore solutions to these issues, assess the long-term impact of community-driven disaster management, and evaluate the applicability of these approaches in various coastal contexts.

In conclusion, harnessing community strengths in disaster management is vital for enhancing resilience in coastal areas. By leveraging local knowledge and practices, and by providing necessary support and resources, disaster management efforts can be significantly improved, better preparing communities for future challenges.

References

Adger WN, Hughes TP, Folke C, Carpenter SR, Rockström J (2005) Social-ecological resilience to coastal disasters. Science 309(5737):1036–1039

Bankoff G (2003) Constructing vulnerability: the historical, natural and social generation of flooding in metropolitan Manila. Disasters 27(3):224–238

Barnett J, Webber M (2012) Rethinking the future of small island developing states: the case of Tuvalu. Climate Dev 4(1):48–56. https://doi.org/10.1080/17565529.2011.631227

Basheer Ahammed KK, Pandey AC (2022) Assessment and prediction of shoreline change using multi-temporal satellite data and geostatistics: a case study on the eastern coast of India. J Water Clim Chang 13(3):1477–1493

Bautista J (2019) Community-based early warning systems: a case study in the Philippines. Int J Disaster Risk Reduct 40:101–109. https://doi.org/10.1016/j.ijdrr.2019.101109

Cutter SL, Boruff BJ, Shirley WL (2003) Social vulnerability to environmental hazards. Soc Sci Q 84(2):242–261

Das S, Crépin AS (2013) Mangroves can provide protection against wind damage during storms. Estuar Coast Shelf Sci 134:98–107

Das S, Vincent K (2009) Mangrove restoration and community-based disaster risk reduction in the Sundarbans, India and Bangladesh. Glob Environ Chang 19(3):274–282. https://doi.org/10.1016/j.gloenvcha.2009.02.002

Emanuel K (2005) Increasing destructiveness of tropical cyclones over the past 30 years. Nature 436(7051):686–688

Enarson E, Morrow BH (1998) The gendered nature of disaster impact and recovery. Gend Work Organ 5(3):207–218. https://doi.org/10.1111/1468-0432.00050

Fordham M (2011) Gender, sexuality and disaster. In: Handbook of disaster research, Springer, pp 357–374

Foster S, Ritchie H (2018) Financial constraints in community-led disaster management. J Disaster Stud 12(2):223–240. https://doi.org/10.1007/s11069-018-3297-3

Gaillard JC, Mercer J (2013) From knowledge to action: bridging gaps in disaster risk reduction. Prog Hum Geogr 37(1):93–114

Gaillard JC et al (2008) Traditional societies in the face of natural hazards: the 1991 Mt. Pinatubo eruption and the Aetas of the Philippines. Int J Mass Emergencies Disasters 26(1):29–58

Huq S, Adger WN, Jones L (2004) Enhancing resilience of coastal communities to climate change: lessons learned from the Bangladesh coastal zone. Clim Policy 4(3):235–251. https://doi.org/10.1080/14693062.2004.9685484

IPCC (2014) Fifth assessment report: impacts, adaptation, and vulnerability. Intergovernmental Panel on Climate Change

IPCC. (2021). Sixth assessment report: impacts, adaptation, and vulnerability. Intergovernmental Panel on Climate Change

Kuriakose AT, Naik S, Bapat D (2018) Adapting disaster management policies to local contexts: evidence from coastal India. Disasters 42(4):823–843. https://doi.org/10.1111/disa.12325

Lewis J (1999) Development in disaster-prone places: studies of the vulnerabilities of the poor. Int J Disaster Res 13(1):15–28. https://doi.org/10.1016/S0305-750X(98)00035-6

Mastrorillo M, McNamara KE, Seneviratne D (2016) Enhancing community resilience through cross-sectoral collaboration: evidence from coastal disaster management in Sri Lanka. Environ Sci Policy 66:118–127. https://doi.org/10.1016/j.envsci.2016.08.002

Nicholls RJ, Cazenave A (2010) Sea-level rise and its impact on coastal zones. Science 328(5985):1517–1520

Nunn PD (2015) The role of indigenous knowledge in disaster risk management: lessons from the Pacific. J Pac Stud 35(1):37–50. https://doi.org/10.21313/jps.2015.35.1.37

O'Brien K, St. Clair AL, Kristoffersen B (2019) Capacity building for community-based disaster risk reduction: a global perspective. Disaster Prev Manag 28(3):402–416. https://doi.org/10.1108/DPM-09-2018-0315

Paton D, Johnston D (2001) Disasters and communities: vulnerability, resilience and preparedness. Disaster Prev Manag 10(4):270–277
Pelling M (2007) Learning from others: the scope and challenges for participatory disaster risk assessment. Disasters 31(4):373–385
Pelling M (2011) Adaptation to climate change: from resilience to transformation. Routledge
Santos MA, van den Berg M (2020) Community-led early warning systems and disaster preparedness: lessons from the Philippines. Disaster Recover J 15(2):65–80. https://doi.org/10.1038/s41598-020-70252-5
Schrader K, Tschakert P, McMichael C (2020) Community-based disaster resilience: bridging the gap between local knowledge and policy. Environ Res Lett 15(6):064006. https://doi.org/10.1088/1748-9326/ab7c3c
Shaw R, Okazaki K (2004) Sustainable community-based disaster management practices in Asia. United Nations Centre for Regional Development
Shaw R, Takeuchi Y, Shiwaku K (2014) Disaster education: community, environment, and disaster risk management. Emerald Group Publishing
Twigg J (2015) Disaster risk reduction. humanitarian policy group, Overseas Development Institute
United Nations (2020) World Population Prospects 2020: highlights. United Nations
Wickramasinghe N, Stoll J (2021) The role of mangrove forests in disaster risk reduction: evidence from the Sundarbans. Ecol Soc 26(2):15–28. https://doi.org/10.5751/ES-12432-260215
Wilhite DA, Glantz MH (1985) Understanding the drought phenomenon: the role of definitions. Water Int 10(1):111–120. https://doi.org/10.1080/02508068508686328

MIX
Papier aus verantwortungsvollen Quellen
Paper from responsible sources
FSC® C105338

If you have any concerns about our products,
you can contact us on
ProductSafety@springernature.com

In case Publisher is established outside the EU,
the EU authorized representative is:
Springer Nature Customer Service Center GmbH
Europaplatz 3, 69115 Heidelberg, Germany

Printed by Libri Plureos GmbH
in Hamburg, Germany